T0263837

Analytical Spectroscopy Library — Volume 5

Atomic Absorption Spectrometry

Theory, Design and Applications

Analytical Spectroscopy Library

A Series of Books Devoted to the Application of Spectroscopic Techniques to Chemical Analysis

Analytical Spectroscopy Library — Volume 5

Atomic Absorption Spectrometry

Theory, Design and Applications

edited by

S.J. Haswell

School of Chemistry, The University of Hull, Hull HU6 7RX, U.K.

ELSEVIER

Amsterdam — Oxford — New York — Tokyo   1991

ELSEVIER SCIENCE PUBLISHERS B.V.
Sara Burgerhartstraat 25
P.O. Box 211, 1000 AE Amsterdam, The Netherlands

Distributors for the United States and Canada:

ELSEVIER SCIENCE PUBLISHING COMPANY INC.
655, Avenue of the Americas
New York, NY 10010, U.S.A.

Library of Congress Cataloging-in-Publication Data

Atomic absorption spectrometry : theory, design, and applications /
   edited by S.J. Haswell.
       p.   cm. -- (Analytical spectroscopy library ; v. 5)
   Includes bibliographical references and index.
   ISBN 0-444-88217-0
   1. Atomic absorption spectroscopy.   I. Haswell, S. J. (Stephen
John), 1954-   . II. Series.
QD96.A8A86   1991
543'.0858--dc20   91-34690
                                                                  CIP

ISBN 0-444-88217-0 (Vol. 5)

© 1991 Elsevier Science Publishers B.V. All rights reserved.

No part of this publication may be reproduced, stored in a retrieval system or transmitted in any form or by any means, electronic, mechanical, photocopying, recording or otherwise, without the prior written permission of the publisher, Elsevier Science Publishers B.V./Physical Sciences & Engineering Division, P.O. Box 330, 1000 AH Amsterdam, The Netherlands.

Special regulations for readers in the U.S.A. – This publication has been registered with the Copyright Clearance Center Inc. (CCC), Salem, Massachusetts. Information can be obtained from the CCC about conditions under which photocopies of parts of this publication may be made in the USA. All other copyright questions, including photocopying outside of the USA, should be referred to the publisher.

No responsibility is assumed by the publisher for any injury and/or damage to persons or property as a matter of products liability, negligence or otherwise, or from any use or operation of any methods, products, instructions or ideas contained in the material herein.

Although all advertising material is expected to conform to ethical (medical) standards, inclusion in this publication does not constitute a guarantee or endorsement of the quality or value of such product or of the claims made of it by its manufacturer.

This book is printed on acid-free paper.

Transferred to digital printing 2005

Printed and bound by Antony Rowe Ltd, Eastbourne

PREFACE

Atomic absorption spectroscopy is now a well established and widely used technique for the determination of trace and major elements in a wide range of analyte types. Some nine years ago a very successful and popular practical text on the technique was produced under the editorship of John Cantle [1]. There have been many advances in the atomic spectroscopy over the last decade and for this reason and to meet the demand, it was felt that there was a need for an updated version of this earlier book. When the fully revised chapters were produced by both original and new contributing authors experienced in the field, it was found that they represented not just a revised text but a new book on its own right, with over 80% of the original text having been replaced and new sections added.

Whilst interest in instrumental design has tended to dominate the minds of spectroscopists, the analyst concerned with obtaining reliable and representative data, in diverse areas of application, has been diligently modifying and developing sample treatment and instrumental introduction techniques. Such methodology remains a fundamental part of our analysis, and so it is fitting that it should form the basis of the fourteen application chapters of this book. The text focuses in the main on AAS; however, the sample handling techniques described are in many cases equally applicable to ICP-OES and ICP-MS analysis. Despite the increasing use of emission techniques AAS still remains a major workhorse in many analytical laboratories, with electrothermal atomization representing a very important tool for trace element determinations, particularly in the environmental and clinical field.

Whilst offering a detailed and critical overview of applications and sample methodology, in which I hope there is something for everyone, the book also represents a complete text on AAS in the sense that the basic theory and fundamental instrumental design are covered in the first three chapters.

The numerous computer enhancements that have rapidly changed the appearance and use of modern instrumentation have not been addressed in this book, as they tend to be specific to particular manufacturers and would by their nature, very soon be out of date. Whilst accepting the valuable role played by microprocessors in modern instrumentation, these often superficial enhancements contribute little in obtaining a reliable analysis by AAS, which still relies on good sample methodology.

I and my co-authors have made considerable efforts to make this book a valuable practical compendium for users of atomic spectroscopy, and I hope you find it of interest and use in the practices of your science.

S.J. HASWELL

Reference

1   J.E. Cantle (Editor), Atomic Absorption Spectrometry, Elsevier, Amsterdam, 1982.

CONTENTS

Chapter 1

Basic Principles

JULIAN F. TYSON
Department of Chemistry, University of Massachusetts, Amherst, MA 01003 (U.S.A.)

I. INTRODUCTION

Spectroscopy is the study of the interaction of electromagnetic radiation with matter. In particular, it is the study of how the magnitude of the interaction is a function of the energy of the radiation. Many sciences, including chemistry, could not have reached their present advanced state were it not for the challenge of interpreting spectroscopic experiments. Some sciences, such as astronomy, owe their very existence to spectroscopy and today there are many scientific disciplines which rely extensively on the experimental results of spectroscopic studies for their continued development. The results of such studies are very often presented in the form of a two-dimensional plot of the extent of interaction vs energy, known as a spectrum (see Fig. 1). If the object of the exercise is to extract quantitative information about the amount of matter involved, the experiment is usually referred to as spectrometry. This is an abbreviation of spectrophotometry, which is what is carried out with a spectrophotometer, an instrument capable of measuring the intensity of radiation at different energies.

A. Development of analytical atomic spectrometry

There are many different types of spectroscopy and many ways of classifying the various techniques (see Table 1). One broad classification is

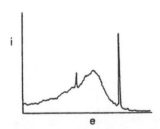

Fig. 1. An example of a spectrum, i.e. a plot of the intensity of interaction of the electromagnetic radiation, i, as a function of the energy, e. The spectrum is, in fact, part of a flame atomic emission spectrum and shows both line and broad-band emission features.

References p. 20

TABLE 1

CLASSIFICATION OF SPECTROSCOPIC TECHNIQUES

Name for region of spectrum:	λ-rays	Hard, soft X-rays	Ultraviolet visible	Infrared	Microwaves	Radiowaves
Origin of radiation:	Nuclear transitions	Inner electron transition	Outer electron transitions	Molecular vibrations	Molecular rotations	Nuclear precessions
Type of interaction[a]:	A, E	A, E, F	A, E, F	A	A	A
Sample phase[b]:	S	S, L	L, G	L, G	G	L
Species[c]:	At	At	At, Mol, Ion	Mol	Mol	Mol

[a] A, absorption; E, emission; F, fluorescence.
[b] S, solid; L, liquid; G, gas or vapour.
[c] At, atomic; Mol, molecular; Ion, ionic.

based on the nature of the interaction. Matter may either scatter, emit or absorb radiation. Each of these three categories may be further subdivided into many sub-categories according to a number of different schemes. In one such scheme there would be a group of techniques associated with the absorption of radiation by free, uncombined gaseous atoms. This particular state of matter is capable of absorbing radiation over a wide range of energies corresponding to gamma-radiation through to near infra-red radiation. A further sub-section would concern the interaction of atoms with near ultra-violet (UV) and visible light.

Although this is a very small part of the electromagnetic spectrum from which scientists extract information nowadays, this region of the spectrum, particularly the visible part, is of immense historical importance. Early spectroscopists such as Newton, Wollaston, Fraunhofer, Brewster, Bunsen and Kirchhof had only their eyes as spectroscopic detectors with which to work. Our present theory concerning the electronic structure of atoms, a basic building block in modern chemistry, is accepted because it explains the observed spectroscopic phenomena.

These early spectroscopists worked mainly with emission spectroscopy, though the explanation of the dark lines in the solar spectrum (absorption of the solar radiation by atoms in the sun's atmosphere) was advanced by several workers around 1830–1850. From this time it was recognised that certain emission lines were characteristic of a particular element and that absorption and emission occurred at the same wavelength. Although a quantitative determination by atomic absorption, that of mercury, had been described in 1939, it was atomic emission spectrometry which was used for quantitative analysis in laboratories. The atom sources used being either electrical discharges (for solid samples) or flames (for liquid samples).

The possibility of the general quantitative potential of atomic absorption spectrometry (AAS) was not raised until the mid-nineteen fifties. It was proposed by Walsh and shown by Alkemade and Milatz that the use of an atomic emission light source allowed measurements of the absorption of radiation by atoms produced in a flame to be made and related to the concentration of the element introduced into the flame. Walsh enthusiastically pursued the development of the technique and from the early nineteen sixties, when instrumentation began to be commercially available, a new technique was available to the analytical chemist for trace metal analysis.

Since then there has been an extremely rapid growth in the use of the technique. At present there are some ten or so manufacturers of atomic absorption instruments selling in various parts of the world and many thousands of instruments must be in routine use. With the air/acetylene or nitrous oxide/acetylene flame it is possible to determine about 65 elements, mainly metals but including some semi-metals, at the ppm level in solution with a robust, compact and inexpensive, bench-top instrument. With the addition of a hydride generation apparatus and an electrothermal atomizer, the detection power for the same elements is pushed down to the ppb level. All of the atomization techniques may be automated and a modern computer-controlled instrument is capable, in principle, of determining sequentially up to sixteen elements in fifty samples without any operator intervention.

II. MODERN ANALYTICAL AAS

This does not mean that the analytical spectroscopist has been made redundant by modern instrumentation. Nearly all analyses will suffer interference effects of one kind or another, and the analytical chemist developing an AAS method must understand such effects and how they can be overcome. In turn this means a thorough understanding of (a) the principles on which the instrument is operating, (b) the need for correct calibration and (c) what is being done with the raw data before the digits appear on the print-out. Of equal importance is a knowledge of how to deal with solid samples. Although considerable developments have been made in recent years in the direct analysis of solids by AAS, this aspect of the technique is really only at the semi-quantitative stage. Solid samples must therefore be brought into a solution that may be presented to the instrument without either losing the analyte or contaminating the sample.

The reminder of this chapter is concerned with the first of these requirements, namely an understanding of the principles on which the instrument is operating. Later chapters will be used to give more specific details on the practical application of the technique.

III. BASIC IDEAS ON LIGHT AND ATOMS

A. The nature of light

In some circumstances light behaves as though it consisted of an oscillating electric field with an oscillating magnetic field at a right angle to it. This wave nature of light is seen in interference phenomena, such as produced by a diffraction grating in a monochromator (see Chapter 2, Section II.C). In other circumstances light behaves as though it consists of a stream of discrete particles (known as photons). The particulate nature of light is seen in the photo-electric effect, such as the emissions of electrons from the photo-cathode of a photomultiplier tube (see Chapter 2, Section II.D). A sharp cut-off in the response curve with decreasing energy is observed as the individual photon energy falls below that necessary to cause the ejection of an electron.

This wave–particle duality is a feature of events on the atomic scale. Very small particles do not behave as we would expect from extrapolating our experiences with large, everyday objects. Nor is their internal structure to be modelled by ideas based on the everyday, macro physical world.

B. The nature of atoms

All the experimental evidence, a considerable amount of which is spectroscopic, suggests that atoms consist of a small central, positively charged nucleus surrounded by as many electrons as are needed to provide overall electrical neutrality. Each electron moves in a specific way unique to that electron, though groups of electrons move in very similar ways. The electrons occupy discrete energy levels, whose values depend on the interaction of the electron with the others in the atom and the interaction with the nuclear charge. Each element thus has a unique set of electronic energies and these can be represented on an energy level diagram. Such diagrams would be very complicated if the energies of all the electrons were shown, and difficult to draw as the energies vary over several orders of magnitude.

At the temperatures available in atom sources used for analytical AAS, most atoms exist in the lowest possible energy state, known as the ground state. It is possible to move electrons around inside an atom by perturbing the electrons by some means. This could be by collision with other atoms (i.e. with the electrons bound to other atoms), by collision with free electrons or by collision with photons. An electron will only be moved from one energy level to another if the energy available from the collision is exactly equal to the energy difference between the initial and final levels.

Analytical atomic spectroscopy is concerned only with the movement of the outermost electrons, as the difference in energies between the initial and final states for these electrons corresponds to photon energies of UV and visible radiation.

IV. PHOTON–ELECTRON INTERACTIONS

The energy of a photon, E, is proportional to the frequency, ν, of the corresponding wave-motion. The constant of proportionality is known as the Planck constant, h, which has the value 6.626×10^{-34} J s, i.e.:

$$E = h\nu \tag{1}$$

As frequency and wavelength, λ, are simply related by:

$$\nu = c/\lambda \tag{2}$$

$$E = hc/\lambda \tag{3}$$

where c is the velocity of light, 2.998×10^8 m s^{-1}.

Although eqns. (1) and (3) would give the energy differences corresponding to particular wavelengths in joules, this unit is not often used in energy level diagrams which are more commonly presented with the energy axis scaled in either electron volts or cm^{-1}. An electron volt, eV, is the energy of

References p. 20

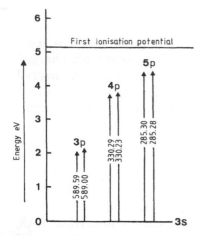

Fig. 2. A partial energy level diagram for sodium. The numbers against the absorption transitions are wavelengths in nm.

an electron (charge 1.602×10^{-19} C) after acceleration from rest across a potential difference of 1 V, i.e. 1 eV is equivalent to 1.602×10^{-19} J.

Part of the energy level diagram for sodium is shown in Fig. 2. Only the possible electronic transitions which may be produced by collision between the ground state atom (i.e. the outermost electron in the case of sodium) and photons, are shown. These transitions are known as resonance transitions. Atoms which have been electronically excited by some means, may emit radiation as electrons move from higher energy levels to lower. The number of transitions which may be observed by emission spectroscopy depends on the energy available for excitation. In flames and hollow-cathode lamps (see Chapter 2, Section II.B), only the resonance transitions will be seen with any prominence, but in arcs, sparks and plasmas many more transitions will be observed.

A. Qualitative analysis

The energy levels are unique for each element so, in principle, it is possible to perform qualitative analysis by atomic spectrometry by examining the wavelengths in the emission spectrum of an unknown. The visible radiation emitted by certain atoms from air/natural gas flames can be used to identify a number of elements, provided the matrix is not too complicated. An emission spectrograph can be used for an initial look at complex unknown samples, but the technique is less widely available nowadays. Direct reading emission spectrometry is less useful because of the dangers of spectral overlap resulting in incorrect deductions being made. Atomic absorption spectrometry is rarely used for qualitative purposes because the technique is essentially single-channel.

B. Relative population of states

For a population of atoms in thermal equilibrium (i.e. the relative populations in the various electronic energy levels are governed only by temperature), it is possible to calculate the ratio of the number of atoms in any one state, $N(1)$, to the number in any other, $N(2)$, by means of the Boltzmann equation:

$$N(1)/N(2) = [g(1)/g(2)] \exp(-E/kT) \tag{4}$$

where k is the Boltzmann constant (1.381×10^{-23} J K^{-1}) and T is the absolute temperature. The ratio $g(1)/g(2)$ is the ratio of statistical weights. These are terms arising from the description of the energy levels, often called spectroscopic states, in quantum mechanical terms and are given by:

$$g(i) = 2J(i) + 1$$

where $J(i)$ is the total angular momentum quantum number. The J-value for a particular energy level is the subscript in the notation used to describe the level. For example, in the transition $^2P_{3/2}$ to $^2S_{1/2}$, an emission transition, the values of J for the two states are 3/2 and 1/2, respectively.

The ratio of the numbers of atoms in the first excited state to the number in the ground state for the temperatures 2000 K and 3000 K are given in Table 2. These temperatures are approximately the values encountered in the air/acetylene and nitrous oxide/acetylene combustion flames, respectively. It can be seen from the table that even for the relatively easily excited elements (such as the alkali metals), the vast majority of the

TABLE 2

RATIO OF NUMBER OF ATOMS IN THE FIRST EXCITED STATE TO THE NUMBER IN THE GROUND STATE

Element	Wavelength of transition (nm)	N (excited)/N (ground)	
		2000 K	3000 K
Cs	852.1	4.44×10^{-4}	7.24×10^{-3}
Na	589.0	9.86×10^{-6}	5.88×10^{-4}
Ca	422.7	1.21×10^{-7}	3.69×10^{-5}
Fe	372.0	2.29×10^{-9}	1.31×10^{-6}
Cu	324.8	4.82×10^{-10}	6.65×10^{-7}
Mg	285.2	3.35×10^{-11}	1.50×10^{-7}
Zn	213.9	7.45×10^{-15}	5.50×10^{-10}

References p. 20

atoms are in the ground state. This is one of the reasons for the sensitivity of atomic absorption spectrometry at these temperatures. The use of much higher temperature sources, such as plasmas, may be attractive in terms of removing chemical interference effects and producing a high degree of atomization, but such sources produce a high degree of excitation as well and are less useful for absorption spectroscopy. On the other hand, these properties make the sources potentially useful for emission spectroscopy. Many more spectroscopic states are accessible at these higher temperatures and so the emission spectra are much more complex, containing many non-resonance as well as the limited number of resonance lines.

C. Transition probabilities

Not all excited atoms will return to a lower energy level by the spontaneous emission of a photon of light. It is possible that the atom could lose energy through a collision. There is thus a probability of less than one that the transition will be by photon emission. This probability is a function of the particular element and is referred to as the transition probability for emission. There is an analogous probability for the absorption process, which is also a function of the element. Each particular transition will have its own transition probability.

Tables of spectroscopic data for analytical spectroscopy often give values of transition probabilities quoted as the so-called f-value. The f-value is the oscillator strength associated with the transition and is a term arising from one theoretical treatment of the absorption and emission of radiation by atoms in which the ability of an atom to absorb and emit a particular wavelength is related to the equivalent ability of a classical oscillating electron. The oscillator strength values are largest for the transition between the ground state and the first excited state and this transition is normally used for atomic absorption spectrometry when the highest sensitivity is required.

Values of oscillator strength are only useful for assessing the relative sensitivities of different transitions for a particular element. They cannot, in general, be used to compare the likely sensitivity of one element with another. This is because the main factor governing the sensitivity of AAS is the partial pressure of atoms (i.e. the concentration of atoms in the vapour phase). This in turn depends mainly on the extent to which the molecular precursor is dissociated at the temperature of the source and to what extent the atoms are ionised. The extent of ionisation may be considerable, especially in the hotter sources. This ionisation and the reverse of the former effect, namely the formation of stable compounds, are two effects which are of particular importance in practical atomic absorption spectrometry and due consideration must be given to controlling these effects when a method is being developed.

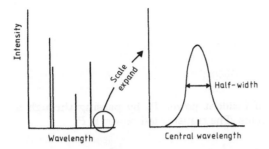

Fig. 3. The shape of atomic spectral "lines".

D. Line shapes

When a recorded atomic spectrum is examined it is observed that if the wavelength scale covers several tens or hundreds of nm, the spectral features are very narrow. For this reason atomic spectral transitions are often referred to as lines. If the spectrum is observed visually the terminology is still appropriate as the instrument which resolves the various component wavelengths will produce geometric images of a narrow rectangular slit mounted between the atom source and the dispersing device. Thus an atomic absorption spectrum would consist of the spectrum of the light source crossed by dark lines corresponding to wavelengths at which the atoms absorbed.

If the lines in the recorded spectrum are examined very closely it is seen that they are not infinitely narrow but have a definite shape (see Fig. 3). The first stage in characterising the line shape is to assign the central wavelength, λ_0, and the width at half maximum height (the so-called half-width, $\Delta\lambda$). The value of the half-width is found to be a function of a number of characteristics of the population of atoms being observed. These parameters include temperature and partial pressure of the various gas-phase species. This latter parameter is also found to affect λ_0 in some cases. In the case of emission lines, the shape is found to depend on the dimensions of the atom source and the temperature gradients in the direction of the line of observation.

V. QUANTITATIVE BASIS FOR AAS

A. Absorbance

An atomic absorption spectrometer measures absorbance. Absorbance is the logarithm (to base 10) of the ratio of the incident light power, P_0, to the transmitted light power, P (see Fig. 4), i.e.:

$$A = \log(P_0/P) \tag{5}$$

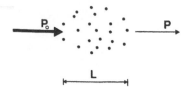

Fig. 4. The attenuation of a light beam of incident power P_0 by passing through a distance L in an absorbing population of atoms. The transmitted power is P.

In general, when a light beam is attenuated by the medium through which it passes, the relationship between P and P_0 is given by:

$$P = P_0 \exp(-kL) \tag{6}$$

where k is the absorption coefficient and L is the path length in the absorbing medium (see Fig. 4). The absorption coefficient is a function of a large number of system parameters such as the oscillator strength for absorption, the wavelength of the light and the number of atoms in the ground state per unit volume.

The detector will measure intensity over the entire spectral range reaching it and thus the values of P and P_0 used in the calculation of absorbance by eqn. (5) will be the integrals of the functions $P(\lambda)$ and $P_0(\lambda)$ over the wavelength interval falling on the detector, i.e.:

$$A = \log \left[\int P_0(\lambda) \, d\lambda \, \Big/ \int P(\lambda) \, d\lambda \right] \tag{7}$$

substituting from eqn. (6):

$$A = \log \left[\int P_0(\lambda) \, d\lambda \, \Big/ \int \exp(-kL) P_0(\lambda) \, d\lambda \right] \tag{8}$$

The difficulty is now to evaluate the integrals in the numerator and denominator. It must be remembered that k is also a function of wavelength as well as P_0. Examination of eqn. (8) shows that it could be simplified considerably if k could be considered a constant. Atomic absorption spectrometers are designed to achieve this.

B. The absorption coefficient

The first step is to consider the way in which k varies with wavelength and such a plot is shown in Fig. 5. This plot is one way in which the absorption line-shape may be represented. It can be seen from Fig. 5 that for wavelength values close to the absorption maximum, the value of k does not change much. Thus the requirement for the instrument design is that the range of wavelengths reaching the detector, over which the absorbance

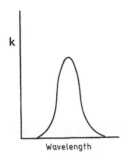

Fig. 5. The variation in absorption coefficient, k, with wavelength.

is measured, is a small range centred around the wavelength maximum of the absorption profile so that over the interval of integration, the value of k may be taken as constant.

In other words the requirement is for a light source which (a) emits radiation at exactly the right wavelength to be absorbed by the atoms of the particular element in the atom source, and (b) whose emission profile is considerably narrower than the absorption profile.

The first of these requirements is met by having an atomic emission source which produces the emission spectrum of the element to be determined. The second requirement requires control over the factors which affect the widths of atomic spectral profiles.

C. Broadening processes

1. *Doppler broadening*

A population of atoms is in random thermal motion and the individual gas phase species are continuously in collision with each other. When the radiation emitted by such a population is observed, the observer looks along a particular line of sight. Due to the random motion of the atoms some of the emitting species will be travelling away from, some towards and some will have no velocity component with respect to the observer (see Fig. 6). To the observer it will appear that the frequency of the radiation emitted by an atom travelling towards the detector is increased compared with the frequency of the radiation emitted by an atom travelling at right angles to the line of observation, while an atom travelling away is emitting radiation of decreased frequency. This effect is known as the Doppler effect and the observed frequencies are said to be Doppler-shifted. In wavelength terms the atoms travelling away from the observer have their emitted radiation shifted to longer wavelengths, and this is often referred to as the red shift. It is a well-known phenomenon in astronomy and it is the reason for considering that the universe is expanding.

References p. 20

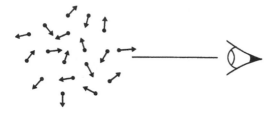

Fig. 6. Random thermal motion in an atomic vapour gives rise to a range of velocity components along the line of observation.

It is possible to treat the effect theoretically and predict its magnitude. It turns out that the Doppler half-width, $\Delta\lambda_D$, is directly proportional to the square root of the absolute temperature, T, and inversely proportional to the square root of the relative atomic mass of the atom, M. The full equation is:

$$\Delta\lambda_D/\lambda_0 = 7.162 \times 10^{-7}(T/M)^{1/2} \tag{9}$$

Values of $\Delta\lambda_D$ for some common elements and temperatures are given in Table 3 from which it can be seen that for temperatures encountered in atomic absorption spectrometry, Doppler half-widths are of the order of 10^{-3} nm. As the spread of velocities in a gas follows a Gaussian distribution, it would be expected that Doppler broadening is a symmetrical effect producing a Gaussian shaped atomic spectral profile.

2. Pressure broadening

Spectral line-widths are also affected by collisions. The energy levels involved in a transition are perturbed by the close approach of another gas phase species. Two types of pressure broadening, as this effect is called,

TABLE 3

DOPPLER HALF-WIDTHS

Element	Atomic weight	Wavelength of transmission (nm)	$\Delta\lambda_D \times 10^3$ nm	
			2000 K	3000 K
Cs	133	852.1	2.4	2.9
Na	23.0	589.0	3.9	4.8
Ca	40.1	422.7	2.1	2.6
Fe	55.8	372.0	1.6	2.0
Cu	63.6	324.8	1.3	1.6
Mg	23.3	285.2	1.9	2.3
Zn	65.4	213.9	0.85	1.0

may be distinguished. The first type is due to collisions with atoms of the same element. This is known as resonance or Holtsmark broadening. It is a symmetrical effect but is swamped by the effect due to collisions with other gas-phase species in real atom sources for atomic spectrometry. In flames, furnaces and plasmas the analyte atoms are present at low partial pressure compared with the combustion gases, sheath gas, plasma gas etc. This second type of effect is known as Lorentz broadening. Lorentz broadening is accompanied by a shift in the wavelength maximum to longer wavelengths and the resulting profile is no longer symmetrical.

These effects are more difficult to treat theoretically than Doppler broadening, as classical theories of collisions do not account for all the observed phenomena and quantum mechanical effects need to be incorporated. It turns out that for the sorts of situation analytical atomic spectroscopists are concerned with, the Lorentz half-width, $\Delta\lambda_L$ is of the same order of magnitude as the Doppler half-width for atoms in flames and furnaces but is negligible at temperatures of a few hundreds of degrees Kelvin.

The ratio of the Lorentz and Doppler half-widths may be derived from the so-called a-parameter where a is given by:

$$a = (\log 2)^{1/2} \Delta\lambda_L / \Delta\lambda_D \tag{10}$$

The value of a does not vary greatly with temperature but depends on the composition and pressure of the gas. A large number of experimentally determined a-parameter values is available most of which are for combustion flames in which the values lie between 0.3 and 2.0.

3. Self-absorption

For emission sources containing a temperature gradient in the direction of observation a third type of process occurs which affects the shape, and hence width of the profile. For such sources, radiation emitted from a hot part of the source must traverse an atom population in a cooler region before being observed. Some of this radiation will be absorbed by these atoms. In general, regardless of the temperature gradient, this effect will occur with sources of finite thickness and the process is known as self-absorption. However, the presence of a temperature gradient means that the absorbing atoms in the cooler part of the source absorb over a narrower wavelength range because they have a narrower Doppler half-width than the emitting atoms in the hotter region. As is shown in Fig. 7a–c, the net result is that more intensity is removed from the centre of the profile than the wings and a distorted profile with an increased half-width is obtained. This process is referred to as self-absorption broadening. Under extreme conditions intensity may be removed to such an extent from the line centre that the profile has the shape shown in Fig. 7d. The profile is said to be self-reversed.

References p. 20

14

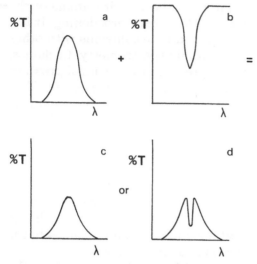

Fig. 7. The effect on the emitted profile (a) of the absorption by atoms with a narrower Doppler half-width (b) to give a resultant profile with an increased half-width (c) or, in severe cases, a self-reversed profile (d).

4. *Other broadening processes*

There are other processes which will increase the width of an atomic line profile. These are the splitting of the energy levels involved into multiplet components due to (a) the presence of a strong magnetic field (the Zeeman effect), (b) the presence of a strong electric field (the Stark effect) and (c) interaction between the electron spin motion and the nuclear spin motion, for those atoms possessing nuclear spin (the hyperfine structure effect). Only the first of these effects is of importance in analytical atomic absorption spectrometry. The Zeeman effect is exploited in a method of background correction.

D. **Design of light source for AAS**

The design criteria outlined above are for an atomic emission source with a narrow emission profile compared with the absorption profiles of analyte atoms. To a first approximation, these will be at atmospheric pressure (10^5 Pa) and at a temperature of between 2000 and 3000 K, whether produced in a flame or a furnace. The light source must therefore be at lower pressure, to reduce the Lorentz broadening and at lower temperature, to reduce the Doppler broadening. In addition the source must not significantly distort the profile due to self-absorption.

All of these three criteria are met by several designs of light source of which the most widely used is the hollow-cathode lamp (HCL). The success

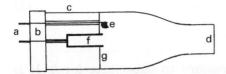

Fig. 8. Schematic diagram of a hollow-cathode lamp. Electrical connections (*a*) with the modulated power supply are made through the base (*b*) into which is sealed the glass envelope (*c*) with a quartz end-window (*d*). The discharge is between the anode (*e*) and the hollow cathode (*f*). A mica disc (*g*) confines the discharge to between the anode "flag" and the interior of the cathode.

of AAS as a laboratory analytical technique is really due to the availability of reliable, inexpensive, long-lasting HCLs which give a sufficiently intense output of the desired spectral shape to give (a) usable calibration functions, and (b) acceptable signal-to-noise ratios. Much research and development work has gone into the design of these sources.

A schematic diagram of the HCL is shown in Fig. 8. The lamp is a vapour discharge lamp with a special electrode geometry. The cathode, a hollow cylinder, is lined with the element whose emission spectrum is to be produced. The interior of the lamp contains an inert gas (argon or neon) at a pressure of approximately 10^3 Pa. A simplified operating mechanism may be considered to be based on the initial release of electrons from the cathode surface when the striking potential of several hundred volts is applied. On their way to the anode these electrons cause impact ionisation of the inert fill-gas atoms. These ions are accelerated to the cathode where the energy of impact causes material of the cathode lining to be vaporised by a process of sputtering. This atomic vapour is collisionally excited by electrons, neutrals and ions and the characteristic emission spectrum is produced.

In operation, the voltage may be between 100 and 200 V and the current between 2 and 20 mA. The value of the operating current will be under the control of the instrument operator. The operating temperatures are only a few hundred degrees Kelvin and thus the operating pressure is still considerably less than atmospheric. The shape of the cathode confines the atomic vapour and thus the partial pressure of atoms outside the cathode, capable of absorbing the emitted radiation and causing self-absorption broadening is minimised.

E. Basic equation

All of these design features mean that the emission profile from the lamp is narrower than the absorption profile in the flame, and for many elements, the use of the HCL means that over the interval of integration in eqn. (8), the value of the absorption coefficient, k, may be considered constant. Thus the equation reduces to:

$$A = \log[1/\exp(kL)] \tag{11}$$

$$= 0.434kL$$

To complete the equation it is now necessary to substitute for k. It is not possible to write an exact equation for k in terms of all of the various broadening processes and so the approach adopted is to use the equation for a Doppler broadened value, k_D, and to modify this by the inclusion of a factor to account for other broadening processes. This factor may be calculated, with an error of about 10%, from the a-parameter (see section above) and the following equations:

$$k = k_D/(1 + 1.2a) \quad 0 < a < 2$$

and:

$$k = k_D/a\pi^{1/2} \quad a > 2$$

$$k = b2(\pi \ln 2)^{1/2} e^2 \lambda_0^2 n_j f / \Delta\lambda_D mc^2 \tag{12}$$

where b is the factor calculated from the appropriate equation, m and e are the mass and charge of the electron, respectively, f is the oscillator strength for absorption and n_j is the number of atoms per unit volume in the lower of the two electronic states involved in the absorption process. Substituting for k in eqn. (11) gives:

$$A = KLn_j \tag{13}$$

where K is a constant incorporating all the relevant terms (these will include temperature as $\Delta\lambda_D$ is a function of temperature).

Thus absorbance, measured under the conditions described above, is linearly related to the number of atoms per unit volume in the lower energy level. For analytical AAS, this level is the ground state and for the temperatures achievable in flames and furnaces, the number of atoms per unit volume in the ground state is, to a good approximation, equal to the total number of atoms per unit volume, N. The fraction of the atoms in the first excited state is shown for some elements and temperatures in Table 2. Thus a further modification may be made to the basic quantifying relationship, namely:

$$A = KLN \tag{14}$$

and absorbance is directly proportional to the total number of atoms per unit volume.

VI. PRODUCTION OF ATOMS

For a physico-chemical phenomenon to be useful as the basis of an analytical technique, it is generally necessary that the magnitude of what is measured, is linearly related to some simple function of concentration or amount of the determinant species in the sample for analysis. As most samples for analysis will be either solids or liquids, a very important step in performing analyses by atomic absorption spectrometry based on eqn. (14) is to convert the elements to be determined into atomic vapour.

Although there are many ways of generating such vapour from solids or liquids, for a variety of practical reasons only two such methods are used in analytical AAS. These are the combustion flame, into which the sample is introduced as a finely-dispersed aerosol solution, and the electrically heated furnace, into which the sample is mainly introduced as a solution but occasionally as a solid. For a limited range of elements, there is a third method in use, namely the direct generation and separation of a chemical vapour from the solution phase with subsequent atomization for which a variety of atomizer devices are used.

A detailed discussion of atomization mechanisms is beyond the scope of the present book. There are several texts available [1, 2] dealing with flames, but atomization mechanism in furnaces is still a hot topic of research in analytical atomic spectroscopy around the world. Some basic concepts have been described [3] and the interested reader can follow developments by consulting the regular review literature [4].

For flames, as may be readily imagined, there is a complex sequence of physical and chemical events between the initial stage of sample solution at room temperature and a population of atoms (at a temperature of between 2000 and 3000°C). Leaving aside some of the problems associated with the physical processes of aerosol generation (nebulization) and transport, it is important to realise that in general, of the vapour phase species containing the analyte element there is a significant proportion which are not atoms. These species, which are predominantly molecules and ions, will not absorb radiation from the hollow-cathode lamp.

A. Molecule formation

Two different types of molecule containing the analyte element may be identified, those involving a component of the sample solution (such as metal chlorides, MCl) and those involving a constituent of the flame gas (such as metal oxides, MO, and hydroxides, MOH). Of course, the presence of molecules may be due to incomplete dissociation, as well as due to reaction between free atoms and flame gases. Most of the experimental evidence suggests that, while a combustion flame is obviously not in a state of equilibrium over the entire physical space occupied by the reacting

References p. 20

gases, local thermodynamic equilibrium can be assumed for regions such as that sampled by the beam of a hollow-cathode lamp. Thus the law of mass action may be applied and the "concentration" (partial pressure or number density) of the various species may be calculated from the relevant equilibrium constant (a function of temperature) and some initial concentration data. The equilibrium constant may be derived on theoretical grounds and an equation formulated from which its value may be calculated as a function of temperature, bond strength etc.

The application of these relationships shows that, for example, about 90% of the lithium in an air–acetylene flame exists as LiOH, $CaCl_2$ is only about 30% dissociated and that there is a negligible degree of oxide dissociation for B, Re, Sc, Si, Ta, Ti, U, V, W, Zr and the lanthanoids.

B. Ionisation

Despite the fact that for many elements the atom sources commonly used for AAS are not energetic enough to produce complete molecular dissociation, there are considerable problems associated with ionisation for other elements. The process may be considered in an analogous fashion to that of diatomic molecular dissociation namely as the reaction:

$$M \rightleftharpoons M^+ + e^-$$

for which an ionisation constant can be written in terms of the numbers of the various species per unit volume in the source. This constant depends only on the temperature and the element in question. The equation governing the variation of ionisation constant with these two parameters is known as the Saha equation [5]. As may be imagined, it is the elements with lowest ionisation potential, such as the alkali and alkaline earth metals, which are ionised to an appreciable extent in the flame. At a temperature of 2500 K potassium is about 50% ionised (at a total K number density of 3×10^{12} cm^{-3} which approximates to a solution concentration of 500 ppm) whereas sodium is only about 7% ionised under the same conditions. Ions of an element, having a different electronic structure, absorb and emit at wavelengths different from those of the neutral atom and thus ionisation produces a reduction in sensitivity. For some elements the ionic spectrum can be used for analytical purposes, though this is usually the case for the emission from hot sources such as the inductively coupled plasma. The extent of ionisation will be influenced by factors which affect the electron number density in the flame. These include the presence of other easily ionised elements (which will tend to suppress the ionisation) and the presence of electronegative species such as chlorine (which will tend to enhance the ionisation).

Any phenomenon which affects the ground state population of atoms has the potential to act as an interference. Solute volatilisation and ionisation effects are commonly encountered in real analyses by AAS. They are discussed further in later chapters.

Both effects are functions of temperature and the nature of the other sample components. This makes it somewhat difficult to predict what will be the result of changing experimental parameters due to the balance of effects.

It is therefore something of an act of faith to accept that the number of atoms formed per unit volume, N, is directly proportional to the concentration in the solution. However, this is the relationship which has to be assumed to produce the quantitative basis for AAS, namely that the absorbance is directly proportional to analyte concentration in the solution.

Fortunately there is plenty of practical evidence that this relationship is valid, making AAS a viable analytical technique.

C. Limitations to the basic linear relationship

It is obvious that the number density of atoms in the light path is a complex function of flame conditions and sample matrix composition as well as the operating characteristics of the nebulizer. The sensitivity obtained depends critically on the operating parameters selected and thus the instrument must be calibrated at the time the analysis is performed. In general the calibration curves are linear at low absorbance values but exhibit an increased curvature towards the concentration axis at higher concentrations. One of the reasons for this is the breakdown of the underlying assumption that the source profile is narrow compared with the absorption profile in the flame.

D. Interference effects

Another reason for increased curvature at higher absorbance values is the presence of unabsorbed or stray light. This is discussed in more detail in Chapter 3, Section II. It should be borne in mind that when a solution of an element is introduced into a flame, some atomic emission will be produced. This emission will be at exactly the same wavelength as that produced by the hollow-cathode lamp and thus will cause measured absorbance values to be inaccurate (too low). To circumvent this problem, the radiation from the hollow-cathode lamp is modulated and the detector electronics are locked into this modulation frequency, thus discriminating against the steady, dc, signal produced by the emission from the flame. This is a basic design feature of atomic absorption spectrometers, which are discussed in more detail in Chapter 2, Section II.D.

Another very commonly encountered interference with real samples is the overlap of a molecular absorption band (see Section VI.A) and the atomic line of the determinant. This background absorption may be compensated for by

References p. 20

use of a background correction device, of which there are three designs being offered in commercial instrumentation. These devices are also discussed in more detail in Chapter 2, Section III.

REFERENCES

1   C.Th.J. Alkemade, Tj. Hollander, W. Snelleman and P.J.Th. Zeegers, Metal Vapours in Flames, Pergamon, Oxford, 1982.
2   C.Th.J. Alkemade and R. Herrman, Fundamentals of Flame Spectroscopy, Adam Hilger, Bristol, 1979.
3   C.W. Fuller, Electrothermal Atomization for Atomic Absorption Spectrometry, Chemical Society, London, 1977.
4   B.L. Sharp, N.W. Barnett, J.C. Burridge, D. Littlejohn and J.F. Tyson, Atomic Spectrometry Update—Atomization and Excitation. J. Anal. Atom. Spectrom., 3 (1988) 133R–154R.
5   M.N. Saha, Phil. Mag., 40 (1920) 472–478.

Chapter 2

Instrumental Requirements and Optimisation

S.J. HASWELL

School of Chemistry, University of Hull, Hull HU6 7RX (Great Britain)

I. INTRODUCTION

An atomic absorption spectrometer can be considered as having two fundamental component parts: an atom cell to generate atoms in the free gaseous ground state, and an optical system for signal measurement. In addition a growing feature of modern instruments is a signal processing system or data station which can be either integrated or externally fitted to the instrument.

A schematic of the basic instrumental components of an atomic absorption spectrometer is shown in Fig. 1. The atom cell (B) can be seen situated between the light source (A) and the monochromator (C). As described in Chapter 1 the role of the atom cell is to primarily dissolvate a liquid sample and dissociate the analyte elements into their free gaseous ground state form. In this form the atoms will be available to absorb radiation emitted from the light source and thus generate a measurable signal proportional to concentration. The design of the atom cell must therefore create the conditions described above in the most efficient way possible. It is interesting to compare for a moment the role of the atom cell in atomic absorption with that used in atomic emission. In atomic emission the maximum number of thermally excited atoms are produced by the atom cell to give an intense emission measurement. From the Boltzmann distribution (Chapter 1) we see that at the typical atomization temperature achieved by both the flame and furnace in atomic absorption methods (2500–3100°C) the majority of the atoms will remain in the ground state. It is this factor which gives atomic absorption its characteristically good sensitivity as an analytical technique for elemental determinations.

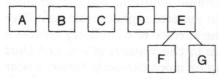

Fig. 1. A schematic diagram of the basic instrumental components of an atomic absorption spectrometer: A, light source; B, atom cell; C, monochromator; D, detector; E, amplifier; F, signal display; G, data station.

The optical system essentially consists of the light source (Fig. 1, A) which emits the characteristic narrow-line spectrum of the elements of interest and a monochromator (Fig. 1, C) which enables, through spectral dispersion, the selection of the specific emission lines produced by the light source. The monochromator may also be fitted with a variable exit and entrance slit to reduce background radiation and aid the selection and isolation of the analytical emission lines. Finally a photomultiplier detector (Fig. 1, D), whose function it is to convert photons into an electrical signal, is mounted close to the exit slit of the monochromator.

The processing of signals produced by the photomultiplier is carried out by an amplifier (Fig. 1, E) with the signal being displayed (Fig. 1, F) to the operator, and as is more common in modern instrumentation, the signal can be fed to a data station (Fig. 1, G) from where the data can be printed out on request in a report format.

Atomic absorption spectrometry as an analytical technique has in general two main limitations: firstly, it is restricted to single elemental determination, and secondly, it has a limited working calibration range. Both these limitations will be considered more fully later in this chapter.

The following sections in this chapter address each of the major instrumental components in turn, highlighting its function and discussing the procedures that may be necessary for its optimization with respect to the other components, to produce the best possible instrumental performance commensurate with the analytical need.

II. THE COMPONENTS OF THE SPECTROMETER

A. The atom cell

1. *Atomization in flames*

The atom cell in flame atomization can be considered to have two basic steps: firstly, nebulization to generate a fine aerosol of the sample solution, and secondly, dissociation of the analyte into gaseous ground state atoms. The most commonly used atomizer in atomic absorption is the chemical flame, based upon the combination of a fuel gas (e.g. acetylene) with an oxidant (e.g. air or nitrous oxide). The sample solution, usually aqueous based, is introduced into the flame using in general a pneumatic nebulizer in which the passage of the oxidant over the surface of a capillary creates a partial vacuum at the capillary surface due to the venturi effect, and thus the sample solution is effectively sucked up or more correctly forced under pressure up through the capillary by the pressure difference created.

In the nebulizer the sample solution emerges from the end of the capillary at a high velocity and is usually caused to collide with an impact bead situated in close proximity. This process produces an aerosol of the solution consisting of a wide range of droplet sizes. It is only the fine droplets however (<10 μm) which eventually reach the flame and these represent at best only 10% of the original solution. Although nebulizer efficiency may appear to be low, it has been found that this level of aerosol introduction into the flame does not significantly increase flame noise. It seems unlikely therefore that little benefit will be derived from increasing the nebulizer efficiency, when using conventional instrumentation, as a subsequent loss in signal stability will occur.

A schematic diagram of a concentric pneumatic nebulizer in which the sample liquid is surrounded by the oxidant gas, as it emerges from the uptake capillary, is shown in Fig. 2. Note that the trap in the drain must be full before the flame is lit to prevent flashback due to loss of pressure in the burner, and to prevent the fuel–oxidant mixture escaping into the laboratory.

The nebulizer is normally made of a corrosion-resistant polymer fitted with a platinum iridium alloy (90:10) capillary. Sample solutions are typically taken up by a narrow bore flexible chemically resistant polymer tube attached to the capillary. The impact bead if present is usually made of borosilicate glass, a chemically resistant polymer or an alloy similar to the capillary.

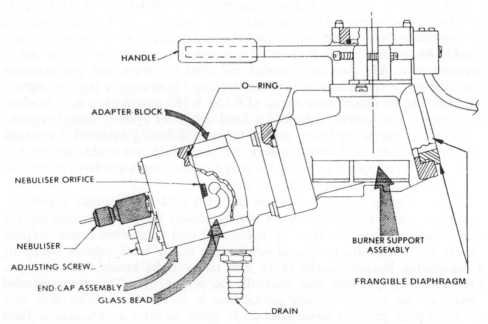

HANDLE

O—RING

ADAPTER·BLOCK

NEBULISER ORIFICE

NEBULISER

ADJUSTING SCREW

END·CAP ASSEMBLY

GLASS BEAD

DRAIN

BURNER SUPPORT ASSEMBLY

FRANGIBLE DIAPHRAGM

Fig. 2. Cross-section of premix chamber of a pneumatic nebulizer (Thermoelectron).

The use of organic-based solvents as opposed to aqueous solutions will have an effect on nebulizer efficiency. This results from different surface tension, density, viscosity or saturated vapour pressure of the solvent, which all influence the aerosol production and hence nebulizer transport. Solvents of this type can also act as secondary fuels in the flame and so modify the flame conditions for atomization. In general, organic solvents do not find a wide application in flame atomic absorption, but it is worth noting however that whilst more care is required when using organic solvents, they do tend to increase atom cell efficiency.

Some nebulizers are designed with an adjustment for sample uptake flow rates, which enables them to be optimised for sample solutions of varying viscosities and surface tensions. Where an adjustable nebulizer is fitted, the optimization is quite straight forward. The operator should set up the spectrometer for the determination of an "easy" element such as copper, which shows little dependence on flame stoichiometry. A sample solution of representative viscosity containing one or two parts per million of copper should then be aspirated. By carefully turning the adjusting screw on the nebulizer anti-clockwise, a point will soon be reached where no nebulization occurs and bubbles of air are seen in the sample beaker. At this point the instrument's absorbance display will show zero. By slowly turning the nebulizer screw clockwise, the aspiration rate is increased. The display should be set to be sufficiently responsive with a time constant of no longer than one second. As the uptake rate increases, so will the absorbance displayed on the readout. When a maximum is reached, the control should then be gently reversed, i.e. some slight anti-clockwise movement again, and by combining these adjusting motions the optimum signal-to-noise ratio will quickly be achieved. The highest absorbance obtained for a certain set of parameters should then be recorded and used for day-to-day performance comparisons. As a guide, a solution containing 1 ppm copper is quite capable of producing an absorbance signal of 0.150–0.180 absorbance on a modern instrument. For instruments with a fixed nebulizer system, sample uptake rates can be varied by using uptake tubes of differing internal diameters. This approach does not offer the flexibility of the fully adjustable system, but is sufficient to compensate for viscosity and surface tension effects associated with, for example, organic-based solvents.

Proper alignment of the flame in the light path is obviously important. The instrument should provide a sturdy mount for the sampling system and should provide for vertical, horizontal and rotational control. Quite clearly the horizontal or lateral adjustment is the most critical assuming a reasonable burner height (4–10 mm below light beam) has been set. The rotational position can generally be set by eye. The recommended procedure for correct burner positioning is to insert a white card into the light path prior to flame ignition in order to identify the source light path. The burner position can then be quickly adjusted such that the slot is a few millimetres squarely below the hollow-cathode lamp beam. Final burner position optimisation must be done with the flame lit and whilst aspirating an appropriate standard solution. Once again, a fairly responsive instrument display of absorbance is maximised by moving the burner head in the available planes, paying particular attention to the lateral adjustment for obvious reasons.

The flame is a chemical reaction which takes place in the gas phase. The ideal flame for atomic absorption would generate the correct amount of thermal energy to dissociate the atoms from their chemical bonds. The most commonly used flames are air–acetylene and nitrous oxide–acetylene. The choice of oxidant depends upon the flame temperature and composition required for the production of free atoms. These temperatures vary the molecular or chemical form of the element. Air and acetylene produce flame temperatures of about 2300°C and permit the analysis by atomic absorption of some 30 or so elements. The more reducing nitrous oxide–acetylene flame is some 650°C hotter and extends the atomic absorption technique to around 66 elements. It also permits the successful analysis of most elements by flame atomic emission, in many cases at fractional part-per-million levels, providing adequate spectral resolution is available.

The fuel and oxidant mixture must be controlled to provide the proper flame conditions for the element being analysed. A modern spectrometer should have a gas control system providing the precise and safe regulation which is important if reproducible results are to be obtained, particularly for those elements that show great dependence on flame stoichiometry.

When a fuel gas such as acetylene and an oxidant such as air are mixed and ignited, a flame is produced, the structure of which is well defined. The mixture of gases enters at the base of the flame, forming the inner cone, ignites and generates the very hot reaction zone. Most of the chemical reaction of combustion takes place in the reaction zone and is completed in the outer zone of the flame. It is worth recalling that several steps occur as the sample leaves the container, enters the flame, is atomized and has its absorption measured. Firstly, the sample solution is nebulized into a fine cloud-like mist. In the premix system the mist mixes with fuel and oxidant before entering the base of the flame where solvent evaporation takes place. The efficiency of solvent evaporation depends upon four factors. Firstly, the drop size; small uniform droplets give rapid evaporation. Secondly, the chemical nature of the solvent affects the evaporation; obviously volatile solvents evaporate more quickly than less volatile solvents such as water. Thirdly, flame temperature will affect evaporation; high temperatures, for example, result in high evaporation rates. Finally, solvent flow rate can be either too slow and thus limit performance or too fast, thereby flooding the flame and causing cooling.

Rapid and complete solvent evaporation is required for optimum performance. Atomization occurs in the flame reaction zone, i.e. the conversion of sample molecules into gaseous ground state atoms. Three factors influence the number of atoms formed. Firstly, the anion with which the metal atom is combined; calcium chloride, for example, is more easily dissociated than calcium phosphate. The second factor is flame temperature; higher temperatures cause more rapid decomposition and, indeed, are often specifically required for elements which form refractory oxides. Finally, gas composition may affect the rate of atomization if the constituents in the gas react with the sample or its derivatives. In the outer zone of the flame the atoms recombine to form oxides. In this form they no longer absorb radiation at the wavelength of the uncombined gaseous ground state atoms.

Thus, it can be seen that there are several variables associated with the flame atomizer that must be optimised to achieve the best sensitivity and detection limit. The flame must be correctly positioned with respect to the light path. The fuel/oxidant ratio should be investigated to establish the optimum chemical environment for atomization. The nebulizer and impact bead (where fitted) must be optimised to produce, overall, the best signal-to-noise ratio.

It is clear then that the chemical flame is an effective means by which a free, gaseous, neutral ground state atom population may be produced from a sample solution for analysis by atomic absorption spectrometry. The fact that flames were inherited from the older technique of flame emission spectrometry, may account in part for their popularity, although they also have the following advantages for use in AAS:

(a) They are convenient to use, reliable and relatively free from a tendency to memory effects. Most flames in common use can be made noiseless and safe to operate.

(b) Burner systems are small, durable and inexpensive. Sample solutions are easily and rapidly handled by the use of relatively simple nebulizer assemblies.

(c) A wide variety of flames is available to allow the selection of optimum conditions for many different analytical purposes. Table 1 lists various fuel–oxidant combinations that have received attention.

(d) The signal-to-noise ratios obtainable are sufficiently high to allow adequate sensitivity and precision to be obtained in a wide range of analyses at different wavelengths in the range 190–900 nm.

Flame atomization systems have some disadvantages, however, which limit their potential and convenience in use. These drawbacks have led many workers to devise techniques for the atomization of samples for analysis that are not based entirely on nebulizer–flame systems. Some of the possible drawbacks of flames for analytical work are:

(a) The sample volume available may be less than that required for continuous use with an indirect nebulizer system (e.g. 3–5 ml). For low analyte concentration it may not be possible to dilute the solution to circumnavigate this limitation. The discrete volume method of nebulization

TABLE 1

CHARACTERISTICS OF VARIOUS FLAMES

Oxidant	Fuel	Approximate temperature (°C)	Typical burning velocity (cm s^{-1})
Argon/diffused air	Hydrogen	400–1000	
Air	Coal gas	1840	55
Air	Propane	1930	45
Air	Hydrogen	2050	320
Air	Acetylene	2300	160
Nitrous oxide	Hydrogen	2650	390
Nitrous oxide	Acetylene	2950	285
Oxygen	Acetylene	3180	1130[a]

[a] Because of the high combustion velocity the flame cannot be burned on a premix burner system.

overcomes this limitation to a certain extent whereby tiny volumes of 50–200 μl can be aspirated to produce a series of transient peaks. Where sample dilution is not possible because of sensitivity problems, this technique is certainly the best way to get the most analytical data from a limited volume of sample.

(b) Pneumatic nebulizers used in premix flame systems are only approximately 10% efficient in getting the sample into the flame.

(c) Flame cells are only rarely able to atomize solid samples and viscous liquid samples directly.

(d) Flame background adsorption and emission at the wavelength of the analysis or thermal emission from the analyte or concomitant matrix at this wavelength, may give rise to unacceptable signal noise with consequent loss of precision.

(e) In some locations it may be inconvenient to use high-pressure cylinders of support and fuel gas, and in automated systems with no operator in attendance it may not be desirable to use a flame as the atom cell.

In addition to these practical disadvantages, other more fundamental factors act in flames to limit the sensitivity and selectivity that may be achieved. These are:

(a) The attainable atomic density in flames is limited by the dilution effect of the relatively high flow-rate of unburnt gas used to support the flame and to transport small volumes of sample solution to the flame. These factors have a limiting effect on the relative sensitivity of a direct flame technique. The optical density is also limited by the flame gas expansion that occurs on combustion. This factor has been estimated to be about 10^4 when a unit volume of acetylene burns in air.

(b) Precise control over the chemical environment of the analyte and concomitant atoms in flame cells is not possible. The degree of control of chemical composition that can be achieved by variation of the fuel/oxidant ratio is accompanied by simultaneous changes in the flame temperature and its spectral absorption and emission characteristics. For many elements, particularly those that form thermally stable oxides, the efficiency of free-atom production from the sample introduced into the flame is low. Clearly, for the analysis of small liquid or solid samples and for the determination of trace amounts of many elements in larger samples, it would be advantageous to achieve a higher concentration of atoms in a small cell volume than is possible with flames when solution nebulization systems are used.

2. Modified flame introduction techniques

Various approaches have been made to improve the sensitivity of flame atom cells. These techniques have been concerned with either trapping the ground state atoms longer in the optical path of the spectrometer (e.g. Delve's cup, STAT) or on preconcentrating the analyte and so releasing it for atomization in effectively a more concentrated form, observed as a transient signal (cold finger).

With the Delve's cup atomizer a silica or alumina tube is orientated along its axis in the optical path of the spectrometer, heated externally by the flame from a normal AA burner. A small sample volume (typically 10 μl) is placed in a nickel cup which is mounted directly under a hole in the trapping tube. When the sample cup is placed in the flame, normal thermal dissociation processes take place and gaseous ground state atoms are generated, but in this case, unlike a normal flame system, the atoms pass into the heated tube and become trapped longer in the optical path, increasing their residence time and consequently instrumental sensitivity. The apparatus does not find wide usage, but has been very effective for the determination of volatile metals such as lead and cadmium at sub-part-per-million levels in biological materials such as whole blood.

The delaying of atoms in the optical path has more recently found popularity with the so-called Slotted Tube Atom Trap (STAT) system (Fig. 3). This simple device can be fitted to a conventional burner, trapping atoms longer in the optical path much in the same way as the Delve's cup. In this practical device a heated quartz tube with a slot along its upper and lower edges is placed in a conventional flame, using a pneumatic nebulizer system. Trapping is achieved as the dissociated ground state atoms pass into the tube and are delayed from passing out of the optical path. Once again the increase in residence time results in an increase in instrumental sensitivity by a factor of up to five times for a wide range of elements.

The concept of atom-trapping is however not new and the cold finger has found some uses in flame application. In this case the gaseous ground state atoms produced by a conventional nebulizer and flame system are allowed to condense onto the outer surface of a cooled silica tube positioned just

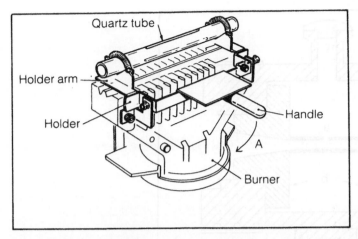

Fig. 3. Schematic diagram of the slatted quartz tube (STAT) (Philips Analytical).

below the optical path of the spectrometer. The tube is normally cooled by a rapid flow of cold water through the silica tube. On switching off the flow of coolant, the silica surface undergoes rapid heating and the condensed atoms are once again released as ground-state free atoms which absorb the light from the source. Thus a transient more concentrated signal is obtained than could be achieved prior to preconcentration of the analysed atoms. Both the atom-trapping and cold-finger techniques described represent relatively simple modifications to a conventional flame spectrometer and yet offer substantial signal enhancement for many application areas.

3. Atomization without flames—electrothermal atomization

Various electrically heated furnaces have been described in recent years, but the most common commercially available systems are based on a Massmann furnace, in which a fixed sample volume is introduced into the furnace and after thermal pretreatment is rapidly atomized (Fig. 4). This results in a transient signal whose height or area is proportional to the quantity of element under study. It has been demonstrated that in a graphite furnace atom cell a substantially higher peak concentration of atoms may be expected compared with a flame. This gain results directly from avoidance of the dilution and expansion effects that occur in flame cells and the longer residence time of the atoms in the optical path. To assist the formation and maintenance of a dense gaseous free ground-state atom fraction of the element for atomic absorption analysis, it is also an advantage that the chemical environment can be controlled by the use of an inert gas atmosphere, generally argon or nitrogen. The more recent designed electrothermal atomizers enable the addition of ancillary gases to,

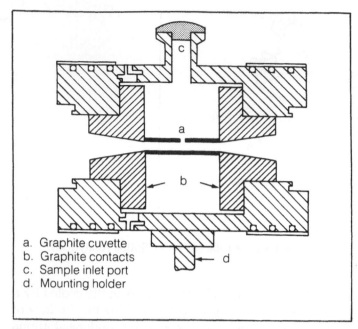

a. Graphite cuvette
b. Graphite contacts
c. Sample inlet port
d. Mounting holder

Fig. 4. The Massmann graphite furnace system (Philips Analytical).

for example, aid ashing (oxygen) and maintain a pyrolytic coating on the tube (methane).

A typical electrothermal atomizer atomic absorption set-up is shown in Fig. 5. The graphite tube or cuvette is mounted in the furnace workhead which replaces the burner in the spectrometer, and must be carefully aligned in the light path. It is after all, performing the same function as the flame but 100–1000 times more efficiently, in terms of sensitivity an improvement. The power supply or programmer is a means of developing a series of carefully controlled voltages across the ends of the cuvette, the purposes of which are: (1) drying liquid samples; (2) melting soil samples; (3) removing or partially removing the matrix; (4) catalysing chemical changes to enhance atomization; (5) catalysing chemical changes to ensure that only one form of the analyte is present at atomization; (6) catalysing reactions intentionally used for matrix modification; (7) atomization; (8) cleaning the furnace.

The selection and optimisation of furnace operating conditions will be dealt with in more detail in Section VI of the next chapter.

As indicated, electrothermal techniques are very sensitive. For example, a 10 ppm lead standard produces an absorption signal of about 0.44 absorbance at 217 nm by flame AAS; this same instrument deflection can be achieved from a 20 μl injection of only 10 ppb lead solution when using the electrothermal atomizer, i.e. a thousand-fold lower in concentration.

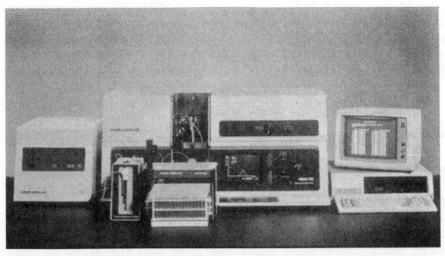

Fig. 5. An atomic absorption spectrometer fitted with electrothermal atomizer and autosampler (Thermoelectron).

However, furnace atomic absorption has certain disadvantages. The flame is more precise, faster, with much less trouble, and should always be used if sensitivity is adequate. The electrothermal method takes longer and requires more skill. Precision for comparable absorbances will generally be poorer. Moreover, chemical reactions in the graphite cuvette and contamination can be a real problem at these ultra-trace levels and all possible sources of analyte contamination have to be scrutinised.

Recent developments in the field of automatic sample injectors have done much to increase the ease of use, reliability, and performance levels of electrothermal atomizers. The sampling process may now be automated, thereby minimising the tedium associated with manual sample injection. The furnace auto samplers presently available fall into two categories; firstly and most common are those that transfer a volume of solution into the furnace mechanically and then commence the normal temperature programme, and secondly, those that produce an aerosol from the solution using a nebulizer and direct the spray into the cuvette which is maintained at elevated temperature so that solvent evaporation occurs continuously with sample injection. Liquid samples (5–50 μl) are placed in the furnace via an injection hole in the centre of the graphite tube. This is normally achieved by a simple robotic arm controlled by an auto sampler unit. The auto sampler enables repeated sample injections, the addition of matrix modifiers and full calibration including the method of standard additions to be carried out. With the aerosol injection method the amount of sample deposited is controlled by time (typically the rate is about 1 μl s^{-1}); accurate depositions of very small volumes are possible. This allows samples of relatively high concentration to be analyzed without dilution. At the same time, when very dilute solutions are being analyzed, long deposit times allow practical injection of volumes much larger than tube furnaces can normally accommodate, thereby greatly improving sensitivity. Using the discrete volume injection technique, sensitivity can be improved in a similar way by a multi-injection technique in which samples are dried between subsequent injections. The aerosol injection technique is particularly well suited to the analysis of oils and other organic liquids. The fine spray is dried instantly on contact with the tube, thereby overcoming the tendency of these materials to creep out of the system. In addition, multipoint calibration curves may be generated from a single standard by dispensing different volumes into the furnace by varying the aerosol deposition time. The aerosol injection technique does however require quite large sample volumes (>5 ml) and may be unsuitable for biological-type samples.

In recent years considerable effort has gone into the study of both the chemistry and physical processes associated with furnace atomization. This work has had a profound influence on the maturing of ETA-AAS as a reliable technique for trace-element analysis. As indicated above, much of this reliability has been attributed to the use of auto-samplers and more sophisticated temperature operating systems, controlled by an onboard microprocessor accessed through a data station. However, the modern electrothermal atomizer and its practical use have been significantly changed by modification in cuvette design. Pyrolytically coated and more recently totally pyrolytic graphite tubes have significantly reduced surface porosity and carbide formation, formally a severe problem with certain elements. Understanding the thermal processes within the graphite tube has also had an important bearing on the design of cuvettes. In simple terms, during the normal use of a Massmann furnace samples are vaporised from a hot tube wall into a cooler and rapidly changing vapour phase, which has been shown to favour the formation of analyte molecules. The early work by L'vov in 1959 indicated that interference effects in the cuvette could be significantly reduced if the atoms were vaporised into an isothermal atmosphere at the atomization temperature of the analyte atmosphere. Simply increasing the mass on which the sample is placed by the introduction of a platform, for example, would delay the thermal atomization of the sample sufficiently to allow the atmosphere within the cuvette to gain stable isothermal conditions before the atoms are vaporised. Today's instruments offer the option of a platform or delayed atomization cuvette as standard. One of the more recent developments in the area of platform atomization has been probe atomization (Fig. 6). Essentially four steps are involved in this process: (1) the sample is introduced onto the probe inside the cuvette, dried and ashed in the usual way; (2) the probe is then withdrawn and the furnace which is then heated and allowed to stabilize at

Fig. 6. The Massmann furnace with probe attachment (Philips Analytical).

the atomization temperature selected; (3) the probe is then reinserted into the isothermal environment and atomization takes place; and finally (4) the probe undergoes a thermal clean stage.

The chemical modification of electrothermal atomization has focused on the use of matrix modifiers, which although normally used with conventional cuvettes, can be used in conjunction with delayed atomization techniques also. Matrix modifiers, e.g. magnesium nitrate or ammonium dihydrogen phosphate, have been widely used to chemically delay atomization again in an attempt to reduce interferences by allowing closer isothermal conditions to be achieved. Recent developments in this area suggest that palladium may be used as a universal matrix modifier.

A greater understanding of the processes associated with electrothermal atomization, which has occurred in recent years, has stimulated much thought on suitable methodology for ETA-AAS. The techniques described above, together with important developments in background correction methods (see Section III), have been combined to create a more reliable, suitable analytical protocol. Today a more reliable robust methodology exists for the determination of trace elements in a wide range of both liquid and solid samples.

4. *Vapour generation techniques*

The chemical properties of some elements are such that special methods may be used, both to separate them from the sample matrix before their introduction into the light path and for conversion into an atomic vapour once there. These elements include antimony, arsenic, bismuth, lead, germanium, selenium, tellurium and tin, which readily form volatile hydrides upon reduction with sodium borohydride, and mercury, which is unique among the elements in possessing a high vapour pressure at room temperature and can readily exist as a monatomic vapour. In this last example no atomizer is required with the sample under examination being reduced, generally by stannous chloride and hydrochloric acid, and the vapour swept into the light path of a mercury hollow-cathode lamp. Most commonly a closed loop system is employed where an electric pump circulates the mercury vapour, and atomic absorption measurements are made using a quartz-ended gas cell in the light path.

In the former case, that of hydride generation, the evolved gas is swept to the atomizer by a stream of argon, nitrogen or air depending upon the type of atomizer employed. A number of flames can be used, the most successful being those burning hydrogen by entraining air from the atmosphere, thereby creating a diffusion flame. A support gas of argon or nitrogen is also required. An air–acetylene flame may be used but severe reduction in analytical sensitivity will occur due to low energy transmission of the elemental emission lines from the hollow-cathode lamp, these being at low wavelengths. Alternatively the hydride(s) may be swept into a heated silica tube mounted in the optical path above an air–acetylene burner supporting its usual air–acetylene flame. This method of atomization has very high sensitivity, and eliminates certain types of molecular absorption observed when the liberated hydrides are passed directly into the flame.

Where appropriate, these methods of atomization will be discussed again in the later chapters describing applications. These elements may be determined by conventional nebulization systems, but with severely reduced sensitivity in most cases.

B. Light sources

The most familiar type of light source is the "continuum" or white light source. The domestic filament bulb is an example, having an emission spectrum over a wide wavelength range starting at about 300 nm and extending into the infra-red. A continuum source which has a useful function in atomic absorption analysis is the hydrogen or deuterium-filled hollow-cathode lamp. This source emits strongly in the UV part of the spectrum. The spectral characteristics of these continuum lamps are illustrated in Fig. 7.

However, since the absorption phenomenon being measured is occurring over an extremely narrow part of the spectrum (0.01 nm), it would require very high resolution to measure any significant absorption from a continuum lamp. This is why atomic absorption spectrometry as an analytical tool made very little progress until Walsh described the hollow-cathode lamp line source in 1955. A line source emits only at discrete wavelengths. The spectral lines are narrower than the absorption lines being measured, thus high resolution is not required. A typical hollow-cathode lamp is shown in cross-section in Fig. 8. The lamp consists of a glass envelope containing a cathode and an anode.

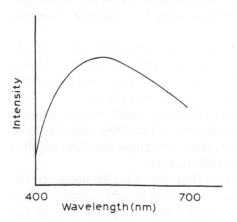

Fig. 7. Spectral characteristics of a continuum source.

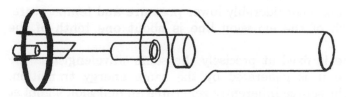

Fig. 8. Cross-section through a hollow-cathode lamp.

The cathode is a metal cup or cylinder which is made of the chemical element for whose analysis it will be used. Thus, for a copper analysis, a lamp will be used having a pure copper or brass cathode. In the second case the lamp spectrum will also contain zinc lines. The sealed glass envelope contains an inert gas, usually neon at low pressure. When a high voltage, 300–600 V is applied across the electrodes, positively charged neon ions bombard the cathode and dislodge (sputter) atoms of the cathode element. These atoms are excited by collisional processes and the emission spectrum of the element or elements present in the cathode are produced. Thus, by choosing the appropriate chemical element for the cathode, the atomic spectrum of that element may be readily generated. The cathode is usually surrounded by an insulating shield of mica, ceramic material or glass. This ensures that the discharge is confined to the interior of the cathode and results in an improvement in the intensity of the emitted lines.

The anode can be of almost any shape. It can be formed as an annular ring around the mouth of the cathode, or as a "flag" near the mouth of the cathode, or as a wire or rod located in a convenient position. In some designs the anode serves the dual purpose in that it also acts as a mechanical support for the insulating shield.

The material used for the lamp window is important since it must transmit the spectral line(s) of the element being studied. Because quartz glass transmits over the full wavelength range, it is suitable for all lamps. The other glasses are less expensive, and they can be used for elements where resonance lines lie above 300 nm.

The filler gas is usually argon or neon at a pressure of 10–15 Torr, i.e. about 1/50 of atmospheric pressure. Neon is preferred because it produces a higher signal intensity than argon, but where a neon line occurs in close proximity to the element's resonance line, argon is used instead.

Hollow-cathode lamps are not, of course, the only light source capable of producing the line spectra of chemical elements, but they are the most universally accepted source for atomic absorption instruments. The reasons for their popularity are to be found in their attributes:

(a) They generate a very narrow emission line. In order to measure the peak absorption, the width of the source must be very much narrower than the width of the absorption line. The width of the absorption line of the atoms in the flame is largely determined by temperature and pressure, and is normally about 0.01 nm. The atoms in the hollow-cathode lamp are in an environment having a considerably lower pressure and temperature. Consequently, the width of the emission line is about one tenth of the absorption line width.

(b) The emission line is fixed at precisely the same wavelength as the absorption line because it is generated by the same energy transition. This means that the light source inherently generates an emission signal at precisely the correct wavelength for optimum absorption. The unique overlap of emission and absorption lines is sometimes known as the lock-and-key effect.

(c) Hollow-cathode lamps can be made for all chemical elements that can be determined by atomic absorption. All of these elements are suitable for inclusion in the cathode by one means or another (machining or powder metallurgy)

(d) They are simple to operate. All that is necessary is to connect the electrodes to a suitable power supply and adjust the current to the prescribed value. No complicated adjustments are involved, and control of environmental conditions is unnecessary.

(e) They are both stable and intense. Adequate intensity of signal is available for most elements such that intensity is not the most common limiting factor. Absorption is independent of source intensity so that increasing the intensity of the signal will not increase the degree of absorption. It must be pointed out, however, that intensity and noise level are inextricably linked in atomic absorption, as in most spectroscopic techniques. An increase in intensity will often yield a reduction in noise level, or provide an improvement in signal-to-noise ratio. This is usually because less electronic amplification of the signal is required and therefore less noise is generated in the detector–amplifier system. But, in flame absorption spectrometry, the flame itself can be a source of noise. Since the flame is a dynamic system, it presents an absorption which is not constant but fluctuating. These fluctuations in absorption are caused by flame movement due to drafts or imperfect gas mixing. Even very minor fluctuations will affect the intensity of the lamp emission reaching the detector and cause a varying or "noisy" signal. This noise component is independent of lamp intensity. For many elements, the flame absorption fluctuation contributes most to the total noise. For these elements, therefore, any increase in source intensity would offer little advantage. Noise is only one aspect of stability. The other is drift, either during the warm-up period or during operation. Most hollow-cathode lamps are sufficiently stable for routine analysis after about five minutes warm-up, but for high precision analysis a longer warm-up period is desirable.

(f) Hollow-cathode lamps are economical. Most manufacturers guarantee their lamps for a life of 5000 milliampere hours or two years. This means that if a lamp is operated at 5 mA, it is guaranteed to last at least 1000 hours of operation. In routine use this would be adequate for many thousands of determinations. The introduction of Smith-Hieftje background correction has had an important influence on the design of certain hollow-cathode lamps (e.g. cadmium). The lifetime of such lamps is still guaranteed however by the manufacturer.

1. *Operating conditions*

One particularly useful feature of the hollow-cathode lamp is that only one parameter needs attention, i.e. the operating current. Manufacturers generally recommend a suitable operating current. This is seldom highly critical and small departures from it will have relatively small effects on sensitivity. The only absolute limitation is that the maximum current specified must not be exceeded. The current actually recommended for a particular lamp is always a compromise since there is no clearly defined optimum and no specific lamp current corresponding to peak performance.

The most obvious effect of altering lamp current is the effect on intensity. Lamp intensity by itself is not important in absorption spectrometry (because the absorption signal does not depend on absolute intensity), but the secondary effect of improving the signal-to-noise ratio seems desirable. But this is not the only consideration. If it were, lamps would always be operated at their maximum rated current. There are two adverse effects of operating a lamp at high current and these require that a compromise be made in the selection of operating current. The first of these effects is that at higher currents the resonance line becomes broadened and distorted. This phenomenon is known as "self-absorption broadening" or "self reversal" and is caused by gaseous ground state atoms in the discharge absorbing at the resonance wavelength emitted by excited atoms. The result is that the absorption sensitivity is degraded and the calibration curvature increases. Figure 9 illustrates the change in curvature for magnesium when the lamp current is increased. The second adverse effect is that lamp life is shortened as the operating current is increased. The effect of self-absorption has, of course, become an important feature of the Smith-Hieftje background

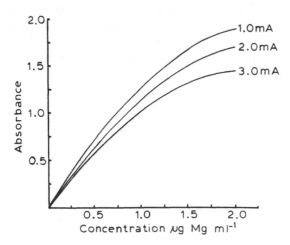

Fig. 9. Graph showing absorbance as a function of lamp current.

correction technique. The pulsing of the lamp at high current to induce self-absorption may reduce the life of lamps but manufacturers will supply at slightly higher cost special lamps for this mode of background correction.

With these factors in mind, it is good operating practice to use the manufacturer's recommendation as a starting point and then determine empirically the current which gives the optimum combination of signal-to-noise ratio and calibration linearity. The major consideration is not to exceed the maximum rated current.

2. *Multi-element lamps*

By combining two or more elements of interest into one cathode, it is possible to produce a hollow-cathode lamp that can be used for the analysis of more than one element. Such a lamp is obviously convenient for the analyst, but there are some limitations to this approach. Some combinations of elements cannot be used because their resonance lines are so close that they interfere with each other. This makes it impossible to resolve the line required and the combination is therefore unworkable. Other combinations cannot be used simply because of manufacturing difficulties in trying to incorporate elements of widely differing physical characteristics into a common cathode. However, for chemical elements to which such limitations do not apply, multi-element lamps do provide the analyst with a convenient source for the routine determination of several elements.

3. *Electrodeless discharge lamps*

For a number of years electrodeless discharge tubes excited by microwave frequencies have stimulated interest. The biggest claimed advantage of EDLs is the increased intensity of the line spectrum, by several orders of magnitude compared with the hollow-cathode lamp. They consist of a sealed quartz tube several centimetres in length and about 5–10 mm in diameter, filled with a few milligrams of the element of interest (as pure metal, halide or metal with added iodine) under an argon pressure of a few Torr. The tube is mounted within the coil of a high-frequency generator at around 2400 MHz and excited by an output of a few watts up to 200 W. Various views exist on the advantages or otherwise of electrodeless discharge lamps in AAS. Higher radiation intensity does not influence the sensitivity but the signal-to-noise ratio can occasionally be improved leading to better precision. Radio frequency EDLs operating at 27.12 MHz are now available which, although having lower light output, have proved to give better long- and short-term stability than their microwave equivalents.

4. *Continuum source*

As indicated earlier the emission profile from a deuterium or hydrogen lamp has a strong emission in the UV part of the spectrum. This spectrum has an important role in quantifying non-atomic absorption and represents one of the most common forms of background correction. More serious attempts at performing simultaneous multi-element atomic absorption measurements has concentrated on using a more intense continuum source, such as the Xenon lamp. The high resolution required to achieve individual elemental quantification lies on the monochromator used and this aspect of spectral resolution will be discussed briefly in the subsequent section.

C. Monochromator and optics

The monochromator separates, isolates and controls the intensity of radiant energy reaching the detector. In effect it may be seen as an adjustable filter which selects a specific, narrow region of the spectrum for transmission to the detector and rejects all wavelengths outside this region (Fig. 10). Ideally, the monochromator should be capable of isolating the resonance line only and excluding all other wavelengths. For some elements this is relatively easy, for others more difficult. The copper spectrum, for instance, is relatively uncluttered, the nearest line being 2.7 nm from the 324.7 nm line. Nickel on the other hand has quite strong lines at 231.7 nm and 232.1 nm, one each side of the 232.0 nm primary line. The ability to discriminate between different wavelengths (i.e. resolution) is thus a very important characteristic of the monochromator.

Several methods may be used to isolate the required part of the spectrum, but the diffraction grating monochromator is now universally used in atomic absorption instruments. A schematic of a typical diffraction grating monochromator is shown in Fig. 11. The geometry of such a monochromator may vary slightly, however, between different instrumental manufactures. The radiation from the hollow-cathode lamp enters the monochromator through the entrance slit and is focused by the first mirror onto the grating.

SPECTRUM ON PLANE OF EXIT SLIT

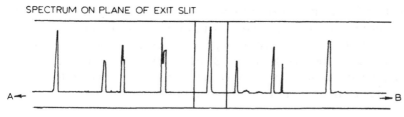

Fig. 10. In the monochromator the resonance line is isolated from unwanted nearby radiation.

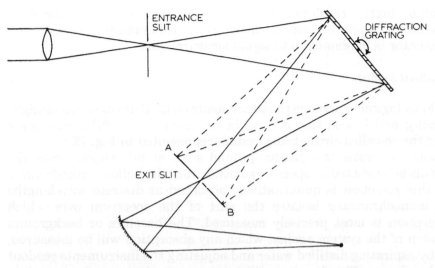

Fig. 11. Schematic of a monochromator as used in an atomic absorption spectrometer.

The grating disperses the radiation into individual wavelengths. By rotating the grating, the analytical wavelength of interest will pass through the exit slit and be focused onto the detector.

The entrance slit, occurring as it does prior to dispersion, controls the amount of light entering the monochromator and should ideally be as large as possible. The exit slit determines the spectral bandwidth, i.e. the width of the tiny part of the spectrum transmitted to the detector. As we have seen already, this can be larger in some cases (Cu) than in others (Ni). In practice the two slits are ganged together so that the choice of slit width is always a compromise between high light throughput corresponding to excellent signal-to-noise characteristics, and having the required degree of line separation to prevent the detector "seeing" more than it should.

Since the hollow-cathode lamp spectra used in AAS are relatively simple, spectral bandwidths narrower than 0.1 nm are seldom if ever used. In atomic emission analysis, however, higher resolving power is often essential, particularly when the excitation source (e.g. the nitrous oxide–acetylene flame) is producing a complex spectrum. The instrument should, therefore, provide a wide range of slit settings and a convenient digital display of the wavelength in use for the operator. The isolation of absorption lines from within a continuum spectrum calls for a higher resolution monochromator than the type described above. In this case an Echell monochromator which carries out dispersion in two dimensions can be used. The requirement for high resolution for other than continuum atomic absorption is not, however, normally required, as the mutual lock-and-key effect which occurs in the atom cell, represents the important mechanism in atomic absorption.

42

As indicated, higher resolution is, of course, of greater importance in atomic emission, where isolation of interference free emission lines by the monochromator is fundamental to signal measurement.

1. *The optical layout*

It would be logical to arrange the components so far described in a straight line optically, and indeed this is done in the most successful instruments producing the so-called single-beam system represented in Fig. 12.

The flame generates the gaseous ground state atomic vapour from the sample. This is irradiated by specific radiation from the hollow-cathode lamp. Some of this radiation is quantitatively absorbed at discrete wavelengths and the monochromator isolates the part of the spectrum over which that absorption is most precisely measured. The baseline or background transmission of the system against which any absorption will be measured, is set up by aspirating distilled water and adjusting the instruments readout to zero absorbance. This baseline will have various degrees of uncertainty associated with it. There are various contributions of noise in the system and certain variables can cause long-term drift. For the most precise and accurate results the baseline zero should be as stable as possible, and on the whole modern spectrometers satisfy this condition. The stability of the instrument is however dependent on the lamp current, flame conditions and photomultiplier voltage selected. It is therefore important that the operator ensures that suitable conditions are selected if stable conditions are to be achieved.

Instabilities in early instruments led to the introduction of a double-beam optical configuration, which attempted to overcome the limitation imposed by contemporary lamps. In this arrangement the radiation from the source is split into two beams, a sample beam going through the atomizer

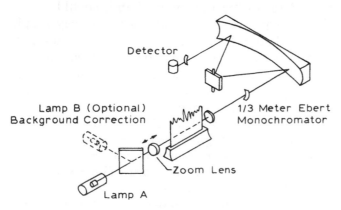

Fig. 12. Single-beam optical arrangement.

and a reference beam bypassing the atomizer and therefore of unvarying magnitude. The beams would be similar in intensity when the baseline zero was being established. Upon introduction of a sample into the flame, the sample beam would be attenuated and the instrument would compute log I_o/I to measure the absorbance. Any variation in output from the hollow-cathode lamp would be equivalently translated in both beams and the end reading would be unaffected. It must be noted, however, that the double-beam configuration will not compensate for fluctuation in atomizer efficiency, or in general for any variation in the flame/premix chamber/nebulizer areas.

Most instrumentation today is of the single-beam type; however, double-beam instruments may be used in a single-beam mode with the reference beam not being sampled by the detector. Statistically, improved precision should result from the selection of single-beam operation if the source lamp is free from drift. In terms of light throughput, of course, the single-beam arrangements uses light most efficiently. However, it must be pointed out that ideally the light throughput for a double-beam system will be no worse than the equivalent single-beam instrument when equipped with a deuterium background corrector, which requires insertion of an extra source to emit a continuum (see Section III). The incorporation of the background correction source into the instrument will always result in some loss of transmission. This loss would be 50% when static beam splitters are employed.

2. *Dual-channel optics*

Figure 13 represents an instrument having dual-channel optics. This means that there are facilities for two line sources and one continuum

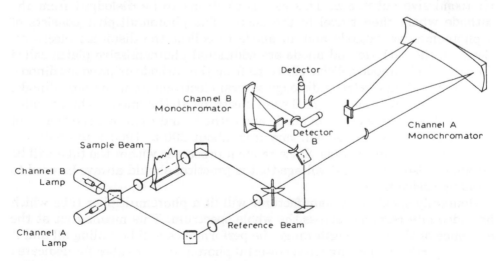

Fig. 13. Dual-channel optical arrangement.

source, each line source having its own monochromator and detector. Many workers have advocated the usefulness of this type of instrument. Certainly the recently introduced microcomputerised dual-channel models are easier to use than their earlier counterparts, so it is worthwhile to summarise the possible attractions of simultaneous dual-element analysis. One obvious possibility is the analysis of two elements in a sample at the same time, thus halving analysis time. A less obvious attraction is the possibility of analysing via an internal standard. This is where a second element, either already present or added to the sample, is measured and ratioed to the analytical element. The analyte display is now independent of physical variations in atomizer characteristics. Additionally, if the internal standard is already present within the sample homogeneously, it means that the sample preparation need not be quantitative. Weighings and dilution ratios may be quite approximate since the two elements' concentrations will always be fixed with respect to each other. Some examples of internal standard analysis will be presented in the application chapters.

D. Electronics and readout

The radiant energy from the source must be converted into an electrical signal for amplification and measurement by the readout or data processing system. This conversion is undertaken almost universally by a photomultiplier tube.

1. *Photomultiplier*

Photons from the radiation source bombard a cathode containing a photoemissive substance. This causes electrons to be dislodged from the cathode which then travel to the anode. The photomultiplier consists of a photoemissive cathode and an anode to collect the displaced electrons. Between the cathode and anode are additional photoemissive plates called dinodes. Each dinode collects electrons from the cathode or previous dinode. The bombarding electrons dislodge several electrons from the next dinode, producing an extremely high flow of electrons at the anode. The operator controls the voltage between anode and cathode and thus sets the "gain" of the detector. This voltage will vary from about 200 to 1000 V to produce a wide range of gain settings. Since random processes within the tube will be expanded also, the lowest voltage that is practical should always be used to avoid excessive noise.

Generally speaking a manufacturer will fit a photomultiplier tube which has adequate response across the whole spectrum. This means that at the extremes of the wavelength range the performance will be tailing off and it may be worth considering more powerful photomuliplier tubes for dedicated work at the red end, for cesium and rubidium for instance. Photomultiplier response curves are available from manufacturers such as Hammamatzu and Philips.

2. *Source modulation*

It is necessary that the hollow-cathode lamp source be pulsed or modulated at a certain frequency, and for the amplifier to be locked into this frequency to permit discrimination between the continuous emission signal coming from the light source and that associated with excited atoms in the atom cell returning to their ground state, after absorbing radiation, producing an emission in the atomizer. Only the resonance radiation from the lamp must be seen by the detector. In modern digital electronic instruments the lamp cycle is controlled by a sophisticated electronic clock which is sampled to provide the short pulses of power to switch the lamps on and off in phase with the amplifier. The flame background emission is measured during periods of time when the lamp is off. It should be remembered, however, that the discrimination occurs at a point after all the light has fallen upon the photomultiplier tubes. This can result in saturation or breakthrough, and conditions must be modified in this case to reduce the level of flame emission. This problem will be considered in some detail in the section on interferences in Chapter 3.

3. *The readout*

The modern practice in scientific instruments is to present the results digitally. This avoids errors in scale readings on meters through parallax, misinterpolation between scale divisions, etc. However, before the amplified output from the photomultiplier is displayed, it must be converted to read in absorbance units, the logarithmic function of percent absorption. A logarithmic amplifier is used to make this conversion. Most current instruments will integrate the signal over a selected period, typically 0.1 s up to 99.9 s, so that an unvarying result appears on the display after the selected time. Clearly, the precision of successive integrated results will improve as the integration period lengthens, so the operators choice will be a compromise. With an update time of 0.1 s the output fed to a chart recorder will appear equivalent to the analogue output, which was a characteristic of older equipment. It is important that these very fast times are available for use with electrothermal atomizers which produce very short-lived transient absorption signals.

Very small signals may be expanded continuously using the instrument's scale expansion control. This facilitates the reading of small absorbances, but it should be remembered that any fluctuations in the signal will be scale-expanded also. It is not always immediately grasped that readout directly in concentration units is achieved via the scale expand functions.

After all, if 10 ppm lead is aspirated versus a distilled water blank, then an absorbance of about 0.500 would result. The readout could then be made to read directly in concentration units (ppm) by employing a scale expand factor of 2 and shifting the decimal point one place to the right. This is what concentration readout consists of. However, to use this facility usefully, deviation from linearity must be compensated for. The practicality of this will be discussed in Chapter 3; suffice it to say at this stage that non-linearity in the calibration graph can be corrected for either using some manual device or automatically using the microprocessor-controlled instruments now widely available on the market.

4. *The use of a microprocessor*

The introduction of a simple on-board microprocessor or external data station in atomic absorption spectrometry has undoubtedly increased flexibility and ease of operation. The benefits can be summarized as follows: (a) easier operation via a keyboard to set up instrumental parameters; (b) a greater control on parameters, especially integration periods and furnace conditions; (c) automatic linearization of calibration graphs using a blank and up to five standards, using various well proven equations; (d) statistical interpretation of data, including percent relative standard deviation on a series of measurements, thus providing important information about the analytical precision; (e) operating conditions and calibration graphs once set up may be stored in the memory and recalled for use when required; one standard can then used to set the slope for that particular occasion; (f) easy visual presentation of instrumental conditions and result reporting.

It is now common to use the microprocessors to selected functions both on the main instrument and for peripheral devices such as an electrothermal atomizer or autosampler. Programmes for instrument setting and data processing can be stored, for example, on floppy or hard disk. Although, as already indicated, the actual speed of analysis may not be vastly improved, the advantages lie in automation and the better reliability and accuracy achievable. The current trend in computer-controlled instrumentation is towards a self-optimising fully automated analysis. In such a system the operator would load the autosampler with samples, select mode of calibration (i.e. aqueous, matched matrix or standard addition) and insert the hollow-cathode lamps (in a multiturret) required. The instrument would then run automatically in the flame or furnace mode and report the data in a suitable format. Problems with samples or instrumental stability could be identified and logged by the instrument. The modern automated computer-controlled atomic absorption spectrometer now represents a truly reliable and efficient analytical instrument.

III. BACKGROUND CORRECTION

We have seen that the atomic emission lines from the hollow-cathode lamp are uniquely different for each element and that the atomic absorption of these emission lines is therefore very specific, producing what we know as the lock-and-key effect. One might imagine therefore that because of this very specific relationship that takes place in the atom cell, that once gaseous ground state atoms had been produced, there would be few problems remaining. This, however, is most certainly not the case and non-specific absorption and/or scattering of the hollow-cathode emission lines by molecular species present in the atom cell can account for quite a substantial absorbance signal or loss of transmission of the atomic emission lines. This problem is most noticeable at shorter wavelengths (<300 nm) and when an electrothermal atomizer is being used, but it can and does also occur in flames. For accurate determinations we therefore require some mechanism for distinguishing between the atom-specific absorption (the real signal) and the non-specific absorption or background signal. One important consideration in dealing with this problem is to remember that the profile of the atomic absorption line is always sharp (10^{-3} nm wide), whilst that of a non-specific background effect is very broad, perhaps tens of nm wide. Several ways of achieving background correction have been developed and all of the methods are common in that they determine the total absorption at the analytical line (i.e. specific and non-specific absorption) and subtract the background from the analyte signal to give a corrected value for the element in question. The three most common techniques used for background correction are considered below.

A. Continuum hydrogen (deuterium) background correction

This represents the oldest and perhaps the least effective of the three techniques, due to the fact that a second light source, the hydrogen lamp is required, and so beam alignment and intensity balancing can give spurious results. The continuum radiation from the hydrogen lamp fills the spectral slit of the monochromator, giving a broad molecular band signal. When a sample is atomized in the atom cell, broad-band non-specific (background) absorption will occur across the entire spectral slit and the absorption of the hydrogen emission will take place. The analyte atoms will, however, also be atomized in the atom cell and these atoms will absorb a proportion of the continuum signal. The atomic lines are, however, very narrow (10^{-3} nm) and so the total amount of the energy absorbed by them across the whole slit (0.1 nm) is very small compared with the molecular energy absorbed. If an atomic line were completely absorbed from the 0.1 nm waveband, then the total absorption would only represent 1%. In this method of background correction this small proportion is essentially ignored. Consequently the absorption of the hydrogen lamp is a measure of the background absorption, whilst switching, using a beam splitter, to the absorption of the emission line from the element-specific hollow-cathode lamp will represent the specific and non-specific or the analyte and background signal. Thus fast switching between the two absorption signals allows a simple subtraction to be made between the background (hydrogen lamp) and the analyte plus background signal (elemental hollow-cathode lamp) to give the corrected analyte signal. Some electronic correction of the element-specific signal using the continuum background measurement will be required. The three main problems with this approach are:

(a) The hydrogen emission profile must be optically aligned with the element-specific emission line.

(b) The background absorption profile is considered uniform across the monochromator slit, i.e. it has no structure.

(c) When the level of background absorption increases above about 0.4 absorbance units, then electronic subtraction of the background signal becomes less accurate.

B. Zeeman background correction

In the Zeeman effect the energy levels of an atom are split under the influence of a strong magnetic field to produce fine structure with wavelengths greater than and less than the elemental resonance line. If the magnetic field is strong enough, the resonance line is eliminated entirely, but such an approach requires a strong varying magnetic field, a technique not favoured by the instrument manufacturers.

In practice a field of around 1 tesla is applied either to the source or, more commonly, to the atom cell. This field will cause the normal Zeeman effect to occur and three spectral components will be produced: the π component at the original wavelength and the σ^+ and σ^- components at wavelengths on either side of the original resonance wavelength. Under a strong enough magnetic field the σ components will lie equal distance either side of the resonance line out of the wave band of the resonance line. In addition to this shift the σ components are polarised perpendicular to the direction of the magnetic field, whilst the π component is polarised parallel with the field. The use of a polarizer thus enables the detector to observe the π component (specific and non-specific absorbance) and the σ components (non-specific, background absorption) alternatively. This switching can be achieved by a rotating device suitably placed in the optical path. Background correction is achieved therefore simply by subtracting the background absorption (σ components) obtained from either side of the resonance line from the resonance line absorption which is subject to interference.

This form of background correction overcomes the problem of requiring two light sources as in the case of continuum correction, and can compensate for structural effects where background absorption may be more significant on one particular edge of the resonance line. Zeeman background correction is therefore regarded as a more powerful correction technique, able to operate for up to 0.8 absorbance units of background. The main limitations are, however, that not all elements undergo normal simple Zeeman splitting and that by placing the magnet, as is common, around the atom cell, it precludes the use of the technique in flame AAS.

C. Smith-Hieftje background correction

This is the most recently introduced method of background correction, which exploits an effect observed for some time in atomic absorption spectroscopy, that of hollow-cathode lamp emission self-reversal. If the hollow-cathode lamp is operated not at a normal current, of say 3–10 mA, but at a few hundred mA, then the sharp emission line changes to a broad profile with the peak centre substantially reduced. This gives the appearance of two peaks either side of the primary resonance line. Thus, at low current a normal sharp resonance line is observed, but when pulsed at a high current the resonance line diminishes and two peaks at the edges of the resonance line are observed. This effect occurs as a result of ground state atoms being generated inside the hollow-cathode lamp, as a result of severe cathode inert gas bombardment from the inert filler gas when using high currents. These gaseous ground state atoms which are generated a short distance from the cathode surface, will absorb the emission lines being generated at the cathode surface and produce an emission profile with the resonance line effectively missing.

In practice, if the lamp is rapidly pulsed, then at low current the normal specific and non-specific absorption takes place whilst at high current due to a significant decrease in the presence of the resonance line, the edges of the profile are measured where only the background absorption is occurring. Correction is achieved therefore by subtracting the background from the elemental-specific line absorption to obtain a corrected signal.

This approach to background correction can be carried out, as in the case for Zeeman, with one light source and so no problem of alignment occurs. In addition any structural differences in the background either side of the resonance line will be compensated for. The major problem with the Smith-Hieftje technique is that the lamp can only be pulsed at a fairly low frequency, and this means that resolution of the corrected signal becomes a problem with transient signals of elements at a low concentration. The technique can, however, be used with either flame or furnace techniques and can correct severer interferences of the order of 0.7–0.8 absorbance units.

Chapter 3

Practical Techniques

S.J. HASWELL

School of Chemistry, University of Hull, Hull HU6 7RX (Great Britain)

I. INTRODUCTION

Atomic absorption spectrometry (AAS) is a virtually universal method for the determination of the majority of metallic elements and metalloids in both trace and major concentrations. In general the form of the original sample is not important, provided that it can be brought into either an aqueous or a non-aqueous solution. Understanding more fully the processes involved when solutions are introduced into an atomizer, inevitably has led to better methodology for the removal of chemical interferences and the development of more suitable calibration procedures. In addition to solution phase sample introduction techniques, some interest has focused on direct solid sample introduction, particularly in ETA-AAS. This interesting development in sample introduction will be discussed later in this chapter.

Atomic absorption methods therefore combine the specificity of other atomic spectral methods with the adaptability of wet chemical methods. High specificity means that elements can be determined in the presence of each other. Separations, which are necessary with almost all other forms of wet analysis, are therefore reduced to a minimum and often avoided altogether, making typical atomic absorption analytical procedures attractively simple. In practical terms elements can be considered as groups rather than individuals, and so in preparation of samples for atomic absorption as many elements as possible are brought together for determination in the final analysis solution. Thus relatively simple sample preparation coupled with the ease of handling available with a modern automated atomic absorption spectrometer, makes it possible for routine analysis to be carried out quickly and economically with minimal instrumental supervision. This together with simple rapid but accurate sample preparation techniques should always be the aim in method development.

II. CALIBRATION

It is worth considering, before a detailed description of sample preparation is given, that atomic absorption like the majority of analytical techniques is only relative and requires careful calibration.

A. Beer's law

Beer's law states that the absorbance of an absorbing species is proportional to its concentration and is represented in fairly simple terms in the idealised calibration graph in Fig. 1, which shows a plot of absorbance versus concentration. The straight line joining the origin with a series of calibration points is referred to as the calibration graph or curve. The concentration of the unknown solution which has absorbance A is read off the graph and produces an answer corresponding to concentration C. Whilst a simple calibration curve can be constructed with a pen and ruler, it is more proper to use a simple algorithm such as linear regression now available on most calculators and a common feature of the computer enhancement of the spectrometer. Using such a technique not only can concentrations be determined directly but a limited amount of information on the quality of the calibration can be obtained. For example, how well the calibration line fits the real data points, details on the slope of the line and where the intercept on the absorption axis occurs, indicating possible errors and potential limits of detection. This assumes that the calibration is linear between the origin and the absorbance/concentration of the highest standard used to produce the graph; however, this is only strictly true at low absorbance values. Deviations from linearity are usually apparent as the absorbance increases and this is apparent by the calibration curve bending towards the concentration axis. The reasons for this are discussed in the following section.

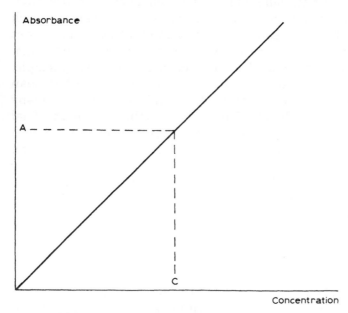

Fig. 1. Idealised calibration graph.

B. Reasons for non-linearity of calibration graphs

The main reasons for non-linearity in atomic absorption can be summarized as follows:

(1) Unabsorbed radiation, stray light. All light must be absorbable to the same extent.

(2) Hollow-cathode lamp linewidth broadening due either to the age of the source or the use of high lamp currents.

(3) If the monochromator slit is too wide, more than one line may be transmitted to the detector. In this case the calibration graph will show greater curvature than would be the result if only the desired line were transmitted to the detector.

(4) Disproportionate decomposition of molecular species at high concentration. This results in a lower proportion of free atoms being available at higher concentrations for a constant atomization temperature.

Clearly these factors must be considered if the calibration is to be as linear as possible in the desired concentration range. An example of point (3) is presented in Fig. 2, which shows a series of calibration graphs for nickel.

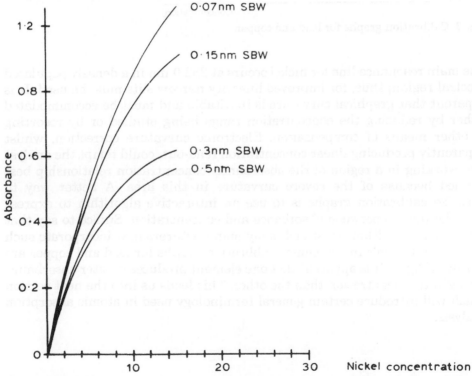

Fig. 2. Improvement in linearity and sensitivity of nickel determination at higher resolution settings.

54

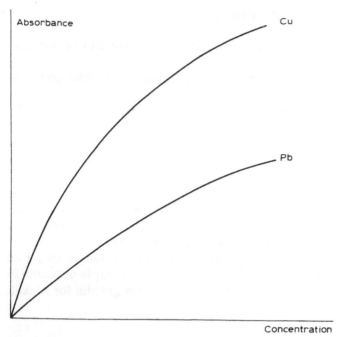

Fig. 3. Calibration graphs for lead and copper.

The main resonance line for nickel occurs at 232.0 nm in a densely populated spectral region; thus, for improved linearity narrow slits must be used. It is apparent that graphical curvature is inevitable and must be accommodated either by reducing the concentration range being studied or by resorting to other means of compensation. Electronic curvature correction, whilst apparently producing linear concentration read-out, could tempt the analyst into working in a region of the absorbance–concentration relationship best avoided because of the severe curvature in this area. A better way to linearise calibration graphs is to use an interactive algorithm to express the relationship between absorbance and concentration. Suffice to say that many such algorithms exist and many manufacturers now incorporate such a facility into their instruments. Calibration graphs for lead and copper are shown in Fig. 3. It is apparent that one element produces greater absorbance for a given concentration than the other. This leads us into the next section which will introduce certain general terminology used in atomic absorption analysis.

C. Terminology

1. *Sensitivity*

Sensitivity is defined as the concentration of a solution (typically in ppm) of an element needed to produce a signal of 0.0044 absorbance units. This is equivalent to a 1% decrease in the transmitted radiation. The absorbance (not % absorption) is proportional to concentration. This is given by Beer's Law. The sensitivities of some commonly analysed elements are given in Table 1. The sensitivity figure is a useful performance index in that it gives the analyst information about how well the instrument has been optimised. For example, the sensitivity for copper is 0.03 ppm. This means that a solution containing this concentration of copper would be expected to produce a reading equivalent to 0.0044 absorbance. Since this represents a very small absorbance it is convenient when optimising to use solutions of greater concentration so that, for example, by extrapolation a solution of copper containing 3 ppm would be expected to produce a deflection of 0.440 absorbance units (440 miliabsorbance units). Similarly, the sensitivity for lead is 0.1 ppm, a higher figure than for copper. This means that a higher concentration is required for lead than for copper to give an equivalent absorbance. Thus, when optimising lead, the operator would expect to see a deflection of 440 milliabsorbance units or more from 10 ppm of lead. It can be seen that the sensitivity figure is a measure of the slope of the calibration graph and gives the analyst useful information about how well the atomic absorption instrument has been optimised. The sensitivity figure however gives no clue as to the minimum detectable amount of the element. This requires a second performance index, which will be considered in the next section.

2. *Detection limit*

The detection limit (or limit of detection) is commonly defined as the lowest concentration of an element in solution which can be detected with 95% certainty. It is therefore the concentration which will produce a deflection equal to twice the standard deviation of a series of readings (typically ten). Historically it has also been defined as the concentration which produces a deflection from the base line of a chart recorder, which is equal to twice the peak-to-peak variability or noise of that base line. The important thing to note, however, is that detection limit information can be obtained in many ways and that the method of calculation should be defined when original work is being published. Recent guidelines suggest that the detection limit should be determined directly from a calibration graph constructed using a regression technique. In this case the value for the blank intercept on the absorbance axis is multiplied by three times the

TABLE 1

SOME TYPICAL SENSITIVITY DATA FOR FLAME AAS

Element		Sensitivity ($\mu g \ ml^{-1}$)	Detection limit ($\mu g \ ml^{-1}$)
Ag	Silver	0.04	0.003
Al	Aluminium	0.43	0.04
As	Arsenic	0.63	0.21
			0.001 [a]
Au	Gold	0.1	0.014
B	Boron	8.0	2.0
Ba	Barium	0.17	0.015
Be	Beryllium	0.014	0.0026
Bi	Bismuth	0.22	0.07
Ca	Calcium	0.02	0.001
Cd	Cadmium	0.01	0.002
Co	Cobalt	0.05	0.005
Cr	Chromium	0.04	0.003
Cs	Cesium	0.058	0.019
Cu	Copper	0.03	0.004
Er	Erbium	0.47	0.07
Fe	Iron	0.05	0.003
Hg	Mercury		0.0005 [a]
K	Potassium	0.02	0.003
Li	Lithium	0.023	0.0016
Mg	Magnesium	0.004	0.0002
Mn	Manganese	0.03	0.003
Mo	Molybdenum	0.28	0.046
Na	Sodium	0.005	0.00023
Ni	Nickel	0.05	0.007
Pb	Lead	0.1	0.01
Pr	Praseodymium	28.1	7.3
Pt	Platinum	1.0	0.098
Rb	Rubidium	0.02	0.0034
Re	Rhenium	8.5	0.73
Rh	Rhodium	0.12	0.022
Sb	Antimony	0.28	0.06
			0.006 [a]
Sc	Scandium	0.25	0.03
Se	Selenium	0.11	0.16
			0.001 [a]
Si	Silicon	1.3	0.35
Sn	Tin	0.88	0.16
Sr	Strontium	0.06	0.001
Ta	Tantalum	12.0	0.94
Te	Tellurium	0.21	0.07
Ti	Titanium	1.5	0.12
Tl	Thallium	0.23	0.03
V	Vanadium	0.96	0.05
W	Tungsten	16.9	1.4
Y	Yttrium	1.69	0.3
Yb	Ytterbium	0.077	0.005
Zn	Zinc	0.007	0.0026

standard deviations obtained for the blank. The calculated value is then interpolated from the calibration graph to give the limit of detection.

The detection limit is a theoretical figure and one would never attempt to measure routinely concentrations at the detection limit for real samples. As a guide to whether an element can be routinely analysed, the detection limit figure should be multiplied by a factor, which in practical terms will be as great as ten, in order to estimate whether a successful analysis can be achieved. This slight ambiguity in quoting detection limit has led many analysts to favour the use of minimum detectable levels when evaluating instrumental performance. For example, the detection limit for lead is 0.01 ppm by flame atomic absorption. One could reasonably expect therefore that the minimum detectable level for lead would be in the order of ten times this value, i.e. 0.1 ppm in real samples, using a suitable background correction technique where necessary.

3. Precision

The precision of an analysis is most conveniently defined in terms of percent relative standard deviation (RSD). The standard deviation can be easily calculated following a series of discrete measurements either of absorbance or of concentration. The relative standard deviation is then defined as the standard deviation expressed as a percentage of the mean of the data used to calculate the standard deviation.

One of the attractions of flame atomic absorption is that relative standard deviations are often better than 1% in ideal situations and only marginally poorer than this when lower levels are being analysed. By definition the precision of measurement is 50% relative standard deviation at the detection limit, because it is at this point that the signal-to-noise ratio equals 2.

The standard deviation (sigma) is calculated according to the formula:

$$\sigma = [(\bar{x} - x)2/(n - 1)]^{1/2} \tag{1}$$

where σ = standard deviation, x = analytical value, $\bar{x}$ = arithmetic mean, n = number of values taken. The standard deviation concept is a very useful one for expressing the analytical confidence one has in the answer obtained from the instrument. Most analysts would be content working with an analytical confidence of 95%. Thus if one's analytical result were 5 ppm, for example, and the calculated sigma were 0.1 ppm, then with a 95% confidence the answer would be 5 ppm plus or minus 2×0.1, i.e. an analytical range of

---

TABLE 1 (continued)

Note: All sensitivities and detection limits were obtained using an air–acetylene or nitrous oxide–acetylene flame unless otherwise noted. Potassium was added to the aqueous standards to suppress ionisation for easily ionised elements.

[a] Hydride generator.

4.8–5.2 In order to provide greater confidence of an answer around 5 ppm the analyst would have to improve the precision, that is, make sigma smaller than 0.1 ppm. This would involve a re-examination of the instrumental parameters to check the possibilities for obtaining more stable analytical results.

D. Working with calibration curves

Calibration graph curvature cannot be avoided completely though it can be minimized by paying attention to the points made in Sections A and B above. The analyst should prepare a range of standards covering the concentration range of interest, optimize the instrument, nebulize the standards and note the absorbances and produce a calibration plot. Depending upon the shape of the graph he or she might either decide to reduce the number of standards or to increase the number of standards, in order to define the curve as accurately as possible. If curvature is severe, it may at this stage be worth examining some of the variables that might control the magnitude of curvature, such as lamp current, spectral bandwidth and, of course, the wavelength of analysis.

As a general guide, in the author's opinion, a minimum of five standards and a blank should be prepared in order to give the instrument's microprocessor sufficient information to fit the curve appropriately. If the instrument has a manual curvature correction function then the manufacturer should be consulted for his recommended method of use, but generally speaking the range of standards would be aspirated and their absorbances noted and the amount of curvature from origin to top-standard calculated. The amount of curvature correction introduced by the curve-correct control would then be progressively increased and at each stage a blank and all standards would be aspirated again and the amount of curvature recalculated. This procedure would be repeated until the amount of curvature were acceptable and the level of acceptability, of course, would depend upon the degree of accuracy required by the analyst. This is considered to be a better approach to working with curved calibration graphs than attempting to linearise the graph by a trial-and-error procedure. On most modern instrumentation curve filter calibration algorithms are available and often calibration is automatically carried out. The discerning analysts may find it interesting to examine more closely the effectiveness of some of the algorithms in current use.

E. De-optimization

If the absorbance produced from the sample is too high to permit accurate analysis in the working range of the standards normally employed, then the analyst has to decide which of three options to take. Firstly, the sample can be diluted to bring its absorbance into the optimum working range for that element's wavelength, or secondly, an alternative wavelength having a lower absorptivity may be used. The majority of elements analysed by atomic absorption offer the analyst a range of wavelengths. Of course in most cases, for accurate analysis, the most sensitive wavelengths would be chosen, but for analysis at higher concentrations the use of a less sensitive line offers an alternative to diluting the samples. The third alternative is for the analyst to reduce the path length by rotating the burner head by the required amount to desensitise the analysis, thus reducing the absorbance to whatever extent is necessary to bring the sample into an optimum absorbance range. On most instruments the burner can be rotated through any number of degrees from 0 to 90.

The three methods given above represent the best ways to de-optimize the analytical system and thus reduce curvature. As a general guide, if the top standard is arranged to provide an absorbance of around 0.4 or 0.5 absorbance units, then a successful analysis will result with more than adequate precision. There is seldom need to work with higher absorbances. The absorbance from the sample could, of course, also be reduced by generally de-tuning the instrument by moving the burner head or reducing the nebulizer uptake rate for instance. These methods are not recommended, however, and the analyst should attempt to adhere to the three general techniques given above.

III. SAMPLE REQUIREMENTS AND GENERAL PREPARATION TECHNIQUES

Consideration will be given in this section to the general principles involved in the preparation of various types of sample for atomic absorption analysis. In general, a sample received in the laboratory can be placed into one of the following categories: (A) aqueous solutions; (B) organic solutions; (C) inorganic solids; (D) organic solids; (E) gases. If these materials are to be aspirated, then solids must be solubilised, gas streams filtered and liquid samples must satisfy certain criteria.

(1) The viscosity and solids content of the solution must be such as to permit nebulization without giving rise to problems associated with burner blocking or nebulizer "salting-out". The long path air–acetylene burner head will accommodate 2–3% solids at 40 psi oxidant pressure, and up to 10% solids by progressively reducing the oxidant pressure. The nitrous oxide burner head has a lower tolerance and 1% solids should be regarded as maximum for continuous aspiration. A viscous solution will be nebulized less efficiently than a less viscous solution and should be calibrated by an addition procedure (see Section VI.B).

(2) Solid particles should be removed, preferably by centrifugation, particularly in trace-level determinations where contamination will be more significant.

(3) The acid concentration should be as low as possible. The nebulizer fitted as standard may have a stainless steel capillary tube which will be attacked by acid over a period of time. Solutions containing more than 5% mineral acid should be nebulized using a nebulizer having a platinum/iridium capillary. A corrosion-resistant nebuliser having a plastic throat is required for solutions containing hydrofluoric acid. Additionally, in this instance, a Teflon impact bead must replace the standard glass one, where these are employed.

(4) Organic solvents must be chosen carefully.

(5) Interferences should be removed or compensated for (a general description of interferences is given in Section V).

(6) The metal concentration must not be so high as to fall in a grossly curved part of the calibration graph. Beer's law is obeyed for most elements up to an absorbance of about 0.4; calibration graph curvature has been discussed above.

The five sample types are now briefly discussed in turn, but full details can be found in the relevant chapters of this book.

A. Aqueous solutions

Typical of these samples are raw and treated waters, seawater, biological fluids, beer, wines, plating solutions, effluents, etc. With this type of sample very little preparation is usually required. If the solution is suitable for aspiration then its approximate concentration can be determined to check whether dilution with water is necessary. In general, aqueous solutions can be introduced in a graphite furnace for ETA-AAS, directly with minimal treatment. Matrix modification may however be necessary and some care in calibration may well be required. For both flame and ETA-AAS analysis degassing may be necessary and/or the addition of releasing agents, ionisation suppressants, complexing agents, etc., may be required for interference compensation. Concentration methods will be described later.

B. Organic liquids

These will mainly be petroleum products, many of which can be aspirated directly or following viscosity adjustment with suitable organic solvents, which should be chosen according to certain criteria, i.e. the solvent should: (i) dissolve or mix with the sample; (ii) burn well, but in a controlled manner; (iii) be available in a pure state, and not contain species having molecular absorption bands in the ultra-violet; (iv) be innocuous and produce no harmful by-products upon combustion; (v) be inexpensive.

Some examples of commonly used solvents are p-xylene, n-heptane, cyclohexane, 10% isopropanol-white spirit mixture, methyl isobutyl ketone, methyl ethyl ketone and cyclohexanone.

Standardisation should be via organometallic standards, which are now available for a range of metals from, for example, Merck, as well as from specialised oil-standard organisations such as Conostan. Viscosity makes direct analysis of oil samples by ETA-AAS difficult; however, with a diluent such as described above and with some matrix modification, satisfactory analysis can be achieved for most trace elements if a powerful background corrector is fitted.

C. Inorganic solids

Typically, these will be alloys, rocks, soils, fertilisers, ceramics etc. These materials are taken into solution using suitable aqueous/acid media, according to solubility: hot water, dilute acid, acid mixtures, concentrated acids, prolonged acid digestion using hydrofluoric acid if necessary, alkali fusion (e.g. using lithium metaborate), Teflon bomb dissolution. Fusion and "bomb" methods are usually reserved for complex siliceous materials, traditionally reluctant to yield to solubilisation. The recent introduction of microwave digestion techniques incorporating a pressure vessel or a bomb offers considerable scope in the area of sample preparation. Using this approach a more rapid controlled dissolution can be achieved for materials which are often difficult to digest by traditional methods. In addition the loss of volatile elements, such as mercury, is less likely in a closed system. A note of caution should be given, however, to remind the analyst of the explosive potential of a closed cold vessel exposed to rapid microwave heating. The digestion or solubilisation procedures commonly used in sample preparation are equally applicable for samples analysed by ETA-AAS determination. The introduction of solids directly into the graphic furnace, either directly or as a slurry, has for some samples proved to be a most successful approach, simplifying preparation and reducing potential contamination and dilution effects associated with wet chemical methods.

With such techniques the introduction of ashing aids or matrix modifiers may be necessary and, although aqueous calibration has been shown to be possible, a standard addition technique may be required for satisfactory calibration (see Section VI.B).

D. Organic solids

Typically foods, feedstuffs, leaves, plants, biological solids, tissue, polymers etc. Prior to solubilisation these types of sample generally require destruction via wet digestion or ashing in a muffle furnace. Perchloric acid is often suggested as a suitable acid for wet oxidation of organic materials and despite having a somewhat undeserved reputation for being a highly dangerous substance, it does, when used with suitable precautions, offer a very satisfactory technique.

As with inorganic samples the use of a microwave oven is a very attractive technique for small organic samples and attachments for the technique are now available, which enable dry ashing to be carried out in the microwave cavity. Dry ashing of samples at say 600°C in a muffle furnace, usually with some type of ashing aid added, have traditionally been popular. This approach has tended to fall from favour as potential contamination and loss of analytes can represent quite severe problems. Specific details of current methodology based on both wet and dry techniques can be found in the relevant sections in Chapters 4a to 4n.

E. Gases

Atomic absorption techniques can be used to analyse gases indirectly, as liquid samples. To prepare the liquid sample the metals are removed from the gas stream or atmospheric sample using a filter medium such as a millipore filter disc. This is then either dissolved or washed in nitric acid and the solution analysed by standard additions. These procedures are now extensively used by Health and Safety Executive Inspectors to monitor (particularly) heavy metals in working environments. The reader is referred to Chapter 4c.

IV. METHODS OF ANALYTE CONCENTRATION

Where the concentration of an element in a sample falls below the detection limit for that element or is low enough to make a precise direct measurement impossible, other techniques must be used to pre-concentrate the element or remove the matrix. The possibilities given below are (i) an alternative to using an electrothermal atomizer where the sensitivity is of the order of 100 to 1000 times greater than the flame technique, or (ii) can be used in combination with ETA-AAS for ultra trace analysis.

A. Concentration by evaporation

This is the simplest technique but is prone to contamination or element loss by evaporation. Also the sample matrix may become too concentrated to pass through the nebuliser and burner without deposition. Since the matrix is also concentrated the final sample aspirated may suffer from matrix interference. This should be investigated and if necessary a method of standard additions used; see Section VI.

B. Solvent extraction of trace metals

This is a popular separation and preconcentration technique, as it can be reduced to its simplest form. It is possible and often desirable in atomic absorption to extract and therefore concentrate more than one element at one operation. Specificity resides in the measuring technique, as has been stated previously. The choice of chelating or complexing reagent is therefore not limited, as in colorimetry, to one which gives a strong colour for the metal being determined, and complicated methods involving extractions and back extractions, in order to improve specificity are avoided. A number of advantages result from the extraction of APDC-metal complexes into a suitable organic phase. The metal may be concentrated by as much as a hundredfold if desired. Analyte metals can be separated from a matrix containing, for example, high concentrations of other solutes which may cause difficulties in nebulization and atomization. The atomic absorption signal for nearly all metals is further enhanced by a factor of maybe threefold when aspirated in an organic solvent instead of an aqueous solution. APDC complexes are soluble in a number of ketones. Methyl isobutyl ketone, which is a recommended solvent for atomic absorption, allows a concentration factor of ten times.

Further details of specific extraction systems will be given in the later application chapters.

C. Ion exchange

Ion exchange techniques have been used to separate certain groups of metals from an undesirable matrix quite successfully for many sample types. Adopting continuous flow or flow injection techniques which incorporate on line ion exchange, both cation and anion, coupled to FAAS, offers many potential advantages. Perhaps the most useful separation of this type is of trace heavy metals from higher concentrations of alkali metals. There is a very large number of examples in the literature in which ion exchange has been used to separate an interfering matrix, either by retaining the analyte elements or by retaining the matrix element. In the latter case relatively large amounts of the ion exchanger may be required and in this case no actual concentration of the analyte is achieved in the process. The combined effect of preconcentration and matrix modification makes this approach to sample introduction an easy and attractive method for FAAS, offering simple calibration and improved limits of detection. The reader is directed to the specialist literature for more information on the kind of separation that may be applicable to atomic absorption methods.

V. INTERFERENCES IN ATOMIC ABSORPTION ANALYSIS

This topic will be treated fairly generally in this section because the authors of the subsequent application chapters will be describing more specific examples.

The newcomer to AAS could easily be led into believing that he or she has been misled when informed that this analytical technique is free from interferences. This impression unfortunately arises from early work in the technique when, of course, only a few applications had been studied. With the increase of interest, a wider range of applications was studied and consequently more problems were encountered. However, the interferences encountered in atomic absorption spectrometry are now extremely well documented and many which were reported early in the literature were found to be due to instrumental imperfections and have now virtually disappeared. All interferences can be overcome by the use of simple techniques.

Interferences encountered in AAS can be separated into the following categories: (A) spectral; (B) flame emission; (C) chemical; (D) matrix; (E) non-specific scatter; (F) ionisation. The majority of difficulties that the analyst can expect to encounter arises from chemical, matrix, light-scattering and ionisation interferences.

A. Spectral interference

Spectral interference is rarely encountered in atomic absorption spectrometry. Spectral interferences in the past were experienced typically if, in a given solution, element A was being determined in the presence of element B. If the source contained both elements and the absorption lines of these could not be resolved by the monochromator, element B would cause an interference. In some early hollow-cathode lamps this was a well-known phenomenon. It could be overcome, however, by using an alternative absorption line, the probability of two lines coinciding again being extremely remote. Most of the interferences have now disappeared due to improvements in the purification techniques of the cathodes.

B. Emission interference

Emission interference was common in many early instruments which were accessories to UV/visible spectrophotometers, which operated in most instances on a d.c. system. The interference was caused by emission of the element at the same wavelength as that at which absorption was occurring. All modern instruments use a.c. systems which are of course "blind" to the continuous emission from the flame. However, if the intensity of the emission is high, the "noise" associated with the determination will increase, since the noise of a photomultiplier detector varies with the square root of the radiation falling upon it.

This effect can be reduced by either increasing the source current or by closing down the slit, both methods resulting in an increase in the signal-to-noise ratio.

C. Chemical interference

This is by far the most frequently encountered interference in AAS. Basically, a chemical interference can be defined as anything that prevents or suppresses the formation of ground state gaseous atoms in the flame. A common example is the interference produced by aluminium, silicon and phosphorus in the determination of magnesium, calcium, strontium, barium and many other metals. This is due to the formation of aluminates, silicates and phosphates which, in many instances, are refractory in the analytical flame being used.

In order to overcome this type of interference, two techniques may be emphasised, both of which release the element under investigation. The first relies upon the application of chemistry, in the knowledge that, in many instances, a compound may be added which will lead to the release of the element that we are interested in by the formation of a preferential complex. Thus, a chelate such as EDTA can be added to complex the cation, thus preventing its association with an anion that could lead to the formation of a refractory compound. Alternatively, a reagent can be added that will preferentially form a compound with the interfering anion, again leading to the "release" of the cation; for example, the addition of lanthanum chloride to solutions of calcium containing the phosphate anion. The calcium is "released" due to the preferential formation of lanthanum phosphate.

Secondly, many chemical interferences may be overcome by using the higher temperatures of a nitrous oxide–acetylene flame, which will simply thermally decompose the sample.

D. Ionisation interference

To understand ionisation interferences, it is necessary to appreciate what is occurring in the flame during the aspiration of a sample. The flame is being used as a source of energy to convert elements present in the solution mist, created by the nebulizer, into ground-state atoms.

$$MX \quad \xrightarrow{E} \quad MX \quad \xrightarrow{E} \quad M^0 + X^0$$

solution   salt particles   ground-state atoms

Many determinations require the use of the nitrous oxide–acetylene flame and it is usually under these conditions that ionisation interferences occur. They arise from the energetic nature of the flame which in addition to producing ground-state atoms, also excites some atoms to such an extent that one or more electrons are lost and ionisation occurs.

$$M^0 \quad \xrightarrow{E} \quad M^+ + e^-$$

This effect will obviously be greatest with elements having low ionisation potentials, such as the alkali and alkaline earth metals, e.g. barium is approximately 80% ionised in the nitrous oxide flame. Since the ground state therefore becomes depopulated, the sensitivity will decrease.

An equally important effect arises when an easily ionised element is being determined in the presence of another. There will be an enhancement of sensitivity compared with pure aqueous standards. This arises from the presence of excess free electrons which suppress further ionisation.

$$M^+ + e^- \longrightarrow M^0$$
$$\text{excess}$$

This effectively increases the population of ground-state atoms. In practical applications some use may be made of this phenomenon. By adding an excess of a readily ionised salt to samples and standards, an increase in sensitivity may be achieved. Potassium chloride is usually chosen for this purpose owing to its high purity, low ionisation potential and lack of visible emission in the flame.

E. Matrix interference

This is a general term covering: (i) enhancement of sensitivity due to the presence of an organic solvent in the aqueous solution; (ii) depression of sensitivity due to the sample having a greater viscosity than the standard solutions; (iii) depression of the result due to a high salt content.

These interferences can be readily overcome by using one of the following techniques: (a) the method of standard additions; (b) matching the matrix of the standards with that of the sample; (c) remove the cation to be determined from the interfering matrix by solvent or ion exchange; and (d) relating the erroneous value obtained to an accurate value by using a factor determined by other means.

F. Non-specific interference

This effect causes an enhancement of an analytical result due to the solution containing a high concentration of dissolved salts or to a less extent stable absorbing molecules. The first effect is due to the presence of dried and semi-dried salt particles in the flame which scatter and absorb the incident radiation from the source, whilst the second effect is associated with broadband absorption from thermally stable matrix species. Since the intensity of the transmitted radiation will be effectively decreased, there will be an increase in the absorption signal. This non-specific effect is wavelength-dependent and is more pronounced at shorter wavelengths. It is most significant below 250 nm.

The effect can be overcome by one of the following techniques:

(i) Solvent extraction or ion exchange to remove the element or compound from the interfering matrix.

(ii) Repeating the determination at a nearby non-absorbing line and subtracting the scatter reading from the signal obtained at the absorbing line. This technique has several limitations, one being that the precision at the non-absorbing wavelength will probably be worse than at the absorbing line. Also, some elements do not have close suitable non-absorbing lines.

(iii) By using a background corrector. The use of background correction should now be common place in all atomic absorption analysis, in particular for ETA-AAS. As one can see in Chapter 2, three main types of correction are available: deuterium, Zeeman and Smith-Hieftje (S-H). The deuterium makes use of the fact that the sample beam is ratioed to the reference beam and that a UV continuum source will behave in the same way as a non-absorbing line, in that it will enable scatter-only measurements to be made. By arranging for it to be out of phase with the sample beam, scatter readings will automatically be subtracted from the erroneous absorption measurement at the detector stage. The deuterium system is, however, prone to misalignment problems associated with the two beams required and is generally accepted not to be as powerful as the Zeeman and S-H systems.

VI. ELECTROTHERMAL ATOMIZATION, ETA

In addition to the normal requirements of a good atomic absorption spectrometer two parameters are of paramount importance when employing a furnace as the atomizing source. These are: (a) the response time of the signal handling circuitry must be sufficiently rapid to capture the transient absorption signals which are characteristic of ETA atomizers; and (b) the background correction system must be extremely effective, since non-specific background absorption is more of a problem in ETA than in flame cells.

Certainly all of the development work using these devices was performed with a chart recorder to display the analytical peaks; today the recorder is replaced by a computer with an enhanced visual display unit (VDU). The value of such a dynamic visual display cannot be overemphasised, for the following reasons.

(a) The drying process can be monitored in that very often the droplet will disturb the light path and attenuate it. This will show as a deflection on the baseline when background correction is not being employed. As the solution is gradually dried the baseline will recover to normal. Erratic signals at this stage would be evidence of a hasty drying stage. As a guide a 20 μl aliquot injected into the graphite tube will require a drying time of about 45 s, depending upon its salt content.

(b) At the intermediate stage(s) the pyrolysis temperature will ideally be high enough to decompose the sample matrix without volatilising the

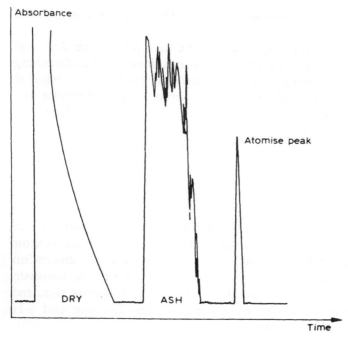

Fig. 4. Chart trace illustrating dry, ash and atomise events with background corrector off.

analyte. Again the effectiveness of this stage can be monitored using a suitable display. Figure 4 shows an ideal trace. The drying stage has been accomplished steadily and the non-atomic peak corresponding to the pyrolysis products of the matrix is clearly resolved from the final atomic signal from the analyte upon atomization.

(c) Changes in peak shape of the atom signal can be spotted with a visual display provided the time constant is adequately small, for example, in the order of 0.2–0.3 s.

It is now common to capture transient signals digitally, and this is possible in two ways, either as a peak height absorbance or an integrated peak area absorbance. Since the exact manner in which these modes of operation function on individual instruments varies, the manufacturer's recommendations should be followed, but it should not be forgotten that the visual display, either on a chart recorder or a VDU is extremely valuable in setting up a new method or monitoring the success of an existing one from time to time.

A. Selection and optimization of furnace operating conditions

The furnace power supply and controller enable the basic steps normally considered to occur in the flame method of atomization to be carried out in a sequential form, each stage governed by the particular phase of the controller programme. In its simplest form the programme will consist of three readily identifiable stages, namely: dry—ash—atomize. Each of these stages has to be carefully optimised to obtain the best results for any particular analysis. The importance of gradual drying has already been discussed. For the intermediate and final temperatures the construction of an ash/atomize curve may be beneficial, and of particular value when dealing with an analytical situation where no comparable studies can be found. The first stage is to prepare an aqueous solution containing only the analyte element at a concentration that should produce an optimised signal of approximately 0.1–0.2 absorbance. The metal should ideally be present as nitrate, sulphate or perchlorate since chloride salts are often volatile and can be lost prior to atomization in molecular form. Having prepared the standard, the drying time can be established quickly and the pyrolysis temperature ignored since no matrix is present. The atomization temperature is then varied and a graph obtained, plotting atomic absorption signal versus atomization temperature. Many elements will produce a graph which increases with temperature and then reaches a plateau, which indicates that there is no advantage in going beyond a certain atomization temperature. Indeed, unnecessarily high temperatures are to be avoided in any case to maximise tube life and to minimise blackbody emission breakthrough to the photomultiplier.

A temperature which is on the plateau part of the graph is selected and fixed and the intermediate temperature gradually raised starting normally from room temperature with an arbitrary time of 30 s. The design of some instruments is such that the sample is aspirated into a warm furnace and in such systems the starting temperature can be typically 150°C. The atomization time will normally be 5 s. A graph of atomize peak height versus ash temperature is now plotted, which should result in something approximating to Fig. 5. The ash curve too has a plateau region, which turns downwards as the temperature at which the analyte is lost is exceeded. At this point the atomization signal will rapidly fall, so the maximum ash temperature will be typically 100°C below this point.

One must bear in mind that should a longer ashing time be required (and this would depend on the sample matrix) then the ash curve might need modifying. Conditions must be rechecked at the longer ash time and the ash temperature changed accordingly. A successful ashing sequence will result in no smoke being evolved upon atomization, but this is not always possible, particularly when dealing with volatile metal analytes. The residual smoke formed during atomization must be compensated for by using background correction; if this effect is found to be excessive (more then 0.5 absorbance) then one of the more powerful background correction techniques should be used or a smaller injection volume should be selected. The more modern designed instruments now offer the facility to add additional gases into the furnace during ashing and atomization. Traditionally an inert gas such as

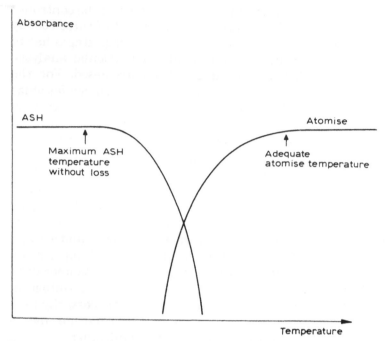

Fig. 5. Ash/atomise curves.

nitrogen or argon is used throughout the ETA cycle. The addition of oxygen to assist ashing offers considerable benefits with organic-based samples. The addition of methane into the furnace chamber after atomization will replenish the pyrolytic coating of the furnace, resulting in longer-term stability of the graphite surface.

B. Calibration

In principle there are no differences between calibration procedures for flame and ETA methods, although the latter case will take longer, as has already been pointed out with a standard ETA system. Calibration standards should match the samples as nearly as possible with respect to major components, otherwise standard additions must be used. The method of standard additions has been shown to be of particular importance where complex solutions and samples are being analysed. Where platform or probe techniques are being used to reduce interference effects, then aqueous calibration has been shown to be quite adequate even for complex samples such as blood. The method of standard additions can, however, be used to check periodically for accuracy or indicate signal interferences and so validate calibration. If the standard additions graph is found to be parallel to the direct calibration graph, then freedom from matrix interferences would be indicated. The standard additions principle as applied to ETA should be carried out as follows.

It is advisable that all readings be obtained in duplicate or triplicate. Various variations of the procedure as described have also been reported. The sample is divided into four or five aliquots and a small but concentrated volume of a standard is added to each aliquot. It is important to always add the same volume at each addition. The additions are made such that standard additions can be made in the furnace tube itself, thus avoiding dilution of the sample and possible contamination from volumetric ware. The selected volume of sample is injected into the tube, dried and ashed, the programme being stopped before atomization. After cooling, a known amount of the analyte element is injected on top of the ashed sample and the complete atomization programme run through. This is repeated for a range of different additions and, of course, for the sample with no addition. The standard addition graph is plotted in the normal way, absorbance versus concentration for the added element. Quantification of the element of interest in the original sample can be achieved by extrapolating the standard addition curve to the point at which it intercepts the x-axis.

The validity of the method of standard additions depends on the forms of the analyte element in the sample and in the added standard responding in the same way during the atomization step. This may not always be so in practice. For example, in determining lead in whole blood by this method, it would have to be proved (or assumed) that lead added as a lead nitrate solution is atomized to the same extent as lead bound organically in the sample. This is most easily checked by running a certified standard material through the procedure. If no such reference standard exists, the sample should also be run after pre-treatment for removal of the matrix, e.g. wet or dry ashing, and the results compared.

Two further limitations apply to the use of the method of standard additions. Simultaneous background correction must be employed because of possibly varying amounts of matrix material present in the tube during the several firings needed to make one determination. Furthermore, all readings must be within the linear portion of the calibration graph in order that meaningful results may be obtained. It is, of course, perfectly permissible to use a linearisation function, either as provided in the atomic absorption spectrometer data processor itself or as may be devised with a desk-top calculator or computer.

C. Sample handling and preparation

The increased sensitivity which is the main feature of electrothermal atomization methods introduces a number of difficulties connected with the handling and preparation of samples. Some practical guidance on the avoidance of errors through contamination and on the choice and use of micro-pipettes and into samplers is set forth in the following subsections. The analyst must appreciate, however, that we are dealing with a technique of trace analysis, and any advice or experience that he or she can make use of on that subject will be entirely relevant here. One obvious procedure is to always prepare reagent and system blanks at all times.

1. *Avoidance of contamination*

The importance of clean-air rooms or at least clean areas and benches, is a point often omitted from the earlier literature.

Atomic absorption with electrothermal atomization typically involves measurement of less than 1 ng of an element during each analytical sequence. At such extremely low levels, contamination of the apparatus with detectable amounts of common elements is a severe problem. Contamination may occur at any stage in the procedure. It may arise from the reagents used in sample preparation, from the vessels used during preparation, or from the laboratory atmosphere at any stage in the procedure, even when the sample is actually situated within the graphite tube. Sodium, magnesium and zinc are often detected as contaminants after conventional laboratory washing procedures, and zinc particularly is often found in cleaning agents. Other elements which frequently figure as contaminants are iron, copper and potassium. If the laboratory deals with a particular type of sample, then the matrix of the sample is a potential contaminant.

The precautions recommended to avoid contamination are detailed below and are divided into two sections: those considered essential to enable electrothermal atomization to be carried out successfully, and those considered desirable.

2. *Essential precautions*

(1) All glass or plastic vessels to be used for electrothermal atomization work should be washed, rinsed and then soaked in 2% v/v nitric acid for at least 24 h and then thoroughly rinsed in high-purity deionised or double-distilled water.
(2) A "clean" bench area should be reserved for solution preparation for use with the furnace. This can now easily be achieved by using a filtered air clean cabinet placed on the bench.
(3) The volumetric and storage ware used for solutions should be kept separate from apparatus used for conventional laboratory work.
(4) Solutions of low concentration should be prepared immediately before use and after preparation should be transferred to a suitable plastic container for storage.

(5) When solutions of the same concentration of an element have to be prepared regularly on a routine basis, it is advisable to keep the same apparatus for the same solutions.

(6) Efficient fume extraction.

(7) A high-purity water supply, either deionised, having a minimum resistivity of 10 $M\Omega$ cm^{-1}, or double-distilled. It is preferable to produce water as it is required rather than to store it for later use. For this purpose a deionising system is usually more convenient.

(8) Micropipette tips used in reagent preparation and sample handling may introduce contamination. If this is excessive it may be necessary to soak the tips in dilute nitric acid and then wash with high-purity water before use. In any case, with each solution, the micropipette tip should be washed through twice with injections which are discarded into a beaker before it is used.

3. *Desirable precautions*

(1) The complete electrothermal atomization system should, preferably, be in a room separate from the general laboratory, and well away from sample preparation procedures.

(2) The room should be under a small positive pressure, supplied from a pump system with filtration for dust particles.

(3) The room should not be used by personnel as an access room for other parts of the laboratory.

Plastic ware

All solutions should be stored in plastic bottles, as glass vessels usually give greater contamination and also adsorb elements from the solution onto the glass surface. Polypropylene and polyethylene are the best general-purpose storage vessels, but even these cause solution deterioration, noticeably with solutions having concentrations below the ppm level. This effect is activated by acidic solutions and soaking in water is preferable to dissolve any contaminating salt. If soaking in acid is necessary then acid concentration no greater than a few percent v/v should be used. Polystyrene vessels should not be used because they are particularly susceptible to adsorption effects. The best storage vessels are those made from PTFE or FEP materials, which give very few problems from contamination or surface adsorption. These are very expensive, however, compared with polyethylene and polypropylene.

Micropipette tips

Micropipette tips may suffer from contamination from materials used in their construction or from packing, but more usually contamination comes from handling in the laboratory. For this reason it is very good practice to put the tip on the micropipette from within the supplier's packet by handling the outside of the packet only. Also, when the pipette is put down between operations it is essential to rest it in such a way that the tip does not come into contact with any bench surfaces or other objects. It is good policy to dispose of the tip when changing solutions so that it is not left for later use, by which time it could be contaminated by contact or by atmospheric dust.

In some cases the dyes used for colouring tips have been found to contain contaminants. Pink tips have been found to contain cobalt, yellow tips may contain cadmium, and white tips may contain lead. This source of contamination is now less common and should not cause any significant problems if the tips are purchased from a reputable manufacturer.

The use of a reagent blank will help keep a check on contamination from this source.

Choice of micropipette for sample handling

The general principles by which pipettes operate are either based on air displacement or a positive plunger displacement and are provided with non-wettable plastic (usually polypropylene) disposable tips to contain the solution, preventing any contamination of the pipette itself. Most micropipettes have a double action plunger system, i.e. calibration and overshoot positions, which ensures that the sample is completely dispensed. There are many manufacturers offering a complete range of volumes; in addition some have available a selection of pipettes with adjustable volumes. Some very sophisticated computer-controlled systems are now available and are very useful where dilution of samples is commonplace.

The tips used with the micropipette contribute greatly to the accuracy of the analysis, and different types are better suited to certain uses and volumes. In general, the "dead" air space between the sample and the plunger seal should be kept to a minimum, since this can seriously affect both accuracy and precision due to the air expanding in response to even very small changes in temperature, such as could be produced by a loose or a firm grip.

Correct operation of a micropipette is essential to enable volumes to be dispensed precisely. The plunger of the micropipette should be depressed slowly but firmly to the first stop position and the tip just inserted in the solution to be tested. The plunger should be allowed to rise carefully over a period of 3–4 s. To inject the sample the plunger is slowly (3–4 s) depressed fully to expel the sample completely.

Standard tips are made from polypropylene which has been treated to be non-wetting for aqueous solutions. When non-aqueous solvents such as ketones, alcohols and chlorinated hydrocarbons etc., are to be handled however, the precision or repeatability is very much poorer, anything from 10% or worse. There are two main reasons for this. The non-wetting coating or finish is rapidly destroyed by organic solvents and, although some tips might work properly, it is likely that droplets of liquid will remain behind on the inner walls of the tip. One recommended procedure is to pre-wash the tips in the solvent to get uniform performance; this not only destroys the coating, however, but encourages droplet formation. Also, when standard tips are used for volumes between 1 μl and 20 μl, there is a comparatively large dead space above the liquid. This results in premature sample ejection, due to the build up of solvent vapour pressure in this space which forces the liquid out. Low boiling point solvents such as chloroform are particularly prone to this problem. The problem can be alleviated to some extent by careful choice of solvent. Chloroform could be replaced, for example, by 1,2-dichloroethane which has a much lower vapour pressure. Avoidance of solvents of low boiling point reduces the problem of expulsion of the sample from the pipette tip, but does not prevent the formation of droplets within the tip. Positive plunger-type pipettes are often useful when oil or organic solvents are being handled.

The autosamplers now supplied with ETA-AAS equipment are normally of high quality to avoid sample contamination. Samples are placed in a disposable covered sample pot and the injector tip is routinely washed between injections. The analyst should be conscious, however, of possible contamination between samples. Where manual injections are necessary the pipette and tip should be completely removed from the tube before the plunger is gently released to avoid taking the sample back into the tip.

Accuracy and precision

Accuracy of pipettes and autosamplers should usually be within 1% of a stated volume, and most manufacturers will specify each pipette to be accurate within this limit. It is, however, important to maintain the instrument in good order according to instructions.

Since atomic absorption is a relative technique, relying on standard solutions for calibration, the absolute accuracy of the micropipette and autosampler is not critical providing the same one is used to dispense both known standards and unknown samples; the prime requirement is therefore the ability to reproduce the volume, i.e. good precision. Precision should be within 1% RSD with volumes down to 20 μl, but will only be within 2% RSD at 10 μl. Volumes below 10 μl should not be used, if at all possible, because of the loss of precision. These figures represent results of sampling very dilute aqueous solutions. As the total dissolved solid content of the solution rises, so the viscosity will increase, and as this becomes significant, the precision will deteriorate.

Pipetting precision of blood or serum is improved by taking up a volatile liquid, e.g. *n*-heptane, between the bottom and first stops of a double action micropipette. The sample is taken up in the calibrated part and both are then injected into the furnace. The heptane washes out the sample completely and is evaporated off at the drying stage.

4. *Sample preparation*

In many analyses preparation will consist either of dilution of an already liquid sample or dissolution of a solid sample followed by dilution. Contamination introduced by the diluents or solvents used can be significant. The need for high-quality deionised water has already been emphasized. Normal analytical reagent-grade solvents and reagents are generally not sufficiently pure and materials of a higher degree of purity, e.g. "Aristar" Spectrolsol, or "Suprapur" grades or their equivalent, must be used. Sometimes solvents even in these grades give rise to a significant blank and this can only be reduced by further distillation.

5. *Chemical separations*

The types of separation procedure described elsewhere in this book for the improvement of sensitivity and for matrix separation in flame atomic absorption analysis can, in principle, be employed for the same purposes before electrothermal atomization.

In solvent-extraction methods only reagents and solvents of the very highest purity can be used and it is most important that blank extractions should be run wherever appropriate. Electrothermal atomizers allow a greater degree of flexibility than the flame, because the solvent does not have to be flame-compatible. The whole range of organic solvents can thus be considered. In particular the separations using APDC and 8-hydroxyquinoline may, under the correct conditions, be more successful using chloroform than methyl isobutyl ketone which is recommended for flame work. Thus, the best solvent for a particular separation can be employed, the only limitations being imposed by the difficulties of handling organic solvents with micropipettes.

It has been confirmed that elements in compounds in organic solvents usually give the same response in electrothermal atomization as the same element in an aqueous standard solution. This is because the solvent is removed at the drying stage and most organic complexes are converted to stable inorganic compounds at low ashing temperatures.

Ion exchange separations may also be carried out in preparing samples for electrothermal atomization. An interesting variant on this method is where the resin itself, containing the bonded analyte element, is subjected to direct analysis in the solid phase. For example one litre of seawater was passed through 500 mg of chitosan (a natural chelating polymer). The resin was then homogenized and 5 mg samples of this were analysed for vanadium. Response of vanadium from the resin and from aqueous standards was shown to be the same. The addition of solids directly or as slurries into the furnace do offer the advantages of not requiring excessive dilution or sample treatment. There are, however, some more serious problems with this approach and this should be attempted only with a powerful background correction technique.

Chapter 4a

Waters, Sewage and Effluents

M. BLANKLEY, A. HENSON and K.C. THOMPSON

Yorkshire Water, Business Services Laboratory, Charlotte Road, Sheffield, S2 4EQ (Great Britain)

I. INTRODUCTION

A. General considerations

The analysis of metals in waters, sewage, trade effluents and sewage sludges is frequently carried out by atomic absorption techniques. Even though most natural water samples contain over 99.9% of water, accurate determination of elements in this simple matrix is frequently not as straight forward as would appear at first sight [1]. Most unpolluted natural waters contain less than 10 μg l^{-1} of the commonly determined toxic metals (e.g. Cd, Cr, Cu, Hg, Ni, Pb etc.) and it is easy to obtain significantly biased results when measuring low concentrations of these elements [2]. The toxic metal levels in sewage and trade effluents and acid digests of sewage sludges are normally significantly higher than those of natural waters. Interlaboratory comparison studies [3–5] have demonstrated that inaccurate results at these higher concentrations can still occur unless meticulous attention to detail is observed.

In most countries sewage sludge is recycled to agricultural land because of the nitrogen and phosphorus content of the sludge. The majority (typically greater than 90%) of the toxic metals present in the crude sewage entering a sewage works end up in the sewage sludge. Consequently accurate analysis of the toxic metals present in sewage sludge is required to ensure that the metal loading being applied to the soil is within safe limits [6, 7].

B. National standards

Table 1 gives the USA, European Community and World Health Organisation (WHO) limits for a range of elements in treated (potable) water [8, 9]. In order to monitor for the elements at concentrations close to these limits and determine compliance, the criterion of detection [10] of the analytical technique (see Section I) should ideally be at least 10 times better than the actual limit. Table 1 also gives the recommended analytical technique and it can be seen that flame atomic absorption spectroscopy (FAAS) can only be used to monitor directly 7 of the 17 listed elements.

TABLE 1

SOME INTERNATIONAL STANDARDS FOR TREATED POTABLE WATER (All concentrations in μg l^{-1})

	USA maximum contaminant level	EEC guide level	EEC maximum admissable concentration	WHO guideline value	Preferred method a
Aluminium		50	200		E
Antimony			10		E, H
Arsenic	50		50	50	E, H
Barium	1000	100			F
Boron		1000			
Cadmium	10		5	5	E
Chromium	50		50	50	E
Copper	1000	100 [b] 3000 [c]			F
Iron	300	50	200		F
Lead	50		50	50	E
Manganese	50	20	50		F
Mercury	2		1	1	C
Nickel			50		E
Selenium	10		10	10	E, H
Silver	50		10		F, E
Sodium		20	150		F
Zinc	5000	100 [b] 5000 [c]			F

[a] E = electrothermal atomization AAS; H = hydride generation AAS; F = flame AAS; C = cold vapour AAS or AFS.
[a] At treatment plant.
[b] After 12 hours in piping.

Environmental quality standards arising from the European Community dangerous-substances directive for the protection of fish and other freshwater life have been published for some list II metals [11–13]. Table 2 lists these limits for the protection of salmonid freshwater fish at various water hardnesses. (The toxicity for many metals is dependent upon the hardness.) It can be seen that most of the limits shown in Table 2 are significantly lower than the equivalent limits for treated water.

The EC limits for metals in sewage sludge with respect to the safe recycling of sewage sludge to agricultural land [6] are much higher than for waters and for this analysis all of the listed elements except mercury (Cd, Cr, Cu, Ni, Pb and Zn) can be satisfactorily determined by flame AAS after acid digestion of the sludge.

TABLE 2

UNITED KINGDOM RIVER ENVIRONMENTAL QUALITY STANDARDS FOR THE PROTECTION OF SALMONID FISH (All concentrations in $\mu g\,l^{-1}$. Standards expressed as dissolved concentration relative to the annual average except where stated)

	Hardness (mg l^{-1} CaCO$_3$)					
	LT 50	50–100	100–150	150–200	200–250	GT 250
Arsenic, as As	50	50	50	50	50	50
Cadmium (1), as Cd	5	5	5	5	5	5
Chromium (2), as Cr	5	10	20	20	50	50
Copper (3), as Cu	5	22	40	40	40	112
Lead, as Pb	4	10	10	20	20	20
Mercury (4), as Hg	1	1	1	1	1	1
Nickel, as Ni	50	100	150	150	200	200
Zinc (5), as Zn	30	200	300	300	300	500
Tin (inorganic) as Sn (6):			25 $\mu g\,l^{-1\,a}$			
Tributyltin (7):			25 $ng\,l^{-1\,a}$			
Triphenyltin (7):			25 $ng\,l^{-1\,a}$			

[a] Proposed values for all freshwater fish.

Explanatory notes:
(1) Total concentration (i.e. dissolved plus particulate).
(2) Defined as Cr(III) + Cr(VI).
(3) This is a guide value, defined as a 95 percentile. The 112 $\mu g\,l^{-1}$ standard only applies at a hardness in excess of 300 mg l^{-1} as CaCO$_3$.
(4) Refers to the total concentration (dissolved plus particulate) defined as an annual average.
(5) This is a mandatory value, referring to the total zinc concentration (dissolved plus particulate), defined as a 95 percentile. The 500 $\mu g\,l^{-1}$ standard only applies at a hardness in excess of 500 mg l^{-1} as CaCO$_3$.
(6) Total concentration (i.e. dissolved plus particulate).
(7) Defined as 95 percentile value.

(This table has been compiled with permission from data in references 11–13.)

C. Speciation of metals

Dissolved and particulate metals in natural waters, effluents and sludges are present in varying speciation. In waters with a low organic carbon content, dissolved metals can occur in the ionised form as hydrated cations; however, in the presence of significant organic material (e.g. humic or fulvic acids etc.) metals are often complexed with organic compounds. In waters with significant amounts of suspended or colloidal material, trace metal species may be adsorbed on the suspended material.

The speciation with respect to oxidation state of some elements (e.g. As, Cr, Fe, Mn and Se) depends upon the redox potential of their environment.

References pp. 120–123

The relative toxicity of organometallic compounds compared with the commonly occurring inorganic forms of a given element varies considerably. For example, tributytin compounds are orders of magnitude more toxic than inorganic forms of tin [12, 13], whilst alkyl mercury and tetralkyl lead compounds are significantly more toxic than the inorganic forms of these elements. However, many alkyl arsenic compounds are less toxic than inorganic arsenic [14].

The current EC directive on the quality of water intended for human consumption [9] only specifies total metals. Other than organotin compounds [13], all current United Kingdom metal environmental quality standards for the protection of fish and other freshwater life only specify dissolved and/or total metal concentrations. Consequently no attempt has been made in this chapter to describe methods to determine the speciation of metals other than total and dissolved (after passage of the sample through a 0.45 μm membrane filter). Although much work has been reported on the development of methods to determine the speciation of metals in environmental samples [13–16] very few of these are carried out routinely by statutory bodies other than organotin speciation analysis. The main reasons for this are: preservation of speciation in samples between sampling and analysis is very difficult; for many metals relative toxicities of the various species are not known; and the very significant additional analysis time required relative to a total and/or dissolved determination of a given metal.

D. Routine analysis considerations

Most statutory bodies carrying out routine analysis of environmental samples handle large numbers of samples and consequently automated methods of analysis with the minimum of operator involvement tend to be used. For low-level atomic absorption analysis of non-saline samples (concentrations below approximately 10 μg l^{-1}), pre-concentration techniques such as solvent extraction and ion exchange are seldom used routinely even though large numbers of these types of methods have been and continue to be published (see Section IV). For most elements of interest electrothermal atomization (ETA) or hydride generation (HG) methods are used.

When developing new methods the number of manual steps in the procedure should be kept to a minimum; this will increase sample throughput and reduce contamination problems. However, for saline sample analysis it is still necessary to separate and concentrate the analytes from the matrix prior to measurement for many elements. Accurate sampling and analysis of unpolluted deep seawater samples is particularly difficult to carry out because of extremely low concentrations of many elements [17].

II. SAMPLING AND SAMPLE PRESERVATION

It is often difficult to specify a general method of sample collection and preservation which is applicable to a wide range of substances, because of the variety of procedures and tests possible [1, 18–22]. The objective of sampling is to obtain a representative quantity of material which is small enough in volume to be handled and transported to a laboratory. It is imperative that the samples be treated in such a way that contamination is minimised. Another important aspect is to remember that for some sources, representative samples can only be obtained by making composites from a number of sampling points. Consultation between the analyst and end user is often a useful prelude to taking a sample. Analysts are prepared to spend large sums of money on instrumentation, but often fail to spend adequate time and effort in the development of sampling strategies. Accurate and precise results on an incorrectly taken sample are valueless.

A. Sample collection

Whenever possible, the preferred method of sample collection for potable waters is via a sampling tap straight into the container. Enough sample should be run to waste such that all pipework is washed out. Where there is the possibility of disturbing heavy particles which often accumulate (e.g. in water mains), care should be taken to regulate the flow of liquid so that these are not disturbed.

Where sample type and conditions allow, the sample bottle may be clamped into a weighted holder attached to a line and a dip sample collected. In this method of sampling any preservatives have to be added immediately after the sample has been collected. Alternatively a plastic or plastic-coated open top container can be used and the sample immediately transferred to a sample bottle containing a suitable preservative. For larger sample volumes a polyethylene bucket and nylon line may be used. Sampling equipment must be kept scrupulously clean. Regular soaking in 10% v/v nitric acid and rinsing with distilled water prior to use are advised. A comprehensive treatise on this subject has been published [18].

It is good practice, on a regular basis, to send out at least two prepared sample bottles that are filled during the sampling trip from a large acid-washed polyethylene bottle containing distilled-deionised water. They should then be processed as normal samples. A significant concentration of any metal above the normal blank level indicates contamination problems associated with sampling.

B. Sample containers

Low-density polyethylene, polypropylene or PTFE bottles are recommended for all metals other than mercury. Bottle caps should be of an identical material and cap liners other than those formed from inert materials (polyethylene or PTFE) should be avoided. Bottles should be cleaned by soaking in 10% v/v nitric or hydrochloric acids for 1–2 days and rinsed with copious amounts of distilled water. Regular tests should be carried out on random bottles to detect any potential contamination.

For mercury analysis, borosilicate glass containers should be used because plastic containers are permeable to mercury vapour from laboratory atmospheres. These should be cleaned as follows: (i) soak for 24 hours in a 2% v/v detergent solution and rinse thoroughly; (ii) soak for 24 hours in a mixture of 2% v/v hydrochloric acid, 1% m/v potassium bromate and 0.3% m/v potassium bromide (this liberates free bromine) and then rinse thoroughly.

C. Sample filtration

When it is desired to differentiate between the total metal concentration and that which is in solution, then filtration is required. Ideally this should be carried out immediately. If the un-preserved sample can be returned to the laboratory within 6 hours, it is probably better to carry out the filtration there and risk a slight change in speciation rather than contamination from field filtration [22]. A 0.45 μm membrane filter is normally used, the filtrate being accepted as containing all the soluble material. Membranes should be used with caution as some heavy metals may be adsorbed and soluble materials in the membrane may be leached out. Polycarbonate or cellulose ester filters are commonly used. It is a good idea to pre-wash membranes with 1 M nitric acid.

For on-site filtration proprietary units are available based either on a syringe-type apparatus or a filter holder, flask and hand-operated vacuum pump. It is worth reiterating that all sampling apparatus must be thoroughly cleaned before use.

D. Sample preservation

To prevent losses of metals from dilute aqueous solutions it is necessary to acidify the sample as soon as possible after collection. If filtration is required this must be carried out before any preservative is added. Usually the addition of nitric or hydrochloric acids to give a 0.1 M solution (pH = 1) is considered adequate. The acid can be added to the sample immediately after collection or a calculated volume dispensed into the empty container and sample added. Care must be taken when adding sample to a bottle containing acid. Eye protection is essential. When acidifying trade wastes or samples of unknown origin care should be exercised because of the possibility of liberating appreciable quantities of hydrogen sulphide or hydrogen cyanide. If the sample is not to be analysed for several days, storage at 4°C is recommended. For mercury analysis, water samples can be preserved with bromine generated in situ by reaction of potassium bromate/bromide with hydrochloric acid, so an excess of free bromine is maintained (see Section VII).

III. SAMPLE PRE-TREATMENT AND ANALYSIS PROCEDURES FOR FLAME AAS AND ELECTROTHERMAL ATOMIZATION

A. Introduction

The sample pre-treatment of water, effluent and sludges consists of simple acid digestion procedures. Natural water samples (rivers, raw and potable waters), containing minimal amounts of solid material require a relatively mild wet digestion pre-treatment procedure whilst samples that contain significant amounts of solid material such as effluents and sewage sludge, require more vigorous procedures [1].

The acids and other reagents used in the sample pre-treatment procedures should be of sufficient purity to result in a minimal blank contribution to the final result. Most chemical reagent manufacturers supply a range of reagents suitable for low-level atomic spectroscopic analysis and these should be used. Blank values are dependent upon the type of sample pre-treatment but are typically less than the criterion of detection except for ubiquitous elements such as calcium, magnesium, sodium and zinc. Ideally a separate enclosed area should be available for trace metal analysis. Volumetric ware should be reserved solely for metal analysis and all new volumetric ware should be acid washed, allowed to stand in contact with 10% v/v nitric acid (70% m/m) for one week and then tested for absence of metal contamination prior to use. All bench and fume cupboard surfaces should be regularly wiped over with damp disposable cloths. Ceramic top hotplates should be used rather than cast iron or aluminium ones. The influence of the laboratory environment on the precision and accuracy of trace element analysis has been comprehensively addressed by Adeljou and Bond [23].

B. Reagents

(Atomic spectroscopy grade reagents should be used whenever possible.)
- Water: distilled-deionised water is normally adequate.
- Nitric acid (70% m/m).
- Nitric acid (5M): dilute 320 ml of nitric acid with water to 1 litre in a volumetric flask and mix well.
- Hydrochloric acid (36% m/m).
- Caesium chloride solution (5% m/v Cs): dissolve 31.7 g caesium chloride in approximately 400 ml of water, dilute to 500 ml and mix well.
- Ammonium perchlorate (10% m/V).

References pp. 120–123

- *n*-Dodecane.
- Hydrochloric acid (50% v/v): dilute 250 ml of hydrochloric acid to 500 ml with water.
- Acid washed anti-bumping granules: simmer 250 g of fused alumina anti-bumping granules with 5 N nitric acid for 2 hours. Wash thoroughly with water, dry at 105°C, store in a clean container.
- Certified reference materials (waters and sludges): these are available from a number of sources including Community Bureau of Reference (BCR), Directorate General XII, Commission of the European Communities, 200 rue de la Loi, B-1049 Brussels, Belgium; Office of Standard Reference Materials, NIST, Gaithersburgh, MD 20899, U.S.A.; Bureau of Analysed Samples Ltd., Newham Hall, Newby, Middlesborough, Cleveland TS8 9EA, U.K.; and Promochem Ltd., P.O. Box 255, St. Albans, Herts. AL1 4LN, U.K.

C. Apparatus

Normal volumetric glass and plastic ware. Polypropylene flasks are recommended for low-level metal analysis (see Section VI.B for silver analysis). The flasks used for the calibration standards for river and water analysis should have a volume of at least 4 ml above the calibration mark.

Polyethylene bottles (125, 260 and 540 ml capacity) with polypropylene caps (suitable bottles can be obtained from TT Containers Ltd., Otford, Kent, TN14 5JF). *Note*: bakelite caps contain significant quantities of zinc and other metals and must not be used.

Filtration apparatus for determining soluble metals with 0.45 μm cellulose acetate membrane filters. These must not have a grid pattern as this can contain significant quantities of metals. For filtering potable waters or clean rivers, 30 mm diameter 0.45 μm Sartorious Minisart NML disposable filters (Cat. No. 165 5SQ) have been found to be suitable together with 50 ml polyethylene disposable syringes, both with Luer fittings (Sartorious Instruments Ltd., Belmont, Surrey SM2 6JD).

A homogeniser is required for the preparation of sewage sludges.

Disposable aluminium or borosilicate trays suitable for drying sludge at 105°C.

Micro Kjeldahl digestion apparatus. This comprises a rack of six heating mantles suitable for 25–50 ml capacity round bottom flasks. Each mantle has its own temperature controller. A suitable device can be obtained from Electrothermal Engineering Ltd., Southend on Sea, Essex, S52 5PH, Cat. No. MM231.

50-ml calibrated borosilicate tubes with glass or polypropylene stoppers for sewage sludge analysis.

D. Method for trace metals in raw and potable waters and rivers

Applicability
(a) Flame AAS: Cd, Cr, Cu, Fe, Mn, Ni, Pb and Zn. For potable water compliance analysis (see Table 1), the flame AAS technique is only considered sensitive enough for Cu, Fe, Mn and Zn analysis. The ETA technique should be used for other elements.
(b) ETA: all relevant elements. For river compliance analysis (see Table 2), the flame AAS technique is only considered suitable for zinc analysis. The ETA technique should be used for other elements.
(c) Regular tests should be carried out to determine contamination errors from the actual sampling procedure (see Section II.A).
(d) The criterion of detection should be similar to those described in Sections V and VI.
(e) For river samples containing significant amounts of suspended solids, method of Section G should be used.

Calibration
Prepare suitable multi-element standards and blank in 100 ml volumetric flasks made upto volume with distilled-deionised water and then add 2 ml of 5 M nitric acid. For concentrations below 0.25 mg l^{-1} prepare standards and blank daily.

Procedure
(i)   Fill a 540 ml polyethylene sample bottle containing 10 ml of 5 M nitric acid with the sample to the 500 ml mark, replace the cap and mix well. Return the sample to the laboratory.
(ii) The sample bottle cap should be loosened slightly and the bottle placed in an oven at 70°C overnight. The weight loss should be less than 0.5%.
(iii) Select the appropriate technique (flame AAS or ETA) and carry out the analysis (see Sections V and VI).

E. Modified method for total and filtered metals in rivers

This method is a modification of the previous method which allows a filtered sub-sample to be taken from the main sample. If samples can be returned to the laboratory within 6 hours, then filtration there is recommended. This will minimise the risk of contamination from on-site filtration [22]. Potential contamination is considered a greater risk than the change in speciation over a 6-hour period.

Procedure
(i)   Fill a 540 ml clean polyethylene bottle with sample to the 500 ml mark and return sample to laboratory within 6 hours. If this is not possible an aliquot must be filtered on site.

(ii) Use a suitable vacuum or pressure filtration apparatus fitted with a 0.45 μm membrane filter (see Section III.C), filter 100 ± 5 ml of the well shaken sample and transfer it to a 125 ml polyethylene bottle. Alternatively a one piece disposable filter unit and a 50 ml polypropylene syringe (see Section III.C) can be used. Draw 50 ml of the sample into the syringe, attach the disposable filter unit, and pressure filter the contents directly into a 125 ml polyethylene bottle. Then filter another 50 ml aliquot. This procedure is not recommended for highly turbid samples.

(iii) Add 2 ml of 5 M nitric acid to the 125 ml bottle, replace the stopper and mix well.

(iv) Add 8 ml of 5 M nitric acid to the remainder of the sample in the 540 ml bottle and mix.

(v) Slightly loosen the stopper and place the total metals sample in the 70°C oven overnight. (The weight loss has been found to be less than 0.5%.)

(vi) Select the appropriate technique (flame AAS or ETA) and carry out the analysis (see Sections V and VI).

F. Method for calcium and magnesium in raw and potable waters and rivers by flame AAS

The reader is referred to the HMSO method [24], but basically the method is as follows. The acidified sample (0.1 M hydrochloric acid) is diluted ten times and before making up to volume, lanthanum chloride solution is added so that the final concentration of lanthanum is 1000 mg l^{-1}. The diluted samples containing the added lanthanum are then nebulized into an air–acetylene flame.

For barium and strontium in potable waters by flame AAS the reader is referred to the HMSO method [24]. Basically the method is as follows. The acidified sample (0.1 M hydrochloric acid) has lanthanum and potassium (as chlorides) added so their final concentrations are 1000 mg l^{-1}. The treated samples are then nebulized into the nitrous oxide–acetylene flame.

G. Method for metals in trade effluent and sewage by flame AAS

Applicability
Ag, Al, As, Ca, Cd, Co, Cr, Cu, Fe, Mg, Mn, Ni, Pb, Sb and Zn.

(a) This method may also be suitable for some other elements that are released by a hydrochloric–nitric acid digestion and that can be determined by flame AAS.

(b) The nitrous oxide–acetylene flame should be used to determine As, Ca, Cr, Fe, Mg, and Mn in order to minimise chemical interference effects.

(c) The criterion of detection should be similar to those given in Table 3 (Section V).

(d) Silver analysis should be carried out within 24 hours of completing the digestion.

Calibration

Prepare suitable multi-element standards and blank in 100 ml volumetric flasks but before making up to volume with distilled-deionised water, add 3 ml nitric acid, 9 ml hydrochloric acid and 2 ml of 5% m/v caesium solution.

The acid concentrations in the standards make allowance for the acid lost during the digestion step.

Procedure

(i)   Collect the sample in a 150 ml polyethylene bottle.
(ii)   Pour 50 ml of the shaken sample into a 100 ml beaker and add 2 ml nitric acid and 6 ml hydrochloric acid.
(iii)   Run two blanks using 50 ml water as (ii).
(iv)   Add 3 or 4 anti-bumping granules.
(v)   Place a watchglass over the beaker.
(vi)   Digest on hot plate until volume is reduced to 10–15 ml. (The sample must not be allowed to boil dry.)
(vii)   After cooling add 1 ml of 5% m/v caesium solution make upto 50 ml with water and mix well.
(viii) If the digested sample contains appreciable solids, large lumps or oil, filter through a Whatman 541 filter paper.
(ix)   Determine the required elements by flame AAS (see Table 3, Section V).

H. Method for metals in sewage sludges by flame AAS

Applicability

Aqua regia acid soluble metals: Ag, Al, As, Ca, Cd, Co, Cr, Cu, Fe, Mg, Mn, Mo, Ni, Pb, Sb and Zn.

(a) The digestion procedure is suitable for mercury determination by the cold vapour method (see Section VII).
(b) This method may also be suitable for some other elements that are released by a hydrochloric–nitric acid digestion and that can be determined by flame AAS.
(c) Criterion of detection (mg kg^{-1}) should be approximately 200 times those given in mg l^{-1} in Table 3 (Section V).
(d) Ammonium perchlorate is added to minimise chemical interference effects in the determination of molybdenum.
(e) The chromium in the multi-element calibration standards must be in the (III) oxidation state, otherwise slow precipitation of metal chromates will occur.

References pp. 120–123

(f) The nitrous oxide–acetylene flame must be used to determine As, Ba, Ca, Cr, Fe, Mg, Mn, Mo and Sn as chemical interference effects occur in the cooler air–acetylene flame.

(g) Silver analysis should be carried out within 24 hours of completing the digestion.

(h) A certified reference sludge should be analysed with each batch of samples.

(i) Full details of this method have been published [1, 25].

Calibration

Prepare multi-element standards and blank in 100 ml volumetric flasks. Prior to making up to volume with distilled-deionised water, add 6 ml hydrochloric acid, 2 ml nitric acid, 2 ml of 5% m/v caesium solution and 10 ml of 10% m/v ammonium perchlorate.

The acid concentrations in the standards make allowance for that lost during the digestion step.

Procedure

(i)   Dry sufficient homogenised sludge (typically 30 ml for a 5% DS sludge) in a 105°C oven overnight. Borosilicate glass or disposable aluminium trays can be used for this.

(ii)   After cooling in a desiccator grind up the sludge so that it passes through a 2 mm mesh stainless steel sieve.

(iii) Transfer 0.500 ± 0.002 g of the ground dried sludge into a 50 ml calibrated borosilicate tube. Run 1.0 ml of water down the side of the tube to wet the sample.

(iv)   Add 3 or 4 aluminium oxide anti-bumping granules. Carefully run 6.0 ml of hydrochloric acid and 2.0 ml of nitric acid down the side of the tube. Place the tube in a rack and allow any vigorous initial reaction to subside. If excessive foaming occurs, add 2 drops of n-dodecane.

(v)   Place the tube on the heating mantle and adjust the heating control until the sample gently refluxes.

(vi)   Allow the sample to reflux for 10 ± 2 minutes. Run 5 ml water down the side of the tube and reflux for a further 5 ± 1 minutes.

(vii) Allow the tube to cool. Add 1 ml of 50,000 mg l^{-1} Cs solution and 5 ml of 10% m/v ammonium perchlorate solution. Dilute with water to 50 ml. Replace the ground glass stopper, shake vigorously and filter through a suitable filter paper into a polyethylene bottle.

(viii) Two blanks should be run with each batch of determinations. Add 1.0 ml of water and then carry out steps (iv)–(vii).

(ix)   Determine the metals by flame AAS (see Table 3).

(x)   Calculation: metal concentration in dry sludge (mg kg^{-1}):

$$\frac{50 \times C}{W}$$

where C is the measure blank corrected concentration in mg l^{-1} and W is the weight (g) of dry sample taken.

IV. PRECONCENTRATION TECHNIQUES

Preconcentration techniques can be used to improve the detection limits for a number of elements. However, these techniques have several drawbacks; they can be time-consuming and can require meticulous attention to detail to avoid variable blanks and poor recoveries. Also if the metals present in the sample are complexed with organic substances, a digestion step will be required to release them prior to carrying out most preconcentration procedures. It should be noted that many elements can be determined without preconcentration using complementary atomic absorption techniques such as electrothermal atomization and hydride generation.

Preconcentration techniques are useful for the analysis of saline samples where the determinant elements can be concentrated and simultaneously removed from the main matrix elements. Some commonly used techniques are described below.

A. Evaporation

Typically up to 200 ml of sample acidified with nitric or hydrochloric acid to 0.1 M is evaporated until the volume is reduced to 10–20 ml. The sample is then quantitatively transferred to a volumetric flask (20–50 ml) and diluted to volume. Standards are prepared in an identical fashion. Background correction is essential because of the increased matrix concentration in the final solution. If the beakers are allowed to boil dry during the evaporation stage, adsorption losses can occur. Adsorbed material (especially iron) is often difficult to desorb even after prolonged heating with nitric acid. For most natural waters chemical interference effects for Cd, Co, Cu, Fe, Pb, Ni and Zn were found not to be very significant by the authors.

The main advantages of this concentration technique are that it is simple, requires a minimum of reagents, gives 5–10 fold improvement in detection limit and is applicable to most elements. The main disadvantages are that background correction must be used and the procedure is not really suitable for samples having conductivities greater that 2000 μS cm^{-1}.

B. Solvent extraction

In this technique a complexing reagent is added to the sample and the resulting metal complex extracted into an organic solvent. The ratio of sample to organic solvent is usually in the range 10 to 20 : 1. An increase in sensitivity is also observed due to the increased rate of nebulization of organic solvents.

References pp. 120–123

One of the most commonly used complexing reagents is ammonium pyrollidine dithiocarbamate (APDC) with 4-methyl-pentan-2-one (MIBK) as organic solvent [26–28]. A number of elements including Ag, Cd, Co, Cr(VI), Cu, Mn, Ni, Pb and Zn can be extracted and concentrated under optimised conditions. A method for the determination of lead has been published by the Department of the Environment using this technique. The use of diethylammonium diethyl dithiocarbamate (DDDC) together with APDC as chelating agents is also used for extracting Cd, Co, Cu, Fe, Pb, Ni, Ag and Zn into MIBK [29]. A treatise by Cresser [27] gives a detailed appraisal of the uses of solvent extraction in atomic absorption. Blanks as well as samples and standards should be taken through the extraction procedure. This will compensate for contaminants and the fact that the organic phase should be saturated with the aqueous phase so that at wavelengths below 250 nm, any background absorption caused by the water-saturated solvent will be identical for samples and standards.

For river and sewage final effluent samples, direct extraction sometimes gives poor recoveries due to interactions between the metals and natural organic constituents. In these cases the samples must be digested before solvent extraction is used. Advantages of solvent extraction are that under optimum conditions it can be applied to a number of elements, background correction is not necessary and it is applicable to seawater and estuary samples. The main disadvantages are that it is time-consuming, blanks can be significant and variable and extraction efficiency may be reduced due to other matrix constituents.

C. Ion exchange

Chelating resins may be found useful for concentrating metals [30–33]. Riley and Taylor [30] discuss retention and recovery data for 29 elements including Bi, Cd, Co, Cu, Mn, Mo, Ni, Pb, V and Zn from a seawater matrix. A paper by Biechler [31] discusses the preconcentration of Cd, Cu, Fe, Ni, Pb and Zn from industrial effluents. The main advantages of ion exchange are that large concentration factors (100 times) can be obtained and the bulk of the matrix removed. It is especially useful for analysis of saline samples. Drawbacks include variable blanks and the fact that the technique can be time-consuming especially if pre-digestion is also required.

D. Reductive precipitation

Sodium borohydride has been employed in a preconcentration technique for natural water and seawater [34, 35]. In the presence of 2–5 mg l^{-1} of iron and palladium and under alkaline conditions, sodium borohydride reduces metal species to the element and/or forms metal borides. These are co-precipitated on the black granular precipitate formed from the added iron and palladium and after a suitable length of time (15–20 hours) may be collected on a 0.45 μm membrane filter and re-dissolved in acid. Using this technique, a 40 times concentration may be achieved. Good recoveries (greater than 90%) are reported for Ag, As(V), Bi, Cd, Co, Cr(VI), Cu, Mn, Ni, Pb, Sb, Se(IV), Sn, Te, Tl and Zn in seawater [35]. Analysis of open-ocean water was feasible using this technique. Detection limits (3σ) of 1.0, 1.3 and 0.3 ng l^{-1} for Cd, Cr and Pb, respectively, and impressive detection limits for the other 12 elements tested were reported.

V. FLAME ATOMIC ABSORPTION SPECTROSCOPY

A. General considerations

This simple technique can be used to determine a wide range of elements relatively simply, although the poor sensitivity limits its direct application in potable and river water analysis [1, 36]. In the U.K., aluminium, iron and manganese are the most commonly determined elements in raw and treated potable waters. Although iron and manganese can be determined satisfactorily by flame AAS, the criterion of detection for aluminium is not quite adequate and alternative techniques are normally employed [1] (e.g. ETA, plasma emission or colorimetry).

For flame AAS analysis, samples are subjected to suitable acid pre-treatment to ensure that all the relevant elements are released into solution. Where necessary, reagents are added to minimise interference effects.

B. Optimum operating conditions

Table 3 gives the optimum wavelength(s), type of flame, characteristic concentration (sensitivity for 1% absorption) and criterion of detection (2.33σ) for a wide range of elements. It also gives some indication of lamp current; spectral bandpass and flame conditions. It should be possible to obtain a coefficient of variation of less than 2.5% for concentrations greater than 20 times the criterion of detection. It is important to realise that the performance characteristics are dependent upon the actual instrument and also on the care with which it is set up: e.g. burner and optical alignment; good housekeeping procedures such as regularly washing out the burner and spray chamber; optimising the operation of the nebulizer and impact bead; and generally maintaining the instrument in a good working condition.

C. Interferences and methods for their minimisation

It is sometimes assumed that interference effects in flame AAS are not very significant for environmental analysis. However, there are a number of interference effects that can be encountered if inappropriate conditions are employed.

References pp. 120–123

TABLE 3

SENSITIVITY, CRITERION OF DETECTION AND SUGGESTED OPERATING CONDITIONS

Element	Wavelength (nm)	Suggested lamp current (% of maximum current)[a]	Spectral bandpass (nm)	Flame type	Flame condition[b]	Sensitivity for 1% absorption (characteristic concentration) (mg l^{-1})	Criterion of detection (mg l^{-1})
Al	309.3		0.3	NA	4	0.4	0.1
Sb	217.6		0.3	NA	1	0.2	0.1
As	193.7	90	1.0	NA	4	0.4	0.25
Ba	553.5	90	0.2	NA	4	0.1	0.1
B	249.7		0.5	NA	5	12	5
Cd	228.8	30–40	1	AA	1	0.01	0.005
Ca	422.7	50	0.5	AA	2	0.05	0.01
Ca	422.7	90	0.5	NA	3	0.02	0.005
Ca	430.2	90	0.3	NA	3	12	5
Cr	357.9		0.5	NA	3	0.1	0.1
Cb	240.7		0.5	NA	1	0.05	0.02
Cu	324.7		0.5	NA	1	0.03	0.01
Fe	248.3		0.3	AA	1	0.1	0.05
Fe	248.3		0.3	NA	3	0.3	0.2
Fe	372.0		0.5	NA	3	1.0	0.5
Pb	217.0		1.0	AA	2	0.1	0.05
Li	670.8	75	0.5	AA	1	0.02	0.005
Mg	285.2		0.5	AA	1	0.003	0.001
Mn	279.5		0.5	AA	1	0.02	0.005
Mb	313.3		0.5	AA	4	0.3	0.1
Ni	232.0		0.15	AA	1	0.06	0.04

K	766.5	75	1	AA	1	0.02	0.01
Se	196.0	90	1	NA	4	0.4	0.4
Ag	328.1		0.05	AA	1	0.03	0.005
Na	589.0		0.5	AA	1	0.01	0.005
Na	330.2		0.5	AA	1	3	1
Sr	460.7	90	0.5	NA	3	0.1	0.02
Sn	235.5		0.5	NA	4	2	0.5
Ti	364.3		0.5	NA	4	1	0.5
V	318.4		0.3	NA	5	1	0.5
Zn	213.9		0.5	AA	1	0.01	0.003

AA = Air Acetylene flame; NA = Nitrous oxide–Acetylene flame.

a When no figure is given a current of about 40–50% of maximum current should be used.

b 1 = slightly fuel-lean (AA), burner set to 1 mm below grazing incidence; 2 = slightly fuel-rich (AA), burner set to 1 mm below grazing incidence; 3 = 2–5 mm red feather (NA) burner set to 1 mm below grazing incidence; 4 = 10–13 mm red feather (NA) burner set to 3 mm below grazing incidence; 5 = 20–25 mm red feather (NA) burner set 3 mm below grazing incidence.

(Reproduced with permission from ref. [1].)

There are six main categories of interference:

1. *Chemical interference*

This is the most common form of interference in flame AAS. A chemical interference effect is observed when a matrix constituent enhances or suppresses the formation of ground-state atoms in the flame. A common example is the interference produced by aluminium, silicates and phosphates in the determination of magnesium, calcium, strontium and many other metals in the air–acetylene flame. This is caused by the formation of low-volatility aluminates, silicates and phosphates which, in many instances, are only poorly atomized in the air–acetylene flame. There are two main methods of overcoming this type of interference:

(a) Addition of 1000–2000 mg l^{-1} of a releasing agent such as lanthanum (as the chloride). The lanthanum preferentially reacts with the interfering species.

(b) Use of the hotter nitrous oxide–acetylene flame. This flame is recommended for the determination of iron, manganese and chromium in relatively complex matrices such as effluents and sludges. However, this results in a degradation of the criterion of detection relative to the air–acetylene flame.

2. *Matrix interference*

This effect is caused by variation in the rate and efficiency of nebulization with changes in viscosity and/or surface tension of the samples relative to the standards. Normally the sensitivity of a given element will be found to decrease with increasing amounts of mineral acids. For most applications in the water industry the effect is not very significant but can be observed with sludge and sediment analyses where significant amounts of mineral acids and/or major matrix elements are present. In most instances it can be minimised by ensuring that samples, standards and blanks contain the same concentrations of reagents such as mineral acids, releasing agents and ionisation suppressants.

3. *Non-specific background absorption*

Background absorption is an interference phenomenon mainly caused by absorption of the resonance line by molecular species in the flame from the sample matrix. It is overcome by using automatic background correction, which is available on all modern AAS instruments. For routine flame AAS measurements, automatic background correction should be used at all wavelengths below 360 nm. Conventional background correction using a deuterium lamp has been found to be adequate for flame AAS determinations in environmental analysis.

4. *Ionisation interference*

Elements with low ionisation potentials such as the alkali metals and alkaline earth metals undergo ionisation in the air–acetylene flame, thus reducing the ground-state atom population. In the hotter nitrous oxide–acetylene flame the effect is more pronounced and is significant for a number of other elements, including aluminium, chromium, magnesium, titanium and vanadium. The interference is overcome by the addition of an ionisation buffer to samples, standards and blank. The buffer is another easily ionised metal that will preferentially ionise to the one being determined. Caesium (as the chloride) at a concentration of 1000–2000 mg l^{-1} is commonly used.

5. *Emission breakthrough interference*

All atomic absorption instruments have detection systems that, in principle, do not respond to emission from the flame or sample matrix. However, if a low-intensity source is used in a region of high flame background emission, the intense emission signal will result in increased noise level and in extreme cases can cause erroneous signals. The effect is negligible for most elements if the manufacturer's operating instructions are followed. However, if this form of interference is suspected the lamp current should be increased and the monochromator spectral bandpass decreased. The only common example of this form of interference using flame techniques is the determination of barium in the presence of calcium using the nitrous oxide–acetylene flame. The intense thermal emission from CaOH species at the 553.6 nm barium resonance line wavelength can overload the phase sensitive amplifier for calcium levels above 100–1000 mg l^{-1} and results in very noisy erratic signals. The effect is very instrument-dependent. Some instruments can tolerate 1000 mg l^{-1} calcium whilst others are affected by as little as 100–200 mg l^{-1} calcium [37].

6. *Non-reproducible calibration graphs for chromium and iron*

The determination of chromium in the luminous air–acetylene flame, optimised with respect to acetylene flow for maximum chromium sensitivity, can result in calibration graphs that exhibit regions with near zero or even negative slope [38]. The response for a given standard can also depend upon the age of the standard. The effect is more pronounced for chromium(III) than for chromium(VI) and is thought to be caused by the kinetics of the reactions resulting in the formation of free chromium atoms in the highly reducing luminous air–acetylene flame. A non-luminous flame should be used although this results in a significant decrease in sensitivity. A better alternative is to utilise the hotter nitrous oxide–acetylene flame and to buffer the samples, standards and blank solution with 1000 mg l^{-1}

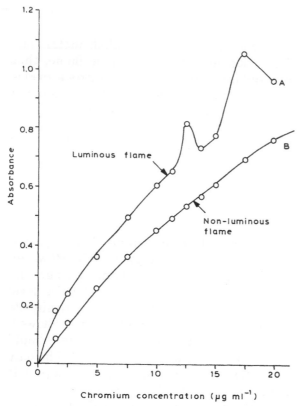

Fig. 1. Chromium(III) calibration graphs in the air–acetylene flame (from [38]).

caesium in order to suppress ionisation. Figure 1 shows typical calibration graphs for chromium(III) in both luminous and non-luminous air–acetylene flames.

Calibration graphs for iron solutions in a fuel-rich air–acetylene flame optimised with respect to acetylene flow for maximum iron response have also been shown [39] to exhibit calibration graphs similar to those observed for chromium. The effect was attributed to traces of silicon slowly leached from glassware (less than 0.5 mg 1^{-1} silicon) and could be eliminated by using a fuel-lean air–acetylene flame although this results in a significant decrease in sensitivity. In both the above examples the anomalous calibration graphs would be unlikely to be detected if only 2–4 standards were used. Most modern AAS instruments would accept the results from these calibration standards and set up a best-fit calibration graph.

D. Notes on interference effects observed in flame AAS for specific elements

(i)   For Cd, Co, Cu, K, Na, Ni and Zn there are few significant chemical interference effects ($\pm 5\%$) for typical environmental samples in the air–acetylene flame.

(ii)   Automatic background correction should be used at all wavelengths below 360 nm.

(iii) For sludges and other samples containing high calcium levels (greater than 200 mg l^{-1} in the final digest), it is advisable to set the burner 3–5 mm below grazing incidence and use a slightly fuel-lean air–acetylene flame when determining lead.

(iv)   Iron and Mn determinations can be prone to interference effects in the air–acetylene flame especially for samples with a complex matrix (e.g. trade effluents, sludges, soils and sediments). Iron can even be subject to non-reproducible calibration graph effects [39]. These effects can be overcome by using the hotter nitrous oxide–acetylene flame, but a significant degradation (4–8 times) in the criterion of detection limit is observed.

(v)   Chromium should always be determined in the nitrous oxide–acetylene flame even though this results in a 2–4 times degradation in the criterion of detection. Chromium exhibits maximum sensitivity in the luminous air–acetylene flame and under these conditions is subject to non-reproducible calibration graph and chemical interference effects [38]. An ionisation suppressor is essential in the hotter nitrous oxide–acetylene flame.

(vi)   Calcium and Mg are prone to significant chemical interference effects in the air–acetylene flame and 2000 mg l^{-1} lanthanum is added to overcome these effects.

(vii) For the determination of Na and K in the air–acetylene flame an addition of 1000 mg l^{-1} caesium (as chloride) to samples, blanks and standards is recommended to overcome ionisation interferences.

(viii) When using the nitrous oxide–acetylene flame it is good practice to add 1000 mg l^{-1} of caesium to all samples, standards and blanks to suppress ionisation.

(ix)   To minimise chemical interference effects on the determination of molybdenum in sludges the addition of ammonium perchlorate to give a final concentration of 1% m/v in samples, standards and blanks has been found to be necessary [1].

(x)   For Ba, Ca and Sr determinations using the nitrous oxide–acetylene flame, the hollow-cathode lamp current should be set to approximately 90% of the maximum rated value. This will minimise the effect of the intense analyte flame emission upon the instrument output.

(xi)   For Ba and Sr determinations, the use of the nitrous oxide–acetylene

References pp. 120–123

flame as well as the addition of 1000 mg l^{-1} of lanthanum and potassium to the samples, standards and blanks has been found necessary to overcome chemical interference and ionisation effects.

(xii) For Ba determinations, a spectral bandpass of 0.2 nm or less should be used to minimise the effect of flame emission from calcium in the sample (see Section V.C.5).

E. Ensuring that a method is not prone to significant interference effects

Many interference effects are instrument-dependent and it is essential that the user should check out any proposed method carefully. Most users select the flame conditions and burner height position for maximum sensitivity when setting up a method. These conditions are often not the best for the minimisation of interference effects. Often reducing the fuel richness of a flame and lowering the burner height slightly will reduce the sensitivity of the determination but significantly reduce many chemical interference effects. It is often assumed that a published method developed on one instrument, will work satisfactorily on all other instruments if the specified conditions are followed. This must not be assumed, any proposed new methods should be carefully checked and all existing methods periodically rechecked. This can be achieved by: carrying out spiking recovery tests on a range of typical samples; regularly analysing relevant certified reference materials; comparing analysis results with those obtained by another technique (e.g. polarography, X-ray fluorescence, colorimetry); and regularly taking part in interlaboratory comparison exercises. The problems of literature interpolation of methods has been addressed by Cresser [40] and the reader is referred to that excellent paper.

VI. FLAMELESS ATOMIC ABSORPTION

A. Application to waters and effluents

The main advantage of ETA-AAS when compared to flame AAS, is enhanced sensitivity; typically 100 times better. Because of this increased sensitivity, ETA is the preferred AAS technique in order to ensure compliance with the various international standards [8, 9] for several trace metals in potable water (see Table 1). Similarly, for river compliance analysis [11–13], the ETA technique should be used for all elements except zinc to achieve the required limits of detection (see Table 2).

Although ETA is significantly slower than the flame technique (typically 15 samples/hour), reliable instrumentation is now available to automatically analyse a large batch of samples (45–200). Furthermore, some ETA-AAS systems can be programmed to sequentially analyse a batch of samples for several elements, while other systems are capable of analysing two elements simultaneously (e.g. Pb/Cd, As/Se, Cu/Ni). These systems can complete the analysis after normal working hours and then close down in a safe manner, thus saving a significant amount of operator time. The other advantage of using an autosampler is that better and more consistent precision can be obtained (typically better than 2% RSD) compared with manual sample introduction.

Further advantages offered by ETA are small sample volumes (5–100 μl) and minimal sample pretreament (because the sample is dried and ashed in the graphite furnace).

This technique does however suffer from one main disadvantage: chemical interference. Methods for minimising these effects are discussed later.

B. Sample collection, preservation and pre-treatment

Samples are collected in pre-washed low-density polyethylene, polypropylene or PTFE bottles as outlined in Section II of this chapter. Great care should be exercised to minimise the level of contamination by using distilled-deionised water and ultra-pure reagents. If a sample is particularly turbid, it may be necessary to further pre-treat it by boiling with acid (see Section III.G). This will, however, increase the risk of contamination with subsequent degradation of detection limit.

If samples are being collected for silver analysis, they should be stored in the dark and analysed within 24 hours. To further minimise the loss of silver, the use of polyethylene rather than polyproplyene bottles for both samples and standards is recommended [41].

C. Interferences and methods for their minimisation

Although much literature has been devoted to interference effects in ETA [42, 43], many of the interfering mechanisms have not been completely elucidated. Interference effects are generally more pronounced with the ETA-AAS technique because the efficient generation of a dense atomic vapour cloud is also accompanied by an efficient generation of the interfering species. The main interferences are discussed below.

1. *Spectral interference (non-specific background absorption)*

The most common causes of this interference are molecular absorption and light scattering. Molecular absorption is caused by matrix components absorbing the analyte atomic resonance line at the atomization stage and alkali halides are the main cause of this interference in environmental analysis. Light scattering is caused by particles condensing at the cooler ends of the graphite furnace, or by the release of graphite from the furnace wall at high atomization temperatures. Direct overlap between atomic lines only exists for a few elements and is of little consequence in environmental analysis (e.g. Hg 253.652 nm and Co at 253.649 nm).

2. *Chemical interference*

This is a broad heading which covers both solute volatilisation interferences (i.e. before atomization stage) and vapour phase interferences (i.e. after atomization). In both cases, the free atom population is altered by the presence of matrix components. These interferences can arise in several ways; the more common ones are discussed below:

(a) Loss of analyte due to volatile compound formation during the drying and ashing stages. This occurs mainly when a high halide concentration (particularly chloride) is present. A reduction in sensitivity is then observed.

(b) Some elements can form stable refractory carbides by reacting with the graphite substrate. When atomization occurs, a proportion of the analyte is retained by the graphite. The degree of carbide formation can be matrix-dependent. Incomplete analyte release can also arise due to formation of intercalation compounds, mainly between alkali and alkaline earths with graphite.

(c) When sample atomization takes place, a change in the rate of analyte release can be effected by reaction between analyte and sample on the hot graphite surface. This results in a change in peak shape and depending upon whether or not the rate of release is retarded or enhanced, the peak height absorbance and consequently the sample result will be negatively or positively biased.

(d) After the analyte has been vaporised, it can interact with the matrix components, resulting in decreased absorbance. The major cause of these vapour phase interferences is the formation of gaseous monohalides.

D. Methods of minimising interferences

Because interference effects are far more common in ETA-AAS analysis than in conventional flame analysis, the analyst must exercise great care in the development of any new method of analysis. In general, a nitric acid matrix (typically 0.5–1% v/v nitric acid) is preferred to a hydrochloric or sulphuric acid matrix.

The degree of interference observed for given solutions is often dependent both upon the design and operating conditions of the graphite tube atomizer. It can also be dependent upon the age (number of firings) and history of the graphite tube. If the atomization temperature is not sufficient to completely remove the matrix, slow changes in sensitivity as well as variations in interelement effects can sometimes be observed. It is often difficult to correlate the degree of interference with the sample matrix. For instance, the interferences observed in the determination of cadmium and lead in natural waters are not directly related to calcium or magnesium concentrations. Some soft waters exhibit more significant interference (suppression of signal) than some very hard waters. There are a number of methods for overcoming or minimising interference effects; the most commonly used methods are summarised below. It should be stressed that any proposed interference tests should be carried out at various stages (e.g. 10, 50 and 90%) of the expected graphite tube lifetime.

1. *Background correction*

The continuum background correction system was developed to correct for errors due to non-specific background absorption [44]. The deuterium arc lamp is commonly used at wavelengths below 360 nm. However, this source does suffer from inadequate energy at the longer wavelengths (above 320 nm) and some instruments are now equipped with a tungsten-iodide lamp to overcome this deficiency. Theoretically, the continuum source is capable of correcting for non-specific background for up to 2 absorbance units, but serious errors due to under- or over-correction when the background level is high, can occur [45].

The Zeeman background correction system [46] claims to offer several advantages over the continuum source, including tolerance to high background levels, correction for structured background, and no extra light source which results in lower noise levels. It also ensures that the light path through the graphite tube is identical for both the total (atomic and background) and the background absorbance measurement. Disadvantages of Zeeman are reduced sensitivity for several transition metals and consequently a reduced linear range. This system has been routinely used for water analysis [47], particularly with the STPF concept (see Section 7 below). A recent study evaluated Zeeman and deuterium background correction techniques to determine trace metals in low ionic strength waters [48]. Both techniques gave satisfactory results. Another widely used background correction system is Smith-Hieftje (S-H) [49], which involves alternately pulsing the hollow-cathode lamp with a high current, causing self-reversal. This system also results in lower sensitivity for some elements.

2. *Method of standard additions*

This method does not correct for non-specific absorption but should correct for chemical interferences. When this technique is used, at least two aliquots of the sample should be spiked with known concentrations of the analyte. The spiked samples and original samples are then measured.

Few problems should be encountered using the additions method with waters and waste waters, but it should only be used when interferences cannot be overcome by other less time-consuming means. Although many modern instruments provide facilities to automate sample spiking and result calculation, this technique is not recommended for routine analysis because it is very time-consuming (at least 15 minutes per sample). Also, the calibration range is reduced and only a limited number of samples can be analysed within the lifetime of the graphite tube.

3. *Selective extraction of the determinand*

Most toxic metals can easily be extracted from natural waters after appropriate pre-treatment using a suitable extraction reagent and solvent (e.g. APDC and Freon). The main matrix elements, i.e. calcium, magnesium, potassium and sodium, are not extracted. The method is particularly suitable for saline samples and the concentration factor results in improved detection limits. The analyte can be back extracted into a small volume of dilute nitric acid. Apte and Gunn [50] have developed a rapid semi-micro scale solvent extraction ETA-AAS method based on APDC/1,1,1,trichloroethane for Cd, Cu, Ni and Pb in saline samples with sub μg l^{-1} detection limits. The extraction is carried out in autosampler cups (1.25 ml sample into 0.25 ml 1,1,1,trichloroethane) and the automatic sampler tip is set to directly sample the lower organic phase from these cups. No phase separation is necessary.

4. *Choice of graphite furnace tube*

It has been demonstrated that when analysing elements which tend to form stable refractory carbides, better sensitivity is obtained from a pyrolytic graphite tube. Early batches of pyrolytically coated graphite tubes suffered from severe ageing effects, where the coating was rapidly degraded to the extent that the tube soon performed similarly to the uncoated version. Modern coated tubes have facilitated the determination of Al [51], Cr and Co [47] using SPTF conditions (see Section 7 below). The more refractory metals Mo, Ti and V should be analysed from the wall of coated tubes rather than a platform. Totally pyrolytic graphite tubes are claimed to offer further advantages than the above type: extended tube life, improved sensitivity for refractory and carbide-forming elements, and consistent signal magnitude throughout their operating period [52]. However, to the best of the authors' knowledge only one manufacturer currently supplies this type of tube.

5. *Choice of sample introduction technique*

The vast majority of ETA methods involve the pipetting of a small sample volume (typically 5–30 μl) onto the wall or platform in a graphite tube. An alternative sample introduction technique has demonstrated reduced interferences by depositing the sample into a hot graphite tube in the form of an aerosol mist [53]. A reduction in solid-phase interferences arises because the aerosol dries immediately on contact with the tube surface. This results in small-crystal formation, rather than the large crystals formed during solvent evaporation. This technique has been used in conjunction with the delayed atomization cuvette (DAC) for the determination of trace metals in water [54]. Aerosol deposition also allowed thallium to be determined in the presence of 1% v/v hydrochloric acid with no interference [55].

6. *Graphite furnace pre-treatment*

Non-pyrolytically coated graphite tubes have been pre-treated with suitable carbide-forming elements to minimise a number of interelement effects [56–61]. The addition of lanthanum compounds has also been used to reduce chemical interference effects in the determination of lead and cadmium in natural water [56–59].

Molybdenum coating of L'vov platforms enabled several volatile elements (arsenic, antimony, cadmium, germanium and selenium) to be determined in iron-rich spring water samples [60]. This coating not only reduced interferences but also extended the tube lifetime. Tantalum carbide coated platforms have enabled interferences to be reduced in selenium determinations, and with increased precision [61].

7. *The L'vov platform and the stabilised temperature platform furnace*

L'vov [62] demonstrated in 1959 that if an integrated absorbance signal was used when analyte was vaporised into a heated tube at constant temperature, interferences which altered the rate of vaporisation could be overcome. He later proposed [63] a method of achieving a constant temperature furnace, known as the L'vov platform. The sample is injected on to a platform inside a graphite tube, and the tube is heated at maximum ramp rate. Because the platform is mainly heated by radiation from the tube walls, there is a time lag before the platform is heated. Thus, the analyte is atomized into a hotter and more isothermal environment, which helps to minimise gaseous-phase interferences especially when integrated absorbance signal measurement is used. This important development led to the "stabilised temperature platform furnace" (STPF) concept [64] which is now widely used.

The requirements for STPF are satisfied by the following [65]:

(a) The L'vov platform.

(b) Fast digital electronics (10 to 20 ms per measurement).

(c) Use of integrated absorbance signals.

(d) Matrix modifiers (see Section 8 below) for analyte stabilisation during the char step.

(e) Good quality pyrolytically coated graphite tubes.

(f) Fast heating of the graphite tube, greater than $1500°C \ s^{-1}$

(g) Zero gas flow during atomization stage.

(h) Use of argon as support gas, not nitrogen.

(i) Zeeman-effect background correction for complex matrices.

Some manufacturers claim that adequate results can be obtained without the use of the L'vov platform, and it does not offer any advantages in the determination of refractory metals. The STPF concept has recently gained substantial popularity [66].

The one piece delayed atomization cuvette (DAC) is offered by one manufacturer as an alternative to the L'vov platform design, but based on the same principle of providing a stable temperature environment during atomization. The DAC has been used to determine trace metals in water [54] with aerosol deposition and S-H. Recently, the EPA has stated that the L'vov platform and the DAC tube are equivalent in maximising the use of an isothermal environment.

8. *Matrix modification*

This is the most widely used technique to minimise interferences in ETA and was introduced by Ediger [67]. Matrix modification involves adding a chemical reagent to the sample to form a more stable analyte compound. This permits higher ashing temperatures to be used and allows the interfering matrix components to be volatilised before analyte atomization.

Many different matrix modifiers have been proposed and used successfully. Some of the commonly used ones are discussed below.

If sodium chloride is a major matrix material, it will produce a large non-atomic absorption signal at many wavelengths, including the 228.8 nm cadmium line. However, Ediger found that the sodium chloride could be converted into less volatile sodium nitrate by adding ammonium nitrate to the sample. Ammonium chloride was removed during the ashing stage so that only a small background signal was observed during the atomization step. Ammonium nitrate was also found to be effective in removing magnesium chloride interference in the determination of lead [68].

Ammonium phosphate was used to decrease cadmium volatility [67] and allowed a significant increase in ashing temperature. The phosphate modifier has since been used to determine many elements, including lead and cadmium [48, 69], chromium, aluminium, tin, cobalt [70] and silver [47].

Nickel was found to increase the stability of arsenic and selenium [67] by the formation of nickel arsenide and selenide, allowing ash temperatures to be used of 1400°C and 1200°C, respectively. Subsequently, this modifier has been used in the determination of gold, antimony and tellurium [47].

Magnesium nitrate greatly reduced interference effects on manganese and aluminium when used with the platform technique [71], and was also found to be useful in the determination of Be [72, 73], Al, Cr, Co, Mn and Ni [73] as well as Fe and Zn [47]. Magnesium nitrate has also been used for Ga and In determination [72]. A mixture of magnesium nitrate and ammonium phosphate has been used for Cd, Pb and Sn determinations [47].

Another commonly used matrix modifier is lanthanum, especially for Pb and Cd analysis in natural waters [56–59]. Initially, La was used to coat the furnace tube prior to analysis of a batch of samples. Eventually the coating is degraded which results in poor recoveries of the analyte. However, by adding a small amount of La as a matrix modifier with each sample injection, the above problem was overcome. It was still necessary to coat the tube before an analytical run however.

A relatively new matrix modifier is palladium, and this promises to be the most wide-ranging modifier to date. It was first used in 1979 [74] to raise the ashing temperature of Hg to 500°C. The interference effect of 28 elemental ions at a thousand times the Hg concentration was not significant. The use of Pd was extended to Te when it was found that inorganic and organic-bound Te could be stabilised up to 1050°C. Previous modifiers could only stabilise one form or the other [75]. Bismuth is a volatile element which can be ashed up to 600°C without the addition of a matrix modifier. The ashing temperature could be raised to 1200°C after adding Pd [76]. Shan Xiao-Quan and his research group then utilised Pd matrix modification for the determination of: As in soil, coal fly-ash and biological samples [77]; As in environmental samples [78]; Tl in waste water [79]; and In in river sediments and coal fly-ash [80]. Previously, Ni had always been used to stabilise As in ETA, but many instruments determine low levels of Ni and thus careful de-contamination procedures have to be carried out. The use of Pd instead should obviate this need. The work with Tl was interesting because the Pd was added as the chloride, and this ion causes most interferences in Tl determination. The usual modifier for Tl determination is sulphuric acid which reduces tube life because of its corrosive nature [47].

An attempt to understand the mechanism of palladium matrix modification indicated the formation of Bi–Pd and Pb–Pd bonds for Bi and Pb, respectively. These metal bonds are very stable and therefore allow high ashing temperatures to be used [81].

Investigations with the use of mixed matrix modifiers with Pd has led to the proposal of a universal modifier for ETA analysis [82]. A number of elements were investigated (As, Bi, In, Pb, Sb, Se, Sn, Te and Tl) and it was found that a mixture of Mg and Pd nitrate offered a better degree of stabilisation in real samples than Pd alone. Other workers [83] found that Pd matrix modification gave better results when used in a reducing environment such as ascorbic acid, hydroxylamine hydrochloride or hydrogen gas. The former two reagents offered certain disadvantages and the use of hydrogen gas during the ashing stage was recommended (5% hydrogen in 95% argon). The Mg and Pd mixed modifier has recently been used to determine As, Cd, Cu, Mn, Pb, Sb, Se and Tl in water [84], and As, Sb, Se, Tl in airborne particles [85]. Interference due to chloride in Tl determination was overcome by using Pd and hydrogen [86]. Other elements successfully analysed using a Pd modifier (either alone or in admixture), are Cd in biological materials [87], Au [88], P [89], Co [90], Pb [91] and Se [92–95].

E. Recommended operating conditions

Although ETA-AAS used to be considered as a non-routine technique, advances with respect to instrument design, background correction, matrix modification and automation have enabled it to be used on a routine basis. However, considerable skill is required in setting up the instrumental conditions and it is virtually impossible to give precise details of a proposed ETA method that can be guaranteed to minimise interference effects on all commercial instruments. A set of general guidelines are given below but the analyst must carry out detailed optimisation and interference effect studies (refer to previous sections). Table 4 lists typical operating parameters for several elements.

TABLE 4

TYPICAL OPERATING PARAMETERS USED IN ETA-AAS

Element	Wavelength (nm)	Typical calibration range (μg l^{-1})	Spectral bandpass (nm)	Typical sensitivity for 1% absorption (pg/0.0044 A.s)	Typical criterion of detection (μg l^{-1}) for 20 μl sample
Sb	217.6	0–120	0.3	10	0.5 –2
As	193.7	0–100	1.0	15	1 –4
Be	234.9	0– 4	0.5	0.5	0.1 –0.4
Cd	228.8	0– 5	0.5	0.25	0.05–0.2
Cr	357.9	0– 40	0.7	5	0.25–1
Cb	240.7	0– 50	0.3	10	0.25–1
Cu	324.7	0– 50	1.0	5	0.25–1
Ga	294.4	0– 50	0.5	25	0.5 –2
In	303.9	0– 50	0.5	15	0.5 –2
Pb	217.0	0– 50	0.5	5	0.5 –2
Mo	313.3	0–150	0.5	15	1 –4
Ni	232.0	0– 40	0.3	10	0.5 –2
Se	196.0	0–100	1.0	25	2 –8
Ag	328.0	0– 50	0.5	1.5	0.1 –0.4
Tl	276.8	0–100	0.5	15	0.5 –2
Sn	224.6	0–100	0.5	15	1 –4
V	318.4	0–200	0.3	40	1.5 –4
Zn	213.9	0– 5	0.5	0.25	[a]

[a] Blank dependent.

F. Matrix modifiers

Table 5 shows matrix modifiers which have been used to determine the listed elements in real samples on a routine basis. Refer to the references for the furnace operating parameters and matrix modifiers.

G. General guidelines

1. General housekeeping

Always carefully align the furnace for maximum transmission of radiation. Regularly clean the silica end windows of the furnace. Take care to balance the intensities of the hollow-cathode and deuterium background lamps.

2. Replacing graphite tubes

When carrying out routine analysis, the graphite tube should be replaced on a regular basis after a given number of atomization cycles. It is always better to replace the tube before its physical destruction because the accuracy and precision of the last 20–30 firings prior to destruction are severely degraded. Ensure that the inner surface of the graphite electrical contacts are clean before inserting a new tube. Follow the manufacturers' instructions for periodically cleaning and renewing these contacts. All new tubes must be thermally conditioned by following the recommendations of the manufacturers.

3. Dry ashing of samples

Always use a ramp time for the dry ashing steps (typically 20–50°C s^{-1}). For some samples, faster ramp times can result in rapid fume development. This can lead to poor analytical precision and shortened tube lifetime. For complex samples two ramps are often employed.

4. Maximum atomization temperature

Do not exceed the maximum atomization temperature specified by the manufacturer. Tube lifetime is substantially shortened for temperatures above approximately 2500°C. For involatile elements (e.g. Cr, Mo and V) it is better to use peak height rather than peak area measurement. A tube cleaning cycle after the measurement of the analytical signal should be employed to ensure removal of the analyte and matrix. Follow the manufacturers' recommendations; if a too vigorous cycle is employed, tube lifetime will be significantly shortened.

References pp. 120–123

TABLE 5

RECOMMENDED MATRIX MODIFIER SOLUTION

Element	Sample volume (μl)	Suggested matrix modifier solution	Matrix modifier volume (μl)	References to the use of various matrix modifiers
Sb	15	4% m/V Ca $(NO_3)_2$	2	72, 82, 83
As	20	0.15% m/V Pd $(NO_3)_2$/0.1% m/V Mg $(NO_3)_2$	82–85	
Be	10	1% m/V Mg $(NO_3)_2$	5	72
Cd	20	0.5% m/V La (as La $(NO_3)_3$)	4	42, 50, 59, 82–85, 96
Cr	10	2% m/V Ca $(NO_3)_2$	2	72, 83, 97
Co	20	None	–	72, 83, 99
Cu	20	None	–	50, 72, 82–85
Ga	20	1% m/V Mg $(NO_3)_2$	5	72
In	20	1% m/V Mg $(NO_3)_2$	5	72, 82.85
Pb	20	0.5% m/V La (as La $(NO_3)_2$)	4	42, 59, 82–85, 96
Mb	20	0.005% m/V BaF_2	5	97, 98
Ni	20	None	–	54, 82, 85, 99
Se	20	0.15% m/V Pd $(NO_3)_2$/0.1% m/V Mg $(NO_3)_2$	10	82, 84, 85
Ag	20	0.5% m/V H_3PO_4	5	72, 83, 84
Te	20	0.15% m/V Pd $(NO_3)_2$/0.1% m/V Mg $(NO_3)_2$	10	72, 82–86
Sn	20	0.5% m/V $NH_4H_2PO_4$/1% m/V Mg $(NO_3)_2$	5	47, 82–85
V	20	None	–	72
Zn	20	None	–	72

5. *Acid matrix*

If possible avoid the use of strong acids and oxidising agents (e.g. H_2SO_4, $HClO_4$, H_2O_2). A nitric acid concentration of 1% v/v (0.15M) does not appear to adversely affect tube lifetime. If strong acids or oxidising agents must be employed, use a slow ramp rate during the dry ashing stage, and use pyrolytically coated tubes and/or platform atomization. Rapid ramp rates cause explosive liberation of oxidising gases which result in rapid surface erosion.

6. *Inert gas*

Argon of a purity greater that 99.995% should be used. The use of nitrogen is not recommended. At temperatures above 2300°C, very toxic cyanogen can be formed. Also some elements (e.g. Mo and V) give reduced sensitivity with nitrogen.

7. *Water cooling*

If the temperature of the cooling water is too low and/or the flow is too high, then water condensation from atmospheric humidity can take place on or around the graphite contacts or even on the silica windows.

VII. MERCURY BY COLD-VAPOUR ATOMIC ABSORPTION AND ATOMIC FLUORESCENCE

A. Introduction

The analysis of mercury by flame AAS is reserved for high concentrations because of its very poor detection limit. Although this element can be successfully determined by ETA-AAS, the detection limit is still poor and matrix interferences usually occur because the high volatility of mercury restricts the ashing temperature. Mercury is most commonly determined by cold-vapour atomic absorption (CVAA) because this technique offers simplicity, high sensitivity and low cost.

The method of CVAA was developed by Hatch and Ott [100] and involved the reduction of inorganic mercury by stannous chloride in a reaction vessel. Elemental mercury has an appreciable vapour pressure at room temperature and exists as a monatomic vapour. Therefore, when a carrier gas (usually argon, nitrogen or air) is bubbled through a solution containing inorganic mercury and stannous chloride, the elemental mercury is rapidly volatilised into the gas stream. The mercury vapour is swept into an absorption cell and mercury absorption is monitored at 253.7 nm using a mercury hollow-cathode lamp. For a 50 ml sample volume the detection limit is approximately

$0.2~\mu g\,l^{-1}$. In order to improve the detection limit significantly it is possible to detect the elemental mercury using a cold vapour atomic fluorescence (CVAF) technique. This inherently simple technique [1, 101–105] has a typical detection limit of better than $0.02~\mu g\,l^{-1}$ for a 2 ml sample volume. It is available commercially as a fully automated stand-alone system [1].

B. Sample collection and preservation

It is essential to collect samples for mercury analysis in glass containers because mercury (0) can permeate polyethylene [106]. Also, volatile organics can leach out of plastic containers and absorb at 253.7 nm. Many different preservation techniques have been successfully utilised to stabilise mercury. It has been shown that aqueous samples preserved with 5% v/v nitric acid and 0.01% m/v potassium dichromate are stable for several months [107], even at the $0.1~\mu g\,l^{-1}$ level. The application of a brominating reagent [1, 103–105] also stabilises mercury because of its oxidising environment (see next section). It is important to use a preservation technique incorporating an oxidising agent because mercury can be lost from acidic solutions used to preserve other toxic metals.

C. Sample pre-treatment

The original method for mercury analysis by CVAA only recovered the inorganic species because stannous chloride is incapable of breaking down most organomercury compounds. Trace levels of organomercury compounds are frequently encountered in waters and effluents, e.g. methyl mercuric chloride, phenyl mercuric acetate etc. Thus, it is necessary to pre-treat samples if total mercury content is required, by utilising strong oxidising conditions.

Various pre-treatment methods have been developed, including prolonged digestion with potassium permanganate [108], prolonged digestion with potassium permanganate–potassium persulphate [109] and digestion with bromine for ten minutes [1, 103–105]. The first technique is very time-consuming because the digestion time is 24 hours. Also, permanganate alone does not break down the organomercury compounds mentioned above. The next procedure reduces the reaction time to about 2.5 hours and this does oxidise organomercury compounds to mercury(II) ions. However, if the sample chloride level is high, additional permanganate is required and chlorides are converted to free chlorine which can cause interference at 253.7 nm by non-specific background absorption.

The bromination procedure is recommended because of speed of reaction (10 minutes) and good recovery (greater than 95%) of organomercury compounds [103]. It also stabilises mercury, provided that excess bromine is present, and therefore this reagent serves both to preserve and pre-treat.

Samples of distilled water, seawater and final sewage effluent containing 2 μg l^{-1} mercury(II) showed no significant loss after 30 days when treated with the brominating reagent [104].

Samples are collected in 100 ml calibrated measuring cylinders containing 15 ml 33% v/v hydrochloric acid (36% m/m) and 2 ml 0.1 N potassium bromate–0.1 N potassium bromide. If a distinct yellow colour is not observed, more brominating reagent is added to maintain oxidising conditions.

D. Interferences and methods for their minimisation

To avoid non-specific absorption due to water vapour, it is preferable to place a light bulb (typically 60–100 W) above the absorption cell to prevent condensation on the cell windows. Alternatively, a drying agent such as magnesium perchlorate can be placed in the gas line.

Sample pre-treatment is required to remove sulphide interference, and it has been shown [105] that the bromination method will remove interference from sulphide concentrations as high as 24 mg l^{-1} (as sodium sulphide).

The potential interference from chloride found in the permanganate-based pre-treatment techniques does not arise with the bromination reagent. This is because the Br$_2$/Br$^-$ couple has a half cell oxidation potential which is lower than that of the Cl$_2$/Cl$^-$ couple [105].

Interference from some volatile organic materials which absorb at 253.7 nm is possible. An analysis run with the neat sample only should determine if this type of interference is present, but automatic background correction can be used to overcome it. Interferences from volatile organic material has not proved to be a significant problem when using the fluorescence technique in environmental analysis.

E. Recommended operating conditions

Prior to analysis the bromination step is terminated by removing excess bromine by the addition of 1–2 drops of 12% m/v hydroxyammonium chloride reagent. The solution is then diluted to 100 ml and is ready for analysis. If potassium dichromate has been used as a preservative, sufficient hydroxylamine should be added to reduce the dichromate to Cr(III) before analysis. Standards and blank are prepared in the same way as samples. Suitable reaction conditions are 50 ml sample and 4 ml 5% m/V tin(II) chloride (dissolve 5 g in 15 ml 36% m/m hydrochloric acid and dilute to 100 ml). An absorbance of about 0.1 should be achieved from 50 ml of a 4 μg l^{-1} mercury solution [1]. Details of the much more sensitive atomic fluorescence technique (CVAF) are given in the references [1, 101–105, and 110].

VIII. ARSENIC AND SELENIUM BY HYDRIDE GENERATION

A. Introduction

These two elements can be determined by flame AAS with typical detection limits of 0.25 mg l^{-1} for As and 0.4 mg l^{-1} for Se (see Table 1). For lower levels it is possible to utilise ETA-AAS and reliable methods using this technique have now been developed (see Table 3). However, considerably more operator skill is required for ETA analysis of these elements and hence hydride generation is commonly employed for arsenic and selenium analysis. This technique is fairly simple to use and is generally more sensitive than ETA and offers better precision at low levels. It is also applicable to antimony, bismuth, germanium, lead, tellurium and tin.

The basis of the technique is the formation of volatile hydrides when an acidic solution of the element reacts with a solution of sodium borohydride. The hydride is carried into a flame-heated silica tube by a stream of argon or nitrogen, whereupon atomization takes place. The main disadvantage of this technique is that interference effects can be very severe [111] and consequently the method of standard additions is sometimes necessary.

B. Sample collection and preservation

The reader is referred to Section II for details concerning sample collection and preservation. The addition of 1 ml of 36% m/m hydrochloric acid per 100 ml of sample is sufficient to preserve samples for several weeks. The use of nitric acid should be avoided if possible (see Section D below)

C. Sample pre-treatment

Many naturally occurring arsenic and selenium species (e.g. arsenobetaine, seleno-methionine, seleno-cystine etc.) do not react with sodium borohydride. These compounds require vigorous digestion (e.g. HNO_3–$HClO_4$, ashing with $HNO_3/Mg(NO_3)_2$, ultraviolet digestion etc.). These organic forms are unlikely to be present in significant amounts in potable and river waters. However, these compounds may be present at significant levels in some biota, sludge and soil samples.

For complex samples (sludge, soil, biota etc.) direct flame AAS determination is recommended for high levels and ETA-AAS for low levels (see Section VI). These methods should give complete recovery for all forms of arsenic and selenium.

Many forms of pre-treatment have been developed [112–117] for breaking down organoarsenic and selenium compounds. The persulphate procedure of Nygaard and Lowry [112] is probably the simplest, and gave complete recovery of various organo compounds of arsenic, selenium and antimony. It involves acidifying a 25 ml sample aliquot with nitric acid to lower the pH below 2. Then 8 ml of 4% m/v potassium persulphate solution is added, followed by evaporation at 95°C to approximately 10 ml. Concentrated hydrochloric acid is added (12.5 ml) and the solution is heated at 95°C for one hour before dilution to 25 ml with water. A simple method for recovering selenate and selenomethiomine has been reported by D'Ulivo [118]. For potable waters HBr is added to the sample to give a concentration of 1.8 M HBr. The sample is then heated at 70°C for one hour and any excess Br_2 is removed with a few microlitres of 10% m/v hydroxylamine solution.

D. Interferences and methods for their minimisation

The hydride generation technique is prone to interferences at the reduction stage by transition metals [119–123]. For the majority of natural water samples this type of interference should not be significant except possibly for the effect of copper on selenium in some potable water distribution samples. The user should carry out recovery tests on typical waters and also determine the maximum allowable copper concentration for the determination of selenium. If the copper level in a sample exceeds this concentration, the method of standard additions should be used. However, Bye [119] has found that the addition of 8 g l^{-1} iron(III) to the digested sample minimised the effect of copper on selenium.

Welz and Schubert-Jacobs [122] have found that increasing the hydrochloric acid in the final sample solution to 5 M and decreasing the sodium borohydride concentration to 0.5% m/v has minimised the effect of cobalt, nickel and copper on the determination of both arsenic and selenium. The addition of Fe(III) to samples containing nickel was found to increase the range of interference-free determination for arsenic [121].

The oxidation state of arsenic and selenium is crucial for accurate results. Both elements can occur in two oxidation states as well as the elemental state (As(O), (III) and (V), Se(O), (IV) and (VI)). It has been found that arsenic(V) gives a significantly lower response than arsenic(III), and hence it is usual to convert all arsenic species to As(III) prior to hydride generation [113–114]. Similarly, it is necessary to convert all selenium species to the Se(IV) state because the higher state gives a negligible response compared with the same concentration as Se(IV).

Reduction of As(V) to As(III) can be carried out by the following procedure: add 25 ml of 36% m/m hydrochloric acid to 50 ml of sample, standard or blank; then add 2 ml of reducing agent (50% m/v potassium iodide–5% m/v ascorbic acid) and allow to stand at 50°C for 15 minutes.

To ensure reduction of all Se(VI) species to Se(IV), add 20 ml of 36% m/m hydrochloric acid to 50 ml of sample and boil gently until the volume is just below 50 ml. Cool and dilute to 50 ml with water.

References pp. 120–123

Trace levels of naturally occurring nitrite or nitrite derived from nitric acid digestion of biological material can cause interference with selenium [126–127] and arsenic [127] determinations. The addition of sulphanilamide has been recommended to overcome this interference effect.

E. Recommended operating conditions

Most manufacturers recommend the use of 1% m/v sodium borohydride. It has been found that the addition of one pellet of sodium hydroxide per 100 ml of reagent, followed by filtration, stabilises the reagent for up to three days. Using the 197.2 nm line for arsenic and the 196.0 nm line for selenium, it should be possible to achieve a limit of detection of less than 1 μg l^{-1} and a sensitivity for 1% absorption of approximately 0.5 μg l^{-1} for a 1 ml discrete sample for both elements.

IX. ANALYTICAL QUALITY CONTROL

A. Introduction

One of the main responsibilities of an analyst is to provide data which can be used in decision making. If this data is to be of any use, it must be sufficiently accurate for the intended purpose. Consequently the analyst and end user of the data must have some means of assessing whether or not it can be trusted and the degree of variation to be expected. Quality control is the visible process used to ensure that the accuracy of analytical data is acceptable and of adequate precision for the use to which it will be put. With quality control such a vital part of everyday laboratory practice it should be inconceivable for any batch of samples to be analysed without including control solutions which have been treated as far as is possible in a similar fashion. In a modern high-throughput laboratory with automated equipment and pressure to produce results quickly, it can become very easy to overlook quality control and quality control procedures. The excellent book by Hunt and Wilson [10] gives comprehensive information on this important topic.

Implementation of quality control schemes vary from laboratory to laboratory. The following procedures are considered useful in developing a quality control strategy for a laboratory: (a) use of control charts; (b) use of spiking recovery tests; (c) interlaboratory quality control exercises; (d) analysis of certified reference materials.

B. Quality control charts

After a method has been fully tested and is put into routine use, a continuous check on analytical errors is required. Quality control charts are a simple and effective way of achieving in-house control. Shewhart control charts will be described. Cusum control charts may also be used although these are slightly more complicated. For a detailed discussion of the statistical techniques used, the reader is urged to consult the various technical papers which are available [1, 10, 128–130].

Shewhart control charts

It is assumed that the analytical results follow a normal distribution for these control charts. For each determination requiring a control chart, a standard is prepared whose concentration falls within the normal working range for that estimation. The standard is analysed at least ten times in separate batches over a number of days. Having obtained a set of data, the mean and standard deviation may be calculated and the control chart shown in Fig. 2 constructed. At some later stage when more results are available, the limits of the control chart may be periodically updated after incorporating these in the statistical calculations. It can be shown that only 0.3% of results will be outside the limits of the mean ± three standard deviations. These correspond to the upper and lower action limits (UAL, LAL). Also 5% of results will fall between the mean ± two standard deviations and UAL and LAL. The upper and lower warning limits (UWL, LWL) are set at these levels. In practice control standards are analysed with every batch of analyses and the results plotted on the chart. Those results which are in control, will fall in the areas between UWL and LWL. If a result falls outside either UAL or LAL, then the method is out of control. Similarly if two consecutive results fall between UWL and UAL or between

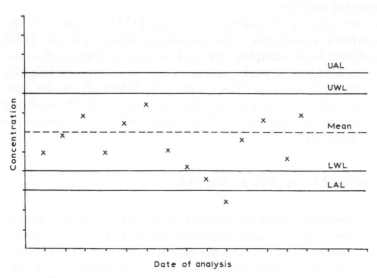

Fig. 2. Shewhart control chart.

LWL and LAL, then the same conclusion can be drawn. Immediate action is required to rectify the problem. All results within that quality control batch are suspect and should be re-analysed.

It is often useful to note on the control chart what action was taken for an out-of-control batch and if possible the reason for the problem. As points accumulate on the chart, other information about the procedure may be gleamed, e.g. the existence of bias. The main advantages of these sorts of charts are that they are very easily constructed and maintained and give instant visual information on a particular method for analysis.

C. Spiking recovery tests

The analysis of control standards gives no indication of bias caused by interference effects. By spiking a second portion of one of the samples in each batch of analyses with a known amount of the determinant, it is possible to calculate the percentage recovery. This can then be plotted on a control chart, the expected value being 100%. To be of value, samples used for recovery tests should have similar determinand concentrations and be spiked to the same concentration.

D. Interlaboratory quality control exercises

There are several schemes available on subscription where a group of laboratories have co-operated and take part in regular analytical exercises involving environmental samples.

The Water Research Centre "Aquacheck" scheme [131] is based on the distribution of synthetic standards. The Yorkshire Water scheme [132] distributes real and synthetic samples (natural waters, sewage, effluents, sediments and sewage sludges) to participating laboratories on a fortnightly basis. The samples reflect all aspects of the water cycle. When all results have been correlated, extremely valuable information becomes available enabling better quality control to be implemented and also helping to overcome problems associated with methodology.

E. Analysis of certified reference materials

A wide range of certified reference materials including water, sludge, soils and sediments are commercially available (see Section III.B). These materials should be incorporated into the analytical quality control protocol of the laboratory. An aliquot of the certified reference material is analysed in each batch of analyses and the results are plotted on a control chart.

F. Standard solutions for quality control

Quality control standards should be prepared from concentrated stock solutions, retained solely for this purpose, by an analyst other than the one performing the test. These standards should be treated identically to real samples, i.e. all preparative procedures should be carried out on them. The concentration of the quality control standard should fall in the normal range for that determination. If the range is large, then several quality standards may be required. When proprietary solutions are used, it is recommended that they be purchased from a different manufacturer to the one supplying other standards used in the laboratory. If this is not possible, the standard must be obtained from a different batch. Mixed quality control standards are useful especially when automated equipment is used. Care should be taken in deciding which mixtures are appropriate due to chemical reactions, matrix problems and contamination, e.g. if a multi-element standard is to contain chromium, this should be in the form of chromium(III). If chromium(VI) is used slow precipitation of metal chromates can occur.

Another important consideration is the stability of the stock and working solutions. Usually acidification and storage at 4°C are considered adequate although regular replacement is recommended for both the stock and working standards after a specified time period or when only one third of the solution remains.

APPENDIX — ABBREVIATIONS

AAS	Atomic absorption spectroscopy
AQC	Analytical quality control
CVAA	Cold vapour atomic absorption
CVAF	Cold vapour atomic fluorescence
DAC	Delayed atomization cuvette
DS	Dry Solids
EC	European Community
ETA	Electrothermal atomization
ETA-AAS	Electrothermal atomization/atomic absorption spectroscopy
HG	Hydride generation
HMSO	Her Majesty's Stationery Office
LAL	Lower action limit
LWL	Lower warning limit
RSD	Relative standard deviation
S-H	Smith-Hieftje
STPF	Stabilised temperature platform furnace
UAL	Upper action limit
UWL	Upper warning limit
σ	Standard deviation

References pp. 120–123

REFERENCES

1 Yorkshire Water Authority Methods of Analysis, 1988, Yorkshire Water, Leeds, 1989, ISBN 0 905057 23 6, Sections 8–10.
2 R. Dybczynski, A. Tugsavul and O. Suschny, Analyst, 103 (1978) 733.
3 R. Smith, Water SA, 5 (1979) 128.
4 K.W. Boyer, W. Horwitz and R. Albert, Anal. Chem., 57 (1985) 454.
5 R.D. Davis and C.H. Carlton-Smith, Water Pollut. Control, 82 (1983) 290.
6 Council of the European Communities, Directive of 12th June 1986 on the protection of the environment, and in particular of the soil, when sewage sludge is used in agriculture, 86/278/EEC, OJL 181/6, 4th July 1986.
7 Code of Practice for the Disposal of Sewage Sludge to Agricultural Land, Yorkshire Water, Leeds LS1 5AA, 1989.
8 I.M. Sayre, J. Am. Water Works Assoc., 80 (1988) 53.
9 Council of the European Communities, Directive of 15th July 1980, relating to the quality of water intended for human consumption, 80/778/EEC, OJL 229, 30th August 1980.
10 D.T.E. Hunt and A.L. Wilson, The Chemical Analysis of Water, General Principles and Techniques, 2nd ed., Royal Society of Chemistry, London, 1986.
11 J. Gardiner and G. Mance, United Kingdom Water Quality Standards arising from European Community Directives, Tech. Rept. TR204, Water Research Centre, Medmenham, 1984, p. 12.
12 G. Mance, A.R. O'Donnell, J.A. Campbell and A.M. Gunn, Proposed Environmental Quality Standards for List II Substances in Water, Inorganic Tin, Tech. Rept. TR254, Water Research Centre, Medmenham, 1988, p. 1.
13 T.F. Zabel, J. Seager and S.D. Oakley, Proposed Environmental Quality Standards for List II Substances in Water, Organotins, Tech. Rept. TR255, Water Research Centre, Medmenham, 1988, p. 1.
14 L. Ebdon, S. Hill, A.P. Walton and R.W. Ward, Analyst, 113 (1988) 1159.
15 T.M. Florence, Analyst, 111 (1986) 489.
16 X. Goenaga and D.J.A. Williams, Environ. Pollut., 52 (1988) 131.
17 G. Topping, Sci. Total Environ., 49 (1986) 9.
18 R. Sturgeon and S.S. Berman, CRC Crit. Rev. Anal. Chem., 18 (1987) 209.
19 R.E. Pellersbarg and T.M. Church, Anal. Chim. Acta, 97 (1978) 81.
20 J.A. Cambell, M.J. Gardner and A.M. Gunn, Anal. Chim. Acta, 176 (1985) 193.
21 M.J. Gardner and D.T.E. Hunt, Analyst, 106 (1981) 471.
22 J.W. Owens, E.C. Gladney and W.D. Purtymun, Anal. Lett., 13.
23 S.B. Adeljou and A.M. Bond, Anal. Chem., 57 (1985) 1778.
24 Lithium, Magnesium, Calcium, Strontium and Barium in Waters and Sewage Effluents by AAS, HMSO, London, 1987.
25 Methods for the Determination of Metals in Soils, Sediments and Sewage Sludge by Hydrochloric–Nitric Acid Digestion, HMSO, London, 1987.
26 Standing Committee of Analysts, Lead in Potable Water by AAS, 1976, HMSO, London, 1977.
27 M.S. Cresser, Solvent Extraction in Flame Spectroscopic Analysis, Butterworths, London, 1978.
28 Standard Methods for the Examination of Water and Wastewater, 16th ed., American Public Health Association, Washington, D.C., 1985.
29 J.O. Kinrade and J.C. Van Loon, Anal. Chem., 46 (1974) 1894.

121

30   J.P. Riley and D. Taylor, Anal. Chim. Acta, 40 (1968) 479.
31   D.G. Biechler, Anal. Chem., 37 (1965) 1054.
32   J. Karkisch and A. Sorio, Anal. Chim. Acta, 76 (1975) 393.
33   J. Dingman, S. Siggia, C. Barton and K.B. Hiscock, Anal. Chem., 44 (1972) 1351.
34   R.K. Stogerboe, W.A. Hanagen and H.E. Taylor, Anal. Chem., 59 (1985) 2817.
35   S. Nakashima, R.E. Sturgeon, S.N. Willie and S.S. Berman, Anal. Chim. Acta, 207 (1988) 291. (Method C, (1980) 253.)
36   K.C. Thompson, Atomic Absorption Spectrophotometry, An Essay Review, HMSO, London, 1980.
37   R.C. Rooney, Analyst, 103 (1978) 1100.
38   K.C. Thompson, Analyst, 103 (1978) 1258.
39   K.C. Thompson and K. Wagstaff, Analyst, 105 (198) 641.
40   M.S. Cresser, Lab. Practice, 26 (1977) 171.
41   A.W. Strumpler, Anal. Chem., 45 (1973) 2251.
42   J.P. Matousek, Prog. Anal. At. Spectrosc., 4 (1981) 247.
43   W. Slavin and D.C. Manning, Prog. Anal. At. Spectrosc., 5 (1982) 243.
44   S.R. Koirtyohann and E.E. Pickett, Anal. Chem., 37 (1965) 601.
45   F.J. Fernandez, S.A. Myers and W. Slavin, Anal. Chem., 52 (1980) 741.
46   H. Koizumi and K. Yasuda, Spectrochim. Acta, Part B, 31B (1976) 523.
47   W. Slavin, G.R. Carnrick, D.C. Manning and E. Pruszkowska, At. Spectrosc., 4 (1983) 69.
48   M.J. Fishman, G.R. Perryman, L.J. Schroder and E.W. Matthews, J. Assoc. Off. Anal. Chem., 69 (1986) 704.
49   S.B. Smith and G.M. Hieftje, Appl. Spectrosc., 37 (1983) 419.
50   S.C. Apte and A.M. Gunn, Anal. Chim. Acta, 193 (1987) 147.
51   W. Slavin, D.C. Manning and G. Carnrick, Anal. Chem., 53 (1981) 1504.
52   D. Littlejohn, I.S. Duncan, J.B.M. Hendry, J. Marshall and J.M. Ottaway, Spectrochim. Acta, 40B (1985) 1677.
53   H.L. Katin, M.K. Conley and J.J. Sotera, Am. Lab., 12 (1980) 72.
54   G.R. Dulude, J.S. Sotera and D.L. Pfeil, Spectroscopy, 2 (1987) 49.
55   J. Fazakas, Spectrosc. Lett., 15 (1982) 221.
56   K.C. Thompson, K.C. Wheatstone and K. Wagstaff, Analyst, 102 (1977) 310.
57   M.P. Bertenshaw, D. Gelsthorpe and K.C. Wheatstone, Analyst, 106 (1981) 23.
58   M.P. Bertenshaw, D. Gelsthorpe and K.C. Wheatstone, Analyst, 107 (1982) 163.
59   Lead and Cadmium in Fresh Waters by AAS, A General Introduction to ETA-AAS, HMSO, London, 1986.
60   A. Criaud and C. Fouillarc, Anal. Chim. Acta, 167 (1985) 257.
61   M.R.A. Michaelis, W. Wegseheider and H.M. Ortner, J. Anal. At. Spectrom., 3 (1988) 503.
62   B.V. L'vov, Spectrochim. Acta, 17 (1961) 761.
63   B.V. L'vov, Spectrochim. Acta, 33B (1978) 153.
64   W. Slavin, D.C. Manning and G.R. Carnrick, At. Spectrosc., 2 (1981) 137.
65   W. Slavin, Graphite Furnace AAS, A Source Book, Perkin-Elmer, Norwalk, Conn., 1984.
66   B.V. L'vov, J. Anal. At. Spectrom., 3 (1988) 9.
67   R.D. Ediger, At. Absorpt. Newsl., 14 (1975) 127.
68   D.C. Manning and W. Slavin, Anal. Chem., 50 (1978) 1234.
69   E.J. Hindeberger, M.L. Kaiser and S.R. Koirtyohaan, At. Spectrosc., 2 (1981) 1.
70   M.L. Kaiser, S.R. Koirtyptiann and E.J. Hinderberger, Spectrochim. Acta, 36B

122

(1981) 773.
71 W. Slavin, G.R. Carnrick and D.C. Manning, Anal. Chem., 54 (1982) 621.
72 The Determination of As, Se, Sb, Be, Cr, Co, Cu, Ga, Ge, In, Ni, Ag, Tl, V and Zn in Raw and Potable Waters by Electrothermal Atomization—AAS, HMSO, London, 1988.
73 D.C. Manning and W. Slavin, Appl. Spectrosc., 37 (1983) 1.
74 Shan Xiao-Quan and N. Zhe-Ming, Acta Chem. Sinica, 37 (1979) 261.
75 G. Weibust, F.J. Langmyhr and Y. Thomassen, Anal. Chim. Acta, 128 (1981) 23.
76 J. Long-Zhu and N. Zhe-Ming, Can. J. Spectrosc., 26 (1981) 219.
77 Shan Xiao-Quan, N. Zhe-Ming and Z. Li, Anal. Chim. Acta, 151 (1983) 179.
78 Shan Xiao-Quan, N. Zhe-Ming and Z. Li, At. Spectrosc., 5 (1984) 1.
79 Shan Xiao-Quan, N. Zhe-Ming and Z. Li, Talanta, 31 (1984) 150.
80 Shan Xiao-Quan, N. Zhe-Ming and Y. Zhi-Neng, Anal. Chim. Acta, 171 (1985) 269.
81 Shan Xiao-Quan and W. Dian-Xun, Anal. Chim. Acta, 173 (1985) 315.
82 G. Schlemmer and B. Welz, Spectrochim. Acta, 41B (1986) 1157.
83 L.M. Voth-Beach and D.E. Shrader, Spectroscopy, 1 (1986) 49.
84 B. Welz, G. Schlemmer and J.R. Mudakavi, J. Anal. At. Spectrom., 3 (1988) 695.
85 B. Welz, G. Schlemmer and J.R. Mudakavi, J. Anal. At. Spectrom., 3 (1988) 93.
86 B. Welz, G. Schlemmer and J.R. Mudakavi, Anal. Chem., 60 (1988) 2567.
87 X. Yin, G. Schlemmer and B. Welz, Anal. Chem., 59 (1987) 1462.
88 J. Egila, D. Littlejohn, J.M. Ottaway and Shan Xiao-Quan, J. Anal. At. Spectrom., 2 (1987) 293.
89 A.J. Curtius, G. Schlemmer and B. Welz, J. Anal. At. Spectrom., 2 (1987) 115.
90 B. Sampson, J. Anal. At. Spectrom., 3 (1988) 465.
91 W. Wendl and G. Muller-Vogt, J. Anal. At. Spectrom., 3 (1988) 63.
92 R.H. Eckerlin, D.W. Hoult and G.R. Carnick, At. Spectrosc., 8 (1987) 64.
93 B.E. Jacobson and G. Lockitch, Clin. Chem., 34 (1988) 709.
94 I. Lindberg, E. Lundberg, P. Arkhammar and P.O. Berozgren, J. Anal. At. Spectrom., 3 (1988) 497.
95 M.B. Knowles and K.G. Brodie, J. Anal. At. Spectrom., 3 (1988) 511.
96 W.K. Oliver, S. Reeve, K. Hammond and F.B. Basketter, J. Inst. Water Eng. Sci., 5 (1983) 460.
97 M. Hoenig, F. Dehairs and A. de Kersabiec, J. Anal. At. Spectrom., 1 (1986) 449.
98 S.P. Ericson and M.L. McHalsky, At. Spectrosc., 8 (1987) 101.
99 P. Kuroda, T. Nakano, Y. Miura and K. Oguma, J. Anal. At. Spectrom., 1 (1986) 429.
100 W.R. Hatch and W.L. Ott, Anal. Chem., 40 (1968) 2085.
101 K.C. Thompson and R.G. Godden, Analyst, 100 (1975) 544.
102 P.B. Stockwell, K.C. Thompson, A. Henson, E. Temmerman and C. Vandecasteele, Int. Labmate, (1989).
103 B.J. Farey, L.A. Nelson and M.G. Rolph, Analyst, 103 (1978) 656.
104 B.J. Farey and L.A. Nelson, Anal. Chem., 50 (1978) 2147.
105 L.A. Nelson, Anal. Chem., 51 (1979) 2289.
106 M.H. Bothner and D.E. Robertson, Anal. Chem., 47 (1975) 592.
107 C. Feldman, Anal. Chem., 46 (1974) 99.
108 S.H. Omang, Anal. Chim. Acta, 54 (1971) 415.
109 J.F. Kopp, M.C. Longbottom and L.B. Lobring, J. Am. Water Works Assoc., 65 (1973) 731.
110 Mercury in Waters, Effluents, Soils and Sediments, etc., Additional Methods 1985,

HMSO, London, 1987.

111   A.E. Smith, Analyst, 100 (1975) 300.

112   D.D. Nygaard and J.H. Lowry, Anal. Chem., 54 (1982) 803.

113   W.A. Maher, Talanta, 30 (1983) 534.

114   L. Hansson, J. Petterson and A. Olin, Talanta, 34 (1987) 829.

115   Arsenic in Potable and Sea Water by Spectrophotometry 1978, HMSO, London, 1980.

116   Selenium in Waters 1984, Selenium and Arsenic in Sludges, Soils and Related Materials 1985, A Note on the Hydride Generator Kits 1987, HMSO, London, 1987.

117   N.G. van der Veen, H.J. Kewkens and G. Vos, Anal. Chim. Acta, 171 (1985) 285.

118   A. D'Ulivo, J. Anal. At. Spectrom., 4 (1989) 67.

119   R. Bye, Anal. Chim. Acta, 192 (1987) 115.

120   B. Welz and M. Melcher, Analyst, 109 (1984) 569.

121   B. Welz and M. Melcher, Analyst, 109 (1984) 573.

122   B. Welz and M. Schubert-Jacobs, J. Anal. Atom. Spectrom., 1 (1986) 23.

123   F.D. Pierce and H.R. Brown, Anal. Chem., 49 (1977) 1417.

124   F.J. Fernandez and D.C. Manning, At. Absorpt. Newsl., 10 (1971) 86.

125   H.W. Sinemus, M. Melcher and B. Welz, At. Spectrosc., 2 (1981) 81.

126   G.A. Cutter, Anal. Chim. Acta, 149 (1983) 391.

127   H.W. Sinemus, D. Maier, M. Schubert-Jacobs and B. Welz, Fortschr. At. Spectrom. Spurenanal., 2 (1986) 571 (in German).

128   U.S. Environmental Protection Agency , Handbook for Analytical Quality Control in Water and Wastewater Laboratories, Cincinnati, Ohio, The Agency, 1972.

129   O.L. Davies and P.L. Goldsmith, Statistical Methods in Research and Production, Longman, London, 1977.

130   D.J. Dewey and D.T.E. Hunt, The Use of Cumulative Sum Charts (CUSUM Charts) in Analytical Quality Control, TR174, Water Research Centre, Medmenham, 1982.

131   Aquacheck Scheme, Water Research Centre, Medmenham SL7 2HD, U.K.

132   Yorkshire Water AQC Scheme, Leeds LS1 5AA, U.K.

Chapter 4b

Application of Atomic Absorption Spectrometry to Marine Analysis

HIROKI HARAGUCHI and TASUKU AKAGI

Department of Applied Chemistry, Faculty of Engineering, Nagoya University, Chigusa-ku, Nagoya 464 (Japan)
Department of Chemistry, Faculty of Science, University of Tokyo, Bunkyo-ku, Tokyo 113 (Japan)

I. INTRODUCTION

In marine chemistry, the concentrations, chemical states, material balances and cycles of various elements and compounds are the main subjects of study for describing the oceans and their environment. The oceans, which cover about 71% of the earth's surface, form a complex multidimensional system with varied constructions and materials. Thus, analyses of the components of seawater, marine plants and animals, and sediments give the fundamental information for the interpretation of material balances and geochemical phenomena in the oceans.

The development of marine chemistry has depended upon proper analytical techniques including sampling methodology. In the 18th and 19th centuries, Boyle and Lavoisier, who were the early pioneers in marine chemistry, investigated various salts in seawater. Dittmar examined the components of seawater using gravimetric and volumetric analytical techniques. Before Dittmar's work, K, Na, Ca, Mg, S, Cl, Br, B, Sr and F had been found in seawater. Dittmar noted that the ratios of the major constituents are almost constant, while the total salt concentration is variable [1].

Various spectroscopic techniques such as flame photometry, emission spectroscopy, atomic absorption spectrometry, spectrophotometry, fluorescence spectrometry, neutron activation analysis and isotope dilution mass spectrometry have been used for marine analysis of elemental and inorganic components [2, 3]. Polarography, anodic stripping voltammetry and other electrochemical techniques are also useful for the determination of Cd, Cu, Mn, Pb, Zn etc. in seawater. Electrochemical techniques sometimes provide information on the chemical species in solution.

Each spectroscopic method has its own characteristic application. For example, flame photometry is still applicable to the direct determination of Ca and Sr, and to the determination of Li, Rb, Cs and Ba after preconcentration with ion-exchange resin. Fluorimetry provides better sensitivities for Al, Be, Ga and U, although it suffers from severe interference effects.

Emission spectrometry, X-ray fluorescence spectrometry and neutron activation analysis allow multi-element analysis of solid samples with quite good sensitivity and precision, and have commonly been applied to the analysis of marine organisms and sediments. Recently, inductively coupled plasma atomic emission spectrometry (ICP-AES) has been increasingly used as a method for simultaneous multi-element analysis of trace metals in seawater [4, 5].

Atomic absorption spectrometry (AAS) is a relatively new analytical technique among the spectroscopic methods, although it is now more than 35 years since the pioneer work by A. Walsh [6]. As described in the previous chapters, AAS gives high sensitivity, precision and accuracy along with experimental convenience and a wide instrumental availability. Thus AAS techniques has been extensively employed for the analysis of marine samples. However, the elemental contents of marine samples are generally extremely low, and some suitable preconcentration procedures are required. The development of electrothermal atomization and gas generation techniques has extended the applicability of AAS to marine analysis. The determination of Cd, Cu, Ni, Pb, Hg, As, Bi, Ge, Sb, Se, Sn and Te has become much easier as a result of the development of the techniques.

In analysis of marine samples by AAS, many problems are encountered in sampling methodology, sample storage, sample treatment and measurement procedures. Analytical difficulties arise from the low concentration of most elements and complex matrices in marine samples. In this chapter, general discussion on marine analysis by AAS will be provided in terms of seawater, marine organisms and sediments.

II. SEAWATER

Seawater contains about 3.5% salts, in which the content of sodium chloride is about 80%. The concentration of dissolved salts, as well as temperature and pressure, influence the physical properties of seawater. The total salt concentration is usually called "salinity". Salinity is generally measured by the electrical conductivity or determination of chloride content. At present, salinity (S) is defined as S = 1.80655 Cl (Cl is the concentration of chloride in seawater) [7]. Dissolved oxygen and silica are usually measured as additional parameters to characterise seawater. The concentrations of nitrogen and phosphorus are the indices of nutrients and the measure of the eutrophication and biological production of the oceans.

The average elemental composition of seawater is shown in Table 1 [8]. The elements which exist at more than 1 mg l^{-1} are the major elements, and those less than 1 mg l^{-1} are the trace elements. Of these, only twelve elements are present at a concentration greater than 1 mg l^{-1}, and most other elements range from 0.5 mg l^{-1} to much less than 1 μg l^{-1} in content.

TABLE 1
MEAN OCEANIC CONCENTRATION

Element	Concentration (μg ml^{-1})	Element	Concentration (μg ml^{-1})
Cl	19353	Ni	0.00048
Na	10781	Zn	0.00039
S	2712	Cr	0.00033
Mg	1280	Se	0.00017
Ca	415	Cu	0.00012
K	399	Cd	0.000070
Br	69	Fe	0.00004
C	26	Y	0.000013
N	8.7	Ga	0.000012
Sr	7.8	Au	0.000011
B	4.4	Mn	0.00001
Si	3.1	Bi	0.00001
F	1.3	Hg	0.000006
Li	0.178	Ge	0.000005
Rb	0.124	Ce	0.000004
P	0.06	Ag	0.000003
I	0.06	Co	0.000002
Ba	0.0117	Pb	0.000001
Mo	0.011	Ti	<0.000001
U	0.0032	W	<0.000001
As	0.002	Th	<0.0000007
Al	0.001	Sn	0.0000005
V	<0.001	Ru	0.0000005
Ni	0.00048		

From M.S. Quinby-Hunt and K.K. Turekian, EOS, 64 (1983) 130–131.

The concentrations of trace elements in seawater vary geographically, spatially, and in depth. Such variations are generally caused by the biological activities, geological phenomena and physicochemical processes in the oceans. Thus, it should be noted that the concentrations of trace elements given in Table 1 are not necessarily representative of those for all oceans. However, the extremely low levels of elements in concentrated salt matrix make accurate and precise seawater analysis difficult. In the interlaboratory investigations of seawater, analysis by skillful marine or analytical chemists, has indicated that a wide range of analytical values can be reported for some trace elements [9, 10]. Chakrabarti et al. [11] summarised the problems in seawater analysis as follows:

(a) Sampling, sample collection, filtration, and storage.

(b) The determination of trace metals in marine samples often involves some preconcentration and/or separation procedures which can cause considerable contamination before analysis.

References pp. 154–157

(c) Contamination from reagents, solvents, laboratory ware and the general laboratory environment.

(d) Inadequate appreciation of the limitations of the analytical techniques used in the determination.

(e) Lack of adequate validation of techniques, methods and procedures.

(f) Lack of assessment on the reliability of techniques, methods and procedures by extensive inter- and intra-laboratory intercalibration studies.

(g) Lack of adequate standard reference materials for calibration purposes, and a standard reference sample that could be certified as blank, i.e. a seawater facsimile.

Recently, Berman and co-workers in the National Research Council, Canada, issued two seawater reference materials (CASS and NASS) and one river water reference material (SLRS). They were carefully produced, i.e. sampled with Go-Flo samplers, filtered through 0.45 μm porosity Millipore filters and stored in a precleaned polyethylene bottle at pH 1.6 containing nitric acid. CASS and NASS were made from nearshore seawater and offshore seawater, respectively. CASS contains trace metals at higher concentrations than NASS. The use of these seawater reference materials will help solve some of the problems outlined above.

In order to obtain reliable analytical data, all the items pointed out above should be carefully taken into account at each step throughout the overall experimental procedures. Sampling, sample collection, filtration and storage are primarily important before instrumental analysis. Table 2 summarises the typical detection limits obtainable by flame and electrothermal atomic absorption spectrometry, and these can be seen to offer quite good sensitivities [12, 13]. The concentration ranges of trace elements in ocean water are also given in Table 2 [8].

For most trace elements, however, the sensitivities (or detection limits) are not high enough to determine them directly without a preconcentration step, where contamination from water, chemicals, solvents, glassware and the laboratory environment is often easily introduced. It should be noted that trace element concentrations in coastal seawater are generally higher than those listed in Table 2. But it is still difficult to determine the concentrations in coastal seawater without preconcentration. To minimize contamination from the laboratory environment, the use of a filtered-air-supply system, a clean room or a clean bench, is strongly recommended.

At the sampling step, contamination from sampling bottles and handling glass and plastic ware should be minimized [11]. Old types of sampling devices such as Nansen and Knudsen are constructed mostly from brass, which may cause contamination of Cu and Zn into seawater. The developed Niskin plastic sampling bottle provides less contamination from the sampler walls. However, the materials used for the end-caps or springs, rubber or Teflon-coated stainless steel, sometimes release significant amounts of Ba, Cu, Sb and Zn [14].

TABLE 2

THE DETECTION LIMITS OF FLAME AND ELECTROTHERMAL ATOMIC AB-SORPTION SPECTROMETRY AND TRACE ELEMENT CONCENTRATION IN SEA-WATER

Element	Detection limit [a]		Trace element concentration [d] (μgl^{-1})
	FAA [b] ($\mu g\ ml^{-1}$)	ETA AAS [c] (ng)	
Ag	1	0.0001	0.1 – 0.7
Al	30	0.001	0 – 7
As	30	0.008	0.8 – 8.0
Au	20	0.001	0.004 – 0.027
B	2500	0.02	–
Ba	20	0.006	4 – 20
Be	2	0.00003	0.001 – 0.03
Bi	50	0.004	0.001 – 0.04
Ca	1	0.0004	–
Cd	1	0.00008	0.07 – 0.71
Co	2	0.002	0.078 – 0.34
Cr	2	0.002	0.1 – 0.5
Cs	50	0.004	0.28 – 0.5
Cu	1	0.0006	0.6 – 20
Fe	4	0.01	0.1 – 61.8
Ga	50	0.001	0.030 – 0.037
Hg	500	0.02	0.01 – 0.075
In	30	0.0004	0.0001– 0.0012
Li	1	0.003	173 –183
Mn	0.8	0.0002	0.2 – 8.6
Mo	30	0.003	2.1 – 18.8
Ni	5	0.009	1.1 – 4.0
Pb	10	0.004	0.02 – 0.04
Pd	10	0.004	–
Pt	5	0.01	–
Rb	5	0.001	119 –125
Rh	20	0.008	–
Sb	30	0.005	0.01 – 0.4
Se	100	0.009	0.052 – 0.11
Si	100	0.000005	–
Sn	0.05	0.02	0.008 – 0.04
Te	50	0.001	
Ti	90	0.04	1.9
Tl	20	0.001	0.0101– 0.019
V	20	0.003	0.2 – 4
Zn	1	0.00003	1.0 – 50

[a] FAA = flame AAS, ETA = electrothermal atomization AAS.
[b] All values cited from Winefordner et al. [11].
[c] All values cited from L'vov [12].
[d] The values cited from Brewer [13].

Bruland et al. [15] compared the sampling methods using the Teflon-coated 20-1 PCV ball-valve samplers (Go-Flo, General Oceanics) and the CIT deep-water sampler, designed and constructed by Schaule and Patterson. Careful sample handling and processing were carried out onboard a ship,

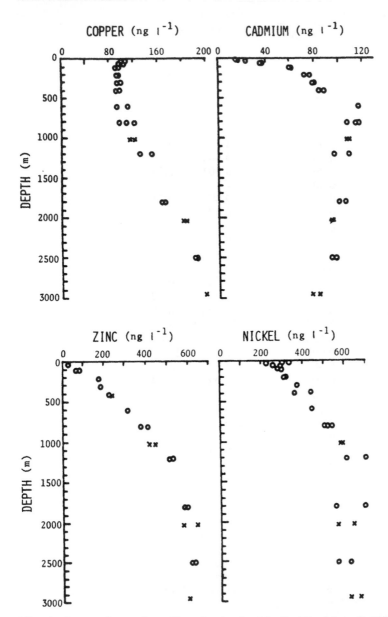

Fig. 1. Comparison of profiles obtained with Go-Flo (o) and CIT (×) samplers. Data points represent replicate extraction values. (From ref. [15].)

inside a modular Portalab equipped with a positive pressure filtered air supply and specially equipped for trace metal analysis. The seawater samples were not filtered, so as to avoid contamination from the filter material, since the primary purpose of their study was the comparison of the sampling techniques. The collected seawater samples were analysed by solvent-extraction/ETA-AAS, which is described in detail in Section II.B.1. Furthermore, they investigated the vertical profiles of Cu, Cd, Zn and Ni in terms of the seawater samples collected with the Go-Flo and CIT samplers. A comparison of the results obtained with the Go-Flo and CIT samplers are shown in Fig. 1 [15].

The results shown in Fig. 1 demonstrate the consistency between the Go-Flo and CIT samplers for four trace metals in deep seawater down to 3000 m. Bruland et al. [15] noted that the samples at depths of 630 m and 1560 m collected with the CIT sampler were contaminated because of a faulty seal. Great care was taken with the methodology with reference to the chemicals, vessels, cleaning and preconcentration procedure, and analytical methods. Consequently it can be concluded that it is clearly very difficult to obtain consistent analytical results for ultra-trace metals in seawater. For more details concerning sampling techniques and contamination from samplers, see ref. [16].

Suspended particulate matter and marine organisms are naturally present in seawater. Thus in seawater analysis, it is desirable to separate the particulate and dissolved fractions, and to analyse them separately. A comparison of analytical results for filtered and unfiltered seawater is shown in Table 3 [15], where upper and lower limits of particulate metal concentrations are also given.

As can be seen in Table 3, the contents of particulate trace metals are less than 1% of the total metal concentrations, which are generally within the range of estimated precision. Thus, filtering deep seawater does

TABLE 3

COMPARISON OF FILTERED AND UNFILTERED SEAWATER SAMPLES ANALYSED BY THE SOLVENT EXTRACTION–ATOMIC ABSORPTION METHOD[a]

(All values in ng l^{-1})

Sample	Copper	Cadmium	Zinc	Nickel
25 m: Unfiltered	110, 109	30.9, 31.5	24, 28	332, 320
Filtered	104, 111	28.8, 25.0	22, 31	325, 321
Particulate	1–9	0.4–1.0	1–7	1–7
1800 m: Unfiltered	173, 170	107, 102	579, 588	556, 693
Filtered	158, 162	107, 109	525, 565	603, –
Particulate	1.5–3	0.05	1.5–2	5–10

[a] Ref. [15].

References pp. 154–157

not cause significant differences. However, the ranges of particulate values at 25 m are 1–9, 1.5–4, 3–27 and 0.2–2% of total Cu, Cd, Zn and Ni, respectively, as calculated from the data for the unfiltered and particulate samples in Table 3. Therefore, differences due to filtration should be taken into account for coastal and surface seawater. Furthermore, contamination from filter materials and filtration apparatus and adsorption of trace metals on the filters, which may produce appreciable error, have to be checked carefully before the analysis of seawater samples. Significant contamination of seawater with Cu [17], Fe and Mn [18], and adsorption of Hg [14] have been reported.

A. Storage of seawater samples

The acidification of samples is required for the storage of seawater in order to prevent precipitation, adsorption of trace metals on the container walls, or volatilization loss from solution. Generally, nitric or hydrochloric acid is added to seawater to bring the pH of the solutions to below 1.5. The addition of acids immediately after collection onboard a vessel may be preferable for sample storage rather than addition before analysis in the laboratory.

Many workers have investigated the significant loss of trace metals from water samples upon storage [19–27]. The extent of loss varies with the pH, the length of storage period, the type of containers, and the concentration of the metals present. Eichholz et al. [19] found losses of trace metals from water samples when stored in polyethylene and borosilicate (Pyrex) glass containers. Robertson investigated losses of Ag, Co, Fe, In, Sc and V from seawater stored in Pyrex and polyethylene containers by using neutron activation analysis, and found significant losses of these metals [20]. Furthermore, Subramanian et al. [26] studied losses of eleven trace metals (Ag, Al, Cd, Co, Cr, Cu, Fe, Mn, Ni, Pb, Sr, V and Zn) from water samples as a function of contact time (days) when stored in pyrex glass, Nalgene-lined polyethylene and Teflon containers, at various solution pH's (1.5–8.0). Electrothermal atomization AAS was used for monitoring metal contents in the solutions. The percent losses of Cd and Pb from synthetic and river-water samples stored in Pyrex glass and polyethylene containers at various pH values as a function of contact time (for 30 days) are shown in Fig. 2.

From these studies it was found that the acidification to pH 1.5 with nitric acid and storage in a Nalgene container is the most effective way of minimizing the loss of trace metals from natural waters. Desorption of trace metals from the container wall should also be considered. Kinsella and Willix [28] have shown that polyethylene and Teflon bottles which have been washed with a 1 : 1 nitric and hydrochloric acid mixture for 2 weeks can still contribute detectable amounts of Cd, Pb and Cu. An ultrasonic treatment

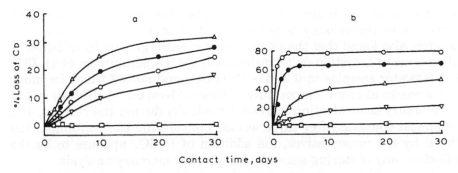

Fig. 2. (a) Loss of cadmium from Pyrex glass and Nalgene containers. Rideau River water sample (0.12 $\mu g\ l^{-1}$ Cd); Pyrex glass: (□) pH 1.6, 2,5, 4.0; (●) pH 6.0; (△) pH 8.0. Nalgene: (□) pH 1.6, 2.5, 4.0, 6.0, 8.0. Synthetic water sample (0.10 $\mu g\ l^{-1}$ Cd); Pyrex glass: (□) pH 1.6, 2.5, 4.0; (▽) pH 6.0; (○) pH 8.0. Nalgene: (□) pH 1.6, 2.5, 4.0, 6.0, 8.0. (b) Loss of lead from Pyrex glass and Nalgene containers. Rideau River water sample (4 μg Pb l^{-1}); Pyrex glass and Nalgene containers: (□) pH 1.6; (△) pH 2.5, 4.0; (○) pH 6.0, 8.0. Synthetic water sample (5 μg Pb l^{-1}); Pyrex glass and Nalgene containers: (□) pH 1.6, (▽) pH 2.5, 4.0; (●) pH 6.0, 8.0. (From ref. [26].)

during acid washing is recommended to remove these elements effectively from the wall.

The loss of mercury from water samples upon storage has been pointed out to be a serious problem by many workers [24–26]. These losses of mercury are caused by rapid adsorption on the container walls [25, 29] and reduction of mercury to the atomic state followed by volatilization from solution [29]. Lo and Wai [29] reported that 81% of mercury in the untreated samples was lost to the walls of polyethylene containers and the remaining 19% was volatilized to the atmosphere. Bothner and Robertson [30] observed mercury contamination of seawater samples due to the diffusion of mercury vapour from the laboratory into the polyethylene containers.

Since polyethylene is one of the most troublesome materials with regard to losses of mercury, Pyrex glass containers should be used with some preservative for storage of seawater samples requiring mercury analysis [24, 29]. Work has yet to be completed on the most suitable preservatives for this application. Feldman [25] proposed the addition of 5% (v/v) HNO_3 and 0.05% $K_2Cr_2O_7$ in the case of a polyethylene container, or of 0.01% $K_2Cr_2O_7$ for a borosilicate glass container. He further recommended that HNO_3 and/or $K_2Cr_2O_7$ should be added to the containers before collection of the samples in order to stabilise the mercury immediately. Keller et al. [31] suggested that a matrix of 5% HNO_3 and 0.1% $K_2Cr_2O_7$ stabilised the mercury. Dokiya et al. [32] showed that L-cystein with dilute HCl is a good preservative for mercury even at the 1 ng ml^{-1} level in water samples. An interesting experiment reported by Kimura and Arikado [33] on the loss of mercury from solution as a result of aeration, indicated that the rate of mercury loss increased with increases in the velocity of aeration and in solution temperature. In addition they also found that the rate of loss decreases greatly by addition of a small amount of L-cystein or tetrachloroaurate(II). Matsunaga et al. [34] acidified seawater samples to 0.2 M with H_2SO_4 and stored them for more than three weeks at room temperature. Using cold-vapour AAS, they noted that the mercury concentration increased gradually during the storage, and became almost constant after three weeks. Considering the contamination introduced by the preservatives, the addition of H_2SO_4 appears to be the most effective way of storing seawater samples for mercury analysis.

B. Preconcentration

The detection limits obtained by flame and electrothermal atomization AAS and the concentration levels of the elements in seawater are summarised in Table 2. In general, ETA-AAS provides better sensitivities for many elements than the flame technique. Using ETA-AAS, several attempts at direct analysis of seawater have been reported [35–38]. In order to avoid severe interference from matrix salts and to enhance the sensitivity of the method, matrix modification and/or platform techniques have been performed. Detection of 10–20 $\mu g/l^{-1}$ can be achieved for Cd, Zn and Pb. Even so, AAS sensitivity is insufficient for the direct determination of most ultra-trace elements. Furthermore, concentrated salts and undissolved particulates cause severe interferences with the determination of trace elements by AAS. Therefore, it is necessary to concentrate the analytes before determination, and, if possible, to separate the analytes from dissolved major matrix constituents and particulates. Solvent extraction, coprecipitation and ion-exchange techniques are the most widely used methods for the preconcentration of seawater. In the following section, these techniques will be reviewed. It should be noted here that the efficiencies of the recovery of the analytes as well as the contamination from reagents and solvents must be carefully examined when preconcentration techniques are employed.

1. *Solvent extraction*

Solvent extraction is one of the methods widely used for preconcentration and separation. Most heavy metals are extracted with a chelating agent into organic solvents [38]. Some of the more common chelating agents used in atomic absorption spectrometry are shown in Fig. 3.

Solvents such as ketones, esters, ethers, alcohols, and other oxygen-containing hydrocarbons are suitable for the flame atomic absorption technique. Of these solvents, MIBK (methyl isobutyl ketone, 4-methylpentane-2-one) is most widely used, together with chelating agents such as APDC, DDC and oxine. Solvent extraction offers the following advantages for AAS:

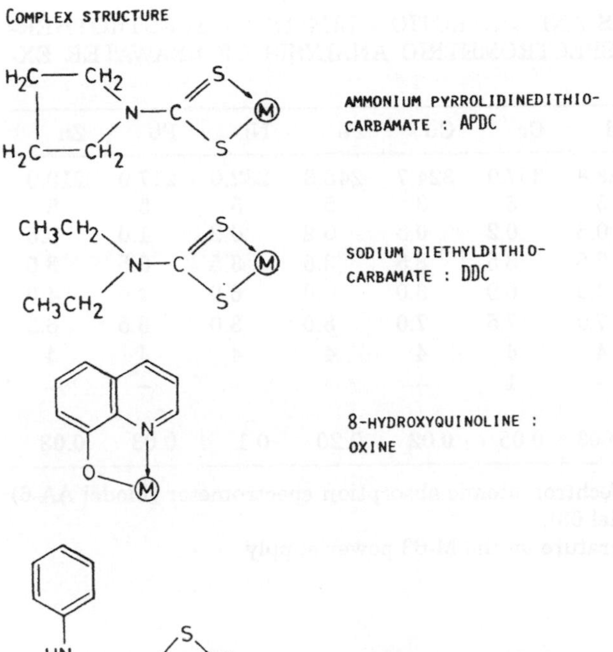

COMPLEX STRUCTURE

AMMONIUM PYRROLIDINEDITHIO-
CARBAMATE : APDC

SODIUM DIETHYLDITHIO-
CARBAMATE : DDC

8-HYDROXYQUINOLINE :
OXINE

DIPHENYLTHIOCARBAZONE :
DITHIZONE

Fig. 3. Typical chelating agents commonly used in atomic absorption spectrometry.

(i) concentration of trace elements; (ii) enhancement of atomic absorption signals; (iii) separation of trace metals from interfering components.

Detailed descriptions of solvent extraction techniques in atomic absorption spectrometry can be found in the book by Cresser [39].

Brooks et al. [40] first investigated solvent extraction for seawater with an APDC-MIBK extraction technique for the determination of six elements in seawater by flame AAS. However, the metal complexes extracted into MIBK were unstable (decomposing within one day), which made the technique inconvenient for routine analysis.

Jan and Young [41] investigated an APDC-MIBK extraction of trace metals (Ag, Cd, Cr, Fe, Ni, Pb and Zn) followed by extraction with 4 M nitric acid, and analysed them by ETA-AAS. They also compared their results with those obtained by the Chelex-100 cation-exchange [42] and APDC-MIBK single extraction methods. The method proposed by Jan and Young offered the advantages of small sample volumes, better recovery efficiency, and

References pp. 154–157

TABLE 4

EXPERIMENTAL CONDITIONS AND DETECTION LIMITS IN ELECTROTHER-
MAL ATOMIC ABSORPTION SPECTROMETRIC ANALYSIS OF SEAWATER EX-
TRACTS [a]

	Ag	Cd	Cr	Cu	Fe	Ni	Pb	Zn
Wavelength (nm)	328.1	228.8	357.9	324.7	248.3	232.0	217.0	213.9
Lamp current (mA)	3	5	5	3	5	5	5	5
Spectral slit (nm)	0.5	0.5	0.2	0.5	0.2	0.2	1.0	0.5
Dry [b] (50 s)	3.5	3.5	3.5	3.5	3.5	3.5	3.5	3.5
Ash [b] (20 s)	5.0	4.0	6.0	6.0	6.0	6.0	4.0	4.0
Atomize [b] (2 s)	6.5	7.0	7.5	7.0	8.0	8.0	6.5	6.5
N_2 gas (l min^{-1})	4	4	4	4	4	4	4	4
H_2 gas (l min^{-1})	–	–	1	–	–	–	–	–
Detection limit ($\mu g\,l^{-1}$)	0.02	0.003	0.05	0.02	0.20	0.1	0.03	0.03

[a] Ref. [37]. Analysed by Varian-Techtron atomic absorption spectrometer (Model AA-6)
with a carbon rod atomizer (Model 63).
[b] Arbitrary dial settings of temperature on the M-63 power supply.

stability of the metal complexes in the acid extracts over the ion-exchange
and APDC-MIBK single extraction methods.

The procedure of the MIBK–nitric acid successive extraction employed by
Jan and Young [41] is as follows. For the APDC-MIBK extraction place a 200
ml seawater sample in a Teflon beaker containing 2 ml of 1% APDC and heat
to incipient boiling at a pH of about 4. After cooling to room temperature
add 7 ml of MIBK to the sample and transfer the mixture into a polyethylene
bottle, which should be shaken for 25 min on a mechanical shaker. After
allowing the layers to separate in a separatory funnel for 20 min, the organic
layer is collected in a polyethylene bottle.

For the acid back-extraction, pipet 5 ml of 4 M HNO_3 into the MIBK
extract obtained from the previous step and shake the mixture for 20 min.
Transfer the mixture into a Teflon separatory funnel and stand for 20 min;
the acid layer can then be drained into a polyethylene bottle and preserved
in a refrigerator until analysed.

The experimental conditions and detection limits for eight metals in
seawater extracts obtained by Jan and Young are summarised in Table 4.

A Varian-Techtron AA-6 atomic absorption spectrometer equipped with an
electrothermal atomizer (Model 63) was used with a background corrector,
and 2.5 μl of the treated sample solution was injected into the furnace
[41]. The analytical results of trace metals in seawater using MIBK single
extraction and MIBK-HNO_3 successive extraction methods are compared in
Table 5.

TABLE 5

TRACE METALS ANALYSIS OF SEAWATER USING MIBK SINGLE EXTRACTION AND MIBK-HNO$_3$ SUCCESSIVE EXTRACTION METHODS [37]

Method	Concentration (μg l^{-1}):					
	Ag	Cr	Cu	Fe	Ni	Pb
MIBK-HNO$_3$	<0.02	0.14	0.78	3.28	0.59	0.29
extraction	<0.02	0.17	0.73	3.57	0.59	0.25
	<0.02	0.16	0.82	3.51	0.59	0.28
	<0.02	0.12	0.82	3.10	0.59	0.31
Mean	<0.02	0.15	0.79	3.37	0.59	0.28
SD [a]	–	0.022	0.043	0.22	0.00	0.025
% RSD [b]	–	7.3	5.4	6.5	0.0	8.9
MIBK single	<0.01	0.11	0.95	3.40	0.51	0.32
extraction	<0.01	0.15	0.75	3.47	0.54	0.40
	<0.01	0.15	0.67	3.15	0.54	0.34
	<0.01	0.19	0.84	4.05	0.56	0.43
Mean	<0.01	0.15	0.80	3.52	0.54	0.37
SD	–	0.033	0.120	0.380	0.021	0.051
% RSD	–	22	15	11	3.9	14

[a] SD = standard deviation.
[b] RSD = relative standard deviation.

As can be seen from Table 5, the precision of the MIBK-HNO$_3$ successive extraction method is improved, in addition to the stability of the extract solution, compared to that of the MIBK single extraction method. In a further experiment mean relative standard deviations ranging from 18 to 25% were obtained for those metals present below 1 μg l^{-1}. The results for the recovery test obtained by the method mentioned above are also summarised in Table 6 [41].

Generally, the recoveries are poor for all the elements, especially for silver. This indicates that the solvent at pH 4 is not suitable or optimal for these elements. It is necessary to examine the optimum pH for each element in order to improve the efficiency of the solvent extraction technique.

Bruland et al. [15] developed a more efficient solvent extraction method for Cu, Cd, Zn and Ni, using a double extraction technique with APDC/DDDC (diethylammonium diethyldithiocarbamate) into chloroform, and back-extraction into 7.5 M HNO$_3$. As this method achieved a great improvement compared with other methods, it is still one of the best methods in ocean seawater analysis. The experimental procedure is described here in detail.

Approximately 250 g of acidified seawater was taken into a 250 ml Teflon separatory funnel and buffered to about pH 4 with 2 ml of ammonium

TABLE 6

RECOVERY OF METALS FROM SPIKED SEAWATER OBTAINED BY SOLVENT
EXTRACTION–ATOMIC ABSORPTION SPECTROMETRY [37]

Element	Added[a] (μg)	Found[b] (μg) mean ± SD[c]	Average recovery (%)
Ag	0.037	0.007 ± 0.0007	19
	0.185	0.024 ± 0.0006	13
Cd	0.020	0.016 ± 0.0025	80
	0.100	0.081 ± 0.0085	81
Cr	0.025	0.020 ± 0.0026	80
	0.125	0.117 ± 0.0279	94
Cu	0.100	0.103 ± 0.0219	103
	0.500	0.373 ± 0.0528	75
Fe	0.100	0.069 ± 0.0095	69
	0.500	0.381 ± 0.0796	76
Ni	0.100	0.080 ± 0.0081	80
	0.500	0.391 ± 0.0099	78
Pb	0.050	0.035 ± 0.0023	70
	0.250	0.163 ± 0.0329	65
Zn	0.050	0.059 ± 0.0064	118
	0.250	0.192 ± 0.0439	77

[a] Filtered island control seawater (200 ml) spiked with metal standard solution.
[b] After correcting for concentration measured in unspiked sample.
[c] SD=standard deviation, $n = 3$.

acetate; 1 ml of a solution containing 1% ammonium hydroxide solution
(purified by chloroform extraction) and 8 ml of chloroform were added, and
the mixture shaken vigorously for 2 min. After 5 min of phase separation,
the chloroform fraction was drained into a 25 ml Teflon separatory funnel,
and 4 ml of 7.5 M nitric acid was added to degrade the dithiocarbamates.
An additional 6 ml of chloroform was added to the original seawater sample
for the second extraction of the chelated species. After phase separation, the
chloroform was combined with the first fraction and shaken vigorously for 2
min. The phases were allowed to stand 5 min for separation, and then the
chloroform phase was discarded.

The 4 ml acid phase containing the back-extracted metals was then
drained into a 10 ml fused quartz beaker, and 2 ml of 7.5 M nitric acid
was used to rinse the 125 ml separatory funnel and stem. This rinse was
combined with the acid in the quartz beaker. The 6 ml back-extract was
evaporated to dryness, and the residue was further oxidised by the addition
of 250 μl or 500 μl concentrated nitric acid. The residue in the quartz beaker
was redissolved in warm 1 M nitric acid and quantitatively transferred with
250 μl rinses (ca. 1.25 ml total) to a 1.8 ml polyvial. This solution was
presented for analysis by ETA-AAS.

After the completion of the procedure described above, the concentration factor was 200-fold, and alkali and alkaline earth elements, which often provide spectral interference in ETA-AAS, have to be removed. In the determination a Varian AA-6 AAS spectrometer with a model 90 carbon-rod atomizer (CRA-90) and a BC-6 hydrogen continuum automatic background corrector, and a Perkin-Elmer 603 spectrometer with a Model 2100 graphite furnace (HGA-2100) and deuterium-arc background corrector were used. The hollow-cathode lamps were used as light sources. In general, the recommended procedures by the manufacturer were followed, except that relatively higher temperatures for shorter durations were utilized during atomization. In addition, with the Varian CRA-90, P-10 gas (90% argon, 10% methane) was mixed with the nitrogen flow (flow rate $1:6$) during the ashing and atomization steps. The use of P-10 gas prolonged the life of the pyrolytic coating of the graphite tubes and increased precision.

The overall recoveries were examined by using seawater samples spiked with ^{115}Cd, ^{64}Cu and ^{65}Zn, and were $98.9 \pm 0.7\%$ ($n = 10$) for Cd, $97.9 \pm 1.2\%$ ($n = 3$) for Cu and $99.0 \pm 0.8\%$ ($n = 5$) for Zn. The experimental results obtained by Bruland et al. [15] are shown in Figs. 1 and 4 and Table 3.

Several authors have developed similar techniques. Danielsson et al. [43] concentrated Cd, Cu, Fe, Pb, Ni and Zn using the same double chelating reagents as used by Bruland et al. They also carried out back-extraction of chelates into nitric acid, but used Freon as the solvent, which was slightly different from that used by Bruland et al. [14].

Chromium as Cr(VI) represents a toxic form of the element, and its environmental pollution should be monitored even in seawater. Whilst Cr(III) also exists in relatively high amounts in seawater, the separation of Cr(VI) from Cr(III) is necessary in the analysis of Cr(VI). For this purpose a solvent extraction technique can also be used, being followed by atomic absorption analysis. Many workers have investigated the solvent extraction of total Cr in seawater, where Cr can be extracted with acetylacetonate, DDC and APDC, and analysed by AAS [4, 41, 44–47]. Hiiro et al. [47] examined in detail the separation of Cr(VI) from Cr(III) in seawater. The effect of pH values on extraction of Cr(VI) is shown in Fig. 5.

Chromium(VI) is most effectively extracted near pH 5, whilst Cr(III) is increasingly extracted above pH 4. Therefore, Cr(VI) must be extracted below pH 4. More than 2×10^{-3} M DDC is desirable for the extraction of Cr(VI), as shown in Fig. 5.

Lo et al. [48] used Hg ion in back-extraction of chelate of APDC and DDTC (sodium diethyldithiocarbamate) with Cd, Cu, Fe, Mn, Pb and Zn in chloroform. The oxine/chloroform system was used for the extraction of Mn [49] and oxine/DIBK (diisobuthyl ketone) for extraction of Mo [50] from seawater. Cd, Cu, Ni and Zn have also been concentrated by extraction of their dithizone complexes into chloroform [51].

References pp. 154–157

140

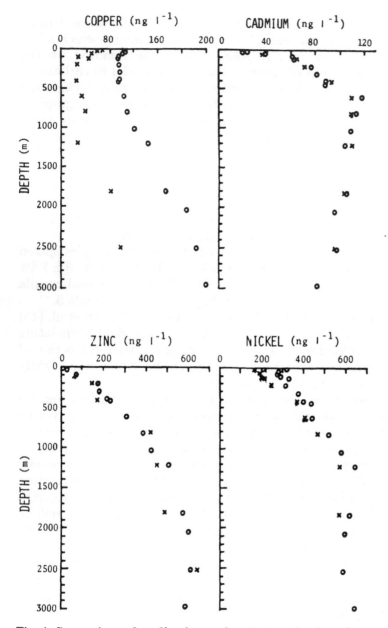

Fig. 4. Comparison of profiles from solvent extraction (o, values expressed as average of two or more replicates), and Chelex-100 (×) techniques. (From ref. [15].)

2. *Coprecipitation*

In the coprecipitation method, trace metals are concentrated by their adsorption onto the surface of a precipitate and/or being incorporated in

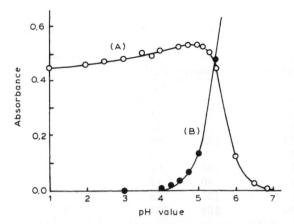

Fig. 5. Effect of pH value on the extraction of Cr(VI) using the flame method. (A) Cr(VI) 0.5 ppm; (B) Cr(III) 20 ppm, V_{aq} 100 ml. (From ref. [42])

a precipitate. Commonly used are the precipitates of hydroxides of Fe(III), Mn(IV), Al(III), Bi(II), and the sulphides of Co(II), Pb(II) and Fe(II). In AAS the use of hydroxide precipitates is more popular than that of sulphides. Coprecipitation using the hydroxide precipitates is generally unspecific, and many trace metals may be concentrated simultaneously. However, some limitations are often found in the coprecipitation techniques [10].

Contamination of the analytes from the carriers (the precipitates) should be first examined, and the blank test carried out carefully. The determination of optimum pH for the precipitation is very important, because the increase of pH leads to the improved recoveries of most analytes as well as interfering matrix salts. Some efficiencies of the recovery for Zr(IV) and Ga(III) coprecipitations are summarised in Table 7, where ICP-AES was applied for simultaneous multi-element analysis [52, 53].

In these experiments, 20 mg of Zr(IV) or 5 mg Ga(III) were added to 1 l of seawater, and the precipitation was made by increasing pH with sodium hydroxide. In the Zr and Ga coprecipitations, the pH was adjusted to 10 and 9, respectively. Poor recoveries of Mo, Sb, As (Zr coprecipitation) and V (Ga coprecipitation) are due to the solubility of the anionic forms.

Sato and Saitoh [54] determined As, Cr and Pb in seawater by ETA-AAS with the Zr-coprecipitation technique. This was done by adding 1 ml of 0.13 M $ZrOCl_2$ solution (about 10 mg of Zr) into 1 l seawater, and adjusting the pH to 9 with ammonia solution. The precipitate was filtered with No. 5 filter paper, washed with 2% ammonia solution, and dissolved with hot 2 M hydrochloric acid. The final solution volume was made up to 50 ml. The treated solution (50 μl) was subjected to ETA-AAS determination (spectrophotometer; Hitachi Model 207, graphite furnace; Perkin-Elmer

References pp. 154–157

TABLE 7

RECOVERIES USING THE ZIRCONIUM AND THE GALLIUM COPRECIPITATION TECHNIQUES

Element	Recovery (%)[a]		Element	Recovery (%)[b]	
	Zr	Ga		Zr	Ga
Al	100	101	Mo	13	1
As	104	19	Ni	92	96
Cd	76	5	Pb	102	94
Co	96	91	Sb	76	5
Cr	100	97	Ti	103	100
Cu	101	96	V	90	17
Fe	103	98	Y	101	88
La	102	82	Zn	101	101
Mn	99	95			

[a] Recovery of 1 mg of each spiked element with 20 mg of Zr at pH 10.
[b] Recovery of 4–200 μg of each spiked element with 5 mg of Ga at pH 9.

Model HGA-70). The recoveries for As, Cr and Pb were almost 100% when the samples containing only a few nanograms of the metals were analysed. The interference effects of the 18 kinds of metal ions and anions examined were almost negligible for the concentrations upto 1000–10000 fold. Sato and Saitoh [54] also suggested that this method might be applicable to Cd^{2+}, Co^{2+}, Cu^{2+}, Fe^{3+}, Al^{3+}, Mn^{2+}, Ni^{2+}, Sb^{3+}, Sb^{5+}, Se^{4+}, Se^{6+}, Zn^{2+}, V^{5+}, and so forth. Other applications of coprecipitation techniques can be found in the references [55–61].

The coprecipitation method combined with colloid flotation using stearylamaine, sodium oleate, octadecylamine etc., has been used to preconcentrate analytes in seawater [59–61]. Combinations of the coprecipitation technique and the hydride generation method have also been reported for the determinations of Te [62] and Bi [63] in seawater by AAS. They used Mg, originally contained in seawater, as a coprecipitation reagent to collect Te and Bi from seawater. After hydride generation with sodium borohydride, the hydrides were directly led to a graphite furnace and trapped there. The detection limits of Te and Bi in seawater were 60 pg l^{-1} and 3 pg l^{-1}, respectively.

3. Resin preconcentration

Various resin preconcentration procedures have been developed in analytical chemistry. Riley [14] has published a detailed review of various ion-exchange techniques for concentration of trace metals in seawater. The extremely high content of sodium ion in seawater makes cation-exchange resins of little use. Anion-exchange resins are useful for elements existing as anions, especially chloro-complexes in seawater. Although Brooks [64] reported large enrichment factors for Au, Bi, Cd and Tl with Amberlite IRA-100 resin, there have not been many applications of anion-exchange resins to seawater analysis [65, 66].

In recent years, the use of chelating exchange resins (e.g. Chelex-100) in preconcentration of trace metals in seawater has become popular [15, 41, 67-76]. Chelex-100 is a copolymer of vinylbenzyliminodiacetic acid, styrene and divinylbenzene, and shows strong chelating activity for various trace metals [14, 67, 68].

Bruland et al. [15] investigated the determination of Cu, Cd, Zn and Ni in seawater by ETA-AAS, carrying out a strict comparison of sampling techniques and preconcentration techniques (solvent extraction and chelating resin methods).

The concentration technique with the Chelex-100 resin can be summarised as follows. The resin (Na form; 100-200 mesh) was cleaned with 6 M HCl by daily suspension, decantation and addition of fresh acid for one week; it was then rinsed with distilled water and 2 M HNO_3, and finally cleaned with distilled water. The resin was changed to the ammonia form and 7.5 ml of resin was packed in disposable polyethylene columns. The excess of ammonia was washed with distilled water to reduce the pH of the effluent to below 9. The concentration procedure of seawater with the Chelex resin was performed immediately after sample collection onboard ship. Seawater (4 l) was taken from the samplers into acid-cleaned polyethylene bottles. It was then pumped through Teflon tubing directly into the resin column; a peristaltic pump was used to maintain the flow rate at 3.5-4.5 ml min^{-1}. Upon completion of pumping, the column was capped at both ends, covered with polyethylene bags, and frozen for storage. After return to the lab, the resins were thawed, washed with distilled water to remove excess of salts and eluted with 30 ml of 2 M HNO_3 into polyethylene bottles. The concentration factor obtained by this concentration procedure was about 120-fold. The recoveries of trace metals for this Chelex procedure was examined with ^{65}Zn- and ^{115}Cd-spiked seawater samples and found to be almost 100%. The treated effluents were analysed by ETA-AAS.

Using the procedure described above, large amounts of Mg and Ca were eluted in the final step; Na and K were also eluted in the final step. These latter components interfered with the determination of Cd, Cu, Pb and Zn by AAS. Therefore, the standard solutions with matched matrices (Ca 188 mg l^{-1}; Mg 1700 mg l^{-1}; Na 150 mg l^{-1} and K 20 mg l^{-1}) were prepared for the analysis.

Bruland et al. [15] compared the results obtained by Chelex-100 concentration with those by solvent extraction (a double extraction with APDC/DDDC into chloroform and back-extraction into 7.5 M HNO_3) described earlier. Vertical profiles obtained by both techniques are shown in Fig. 4.

The profiles for cadmium and zinc show good agreement between the solvent extraction and Chelex results, although the Chelex results are somewhat lower than those by solvent extraction preconcentration. The values for nickel were also lower when the Chelex technique was employed. The mean difference between the two techniques was 65 ± 35 ng l^{-1}.

A large discrepancy between the two concentration techniques was found for the results of copper. The difference was 72 ± 27 ng l^{-1}. Bruland et al. [15] analysed the Chelex column effluent by the solvent extraction technique, and found 63 and 135 ng l^{-1} copper for the samples at 25 m and 2500 m, respectively. These values are almost equal to the difference between the Chelex and solvent extraction results. They concluded therefore, that about 60% of copper in seawater (unfiltered and unacidified) was not removed by the Chelex technique. As Riley et al. [77] suggested, copper in seawater is not liberated by the Chelex resin because of association with colloids and fine particulates. In order to avoid this error, acidification and heating of seawater is necessary prior to the Chelex treatment. According to the results of Bruland et al. [15] acidification and storage followed by solvent extraction appears to be superior to the Chelex resin preconcentration for the quantitative determination of copper in seawater. Similar problems have been pointed out by Eisner and Mark [78] and Florence and Batley [79].

Chu et al. [75, 76] preconcentrated Al, Cd, Cu, Fe, Mn, Ni, Pb, Ti, V and Zn from seawater using Chelex-100 resin followed by ICP-AES measurement. They evaluated and compared two techniques, i.e. column preconcentration and batch preconcentration. The salts remaining on Chelex-100 were removed efficiently by the treatment of ammonium acetate solution. The preconcentration techniques developed were also applicable to AAS.

The preconcentration procedure for the column method [75] was as follows. One litre of seawater sample was filtered and the pH of the solution was adjusted to pH 5.5–6 with 10 ml of 1 M ammonium acetate solution. The seawater sample was eluted at a flow rate of 2 ml min^{-1} through the preconditioned column. For removal of matrix salt, the column was treated with 100 ml of 1 M ammonium acetate buffer (pH 6) and 50 ml of sub-boiled pure water after seawater elution. Trace metals collected on the resin were then eluted with 25 ml of 2 M HNO_3 solution into a Teflon beaker. The eluent in the beaker was heated to dryness on a hot plate, and then the residue was dissolved with 5 ml of 0.2 M HNO_3. Thus a 200-fold concentration was achieved.

The procedure for the batch preconcentration (100 fold) [76] was as follows. Into exactly 1 l of seawater, 0.5 g of Chelex-100 resin and 10 ml of 1 M ammonium acetate buffer solution (pH 6) were added. The pH of the solution was adjusted to 6 with 1 M acetic acid or 1 M ammonia. The solution was stirred with a magnetic stirrer for 2 h. The resin was collected on a glass filter and was washed with 30 ml of 1 M ammonium acetate buffer (pH 6) and 15 ml of pure water. The analyte elements adsorbed on the resin were desorbed with 7 ml of 2 M nitric acid, and finally the solution was made up to 10 ml with pure water. It was concluded that the batch preconcentration was better than the column preconcentration because the batch method required less time (3 h) than the column method (20 h).

Recently a reversed-phase liquid chromatographic technique was applied for preconcentration of trace metals from seawater by Sturgeon et al. [80]. They treated seawater with oxine (8-hydroxyquinoline) to collect Cd, Zn, Cu, Ni, Co, Mn and Fe therein, prior to subsequent adsorption on C_{18}-bonded silica gel. The oxine chelates were quantitatively recovered on the large capacity C_{18}-bonded silica gel. A small disposable cartridge of C_{18}-bonded silica gel is now commercially available. This method has some merit, because naturally occurring metal chelates, which cannot be recovered on Chelex-100 resin, can be collected on the C_{18} bonded silica gel. Here the procedure is introduced in detail.

The glass column was slurry-loaded with 600 mg of the C_{18} gel. Three 10 ml portions of methanol were then passed through the bed under gravity flow for effective cleaning-up. Before passage of the last 200–300 μl of the methanol, distilled deionized water was carefully added in small portions (100–200 μl), thereby changing the solvent from methanol to water, in the manner of a gradient elution, without disturbing the resin bed. The column was further washed with deionized water. A 500 ml seawater sample was loaded into the reservoir and 0.5 ml of 5% purified 8-hydroxyquinoline solution was added. The pH was adjusted to 8.9 with ammonia solution and the sample was drawn through the C_{18} bed with a water-aspirator.

After passage of the sample, the column was washed with 10 ml of demineralised water of pH 8.9 containing 50 μg of oxine per ml. The adsorbed metal chelates were then eluted from the column with 5 ml of methanol. This eluate was collected either in a flask, making up to 10 ml with 2% (v/v) nitric acid, or in a "Vycor" crucible, evaporating to dryness in the presence of 200 μg concentrated nitric acid and 500 μg of 60% perchloric acid and thereafter diluting to 10.0 ml with 1% (v/v) nitric acid. These procedures provide a theoretical concentration factor of 50.

The analytical results of seawaters obtained by Sturgeon et al. [80] are shown in Table 8.

Good agreement with accepted values can be seen. The analytical blank for the methanolic concentrate is smaller than that for the acid-decomposed concentrate. Therefore, they performed all the analyses in the latter method. In the methanolic concentrate no other matrix except the oxine was present. Even so, they made a calibration by spiking a sample of the concentrate with the element of interest, so as to prevent enhancement of signal by oxine.

The same authors [81] developed a preconcentration technique of Se and Sb using a combination of multi-element chelation with APDC and adsorption on C_{18}-bonded gel.

Another chelating reagent (sodium bis(2-hydroxyethyl) dithiocarbamate; NaHEDC) has been used to chelate Ag, Cd, Fe, Mo, Ni, U, V and Zn followed by adsorption on a small column of XAD-4 resin [82].

TABLE 8

DETERMINATION OF TRACE ELEMENTS IN SEAWATER USING C_{18}-BONDED GEL

Element	Concentration (ngl^{-1})			
	Sample A		Sample B	
	C_{18} preconcentration [a]	Accepted value [b]	C_{18} preconcentration [a]	Accepted value [b]
Cd	0.030 ± 0.002	0.027 ± 0.003	0.031 ± 0.002	0.025 ± 0.001
Zn	0.30 ± 0.07	0.41 ± 0.05	0.4 ± 0.1	0.28 ± 0.01
Cu	0.97 ± 0.02	0.96 ± 0.04	0.11 ± 0.02	0.13 ± 0.01
Mn	0.78 ± 0.02	0.68 ± 0.05	0.98 ± 0.05	1.06 ± 0.02
Fe	0.9 ± 0.1	1.03 ± 0.04	7.7 ± 0.5	6.9 ± 0.2
Ni	0.26 ± 0.01	0.31 ± 0.04	0.46 ± 0.06	0.39 ± 0.01
Co	0.019	0.015 ± 0.007	0.017 ± 0.001	0.017 ± 0.002

[a] Mean and standard deviation for triplicate analyses.
[b] Values accepted by the extensive analysis by several methods.

Muzzarelli and Marinelli [83–85] proposed the use of a natural chelating polymer, chitosan, which is de-acetylated chitin containing glucosamine and N-acetyl glucosamine. Sugawara et al. [86] have developed another promising chelating agent: controlled pore glass (GPC) immobilized chelate(s). The properties and feasibilities of the GPC-immobilized chelates have been summarised by Chakrabarti et al. [11].

Combinations with resin preconcentration technique and flow injection (FI) have been reported [87, 88]. Olsen et al. [87] achieved the detection limits of several ng l^{-1} for Pb, Cd and Zn using a microcolumn of Chelex-100.

4. Other preconcentration techniques

Solvent extraction, coprecipitation and resin techniques are the main preconcentration methods used for seawater analysis. Other interesting preconcentration techniques, such as electrodeposition, amalgam trap (for mercury), a cold trap-vaporisation system for hydride generation and recrystallisation, are often used by marine and analytical chemists. The first three methods are briefly reviewed here.

Electrodeposition is an unique preconcentration technique, because the separation of trace metals from interfering matrix species can be easily carried out at the same time. Thus, the detection limits of AAS can be markedly improved without further chemical preconcentration steps, and spectral and chemical interferences due to major components in seawater can be also eliminated easily. Thus, many applications of such techniques to seawater analysis by AAS have been made, especially in ETA-AAS [87–98]. The HMDE (hanging mercury drop electrode [89, 90], gold-foil [91], copper-wire [92], platinum-wire [93], tungsten-wire [94–96] and pyrolytic graphite-coated tube [97] have been used as the electrodes for electrochemical deposition, and successfully applied to the determination of Cu, Cd, Pb, Zn, Hg and so forth. In atomic absorption analysis the electrodes are usually heated directly for atomization of metals.

Flame AAS does not usually have sufficient sensitivity for the determination of mercury. Thus, flameless AAS employing the cold-vapour generation technique [98] has been developed to improve the analytical sensitivity for mercury. In the cold-vapour atomic absorption technique, a long tube absorption cell is commonly used, with detection limits in the order of sub-ppb being obtained. Tanabe et al. [99] reported the detection limit of 2 ng l^{-1} using the 185.0 nm atomic resonance line instead of 253.7 nm. The concentration of mercury in seawater is, however, much lower than the detection limit obtained by such a cold-vapour technique [100]. A few ng l^{-1} of mercury in seawater have been reported by some workers [101, 102], and the sub-ppt level of organic mercury has been reported [103]. Thus, proper preconcentration of mercury in seawater should be explored before atomic absorption analysis. Among others, gold-foil [91], gold-wire [104–106], gold-sponge [107], silver metal particles [102, 108], and a cold trap cooled with liquid nitrogen [109] have been used for the trap-concentration of mercury generated from seawater by proper reduction. These metallic or cold traps are generally heated up to about 400–500°C immediately after the preconcentration process. The quantitative determination of mercury at the sub-ppt level can be achieved by using preconcentration techniques. It should, however, be stressed that the preservation of mercury during the storage of the samples is the critical factor rather than the analytical procedures. Therefore, immediate analysis onboard ship may be desirable [102]. The present authors and co-workers have developed a sensitive non-dispersive vacuum-ultraviolet (185.0 nm) mercury atomic absorption analyser [110, 111], which is a compact instrument easily set even in an onboard ship laboratory [102].

Owing to the development of the hydride generation technique, the analytical sensitivities for As, Ge, Sb, Se, Sn, Te, Bi and Pb have been improved significantly to the ppb or sub-ppb level. The use of sodium borohydride ($NaBH_4$), which has strongly reducing capability, has led to successful improvement of sensitivity and analytical convenience. Recent topics of interest in this field include species analysis, i.e. chemical speciation of inorganic and organic compounds. Braman and Foreback [112] investigated sequential volatilisation techniques for arsines as a result of their interest in environmental chemistry. Arsenic compounds are toxic and they are used as silvicides and pesticides [112]. Talmi and Bostick [113] used gas-chromatographic separation of sodium borohydride-reduced arsenic compounds with a microwave discharge emission spectrometric detection system. Furthermore, Edmonds and Francesconi [114], Andreae [115] and

References pp. 154–157

Tanaka et al. [116] used similar fractional volatilisation techniques as proposed by Braman and Foreback [112], and detected arsines by atomic absorption spectrometric methods.

The properties of arsenic compounds and their reduced forms are summarised in Table 9 [94]. As can be seen, various forms of arsines show different boiling points. Thus, these arsines can be trapped and concentrated in a liquid-nitrogen cold trap, and then separated sequentially by heating

TABLE 9

THE PROPERTIES OF ARSENIC COMPOUNDS AND ARSINES [94]

Molecules	pK_a	Reduction pH	Reduction product	B.p. (°C)
As(III), arsenous acid, $HAsO_2$	9.23	4	AsH_3	−55
As(V), arsenic acid H_3AsO_4	2.25	1–2	AsH_3	−55
Monomethylarsonic acid, $CH_3AsO(OH)_2$	2.60	1–2	CH_3AsH_2	2
Dimethylarsenic acid, $(CH_3)_2AsO(OH)$	6.19	1–2	$(CH_3)_2AsH$	35.6
Trimethylarsine[a], $(CH_3)_3As$	–	1–4	$(CH_3)_3As$	70
Phenylarsonic acid, $C_6H_5AsO(OH)_2$	3.59	1–2	$C_6H_5AsH_2$	148

[a] Or its oxidized form, presumed to be $(CH_3)_3AsO$.

TABLE 10

CONCENTRATIONS OF ARSENIC SPECIES IN NATURAL WATERS (ng As ml^{-1})

Locality and sample type	As(III)	As(V)	MMAA[a]	DMAA[b]
Seawater, Scripps Pier, La Jolla, Calif.:				
5 Nov. 1976	0.019	1.75	0.017	0.12
11 Nov. 1976	0.034	1.70	0.019	0.12
Seawater, San Diego Trough:				
surface	0.017	1.49	0.005	0.21
25 m below surface	0.016	1.32	0.003	0.14
50 m below surface	0.016	1.67	0.003	0.004
75 m below surface	0.021	1.52	0.004	0.002
100 m below surface	0.060	1.59	0.003	0.002
Sacramento River, Red Bluff, Calif.	0.040	1.08	0.021	<0.004
Owens River, Bishop, Calif.	0.085	42.5	0.062	0.22
Colorado River, Parker, Ariz.	0.114	1.95	0.063	0.051
Colorado River, Slough near Topock, Calif.	0.085	2.25	0.13	0.31
Saddleback Lade, Calif.	0.053	0.020	<0.002	0.006
Rain, La Jolla, Calif.:				
10 Sept. 1976	<0.002	0.180	<0.002	0.024
11 Sept. 1976	<0.002	0.094	<0.002	<0.002

[a] Monomethylarsonic acid.
[b] Dimethylarsenic acid.

the trap. The stripping and trapping efficiencies of arsines were examined by Andreae [115] and found to be more than 90%. Andreae determined the arsines with an atomic absorption detector using a hydrogen-rich hydrogen/air flame. The detection limit at 197.3 nm was 0.5 ng l^{-1}. The determined values of arsenic compounds in natural waters, including seawater, are shown in Table 10.

The concentrations of As(V) are higher compared with other arsenic compounds. Since the use of MMAA and DMAA as pesticides is increasing, the concentrations of these compounds should be monitored in relation to environmental pollution of the oceans. Braman and Tompkins [117] reported the determinations of inorganic and organic tin compounds in seawater by means of a technique similar to that for arsenic, although they used a flame emission spectrometric method in a hydrogen/air flame.

Species determination of other hydride generating elements such as Sn [118], Ge [119] and Sb [120] have been carried out using AAS. The separation of species was carried out by either fractional volatilisation [118] or chromatography [119, 120]. Recently, Sturgeon et al. developed a new concentration technique for hydride generating elements (As [121], Sb [122], Se [123]) inheriting the methods developed by Andreae [62] and Lee [63]. In their method the elements were directly collected on the surface of a graphite furnace, following the hydride generation treatment.

The combination of FIA and the hydride generation technique has also been used to realise automatic analysis with AAS [124].

III. MARINE ORGANISMS

The study of the elemental composition of marine organisms (animals and plants) is also an important subject in marine chemistry. Generally, the elemental concentrations in the organisms are much higher than the abundances in seawater [125] (see also Tables 1 and 2). This phenomenon is usually interpreted as biological accumulation of elements from seawater. The marine organisms take up nutrient and trace elements from seawater, and concentrate them inside their bodies. The concentration of trace elements in organisms is sometimes anomalously high. The concentration of elements from seawater is defined as the concentrsion factor [125]; it may be as large as $10-10^6$ [125, 126]. According to the references, the elements concentrated are sometimes specific for the organisms and sometimes not specific [127]. Therefore, the concentrations of elements may be primarily linked to the food chain [125, 128].

On the other hand, marine organisms can influence elemental compositions in seawater. They concentrate the elements from seawater and return them to the ocean after death. Thus, organisms play an important role in the movement, transfer and circulation of elements from surface to deep water or from seawater to organic matter. For the last 20 years AAS has been extensively applied to the analysis of marine organisms. When AAS is employed, the storage of organisms and sample treatments (digestion by ashing, wet acidic digestion, lyophilization, or low-temperature plasma ashing) must be carefully investigated when accurate and precise analytical data is to be obtained. The following three methods are usually employed as pretreatment of marine organisms for storage: (a) freezing the wet sample; (b) drying the wet samples in an oven at 80–120°C to remove water; and (c) lyophilization.

The drying and lyophilization methods may be more convenient than freezing, because the dried samples are easily stored without freezing boxes or rooms.

The samples dried in an oven, can be used for the determination of most elements, except for mercury, which is extremely volatile. Therefore, it is most important to know whether any loss of mercury occurs during the drying or lyophilization process. Pillay et al. [129] reported considerable losses of mercury from plankton/algae during low-temperature oven-drying prior to neutron activation analysis and cold-vapour atomic absorption spectrometry. The results are shown in Table 11 along with the results for sediment.

The samples were dried in the laboratory oven at 60°C for 50 h or until a constant weight was attained. Significant losses of mercury were found even for low-temperature drying. Thus, oven-drying of marine samples is not suitable for mercury analysis.

The possible loss of mercury from biological materials upon lyophilization has also been studied by several workers. Some workers have observed losses of mercury [129, 130], and others no losses from fish samples and animal

TABLE 11

LOSS OF MERCURY FROM PLANKTON AND SEDIMENT SAMPLES DURING LOW-TEMPERATURE OVEN-DRYING [99]

Sample identification[a]	Initial levels of mercury (ppm)	Loss of mercury (%)
A. Plankton/algae:		
PL-Bx	17.86	51.1
PL-By	17.86	71.7
PL-Bz	17.86	60.6
B. Sediment/silt:		
S/S-EA	2.25	23.6
S/S-EB	2.25	12.4
S/S-EC	2.25	12.4

[a] Drying for 50 h in a laboratory oven at 60°C.

organisms [131, 132]. Recently, Ramelow and Hornung [133] investigated mercury losses upon lyophilization of fish and mussel samples, using an AAS technique. The results were carefully compared with the analyses of fresh fish samples. In their experiment, fresh fillets of single or several individual specimens were homogenised in a blender, spread inside small glass bottles and frozen overnight in a freezer. Before placing the samples in the lyophilizer, they were frozen at 70°C in a slurry of ethanol and dry ice and then kept in the lyophilizer for about 24 h. The mussel samples were prepared by grinding the soft parts of several individual specimens in an agate mortar until visibly homogeneous. Then they were treated in the same manner as the fish samples. The fresh fish and mussel samples were prepared for analysis by digesting amounts of wet tissue (0.3–1 g) in Teflon-lined high-pressure decomposition vessels. The homogenised samples were digested in the same manner as fresh samples. The analytical results are summarised in Table 12 [133], where the concentration of mercury in the lyophilized samples has been converted to wet weight.

The average water losses of lyophilized samples were 75.9% for fish and 74.3% for mussel. The results in Table 12 show no significant losses of mercury from the fish samples prepared by lyophilization. As can be seen from these results, lyophilization is more efficient than oven-drying. However, when a large number of samples is analysed, lyophilization is not practically convenient. In such cases oven-drying may be chosen when mercury analysis is not required.

IV. SEDIMENT

The main components of marine sediments are inorganic aluminosilicate minerals, which are usually accumulated on the sea floor by river and other geological activities, and also shells of marine organisms (mainly calcium carbonate and silica) [2]. Of course, some metal oxides/hydroxides or particulates which precipitate from seawater deposit onto the sea floor as sediments. The chemical compositions of the three principal types of sediments in the ocean are shown in Table 13 [134].

Most of the sediments found on the deep-sea floor are the mixtures of these three principal sediments. The study on the sediments in the oceans and seashores can provide important data related to geochemical, oceanographical or biological circulation and deposition of elements, formation and distribution of marine sediments, and exploitation of marine resources.

The major elements of marine sediments are still determined by classical gravimetric and volumetric methods. Atomic absorption spectrometric methods have been applied to sediment analysis, although ICP-AES, X-ray fluorescence spectrometry and neutron activation analysis are also used because of their advantages of direct multi-element analysis [2]. Chakrabarti et

References pp. 154–157

TABLE 12

A COMPARISON OF MERCURY CONCENTRATION IN BIOLOGICAL SAMPLES USING WET AND LYOPHILIZED SAMPLES [133]

Organisms	Hg concentration (μg g^{-1} wet weight)					
	Wet sample			Lyophilized sample		
	Number of samples	Range	Average ± s.d.	Number of samples	Range	average ± s.d.
PISCES (FISH)						
Synodontidae:						
Saurida sp.	9	0.05–0.65	0.24 ± 0.19	9	0.07–0.70	0.27 ± 0.20
Carangidae:						
Trachurus sp.	3	0.06–0.19	0.12 ± 0.07	3	0.07–0.19	0.12 ± 0.06
Mullidae:						
Mullus sp. 1	25	0.02–0.35	0.13 ± 0.06	25	0.03–0.23	0.15 ± 0.05
Mullus sp. 2	1	0.09	–	1	0.13	–
Sparidae:						
Boops sp.	2	0.07–0.08	0.08 ± 0.007	2	0.08–0.08	0.08 ± 0
Pagellus sp.	6	0.07–0.28	0.17 ± 0.07	6	0.08–0.28	0.18 ± 0.08
Sphyraenidae:						
Sphyraena sp.	2	0.08–0.16	0.12 ± 0.06	2	0.08–0.17	0.12 ± 0.06
Triglidae:						
Chelidonichthys sp.	1	0.09	–	1	0.13	–
MOLLUSCA						
Donacidae:						
Donax sp.	8	0.07–0.80	0.38 ± 0.29	8	0.08–0.93	0.37 ± 0.26

TABLE 13

CHEMICAL COMPOSITION OF THREE PRINCIPAL SEDIMENTS IN THE OCEAN (wt.%) [134]

Composition	Red clay	Calcareous ooze	Siliceous ooze
SiO_2	53.93	24.23	67.36
TiO_2	0.96	0.25	0.59
Al_2O_3	17.46	6.60	11.33
Fe_2O_3	8.53	2.43	3.40
FeO	0.45	0.64	1.42
MnO	0.78	0.31	0.19
CaO	1.34	0.20	0.86
MgO	4.35	1.07	1.71
Na_2O	1.27	0.75	1.64
K_2O	3.65	1.40	2.15
P_2O_5	0.09	0.10	0.10
H_2O	6.30	3.31	6.33
$CaCO_3$	0.39	56.73	1.52
$MgCO_3$	0.44	1.78	1.21
C	0.13	0.30	0.26
N	0.016	0.017	–

al. [11] and Van Loon [135] have summarised the use of flame and ETA-AAS. Atomic absorption methods as well as classical analytical methods require acid wet digestion or fusion with adequate salts. These sample treatments are not easy, particularly when trace elements are analysed. In addition, variable and complicated compositions of major elements cause severe matrix interference with the determination of trace metals in marine sediments [136–138].

V. CONCLUSION

This chapter has briefly described the present status of AAS applied to marine analysis, particularly for seawater, marine organisms and sediments. Atomic absorption spectrometry has been increasingly applied to the compositional analysis of marine samples. Problems have been, and will be met in each experimental step, sampling, storage and standardisation of analytical methods, even though AAS is one of the best suited, most sensitive, precise and selective methods for such purpose. The oceans hold limitless future mineralogical and biological resources. Therefore, accurate data related to the oceans from the aspects of marine geology, chemistry, physics and biology is required for exploitation. Atomic absorption spectrometry will no doubt contribute to the future development of marine science and improvement of global environmental problems in estuarine, coastal and open sea.

References pp. 154–157

154

REFERENCES

1 E.C. Goldberg, W.S. Broecker, G.M. Gross and K.K. Turekian, Radioactivity in the Marine Environment, National Academy of Sciences, Washington, D.C., 1971.
2 G. Thompson, in E.L. Grove (Ed.), Applied Atomic Spectroscopy, Vol. 1, Plenum Press, New York, N.Y., 1978, pp. 273–300.
3 J. Van Loon, Selective Method of Trace Metal Analysis, Wiley, New York, N.Y., 1987.
4 A. Montaser and D.W. Golightly (Eds.), Inductively Coupled Plasmas in Analytical Atomic Spectrometry, VCH Publishers, New York, N.Y., 1987.
5 P.W.J.M. Boumans (Ed.), Inductively Coupled Plasma Emission Spectrometry, Parts I and II, Wiley, New York, N.Y., 1987.
6 A. Walsh, Spectrochim. Acta, 7 (1955) 108.
7 T.R.S. Wilson, in J.P. Riley and G. Skirrow (Eds.), Chemical Oceanography, 2nd ed., Vol. 1, Academic Press, New York, N.Y., 1975.
8 M.S. Quinby-Hunt and K.K. Turekian, EOS, 64 (1983) 130.
9 J.B. Bewer, J. Dalziel, P.A. Yeasts and J.L. Barron, Mar. Chem., 10 (1881) 173.
10 K. Sugawara, Deep-Sea Res., 25 (1978) 323.
11 C.L. Chakrabarti, K.S. Subramanian and T. Nakahara, The Application of Atomic Absorption Spectrometry to the Analysis of Trace Metals in Non-Biological Marine Samples, Rept. 6 (NRCC No. 17530), National Research Council of Canada, 1978.
12 J.D. Winefordner, J.J. Fitzgerald and N. Omenetto, Appl. Spectrosc., 29 (1975) 369.
13 B.V. L'vov, Atomic Absorption Spectrochemical Analysis (translated by J.H. Dixon), Adam Hilger, London, 1970.
14 J.P. Riley, in J.P. Riley and G. Skirrow (Eds.), Chemical Oceanography, Vol. 3, 2nd ed., Academic Press, New York, N.Y., 1975.
15 K.W. Bruland, R.P. Franks, G.A. Knauer and J.H. Martin, Anal. Chim. Acta, 105 (1979) 233.
16 D.A. Segar and G.A. Berberian, in T.R.P. Gibb, Jr. (Ed.), Analytical Methods in Oceanography, Adv. Chem. Ser. No. 147, American Chemical Society, Washington, D.C., 1975.
17 K.T. Marvin, R.P. Proctor and R.A. Neal, Limnol. Oceanogr., 15 (1970) 320.
18 W.R. Hirsbrunner and P.J. Wangersky, Mar. Chem., 3 (1975) 55.
19 G.G. Eichholz, A.E. Nagel and R.M. Hughes, Anal. Chem., 37 (1965) 863.
20 D.E. Robertson, Anal. Chim. Acta, 42 (1968) 533.
21 R.A. Durst and B.T. Duhart, Anal. Chem., 42 (1970) 1002.
22 W.G. King, J.M. Rodriguez and C.M. Wai, Anal. Chem., 46 (1974) 771.
23 R.D. Edigar, At. Absorpt. Newsl., 12 (1973) 151.
24 R.V. Coyne and J.A. Collins, Anal. Chem., 44 (1972) 1093.
25 C. Feldman, Anal. Chem., 46 (1974) 99.
26 K.S. Subramanian, C.L. Chakrabarti, J.E. Sueiras and I.S. Maines, Anal. Chem., 50 (1978) 444.
27 H. Scheurmann and H. Hartkamp, Frezenius Z. Anal. Chem., 315 (1983) 430.
28 B. Kinsella and R.L. Willix, Anal. Chem., 55 (1982) 2614.
29 J. Lo and C. Wai, Anal. Chem., 47 (1975) 1869.
30 M. Bothner and D. Robertson, Anal. Chem., 47 (1975) 592.
31 B.J. Keller, M.E. Peden and A. Rattonetti, Anal. Chem., 56 (1984) 2617.
32 Y. Dokiya, H. Ashikawa and K. Fuwa, Spectrosc. Lett., 7 (1974) 551.
33 M. Kimura and T. Arikado, Bunseki Kagaku, 32 (1983) E157.

34 K. Matsunaga, M. Nishimura and S. Konishi, Nature, 258 (1975) 224.
35 R. Guevremont, Anal. Chem., 52 (1980) 1574.
36 R. Guevremont, Anal. Chem., 53 (1981) 911.
37 E. Pruszkowska, G.R. Carnrick and W. Slavin, Anal. Chem., 53 (1981) 182.
38 G.R. Carnrick, W. Slavin and D.C. Manning, Anal. Chem., 53 (1981) 1866.
39 M.S. Cresser, Solvent Extraction in Flame Spectroscopic Analysis, Butterworths, London, 1978.
40 R.P. Brooks, B.J. Presley and I.R. Kaplan, Talanta, 14 (1967) 809.
41 T.K. Jan and D.R. Young, Anal. Chem., 50 (1978) 1250.
42 J.P. Riley and D. Taylor, Anal. Chim. Acta, 40 (1968) 470.
43 L.G. Danielsson, B. Magnusson, S. Westerlund and K. Zhang, Anal. Chim. Acta, 144 (1982) 183.
44 B. Belaughter, At. Absorpt. Newsl., 4 (1968) 273.
45 Y.K. Chau, S.S. Sim and Y.H. Wong, Anal. Chim. Acta, 43 (1968) 13.
46 T.R. Gilvert and A.M. Clay, Anal. Chim. Acta, 67 (1973) 289.
47 K. Hiiro, T. Owa, M. Takaoka, T. Tanaka and A. Kawahara, Bunseki Kagaku, 25 (1976) 122.
48 J.M. Lo, J.C. Yu, F.I. Hatchson and C.M. Wal, Anal. Chem., 54 (1982) 2536.
49 G.P. Klinkhammer, Anal. Chem., 52 (1980) 117.
50 S. Gohda, H. Yamazaki and T. Shigematsu, Anal. Sci., 2 (1986) 37.
51 R.G. Smith, Jr. and H.L. Windom, Anal. Chim. Acta, 113 (1980) 39, 48.
52 T. Akagi, Y. Nojiri, M. Matsui and H. Haraguchi, Appl. Spectrosc., 39 (1985) 662.
53 T. Akagi, K. Fuwa and H. Haraguchi, Anal. Chim. Acta, 177 (1985) 139.
54 A. Sato and N. Saitoh, Bunseki Kagaku, 25 (1976) 663.
55 Y.K. Chau and P.Y. Wong, Talanta, 15 (1968) 867.
56 T. Owa, K. Hiiro and T. Tanaka, Bunseki Kagaku, 21 (1968) 878.
57 K. Fujiwara, T. Morikawa and K. Fuwa, Bunseki Kagaku, 35 (1985) 361.
58 P. Burba and P.G. Willmer, Frezenius Z. Anal. Chem., 324 (1986) 298.
59 N. Rothstein and H. Zeitlin, Anal. Lett., 9 (1976) 461.
60 M. Hiraide, Y. Yoshida and A. Mizuike, Anal. Chim. Acta, 81 (1976) 185.
61 L.M. Cabezon, M. Caballero, R. Cela and J.A. Perez-Bustamante, Talanta, (1984) 597.
62 M.O. Andreae, Anal. Chem., 56 (1984) 2064.
63 D.S. Lee, Anal. Chem., 54 (1982) 1682.
64 R.R. Brooks, Analyst (London), 85 (1960) 745.
65 A.D. Matthews and J.P. Riley, Anal. Chim. Acta, 51 (1970) 455.
66 K. Hiiro, A. Kawahara and T. Tanaka, Bunseki Kagaku, 22 (1973) 1210.
67 J.P. Riley and D. Taylor, Anal. Chim. Acta, 40 (1968) 175.
68 J.P. Riley and D. Taylor, Anal. Chim. Acta, 40 (1968) 479.
69 H.L. Windom and R.G. Smith, Deep-Sea Res., 19 (1972) 727.
70 J.P. Riley and D. Taylor, Deep-Sea Res., 19 (1972) 307.
71 A. Sato, T. Oikawa and N. Saitoh, Bunseki Kagaku, 24 (1975) 584.
72 C.H. Van der Weijden, Chem. Geol., 18 (1976) 65.
73 R. Franche, F. Baffi, A. Dadone and G. Zainicchi, Mar. Chem., 4 (1976) 365.
74 R.E. Sturgeon, S.S. Berman, A. Desaulinier and D.S. Russell, Talanta, 27 (1980) 85.
75 C.J. Chen, T. Akagi and H. Haraguchi, Bull. Chem. Soc. Jpn., 58 (1985) 3229.
76 J.C. Chu, T. Akagi and H. Haraguchi, Anal. Chim. Acta, 198 (1987) 173.
77 M.I. Abdullah, O.A. El-Rayis and J.P. Riley, Anal. Chim. Acta, 84 (1976) 363.

156

78   U. Eisner and H.B. Mark, Jr., Talanta, 16 (1969) 27.
79   T.M. Florence and G.E. Batley, Talanta, 23 (1976) 179.
80   R.E. Sturgeon, S.S. Berman and S.N. Willie, Talanta, 29 (1982) 167.
81   R.E. Sturgeon, S.N. Willie and S.S. Berman, Anal. Chem., 57 (1985) 6.
82   J.N. King and J.S. Fritz, Anal. Chem., 57 (1985) 1016, 80.
83   R.A.A. Muzzarelli and M. Marinelli, Anal. Chim. Acta, 64 (1973) 371.
84   R.A.A. Muzzarelli and M. Marinelli, Anal. Chim. Acta, 69 (1974) 35.
85   R.A.A. Muzzarelli and M. Marinelli, Anal. Chim. Acta, 70 (1974) 283.
86   K.F. Sugawara, H.H. Weerall and G.D. Schucker, Anal. Chem., 46 (1974) 489.
87   H. Brandeuberger and H. Bader, At. Absorpt. Newsl., 6 (1967) 101.
88   C. Fairless and A.J. Bard, Anal. Lett., 5 (1972) 433.
89   C. Fairless and A.J. Bard, Anal. Chem., 45 (1973) 2289.
90   F.O. Jensen, J. Dolezal and F.J. Langmyhr, Anal. Chim. Acta, 72 (1974) 245.
91   J. Olafsson, Anal. Chim. Acta, 68 (1974) 207.
92   M.P. Newton and D.G. Davis, Anal. Lett., 8 (1975) 729.
93   X. Bo-Xing, X. Tong-Ming, S. Ming-Neong and F. Yu-Zhi, Talanta, 32 (1985) 1016.
94   W. Lund and B.V. Larsen, Anal. Chim. Acta, 70 (1974) 229.
95   W. Lund and B.V. Larsen, Anal. Chim. Acta, 72 (1974) 57.
96   E.J. Czobik and J.P. Matousek, Spectrochim. Acta, 35B (1980) 741.
97   G.E. Batley and J.P. Matousek, Anal. Chem., 49 (1977) 2031.
98   W.R. Hatch and W.L. Ott, Anal. Chem., 40 (1968) 2085.
99   K. Tanabe, J. Takahashi, H. Haraguchi and K. Fuwa, Anal. Chem., 52 (1980) 453.
100 S. Yamazaki, Y. Dokiya, T. Watanabe and K. Fuwa, Ecotox. Environ. Safety, 2 (1978) 1.
101 K. Matsunaga, M. Nishimura and S. Kobishi, Nature, 258 (1975) 224.
102 J. Takahashi, H. Haraguchi and K. Fuwa, Chem. Lett., (1981) 7.
103 M. Fujita, K. Iwashima, I. Fukuoka, E. Takabatake and N. Yamagata, Suishitsu Odaku Kenkyu, 1 (1978) 133.
104 R.A. Carr, J.B. Hoover and P.E. Wilkniss, Deep-Sea Res., 19 (1972) 747.
105 S. Nishi, Y. Horimoto and N. Nalano, Bunseki Kagaku, 23 (1974) 386.
106 P. Neske, A. Hellwing, L. Dornheim and B. Triene, Frezenius Z. Anal. Chem., 318 (1984) 498.
107 D.H. Anderson, J.H. Evans, J.J. Murphy, and W.W. White, Anal. Chem., 43 (1971) 1511.
108 M. Nishimura, K. Matsunaga and S. Konishi, Bunseki Kagaku, 24 (1975) 655.
109 W.F. Fitzgerald, W.B. Lyons and C.D. Hunt, Anal. Chem., 46 (1974) 1882.
110 H. Haraguchi, J. Takahashi, K. Tanabe, Y. Akai, A. Homma and K. Fuwa, Bunseki Kagaku, 29 (1980) 348.
111 H. Haraguchi, J. Takahashi, K. Tanabe and K. Fuwa, Spectrochim. Acta, 36B (1981) 719.
112 R.S. Braman and C.C. Foreback, Science, 182 (1973) 1247.
113 Y. Talmi and D.T. Bostick, Anal. Chem., 47 (1975) 2145.
114 J.S. Edmonds and K.A. Francesconi, Anal. Chem., 48 (1976) 2019.
115 M.O. Andreae, Anal. Chem., 49 (1977) 820.
116 S. Tanaka, M. Kaneko, Y. Konno and Y. Hashimoto, Bunseki Kagaku, 33 (1983) 535.
117 R.S. Braman and M.A. Tompkins, Anal. Chem., 51 (1979) 12.
118 O.F.X. Donard, S. Rapsomanikis and J.H. Weber, Anal. Chem., 58 (1986) 772.

119   G.A. Hambrick III, P.N. Froelich, Jr., M.O. Andreae and B.L. Lewis, Anal. Chem., 56 (1984) 421.
120   M.O. Andreae, J.F. Asmode, P. Foster and L. Van't Dack, Anal. Chem., 53 (1981) 1766.
121   R.E. Sturgeon, S.N. Willie and S.S. Berman, J. Anal. At. Spectrom., 1 (1986) 115.
122   R.E. Sturgeon, S.N. Willie and S.S. Berman, Anal. Chem., 57 (1985) 2311.
123   S.N. Willie, R.E. Sturgeon and S.S. Berman, Anal. Chem., 58 (1986) 1140.
124   M. Yamamoto, M. Yasuda and Y. Yamamoto, Anal. Chem., 57 (1985) 1382.
125   H.J.M. Bowen, Trace Elements in Biochemistry, Academic Press, New York, N.Y., 1966.
126   V.T. Bowen, J.S. Olsen, C.L. Osterberg and J. Ravera, Radioactivity in the Marine Environment, National Academy of Science, Washington, D.C., 1971.
127   D.B. Carlisle, Nature, 181 (1958) 922.
128   G.D. Nicholls, H. Curl and V.T. Bowen, Limnol. Oceanogr., 5 (1960) 472.
129   K.K. Pillay, C.C. Thomas, Jr., J.A. Sondel and C.M. Hyche, Anal. Chem., 43 (1971) 1419.
130   R. Litman, H.L. Finston and E.T. Williams, Anal. Chem., 47 (1975) 2364.
131   P.D. LaFleur, Anal. Chem., 45 (1973) 1534.
132   M. Freidman, E. Miller and J. Tanner, Anal. Chem., 46 (1974) 236.
133   G. Ramelow and H. Hornung, At. Absorpt. Newsl., 17 (1978) 59.
134   S.K. El Wakeel and J.P. Riley, Geochim. Cosmochim. Acta, 25 (1961) 110.
135   J.C. Van Loon, Spectrochim. Acta, 38B (1983) 1509.
136   K.W. Bruland, K. Bertine, M. Koide and E.D. Goldberg, Environ. Sci. Technol., 8 (1974) 425.
137   V. Talbot, R.J. Magee and M. Hussain, Mar. Pollut. Bull., 7 (1976) 53.
138   R.T.T. Rantala and D.H. Loring, At. Absorpt. Newsl., 14 (1975) 117.

Chapter 4c

Analysis of Airborne Particles in Workplace Atmospheres

J.C. SEPTON

United States Department of Labour, Occupational Safety and Health Administration, Salt Lake City Analytical Laboratory, P.O. Box 65200, Salt Lake City, Utah 84165-0200 (U.S.A.)

I. INTRODUCTION

In this chapter, trace element analysis of air samples by atomic absorption spectrometry (AAS) will be surveyed. Sampling techniques and analytical methods used to evaluate workplace environments will be presented.

A review of the advantages offered by AAS as an analytical technique in occupational health laboratories was first discussed in 1965. At that time, AAS could be used to analyse 40 elements at concentrations below 1 microgram per millilitre (μg ml^{-1}) [1]. Flame and flameless AAS techniques have been used since the early 1970's to analyse metals in environmental samples [2, 3].

Advantages of AAS include minimal sample preparation, good selectivity and relatively simple operating procedures. The techniques are readily adaptable to the measurement of elements in biologic media and combine good sensitivity with reasonably low cost. Moreover, flame AAS is relatively free from sample carry-over and memory effects. Samples are easily nebulized and a wide range of operating conditions is available [4]. Another advantage of AAS over classical techniques is improved sensitivity. This has been exploited for the analysis of environmental air samples by using flameless AAS techniques such as electrothermal atomization (ETA), hydride generation atomic absorption spectrometry (HG-AAS), or cold vapour atomic absorption (CVAA).

In addition, special speciation techniques have also been developed using chromatographic separations with atomic absorption (AA) detection. Instrumental parameter guidelines for AA analysis of samples taken for industrial hygiene applications may be found in many publications. Examples of such industrial hygiene related manuals published in the United States include those from the National Institute for Occupational Safety and Health (NIOSH) [5, 6] and the Occupational Safety and Health Administration (OSHA) [7]. The majority of the instrumental operating parameters listed in these publications apply to almost any type of sample matrix and may be found in numerous instrument manufacturers' operating instructions.

In addition, NIOSH has "ruggedized" their AAS procedures for Be, Cd,

References pp. 186–190

Co, Cr, Cu, Mn, Mo, Ni and Pb and collaboratively tested their methods for Cd, Co, Cr, Ni and Pb [8]. Therefore, the material presented within this chapter will emphasize special applications unique to the sampling and analysis of environmental air samples, instead of elaborating upon routine instrumental parameters or operations.

II. SAMPLE COLLECTION AND PREPARATION

Sample collection techniques are an integral part of the analytical process for determining analytes in workplace environments. In the United States, OSHA takes workplace air samples in the employee breathing zone to determine compliance. Therefore, sampling devices ideally should be small and portable so as not to interfere with employee duties. Generally, for inorganic particulates and aerosols determined by AAS, the collection medium is a filter contained in a sampling cassette. For analytes that are in a gaseous or vapour state, the collection medium is usually a solid sorbent material.

The type of collection medium determines the analytical preparation required for the instrumental determination. Because of this requirement, specialized collection procedures are included in the specific analytical sections of this chapter. Generalized sample collection and preparation procedures are outlined below.

A. Sample collection

The most commonly used sampling device for the collection of industrial hygiene-related air samples for use in AAS, consists of a 2- or 3-piece filter cassette containing an 0.8-μm pore size mixed cellulose ester membrane filter supported by a cellulose backup pad. Other micro-porous films which have been used contain polymers of polyvinyl chloride or acrylonitrile. A personal sampling pump is connected to the cassette to draw workplace air through the membrane filter. This cassette is usually placed within the breathing zone of a worker. The pump sampling rate is commonly set in the range of 1.5–2.5 litres per minute (l min^{-1}). The sampling pump should be capable of achieving a flow rate throughout the sampling period with an error of $\leq \pm 5\%$.

B. Sample preparation

The most common procedure for removing the analyte from the filter is nitric acid digestion; therefore, most of the samples determined by AA will be in a dilute acid matrix. If necessary, other acids are used to aid in sample solubilisation. Matrix modifiers are also added when determining certain analytes to minimize interferences in the AA analysis.

To determine contamination or background corrections, field and laboratory blanks are analysed along with the samples. The field blank is a filter
that is handled in exactly the same manner as the samples (i.e., placed in a
cassette) except that no air is drawn through it. The laboratory blank is a
clean, unused filter which has not been handled in the field.

III. EXPOSURE LIMIT VALUES

Illness or death can result from hazardous exposure to gases, vapors,
or dusts in the workplace. Exposures to toxic substances are evaluated by
comparison to applicable standards or guidelines. The reference standards
or guidelines vary in different countries. Some of the reference guidelines or
standards used in the United States are: Maximal Allowable Concentrations
(MACS); Threshold Limit Values (TLVs); Permissible Exposure Limits
(PELs); and Recommended Exposure Limits (RELs).

In 1941 the American Conference of Governmental Industrial Hygienists
(ACGIH) set maximal allowable concentrations (MAC) for atmospheric
contaminants in the workplace. These exposure guidelines were eventually
used in industry. However, the phrase "MAC" presented some amibiguity in
the application of these guidelines. For example, these values were based
upon time-weighted averages rather than the implied maximal ceiling values.
In addition, the phrase suggested that such concentrations were "allowable".
To minimize confusion, the term "MAC" was changed to "Threshold Limit
Value" in the United States.

The ACGIH annually publishes a list of TLVs. Threshold Limit Values
refer to airborne concentrations of substances and represent conditions
under which it is believed that nearly all workers may be repeatedly exposed
day after day without adverse effect. These limits are intended for use in
the practice of industrial hygiene as guidelines or recommendations in the
control of potential health hazards [9].

In connection with the enforcement of the United States Occupational
Safety and Health Act, legal standards for exposure to gases, vapors, dust,
noise, and ionizing and non-ionizing radiation are published in the Code
of Federal Regulations (CFR). These legal limits are known as Permissible
Exposure Limits (PELs). Most of the PELs were adopted from ACGIH
TLVs. About 20 PELs were adopted from the American National Standards
Institute (ANSI) Z-37 standards. More recently, the OSHA standards have
been revised to contain most of the current ACGIH TLVs and some RELs
[10]. Many RELs can be found in NIOSH Criteria Documents.

In order to understand the problems unique to the application of AA
analysis of industrial hygiene samples, a background knowledge of applicable
exposure limit values for workplace air is necessary. Based upon the standard
and the air volume of the collected sample, it is possible to determine many
of the analytical parameters required for an analysis.

TABLE 1

EXAMPLES OF EXPOSURE LIMIT STANDARDS

Compounds	PEL	TLV	REL
Aluminium, welding fumes	5	5	
Aluminum, soluble salts and alkyls	2	2	
Antimony and compounds	0.5	0.5	0.5
Arsenic and compounds	0.01 (inorganic) 0.5 (organic)	0.2	0.002 STEL
Arsine	0.2 (as AsH_3)	0.2 (as AsH_3)	0.002 STEL (as As)
Barium, soluble compounds	0.5	0.5	
Beryllium and compounds	0.002 0.025C 0.005 (STEL 30 min)	0.002	0.0005
Cadmium dust and salts	*	0.05	Reduce exposure to lowest feasible limit
Cadmium oxide fume	*	0.05C	Reduce exposure to lowest feasible limit
Chromium (VI), chromic acid and chromates	0.1 (as CrO_3)	0.05 (as Cr)	0.025 (as Cr) 0.05C (non-carc.) 0.001 (carcinogenic)
Chromium (II) and (III) compounds	0.5	0.5	
Chromium (metal)	1	0.5	
Cobalt	0.05	0.05	
Copper dusts and mists	1	1	
Copper fume	0.1	0.2	
Hafnium	0.5	0.5	
Indium and compounds	0.1	0.1	
Iron salts (soluble)	1		
Lead (inorganic compounds, dust and fume)	0.05	0.15	0.1
Manganese compounds	5C	5	
Manganese fume	1 3C	1 3 STEL	
Mercury, vapor	0.05	0.05	0.05
Mercury (alkyl compounds)	0.01 0.03 STEL	0.01 0.03 STEL	
Mercury (aryl and inorganic compounds)	0.1C	0.1	
Molybdenum, insoluble compounds respirable fraction	10 5	10	
Molybdenum, soluble compounds	5	5	
Nickel carbonyl	0.007	0.1	0.007
Nickel, metal and insoluble compounds	1	1	0.015
Nickel, soluble compounds	0.1	0.1	0.015
Osmium tetroxide	0.002 0.006 STEL	0.002 0.006 STEL	
Platinum metal	1	1	
Platinum, soluble salts	0.002	0.002	

TABLE 1 (continued)

Compounds	PEL	TLV	REL
Rhodium, fume and insoluble compounds	0.1	1	
Rhodium, soluble compounds	0.001	0.01	
Selenium compounds	0.2	0.2	
Silver metal dust and fume	0.01	0.1	
Silver, soluble compounds	0.01		
Stibine (as SbH$_3$)	0.5	0.5	
Tantalum	5	5	
Tellurium compounds	0.1	0.1	
Tetraethyl lead	0.075	0.1	
Tetramethyl lead	0.075	0.15	
Thallium, soluble compounds	0.1	0.1	
Tin, inorganic compounds except oxides	2	2	
Tin, organic compounds	0.1	0.1	0.1
Tin oxide	2	2	
Titanium dioxide, respirable fraction	10 (as TiO$_2$) 5	10 (as TiO$_2$)	
Tungsten, insoluble compounds	5 10 STEL	5 10 STEL	5
Tungsten, soluble compounds	1 3 STEL	11 3 STEL	
Uranium, insoluble compounds	0.2 0.6 STEL	0.2 0.6 STEL	
Uranium, soluble compounds	0.05	0.2 0.6 STEL	
Vanadium pentoxide, respirable dust and fume	0.05 (as V$_2$O$_5$)	0.05 (as V$_2$O$_5$)	0.05 (as V) STEL
Welding fumes (total particulate)	5	5	
Yttrium	1	1	
Zinc oxide dust (as ZnO)	10	10	
Zinc oxide respirable	5 (as ZnO)		
Zinc oxide fume (as ZnO)	5 10 STEL	55 10 STEL	15 STEL
Zinc chloride fume (as ZnCl$_2$)	1 2 STEL	1 2 STEL	
Zirconium compounds	5 10 STEL	5 10 STEL	

* = In rulemaking.

C = ceiling; this value should not be exceeded at any time.

STEL = short-term exposure limit; this value should not be exceeded during a specifed period of time. The time limit is 15 min unless otherwise noted.

All limit values are listed as mg m^{-3} and *as the element* unless otherwise noted (i.e. the OSHA PEL for nickel carbonyl is 0.007 mg m^{-3} as Ni).

These limits are listed to illustrate the relative sensitivities needed to properly analyse industrial hygiene samples. These limits are subject to change as more data regarding toxicity is accumulated and/or more sensitive analytical techniques are developed.

Some of these exposure limit values significant in AA analysis of industrial hygiene samples are listed within Table 1.

A. Analytical detection limit related to minimum air volume calculations

The detection limit of a given procedure or analytical technique effectively sets a minimum to the required volume of air which constitutes a sample. As a guideline, a satisfactory combination of sampling method and duration of sampling and flowrate should be chosen to achieve an adequate analytical detection limit. The sampling procedure and method of assessment should be sensitive enough to measure the concentration in air within at least one-tenth of the hygiene exposure limit given in Table 1. The relationship between minimum air volume and detection limit can be defined by:

$$A = \frac{10B}{CD}$$

where: A = minimum duration of sampling (min); B = analytical detection limit (μg); C = TLV or PEL (μg l^{-1}, from Table 1); D = flowrate (l min^{-1}).

The volume of air to be sampled can be calculated by:

$$E = AD$$

where E = volume of air to be sampled (l).

IV. TECHNIQUES FOR IMPROVED SENSITIVITY

Some elements determined in the workplace require the use of flameless techniques, such as electrothermal atomization or hydride generation, which provide improved sensitivity and require a minimum sample size. Some examples cited in the literature for environmental or workplace air samples that use such analytical techniques include those for Sb, As, Se and Bi [11–40].

Solvent extraction is another alternative for increasing analytical sensitivity for Cu and Cd, in addition to those elements with sensitivities improved by flameless techniques [41].

A. Electrothermal atomization

Electrothermal atomization or vaporization (ETA or ETV) is also known as graphite furnace (GF-AAS).

Mitchell et al. [11], in 1975, suggested that ETA-AAS should be used only to exploit its unique characteristic: excellent sensitivity with a small sample size. Otherwise, ETA-AAS methods are reported to be slow, interference-prone, and to require precise operator skills. ETA-AAS is prone to matrix effects; standard addition methods can be used to evaluate such interferences. A number of acids have been used in sample preparation for ETA-AAS analysis; however, Begnoche and Risby [3] have reported that aqua regia caused the graphite tube in their furnace to become unreliable after four or five determinations.

These problems have been overcome through improvements in technique, and ETA-AAS parameters have been sufficiently well characterized for routine determinations.

Air samples containing the elements As, Se, Sb [12] and their hydrides (arsine, hydrogen selenide and stibine) [20–22] can be routinely analyzed by graphite furnace techniques and are frequently reported in the literature.

1. *Arsenic, selenium and antimony*

Ashing temperatures and interferences for As, Sb, Se and Te in surface water and industrial effluents have been analysed using ETA-AAS [12]. A variety of acid matrices was investigated. Using a 1% HNO_3 matrix, ashing temperatures of 700°C for Te and As and 1000°C for Sb were found to cause no significant signal loss. Sensitivities were independent of atomization temperature for many metals. All acids used had a positive and concentration-dependent interference for As, Sb and Te; however, they suppressed Se.

(i) *Antimony*

A flame AAS method for determination of particulate Sb in air has been evaluated; however, the procedure suffered from limited sensitivity [13]. The sensitivity of Sb for flame AAS is typically in the order of 0.3–0.5 μg ml^{-1} in an aspirated solution for 1% absorbance (5–10 μg total Sb in the sample), limiting the lower end of the working range to about 0.05–1.0 mg m^{-3} for a 360-l air sample (approximately 18–36 μg/sample) [12, 16]. Sensitivities about 50–100 times better than this may be obtained with either HG-AAS or ETA-AAS.

The ETA-AAS method for Sb has a detection limit of about 5–15 nanograms (ng) for an ashed air filter sample [12, 14–16]. The results are more reproducible than flame methods if Sb concentrations are below 10 μg ml^{-1} [14].

A 50-l air sample is adequate for analysis as opposed to a 360-l air sample required for equivalent sensitivity with flame AAS.

(ii) *Arsenic*

Inorganic and organic arsenic have been determined through extraction separation techniques [17]. The sensitivity for the determination of arsenite was improved by pre-concentration into a small volume of organic solvent [18]. A graphite cloth ribbon enhanced the sensitivity of As when determined by ETA-AAS [19].

A procedure has been described for the determination of inorganic arsenic and organic arsenic in water and urine by ETA-AAS [17]. Samples of water or urine were heated with HCl and treated with iodide ion. Arsenic-iodide species were extracted into chloroform and then either re-extracted into deionized water for measurement of inorganic arsenic, or re-extracted into dilute dichromate solution for total arsenic determination; the difference provided an indirect measurement of the levels for organic arsenic. The lower detection limit for arsenic in water and urine was 10 parts per billion (ppb). The recoveries were: 87.0 and 93.0% for inorganic arsenic in water and urine, respectively; 92.3% for mixtures of inorganic and methylated arsenic (total arsenic) in water and urine; and 98.7 and 88.4% for dimethylarsenic in water and urine, respectively. This procedure avoids the isolation and transfer of arsine(s) and permits some differentiation between the inorganic and organic (methyl) arsenic content of a sample.

A method for the HG-AAS determination of arsenite from aqueous solutions has been reported [18]. Arsenite was extracted with hexane. The sensitivity for determining arsenite was increased several hundred times when arsenite was extracted preferentially into a small volume of organic solvent such as ammonium sec-butyldithiophosphate.

A method was described for determining As by AAS using a graphite cloth ribbon placed inside various types of graphite tubes [19]. Samples containing As were treated with ammonium pyrrolidine dithiocarbamate (APDC) solution. They were then extracted with a 1 : 1 solution of carbon tetrachloride and chloroform. The organic phase was separated and a 30-μl aliquot was taken for AA analysis. Absorbances were measured for 1.5 ng As atomized from non-pyrolytic graphite (NPG), pyrolytic graphite coated (PGC), or a pyrolytic graphite platform tube with or without a ribbon of graphite cloth placed inside. Matrices used included HCl, HNO_3, H_2SO_4, and APDC. The standard deviation of the measured absorbance values was less than 0.013. Higher sensitivity was generally obtained with NPG than with platform tubes. The graphite cloth ribbon enhanced the sensitivity even further. For each type of tube tested, the HNO_3 matrix gave the best sensitivity for As. The effects of Cu, Ni, Mo and Fe on As absorbance were evaluated using the same experimental protocol as before. Without the ribbon, these metals all gave enhancement effects except the NPG tube and the platform/organic extract combination. When the graphite ribbon was used, improved sensitivity was attained and the effects of the foreign metals were almost eliminated. No enhancement effect was observed for the APDC extract. It was concluded that using the graphite cloth ribbon and the HNO_3 matrix reduced the loss of As from the sample.

(iii) *Stibine or arsine*

Stibine can be collected using either a solid sorbent device containing mercuric chloride on silica gel, or an acidic mercuric chloride solution in an impinger [20]. Arsine can be collected on charcoal tubes or on a material comprised of a synthetic resin activated carbon [21, 22]. The collected gaseous hydrides can then be determined by HG-AAS as described below (see Section IV.B).

A sampling and analytical system has been developed involving the collection of arsine on charcoal in glass tubes followed by ETA analysis utilizing a heated graphite furnace. Arsine gas was measured at the concentration range of 1 to 10 micrograms per cubic meter ($\mu g\ m^{-3}$) [21]. The recoveries at all levels averaged between 85 and 90%. The detection limit (twice the noise level) was $0.3\ \mu g\ m^{-3}$ when using an air volume of 13.15 l and a 15-min sampling time. When using a Ni matrix modification procedure, the mean percent recoveries for arsine loading from 0.01 to 0.2 μg were within 5% for all levels. Thus, the efficiency of the collection desorption process did not vary significantly, and the same calibration procedure was applicable at all levels of arsine loading. A trace metal analysis was performed on 10 ml of 0.01 M HNO_3 solution after desorption of 1000 mg charcoal. Only Mg was detectable and then only at the 50-ng ml^{-1} level, which is unlikely to produce an enhancement or linearising effect.

A further method for determining airborne arsine was described in which air samples were passed through adsorbent tubes containing 150 or 200 mg of a synthetic resin activated carbon [22]. After adsorption, the samples were desorbed with 2 or 10 ml 0.01 M HNO_3 in 10% ethanol and heated at 75°C for 1 hour (hr). The HNO_3 extracts were then analyzed by ETA-AAS. Absorption capacities of the carbon were determined. A 150-mg quantity of carbon could adsorb 0.1 ppm arsine from a 5-l sample at 0 or 50% relative humidity (%RH) without breakthrough. A breakthrough of 2.03% occurred when a 5-l sample containing 0.5 ppm arsine was passed through 150 mg of adsorbent. Desorption efficiencies were on the order of 80 to 94%. Six replicate determinations of spiked solutions yielded relative coefficients of variation of around 3% for samples containing As at concentrations above 20 nanograms per millilitre (ng ml^{-1}). The detection limit was 5 ng ml^{-1}, which had a relative coefficient of variation of 12%. This corresponded to 0.003 ppm l^{-1} arsine in the air. The method was therefore able to determine arsine in air, with an upper limit of 0.5 ppm in a 4-l sample and the lower limit 0.003 ppm l^{-1}.

B. Hydride generation

The sensitivity for elements which form gaseous hydrides can be improved by using HG-AAS. The first application of HG-AAS techniques for metal hydrides used aluminum [23], or magnesium and titanium trichloride, or Zn as reductants [24–26]. Subsequently, sodium borohydride ($NaBH_4$) showed advantages as a reductant because of its wide applicability and usefulness in both solid and solution forms [27–29].

The most common analytes reported in the literature for HG-AAS analysis of workplace air samples include those for Sb, As, Se, Sn and Bi.

1. *Antimony, arsenic, selenium, tin and bismuth*

Various methods for generating and determining covalent hydrides using AAS have been developed [23, 25]. One method, for example, described the determination of sub-microgram levels of Sb, As and Se in natural waters [23]. Stibine, arsine and hydrogen selenide were produced from the samples in an automated system and passed to a tube furnace mounted in the light path of an AA spectrophotometer. The method is reported to be capable of analyzing 40 samples per hour with a limit of detection of 0.1 μg l^{-1} for As and Se, and 0.5 μg l^{-1} for Sb.

Sub-microgram concentrations of As, Se, Sb, Bi and Te have been determined using the TiCl$_3$-Mg method [25], whilst the determination of As, Bi, Ge, Pb, Sb, Se, Sn and Te by means of HG-AAS has been described [28].

Methods for determining As, Se, Sb and Bi using NaBH$_4$ as a reducing agent to convert these elements into their corresponding hydrides have been developed [29, 33]. The control of pH was reported to be critical when determining Se and Bi.

An analytical method has been reported for the determination of As, Se and Sb in urine and in air samples [30]. Using this method, urine and air particulate samples were wet-ashed with a mixture of HNO$_3$, HClO$_4$ and H$_2$SO$_4$ to destroy the organic matrix. Analysis was subsequently made by generating the hydride in an acidic solution with NaBH$_4$ and measuring the evolved gas by background corrected AAS. In the absence of interfering metals, the hydride generation measurement system was highly sensitive. For a 60-l air sample and using optimum conditions, the optimum range extended from 0.4 to 50 μg m^{-3} for arsenic and selenium, and from 4 to 500 μg m^{-3} for Sb. This procedure determined 5 ng of As and Se, and 50 ng of Sb with a relative standard deviation of 20% or better if interferences were not present.

Two techniques to determine Se at nanogram levels were compared [31]. The two techniques, hydride generation and AAS, utilize the same chemical reaction but different methods of atomization. Both techniques were found to be sensitive and precise in determining nanogram quantities of Se. The hydride generation assembly yielded a more rapid analysis. The AAS system was considerably cheaper to operate, slightly more sensitive in detection, less affected by the choice of acid type, and had a wider linear response range.

Manual and automated methods for As, Se, Sb, Bi and Sn were evaluated when using NaBH$_4$ as a reducing agent [32]. The limits of determination were respectively 100, 50, 50, 50 and 50 ng ml^{-1}; the detection limits using the automated system were 2.0, 0.8, 12.0, 10.0 and 1.0 ng for As, Se, Bi, Sb and Sn, respectively.

169

An interference study involving 48 interfering elements was carried out during the determination of As, Bi, Ge, Sb, Se, Sn and Te [33]. This study demonstrated that significant interferences occur in many instances; however, the sensitivity and detection limits for the elements listed, increased when using the HG-AAS technique. An exception was Sn, in which high blank values were noted due to the presence of Sn in the $NaBH_4$ reagent.

(i) *Antimony*

Methods for the HG-AAS determination of Sb have been reported [26–28] with improvement in sensitivity resulting from modification of fuel conditions [27] or increased path-length [28]. The sensitivity for 1% absorption was reported to be 0.004 μg ml^{-1} Sb when using HG-AAS [26]. A sensitivity of 10 ng Sb was reported when using an argon–hydrogen burner system [27]. When increasing the path-length of the atom cell, a sensitivity of 0.6 ng Sb was reported [28].

(ii) *Tin*

The determination of Sn in atmospheric particulate matter has been carried out after sample collection on a Whatman 41 filter paper by continuous hydride generation and AAS [34]. Sulphuric and nitric acids were used to decompose the samples, and hydrofluoric acid was used to dissolve residual silicates. For actual samples, the practical detection limit was 4 ng.

(iii) *Arsenic*

Detailed methods for the generation of arsine gas can be found in the literature [35–37]. Analytical methods for the collection and analysis of inorganic [38] and organic [39] arsenic species in air by hydride generation have been described. A hydride generation system was developed for the AA determination of As [35]. In this method a sodium tetrahydroborate solution was injected into acidified sample solutions held in ordinary boiling tubes. The between-sample reproducibility was reported to be better than 1% RSD at 0.5 μg of As per sample, and the limit of detection was approximately 6 ng.

A method for the generation of arsine using a hypodermic syringe described the determination of As generated as arsine in a 50-ml disposable syringe [36]. A silica absorption cell was heated to between 700 and 800°C using Chromel-C wire. Samples containing siliceous material were fused with potassium hydroxide in Ni crucibles at 550°C, and then made acidic with HCl. Organic samples were digested with concentrated H_2SO_4 at 350°C, made clear with the addition of hydrogen peroxide, and then HCl was added. Samples containing 1 to 500 μg of As in 5-ml volumes of 10% HCl were loaded into a syringe, then 1.5 ml of 2% $NaBH_4$ solution was drawn up. After a 20 second shaking period, the arsine was injected into the furnace.

References pp. 186–190

The optimum hydrogen flow rate was 75 ml min^{-1}; whilst for nitrogen it was 225 ml min^{-1}. Peak heights gave better precision, whereas peak area measurements gave better sensitivity. The most likely interferences were from Cu, Ni, Fe, Sb and Se. It was reported that other cations present were not expected to cause an interference. By using a 1% NaBH$_4$ solution, the tolerance levels for Cr, Cu, Ni, Co, and Ag could be almost doubled. It was concluded that this method was simpler and less expensive than peristaltic pumping systems.

Another arsine generation technique was described which could determine As at levels of 5 μg [37]. This procedure could also be used for the determination of Se and Sb as their respective hydrides. In addition, the generator could be used in the determination of Hg by flameless AA.

The determination of inorganic As compounds in air is described in a method in which arsenic trioxide vapour was collected on a paper filter treated with NaOH solution [38]. Particulate As dust and fume were collected separately on a mixed cellulose ester filter or together with arsenic trioxide vapor on the treated paper filter. Filters were then treated with aqua regia to oxidize and dissolve the As compounds. The solution was then analyzed for As using HG-AAS either directly or coupled with a standard additions method. This method was used to determine As concentrations in inorganic dusts, fumes and vapors in the workplace air. Sampling for as little as 10 min or for as long as 8 hr. The method had a lower analytical detection limit of 0.005 mg m^{-3}. Interference with the evolution of arsenic hydride may result from the presence of Sb, Bi, Ge, Pb, Se, Te, Sn, Cu, Ni and Pt. The precision of the method was greater than 10% for samples of a minimum of 1 hr and concentrations ranging from 0.005 to 0.5 mg m^{-3}.

A method for detecting and identifying organo-arsenical compounds in air described sample collection methods for particulate and vapor forms of methylarsonic-acid (MMA), dimethylarsenic-acid (DMA) and para-aminophenylarsonic-acid (p-APA) [39]. The detection system which utilized continuous hydride generation followed by heated quartz furnace atomization and AAS was described. The method was very precise and accurate for concentrations ranging from 5 to 20 μg m^{-3}, with standard deviations from about 14.4 to 4.7%.

C. Solvent extraction

Enhancement of AA sensitivity has been reported for Cu, Sb, Cd, As and Se by organic solvent extraction of the aqueous solution and aspiration of the organic phase [40]. The organic phase may, however, be back-extracted with an aqueous solution and the aqueous phase aspirated. Detection limits for this type of methodology have been reported to be: Cu 0.01 ng ml^{-1}, Cd 0.1 ng ml^{-1}, Sb 10 ng ml^{-1}, As 20 ng ml^{-1}, and Se 2 ng ml^{-1}.

V. SPECIFICITY

Since AAS instrumental determinations are specifically elemental and many of the hygiene exposure limits are regulated as compounds, some additional techniques may be required during sampling and analysis to aid in the characterization of the sample.

A. Specificity from particle size analysis

These characterizations include such considerations as evaluating the composition of the sample to determine if the analyte is in the form of a dust or fume. This may be accomplished by: (1) observation by the person conducting the sampling of the industrial operation to aid in determining the form of the sample; (2) use of a particle size fractionator (such as a cascade impactor or Anderson sampler). Samples collected according to particle size diameters have been reported to aid in the characterisation of particles. Each sample fraction can then be determined by AAS to evaluate the mass of analyte present for each particle size range. The form of the sample can then be evaluated by comparing the particle size [41].

A method has been developed to separate the fraction ofindustrial dusts less than 10 μm [42]. This was accomplished using a 10-μm sieve which was placed into an ultrasonic transducer container with 50 ml of alcohol. This method reportedly allowed the determination of the true chemical nature of the respirable fraction.

B. Specificity from chromatographic separation–atomic absorption detection

The toxicity of a metal may be dependent upon its chemical form or species in the environment; i.e., whether the metal is an organometallic or an inorganic compound. The determination of this speciation is an important application in the field of industrial hygiene, particularly for ambient-air sampling and analysis. The most prevalent technique for accomplishing this task is the coupling of chromatographic and atomic spectroscopic techniques. This allows for the separation of the sample into its various components and subsequent detection of the metal species by AAS.

General techniques
Overviews of instrumentation reported in the literature include chromatographic coupling to AA spectrometers. The chromatographs include gas (GC), liquid (LC), high pressure liquid (HPLC), and ion separation. The AA spectrometers can be equipped with graphite furnaces and Zeeman background correctors .

References pp. 186–190

The use of Zeeman correction, interfaced with an HPLC, was considered superior to GC, because it is applicable to the separation of polar and thermally unstable high molecular weight compounds [43].

Other chromatographic methods used for metal speciation studies have been reviewed [44]. These methods included LC and GC columns connected to an AA burner interface or to an ETA-AAS interface, and a GC column connected to a cold vapor tube or microwave plasma emission atomizer interface. The detector of choice in most cases was the AA spectrometer.

C. Organometallic compounds

The use of a commercial automated ETA-AAS detector coupled to an HPLC was demonstrated to provide element-specific separation and detection of organometallic compounds at nanogram levels in both polar and nonpolar solvents using conventional columns [45]. Sensitivities of the system for compounds of As, Pb, Hg and Sn were shown to be mainly functions of liquid chromatograph (LC) flow rate and relative AA sensitivity for each element.

Some applications for the use of a chromatographic separation–atomic absorption detection technique have been reported for the environmental analysis of compounds of Pb, Sn, As, Hg, Se, P and Cr.

1. *Organolead and organotin compounds*

Modification of a trace element monitoring technique for the analysis of organolead and organotin using high resolution LC and ETA-AAS, was studied using the following procedures [46]: iodine was used to digest alkyllead compounds; zirconium-treated graphite cuvettes were used in the graphite furnace to analyse organotin compounds. Mixtures of tetramethyllead, tetraethyllead (TEL), trimethylethyllead (TMEL), diethyldimethyllead (DEDML), and triethylmethyllead (TEML) were prepared from stock solutions, chromatographed, and analyzed in the graphite furnace with and without the addition of iodine to the solutions. Solutions of organotin were prepared from tetraphenyltin (TPT), tetrabutyltin (TBT), and dibutyldichlorotin (DBDCT) and analysed with untreated and zirconium carbide-treated graphite cup cuvettes. Incomplete molecular decomposition and metal carbide formation (which produce reduced AA signals) were considerably overcome by the use of iodine in the analysis of organolead compounds and zirconium-treated graphite cuvettes in the analysis of organotin compounds. Improved precision of atomization resulted in both cases.

(i) *Organolead*
Various procedures have been reported for the analysis of alkyllead samples [47–53]. Methods were described for the analysis of TML and TEL compounds in gasoline [47], and gas vapors in air [53]. A procedure for

determining gaseous ionic alkyllead species was described which included the determination of trialkyl and dialkyllead compounds, such as diethyl and trimethyllead (DEL and TML), in air [51]. Other procedures for the chromatographic separation and AAS detection of tetra-alkyllead (TAL) compounds have also been described [49–50, 52].

Two methods for collecting alkylead compounds have been reported, the first used a glass fibre iodized carbon filter [52], whilst the second employed activated carbon [53] as an adsorbent.

A GC interfaced to an AA spectrometer was used for the analysis of individual and total lead alkyls, especially TML and TEL, in leaded or unleaded gasoline [47]. The detection limit was found to be around 0.2 μg ml^{-1} Pb for each alkyl compound and the method was suitable for trace Pb levels in unleaded gasoline. With some modifications, this limit could probably be lowered to levels sufficiently low for pollution monitoring applications. Evidence was given which casts doubt upon measurements obtained by collection of organic lead in iodine monochloride solution when followed by extraction of total lead and AA analysis [48].

A system for sampling TAL compounds in air and analysis by GC-AAS has been reported [49], whilst a method for the determination of TAL compounds by LC and AAS gave detection limits for the LC-AAS system of about 10 ng for each TAL [50]. It was concluded that the LC-AAS hybrid technique could be easily extended to other organometallic speciation analyses.

A procedure for determining gaseous ionic alkyllead species in air has been described [51]. The lead species in the extracts were propylated or butylated and the alkyl derivatives were analyzed by GC/ETA-AAS. Air samples were drawn in at the rate of 1 to 5 min^{-1} through two 125-ml gas bubblers connected in series, with 80 ml of double distilled deionized water. The bubblers were preceded by a 0.45-μm membrane filter. After 24 to 48 hr of sampling, the contents of the bubblers were combined and extracted with sodium diethyldithiocarbamate into hexane. Collection efficiencies for the bubblers, when tested with a variety of alkyllead compounds, ranged from 46% for DEL to 79% for TML. The membrane filter in front of the bubblers successfully excluded inorganic lead and solid phase alkyllead species. Gas phase TAL compounds gave a slight interference of less than 4%. Loss of various trialkyl and dialkyllead compounds from the bubblers was investigated by spiking water with known amounts of alkyllead chlorides. Carbon-filtered air was bubbled through the solutions at 2 min^{-1}. After passage of approximately 3 m^3 of air, the solutions were analyzed. No loss of alkyllead was observed. Vapor from TAL compounds were collected on a glass fiber iodized carbon filter which was subsequently treated with HNO$_3$ containing bromine [52]. If both lead alkyls and inorganic lead were suspected to be in the atmosphere, two filters were connected in series; a membrane or glass fiber filter for inorganic lead and a second filter for the analysis of organic lead. A minimal air gap was used between the two filters

References pp. 186–190

to eliminate the adsorption of lead alkyls onto plastic surfaces. The solution was then analysed for Pb by AAS. The method used an air/acetylene flame at an analytical wavelength of 217.0 nm. The only spectral interference was from Sb. Significant suppression of the Pb absorbance was caused by phosphate, carbonate, iodide, fluoride and acetate when these were present at ten times the concentration of Pb. The method had a detection limit of 0.01 mg m^{-3} for a 120-l air sample. Samples taken over a period of 1 hr at a concentration range of 0.075 to 0.3 mg m^{-3} had an accuracy of better than 10%.

A method for determining alkyllead compounds found in gasoline vapor/air mixtures with activated carbon as the adsorbent has been reported [53]. A 0.80 μm pore size filter was placed in front of an adsorption tube containing 150 mg of activated carbon which was connected to a sampling pump. Air samples were drawn through the tube and filter at a rate of 0.2 min^{-1}. Total sample volumes ranged from 0.2 to 10 l, depending on the expected concentration of gasoline vapor. After sampling, the front and back sections of the adsorption tube were placed in 0.4 ml of 65% HNO$_3$ at 70°C for 30 min. Each extract was diluted with 2.0 ml demineralized water. If the amount of alkyllead adsorbed was above 100 ng, the sample was diluted further. The filter was dissolved in 2.0 ml 65% HNO$_3$ and digested at 70°C for 1 hr, followed by dilution to 10 ml with demineralized water. The Pb in the extracts was determined by ETA-AAS. The detection limit for the procedure was 0.002 μg m^{-3} for a sample volume of 1 m^3. The relative standard deviation for ten replicate determinations of a sample containing 2.1 μg m^{-3} was 9.5%. The recovery of alkyllead from samples spiked with solutions containing TML and TEL stored at 4 or 25°C for up to 10 weeks was 95 to 105%. It was concluded that the method was sensitive enough to determine alkyllead concentrations in air below 1 μg m^{-3}.

(ii) *Organotin*

A technique for the sampling and analysis of organotin compounds has been developed and tested [54]. Samples were collected on a glass fibre filter/XAD-2 resin system. These filter samples were stable under refrigeration or freezing for up to 7 days after collection. Separation of organotin species prior to measurement was accomplished using HPLC with a strong cation exchange column. The test compounds were measured using ETA-AAS connected directly to the HPLC. Atomization was accomplished with a zirconium-coated graphite tube furnace. Recovery and collection tests were performed using tetrabutyltin (TBT), tributyltin chloride and tricyclohexyltin hydroxide. These compounds were chosen as representative of the classes of organotins most important with regard to health effects or large production/use volumes. All recovery values exceeded 95% except for TBT at 0.01 mg m^{-3}, which exceeded 90% recovery. All collection/recovery values exceeded 90%. Overall method precision was 10.8% for TBT, 6.0% for tributyltin chloride, and 6.6% for tricyclohexyltin hydroxide in the 0.05 to 0.2 mg m^{-3} concentration range.

2. *Organoarsenicals*

The trivalent form of As is considered more toxic than the pentavalent [55]; therefore, methods have been developed to speciate the various As compounds by chromatographic separation with AAS detection. For example, commercial filters have been evaluated for collecting particulate organoarsenicals [56]. A review of various approaches to As speciation using coupled chromatography/AAS was reported [57]. More detailed descriptions of chromatographic/AAS coupling were reported including: (1) the GLC or GC/AAS analysis of As in airborne samples [58]; (2) the LC or HPLC/HG-AAS analysis of As [59–60] for pesticides [60].

Three commercially available filtering media were evaluated for the collection of organoarsenical airborne particulates [56]. The organoarsenicals studied included monomethylarsonic acid (MMA), dimethylarsenic acid (DMA), and p-aminophenylarsonic acid (APA). The filtering media tested were 0.8 μm pore size cellulose ester membranes, 1-μm pore size Fluoropore, and 5-μm pore size Mitex. Ion exchange chromatography followed by continuous generation of the arsine derivatives and AA were used for separation and detection of the compounds of interest. The contribution of vapor forms of organoarsenicals to the total atmospheric concentration was also examined. Filters were spiked with MMA, DMA, APA, As(III) and As(V), allowed to dry, extracted ultrasonically in either deionized water, acid or basic medium, stored and then analyzed over a period of 1 week. Filters were then spiked with combinations of the organoarsenicals and analyzed against calibration standards. Mixed aerosols containing the organoarsenicals were generated and collected to test collection efficiency. Effects of high temperature and 95% relative humidity were also examined. Filter extracts changed drastically unless stored under refrigeration in acetate or borate buffers. Percent recoveries from cellulose ester filters were poor, 62 to 78%, while they exceeded 90% for Fluoropore and Mitex. Collection efficiency of the Fluoropore filters was better than 99% while that of Mitex was about 65% due to the large pore size. No significant effects of temperature or humidity were demonstrated. It was concluded that the efficiency of the Fluoropore filters was equal to the traditionally used cellulose ester filters. In addition, Fluoropore filters are inert towards the various organoarsenical species.

Various novel approaches to As speciation have been examined using coupled chromatography/atomic spectrometry techniques [57]. All of the interfaces described were simple, reliable, permitted real time analysis, and produced continuous chromatograms. For GC separation it was necessary to derive the As compounds either as the hydrides or as the methylthio-

References pp. 186–190

glycolates. Cryogenic trapping GC/AAS offered superior detection limits, down to 0.22 to 0.55 ng for different species, but could not distinguish As(III) and As(V) unless a two-stage procedure involving control of the pH of reduction conditions was used. This was the preferred method for analysis of low concentrations of As. A hybrid method of coupling an HPLC, hydride generator, and an inductively coupled plasma as the detector was suggested for multi-element studies, with an AA detector used for more routine determinations.

A method for determining methylated arsenic compounds in airborne particulate matter by GC and AAS describes the extraction of samples with 0.02 M ethylenediaminetetraacetic acid (EDTA) and 1 M HCl. After centrifuging, the supernatant was treated with 10% aqueous sodium tetrahydroborate solution to reduce the As compounds to their hydrides (arsines). The released arsines, after being purged with helium at 50°C, were trapped in a half packed U-tube immersed in liquid nitrogen. On heating the arsines were injected into a gas chromatograph interfaced to an AAS. The separated arsines were quantified by AAS. When the system was used to determine MMA, DMA or TMA, the detection limits were 70, 80, and 100 picograms (pg) As, respectively [58].

Other coupled methods have been based on LC systems in which the As species are converted to their volatile hydrides. These systems often incorporate direct coupling of HG-AAS with HPLC [59–60]. Hydrides are generated from the HPLC effluent and then are detected by the AA spectrometer. Reported separations included the salts or acid compounds of arsenite [62], MMA [58, 60], DMA [58, 60], TMA [58], and arsenate [60].

A commercially available automated HPLC coupled with a GF-AAS detector is available for the analysis of arsenical compounds and other organometallics [59].

For improved separations, the HPLC-HG-AAS systems have been equipped with anion exchange column [60].

For arsenical pesticides, separation can be performed on a low capacity anion exchange column, with a limit of detection of about 5 ng As [62]. Matrix interferences were reported to be minimal. For single atomizations, calibration curves were linear from 0.1 to 2.0 ng. For multiple sequential atomizations, calibrations were linear up to 200 ng.

A ETA-AAS method using matrix modification was developed for determining As. The maximum tolerable charring temperature for As was raised by addition of Pd, and the sensitivity for As also increased [61].

3. Organomercuric compounds

In the past, a GC with an electron capture detector (ECD) has been used for the separation and detection of organomercurials in biological and environmental samples. There have been numerous reports on the analysis of alkylmercury compounds; however, only a few papers have been published regarding the determination of dialkylmercury compounds. When an ECD was used, the analysis of dialkyl compounds required conversion into their corresponding halide salts.

These past techniques have been improved through the use of GC separation with emission [62], CVAA [63], and more recently with ETA-AAS [64–68]. The first analytical improvement was accomplished using a GC with detection by a mercury specific emission spectrometer. This method had the advantage of selectively determining dialkylmercury compounds at the nanogram level without converting the dimethylmercury compounds into their chloride salts [62]. Another technique separated dialkyl compounds by GC with combustion in a flame-ionization detector and determined the resulting free mercury by CVAA [63].

Later, GC separation and reduction of dialkylmercury by a combustion furnace with detection of elemental mercury by ETA-AAS was described [64]. This method reportedly possessed adequate sensitivity and specificity and also was substantially faster and simpler to use than an ECD, which needed extensive clean-up steps [65].

A method was described for the determination of dimethylmercury and methyl mercury chloride in air by GC-AAS. An absorbance of 0.044 was obtained for 3.5 ng of mercury in dimethylmercury and 13–14 ng of mercury in methylmercury chloride [66]. The validity of this method was confirmed using a similar procedure for the determination of dimethylmercury, methylmercury and ethylmercury in water [67].

More recently, the sensitivity of this procedure was improved for the simultaneous determination of nanogram amounts of both alkylmercury and dialkylmercury in air [68]. Dimethyl- and diethylmercury and methyl- and ethylmercury chloride were analysed after separation by GC and the mercury vapor produced after decomposition of the organomercury compounds was measured by AAS. Detection limits of 0.2 ng of mercury for dimethylmercury and diethylmercury and 0.5 ng of mercury for the methylmercury chloride and ethylmercury chloride were obtained.

4. *Organoselenide compounds*

Selenide compounds have been separated by GC [69–71] or by ion chromatography [72] with AA detection. Alkylselenide compounds in air have been determined by GC-AAS [69]. A graphite furnace was used to enhance sensitivity. Dimethyl selenide (DMS), diethyl selenide, and dimethyl diselenide (DMDS) were selected as the analytes of interest. The detection limits of this system were 0.2 ng m^{-3} Se in air. The method allowed for a simple determination of volatile alkylselenides in air with a precision of about 10% at nanogram levels. In addition to air samples, the method was augmented to include the analysis of total dissolved Se and selenite in water [70].

Later, a systematic study of several factors which affected the recovery of alkylselenides from the atmosphere and the sensitivity of determinations by GC/AAS was presented [71]. Flash evaporation was used instead of direct injection for the introduction of heat-sensitive Se compounds into the gas chromatograph. This method decreased as much as possible any thermal decomposition on the chromatographic column. A sampling system was also designed to measure collection efficiency and recovery from a cryogenic trap. The cryogenic trap temperature was maintained at $-140°C$ for quantitative collection of the three compounds investigated. The flow rate was $6 \, l \, min^{-1}$ for these compounds. The surface condition of the trap system had an important effect on the recovery. The best recoveries were obtained when using acid-washed polyvinylchloride and when using Pyrex glass wool. For AA detection, the addition of about 10% hydrogen to the argon carrier gas increased the sensitivity for the organoselenium species. Reproducible measurements were obtained for the alkylselenides with a detection limit of $0.07 \, ng \, m^{-3}$ for DMS, $0.03 \, ng \, m^{-3}$ for diethyl selenide, and $0.15 \, ng \, m^{-3}$ for DMDS. The separation of selenite and selenate by ion chromatography and the use of ETA-AAS as a Se specific detector are reported in a hybrid system which clearly showed the presence of selenite and selenate [72]. The method successfully detected selenite and selenate with no effects from anion concentrations which were 1000 times greater than the selenite or selenate concentrations. The detection limit for the ETA-AAS was 20 ng ml^{-1} for each Se compound. Preconcentration of 4 ml of an anion-rich water sample extended the detection limit to 5 ng Se.

5. Organophosphorus and chromium species

(i) Organophosphorus

An HPLC method for detecting and measuring organophosphorus compounds describes the uses an ETA-AA spectrometer coupled to the LC. The ETA-AAS detection limit for phosphorus was 6 ng [73].

(ii) Hexavalent and trivalent chromium complexes

The application of an anion exchange HPLC procedure to the simultaneous determination of water soluble hexavalent chromium (Cr(VI)) and trivalent chromium (Cr(III)) complexes has been reported [74]. In this work an anion exchange HPLC was interfaced to a visible spectrophotometer, an ultraviolet spectrophotometer, and an AA spectrometer. Hexavalent chromium was separated by HPLC, and was continuously determined by spectrophotometry at 370 nm, and also by AAS. The separation patterns of other species were monitored by AAS. When standard Cr(VI) solutions were tested, the detection limits for visible spectrometry and AAS were 2 and 5 ng, respectively.

D. Specificity from other atomic absorption techniques

1. *Chromic acid/chromates*

An analytical method for the detection of airborne Cr, oxides of Cr, soluble and insoluble salts, and chromic acid in the workplace atmosphere has been described, in which samples from 15 min to 8 hr were taken [75]. The sample was collected on a filter which was then treated with 50% HNO_3 and hydrogen peroxide. Atomic absorption spectrometry was used to analyse the resulting solution. A wavelength of 357.9 nm without background correction was recommended for Cr analysis. Sensitivity for Cr was 0.04 μg ml^{-1}. The detection limit was 0.003 mg m^{-3} for 30 l of air. Elements that may suppress the AA signal for Cr are Co, Fe and Ni; interference may also be caused by Cu, Ba, Al, Ca and Mg. When these cations were suspected to be present, it was recommended to use a nitrous oxide/acetylene flame.

The most common non-chromatographic separation technique for the determination of Cr(VI) uses the principle of chelation with ammonium pyrrolidine dithiocarbamate (APDC) and extraction of the complex with methylisobutyl ketone (MIBK) [76–78]. It was reported that Cr(VI) in welding fumes decays when collected with certain types of filters, therefore a discussion of this problem with possible solutions was included [79–82].

A method was described for the determination of Cr(VI) that was essentially free from interference [76–77]. The sample was collected on a 0.8-μm pore-size cellulose membrane filter. By adjusting the pH prior to chelation and extraction, the effect of Mn interference was minimized. The Cr(VI) was separated from Cr(III) by chelation of the former with APDC and extraction of the complex with MIBK. The limit of detection was 0.2 μg chromic acid per filter or 0.01 mg m^{-3} for a 20-l air sample.

Extraction using APDC/MIBK and subsequent detection by AA was used to determine Cr(III) and Cr(VI) in aqueous media at room temperature [78]. Effect of pH on extraction, optimum extraction time, APDC and phthalate concentrations, time stability of chelates in MIBK, effect of foreign ions, as well as calibration, precision, and accuracy were studied. For both Cr(III) and Cr(VI), optimised parameters and peak absorbance values were identical. No difference in the peak absorbance values of Cr(III) and Cr(VI) was observed in MIBK solutions with or without background correction. Simultaneous and quantitative extractions of Cr(III) and Cr(VI) were possible in the pH range of 2.5 to 4.0. Selective separation of Cr(VI) from Cr(III) occurred when the APDC-to-metal and phthalate-to-metal concentration ratios were $\geq 4 \times 10^4$ and $\geq 2 \times 10^5$, respectively. Hexavalent chromium was quantitatively extracted in the pH range of 2.0 to 4.0. Simultaneous extraction of Cr(III) and Cr(VI) at room temperature was possible without prior oxidation of Cr(III) to Cr(VI). No significant differences were observed for solutions kept in light or dark environments.

The chelates were stable for 30 days. For both Cr(III) and Cr(VI), the linear range was 0 to 50 ng ml^{-1}, the detection limit was 0.3 ng ml^{-1}, and the sensitivity was 0.2 ng ml^{-1} in the MIBK phase. The effect of foreign ions was examined at levels exceeding those found in most fresh water and drinking water samples. When using the described conditions, no problems with the measurement of Cr(III) and Cr(VI) were observed.

It has been shown [79] that Cr(VI) may be reduced to Cr(III) by reaction with cellulose filters, hence such filters are not recommended if samples must be stored for an appreciable length of time prior to analysis. The use of polyvinyl chloride (PVC) filters eliminates this difficulty and collected samples may be stored for at least 2 weeks [80].

Chromium formation and decay in welding fumes, both airborne and collected on filters, were investigated on metal inert gas (MIG) and manual metal arc (MMA) welding operations using stainless steel [81]. Fume samples were collected on glass fiber filters or in an impinger train. MIG fumes collected by impinger showed significant variations in the ratio of Cr(VI) to total Cr after aging. The ratio rose sharply to a maximum at 20 seconds, then fell and reached a steady value within a few minutes. Preliminary field trials were conducted using both collection methods during MIG welding of stainless steel. The amount of total Cr collected by the two methods was very similar, although the impinger method gave concentrations of airborne Cr(VI) 2 to 5 times larger than that measured simultaneously by filter sampling. It was concluded that serious underestimates of exposure for workers close to fume sources could result when using the filter collection technique.

Later in another study, the effects of fume type, collection media, storage time, and type of analytical procedure on measured concentrations of Cr(VI) in welding fumes were investigated on both stainless and mild steel [82]. Fumes generated by manual metal arc welding of stainless steel (MMA/SS), manual metal arc welding of mild steel (MMA/MS), and metal inert gas welding of stainless steel (MIG/SS) were collected either on polyvinyl chloride (PVC), cellulose acetate filters, or in water-filled impingers. Impinger samples showed enrichment of Cr(VI) in the smaller size fractions of MIG/SS and MMA/SS fume particles, and the water-soluble content of Cr(VI) of MIG/SS fume was significantly increased. Collecting and storing MMA/SS fume on PVC filters for up to 15 days did not affect the Cr(VI) content; however, reduction of Cr(VI) occurred when samples were stored on cellulose acetate filters. When basic media were used to process the samples, the amount of Cr(VI) increased, indicating that oxidation of Cr(III) to Cr(VI) had occurred. It was concluded that presently used sampling and analytical procedures could result in reduction or oxidation of Cr(VI), and that this may cause under- or over-reporting of Cr(VI) for industrial hygiene purposes.

2. *Mercury*

Analytical methods for the determination of Hg have been reported [83–85], including a procedure for evaluating the amount of organic Hg from the difference between total and inorganic Hg concentrations [85].

A flameless AA technique has been described and used for analysis of Hg in water, air, fish, urine and organo-mercury compounds [84]. An automatic background correction system was employed for more accurate analysis. An investigation of analytical variables, including behavior of reductants, flow rates, chemical compositions of flushing gases, mixing time, sample volume effect, and chemical interferences were reported.

A detailed study was subsequently made using an ordinary rectangular 4-cm ultraviolet cell for CVAA determinations of Hg [84]. At concentrations from 0 to 30 ng ml^{-1}, the calibration graph was linear. Absorbance at concentrations up to 50 ng ml^{-1} showed only slight deviation from linearity. The detection limit was 0.02 ng ml^{-1} or 0.1 ng. Use of this method allowed for replacement of transient peak atomic absorption and peak area integration measurements for the CVAA method by a steady state AA method.

A procedure to determine the amount of organic mercury concentration from the difference between total and inorganic mercury concentrations was reported [85]. Samples to be analyzed for inorganic mercury were treated with stannous chloride which reduced the Hg to its elemental form. When total mercury was to be determined, cadmium chloride was added to the stannous chloride. The concentration of organic mercury was taken as the difference between the total and inorganic mercury concentrations. During the analysis, Hg became embedded in the glassware in the CVAA apparatus by chemisorption and was effectively removed by cadmium chloride during a determination, giving rise to abnormally high blank values. The author concluded that sequential blank runs should be made with cadmium chloride until appropriate blank values are re-established.

Many articles reported in the literature concern problems associated with Hg sampling [86–99]. Sampling techniques include the use of iodide-impregnated charcoal [86], hopcalite tubes [87–89] (including the use of a commercially developed passive-sampling device [90–91]), and the use of chromatographic type sorbents [94–97]. Comparisons of these various sampling techniques have been evaluated and reported [98–99].

Methods for the AAS determination of Hg vapor for industrial hygiene samples have evolved primarily in the area of sampling procedures. The Hg vapor was first collected on iodine-impregnated charcoal tubes. Another common procedure was the collection of Hg on hopcalite. Currently, a proprietary material known as Hydrar is commonly used for sampling Hg. Sampling devices containing this type of material include packed tubes, or passive samplers which do not require the use of an active pump to sample the material. The evolution of these methods is discussed below.

References pp. 186–190

A method was published in which Hg was collected on iodide-impregnated charcoal. Analysis was performed on this treated charcoal by placing it in a tantalum sampling boat and heating it to drive the Hg vapor into the optical path of an AA spectrometer [86]. This procedure had several disadvantages, including: (a) a high detection limit, (b) a loss of the Hg vapor, (c) poor precision, especially at lower sample loadings.

Also, the analysis of the iodine-impregnated charcoal tubes appeared to be exceedingly operator- and technique-dependent. Subsequently, a method using hopcalite tubes was reported [87–88].

Granules of hopcalite were used to trap vapors of Hg in a small glass tube through which air was drawn by a battery-powered pump, followed by treatment of the granules with HNO_3 reduction of Hg with stannous chloride, and determination of the reduced Hg by CVAA with a spectrophotometer or a detector for Hg vapor [87]. Tests carried out over a wide range of concentrations and sampling rates revealed that the absorption of Hg vapors was nearly 100% efficient for the determination of air Hg levels in the nanogram range. Sampling times ranging from 3 to 365 min were tested. Efficiency of the hopcalite sampler was tested at flow rates of up to 846 l min^{-1}, with results greater than 99%. Efficiencies of about 97% were achieved with an activated carbon filter, but the analytical results obtained with the same method were inconsistent. A small amount of Hg was present in the hopcalite and varied from one batch to another; therefore, it was recommended that a blank obtained from mixing several batches of absorbent should be determined. It was reported that this method could be used for personal monitoring or measurements of larger air volumes for stack or community air monitoring.

The hopcalite procedure was later simplified and shortened such that the error due to incomplete removal of mercury from the hopcalite was eliminated [88].

The performance of hopcalite in another study produced results 26 to 33% less than expected when exposed to 10 ppm chlorine gas in an active sampler; however, at lower concentrations of 0.4 ppm, chlorine had no effect [89].

Later, Hydrar was developed from a manganese dioxide catalyst material which is very similar to hopcalite. The Hydrar sampler collects elemental Hg in the vapor phase and does not collect organic Hg vapors or particulates. The Hydrar sorbent is prepared for analysis by dissolving it in a mixture of concentrated HCl and then HNO_3. This vapor is analyzed by CVAA using excess stannous chloride as a reductant.

A commercial passive sampling device was developed using this type of absorbent material [90]. The results from an evaluation of this device have been reported [91–92] with the following conclusions: (1) the Hg dosimeter had a sampling rate of approximately 20 cm^3 min^{-1} with a sampling time in excess of 8 hr; (2) the collected samples were stable for at least 30 days;

(3) a detection limit of 0.005 mg m^{-3} (0.05 μg for an 8-hr air sample) was achieved for the flameless AAS technique; (4) the recovery was 97.2% with an over-all sampling and analysis precision of 0.038.

In addition to the passive device, the performance of an active sampler containing the Hydrar absorbent material was reported [93]. This sampler had a detection limit of 0.02 μg Hg or 0.0013 mg m^{-3} for a 15-l air volume. The analytical method recovery was 0.945 with an analytical coefficient of variation of 0.081. The sampling rate was 0.2 l min^{-1}, with a recommended air volume range of 3 to 100 l.

The Hydrar contains traces of Hg as a contaminant; therefore, a blank determination should be made on each lot of material used. In comparison with hopcalite [88], Hydrar has a lower background and increased capacity for mercury.

A technique for analyzing vapor phase Hg species in the atmosphere was tested [94]. The purpose was to collect and analyze the principal gaseous species of airborne mercury. Several volatile forms of Hg were used: elemental mercury, mercuric chloride, methylmercuric chloride, and dimethyl mercury. Tests were carried out to determine the collection efficiency, recovery efficiency, release characteristics, and sensitivity of chemically treated and untreated sorbent materials. Performance characteristics of candidate sorbent materials were evaluated. Among the chromatographic type sorbents ultimately selected were Chromosorb W, Tenax GC, and Carbosieve B. Chromosorb W passed virtually all the elemental Hg and dimethyl mercury vapor, whereas Ag- and Au-coated glass beads retained both species. Chromosorb W treated with HCl displayed a 97% collection efficiency for mercuric chloride. The other sorbent materials also retained this compound quantitatively. It was reported that for methylmercuric chloride, Chromosorb W treated with NaOH resulted in a collection efficiency of 46% with an overall recovery efficiency of 95 to 100%.

Another sampling procedure requiring a special complex thermal desorption unit [95] with analysis in an absorption cell of a flameless AA spectrophotometer was reported [96–97]. This method required a sampling tube which contained a trapping section of Carbosieve B to collect the organic Hg vapor fraction. This section was then followed by a section of silvered Chromosorb P to collect the elemental Hg vapor. The tube can be preceded with a glass fiber filter to collect particulate species. This method allowed for the determination (as discrete species) of particulate, volatile organic, and elemental mercury, which were collected simultaneously.

Results from a collaborative NIOSH/OSHA study of the 3M passive monitor, Los Alamos Scientific Laboratories (LASL) tandem sampling tube, hopcalite tube and iodine-impregnated charcoal tube were reported by McCammon et al. [98].

A comprehensive evaluation comparing most of the sampling and analytical techniques for mercury, including a comparison of active and passive devices in the laboratory and field sampling evaluations, has been given in Cee et al. [99].

VI. GENERAL METHODS FOR METAL AEROSOL SAMPLING AND ANALYSIS

Although most metallic aerosols are collected using filters [100], some research has been conducted using solid sorbent sampling procedures [101].

Generally, the analysis of metallic fumes involves sample preparation by digesting the filter in HNO_3 and is sometimes followed with additional reagent pretreatment techniques [102].

A. Metallic fumes

A method for the sampling and analysis, by AAS, of metallic fumes of Pb, Te and Se was reported [100]. Compounds of Pb and Te were dissolved in HNO_3/H_2O_2 solution. Selenium and platinum compounds were dissolved in an aqueous solution. Recoveries of these compounds were greater than 90% using the procedures described in the method.

B. Inorganic lead

1. Sampling

Lead is often collected on a filter; however, the potential use of solid sorbent packed tubes as collection devices for metal aerosol sampling has also been evaluated [101]. Lead aerosols were collected in Pyrex tubes containing four different solid sorbents: alumina, silica gel, Chromosorb 102, or Tenax GC. Lead was determined with an AA spectrophotometer equipped with a Pb hollow-cathode lamp. The average collection efficiencies on the first 200-mg portion of sorbent when sampling a laboratory aerosol were $99.2 \pm 1.0\%$ with alumina, $99.2 \pm 1.1\%$ with silica gel, $99.9 \pm 0.1\%$ with Chromosorb 102, and $100.0 \pm 0.0\%$ with Tenax GC. When sampling an automobile exhaust aerosol, the collection efficiencies were 95.8 ± 5.7, 98.5 ± 3.2, 98.0 ± 3.7, and $99.2 \pm 2.6\%$ with alumina, silica gel, Chromosorb 102 and Tenax GC, respectively. The highest collection efficiency and lowest background absorbance values were obtained with Tenax GC; however, it was also the most difficult sorbent to work with due to its static charge, and was much more expensive than the other sorbents. Similar results were obtained with silica gel which was much easier to use. The mesh size and quantity of sorbent were found to be more important than the type of sorbent when determining the collection efficiency for Pb aerosols.

2. *Analysis*

An AA spectrometric method for determining Pb and airborne Pb compounds reports the analysis of lead dust or fume samples collected on a filter using sampling periods of 15 min to 8 hr [102]. After sampling, the filter was placed in a 100-ml beaker to which 5 ml of a 5% HNO_3/H_2O_2 mixture was added. The beaker was heated to simmering for 15 min. After cooling, the solution was filtered through a cellulose filter and collected in a volumetric flask. The size of the flask depended on the amount of Pb expected. The flask was filled to the mark with 5% HNO_3. The sample was analyzed by AAS using 217.0 nm as the analytical wavelength. The spectrometer was equipped with a Pb hollow-cathode lamp and fitted with an air/acetylene burner. The detection limit was 0.4 mg m^{-3} for a 30-l sample. The relative coefficient of variation was less than 10% for samples containing 0.075 to 0.3 mg m^{-3} Pb.

VII. INDIRECT ANALYSIS

Atomic absorption may be used to indirectly determine the analyte of interest by analysing a component of the compound. One example is the analysis of Ni for the determination of nickel carbonyl [103–104]. Another example is the analysis of Pb for the determination of hydrogen sulphide [105].

A. Nickel carbonyl

The concentration of nickel carbonyl in a test atmosphere was determined by AAS after sampling in midget impingers containing 5% iodine in isopropanol [103]. This new sampling method used a Calgon coconut-based charcoal sorbent which was washed in HNO_3 before use. After sample adsorption, the sorbent was eluted with 3% HNO_3 in an ultrasonic water bath and aliquots were analysed by ETA-AAS. The desorption efficiency was 0.934 for nickel nitrate throughout the concentration range of 0.08 to 0.5 μg Ni ion. The recovery of nickel carbonyl collected in air ranged between 90.8 and 95.0%. The sensitivity of the method makes it possible to determine nickel carbonyl concentrations of 2 to 50 μg m^{-3} of air which corresponds to 0.4 to 7 ppb for a 20-l air sample.

A further method was described for the determination of trace amounts of Ni by AAS after carbonyl generation [104]. After appropriate digestion with mineral acids, organic or inorganic samples were filtered, evaporated to dryness, and reconstituted in HNO_3. The samples were reacted with $NaBH_4$ to reduce Ni in the compounds to the metal, which were then treated with carbon monoxide to give nickel carbonyl. Nickel carbonyl was then decomposed by heating to give elemental Ni which was analyzed by AAS at a wavelength of 232 nm. The detection limit was about 10 ng. The precision was of the order of 2 to 20%, depending on the amount of Ni present. The effects of Pb, Sb, Bi, Se, As and Sn were also investigated. Only Sn, which had an intense absorption line at 231.7 nm, and to a lesser extent As, showed interference. Equal amounts of Ni and Sn gave about the same absorbance. Metallic Fe was found to hinder the formation of nickel carbonyl. This interference could be avoided by conducting the borohydride reduction in 2% HNO_3. Under those conditions, trivalent Fe was reduced only to divalent Fe and not completely to metallic Fe. Although the method was less sensitive than a standard method, it was simpler and well adapted to routine determination of Ni in a wide range of samples.

B. Hydrogen sulphide

An AAS method for the determination of hydrogen sulphide in the atmosphere describes a procedure in which a known amount of buffered lead acetate solution was added to the sample and the excess reagent was measured by AAS [105]. For 1% absorption, 0.5 μg ml^{-1} Pb was equivalent to 0.082 μg H_2S. The method was accurate up to 200 ppm of H_2S. It was suggested that the method was quicker and easier than gas chromatographic or colorimetric methods for H_2S determination.

VIII. CONCLUSIONS

Techniques presented within this chapter are not implied to be "state of the art" solutions for analysis of environmental air samples, but simply some of the applications for AA summarized from the literature.

In the future, as analytical instrumentation becomes more sophisticated, many of these techniques will be improved, and newer ones developed. Previous experience indicates that environmental hygiene exposure limits will be reduced as advances in technology allow for improved sensitivity and speciation of compounds collected in workplace atmospheres.

REFERENCES

1   R.G. Keenan, J. Occup. Med., 7(1965) 276.
2   J.Y. Hwang and F.J. Feldman, Appl. Spectrosc., 24 (1970) 371.
3   B.C. Begnoche and T.H. Risby, Anal. Chem., 47 (1975) 1041.
4   F.W. Sunderman Jr., Hum. Pathol., 4 (1973) 549.
5   NIOSH Manual of Analytical Methods, 2nd ed., Vols. 1–4, U.S. Dept. of Health, Education and Welfare, DHEW(NIOSH) Publ. No. 77-157-A, U.S. Government Printing Office, Washington, D.C. 20402, April, 1977.

6   P.M. Eller, in 1985 Supplement to NIOSH Manual of Analytical Methods, 3rd ed., U.S. Dept. of Health and Human Services, Public Health Service Centers for Disease Control, NIOSH Division of Physical Sciences and Engineering, NIOSH Publ. No. 84-100, Cincinnati, Ohio, February 1984.

7   OSHA Analytical Methods Manual, OSHA Analytical Laboratory, Salt Lake City, Utah, 1985, Vols. I–III, Metal Particulates in Workplace Atmospheres (AAS), ID-121.

8   L.S. Shepard, G.R. Ricci, G. Colovos, W.S. Eaton and H. Wang, Analytical Techniques in Occupational Health Chemistry, ACS Symp. Ser. No. 120 (1980) 267.

9   Threshold Limit Values of Airborne Contaminants and Physical Agents with Intended Changes adopted by the American Conference of Governmental Industrial Hygienists, Secretary-Treasurer, P.O. Box 1937, Cincinnati, Ohio 45201.

10   Code of Federal Regulations (CFR), Title 29, Part 1910.1000, Occupational and Environmental Health Standards. These were first published in the Federal Register, Vol. 26, No. 105, May 29, 1971 and subject to revision. Currently published as OSHA 3112, in Table Z-1-A "Limits For Air Contaminants", effective March 1, 1989.

11   D.G. Mitchell, A.F. Ward and M. Kahl, Anal. Chim. Acta, 76 (1975) 456.

12   G.C. Kunselman and E.A. Huff, At. Absorpt. Newsl., 15 (1976) 29.

13   Antimony and Compounds (as Sb), in NIOSH Manual of Analytical Methods, 2nd ed., DHEW (NIOSH) Publication No. 77-157-B., U.S. Dept. of Health, Education and Welfare, Public Health Service, Center for Disease Control, National Institute for Occupational Safety and Health, 2 (1977) S2-1-S2-7, Cincinnati.

14   B.E. Schreiber and R.W. Frei, Int. J. Environ. Anal. Chem., 2 (1972) 149.

15   G.D. Renshaw, C.A. Pounds and E.F. Pearson, At. Absorpt. Newsl., 12 (1973) 55.

16   J.Y. Hwang, C.J. Mokeler and P.A. Ullucci, Anal. Chem., 44 (1972) 2018.

17   A.W. Fitchett, E.H. Daughtrey and P. Mushak, Anal. Chim. Acta, 79 (1975) 93.

18   D. Chakraborti, K.J. Irgolic and F. Adams, J. Assoc. Off. Anal. Chem., 67 (1984) 277.

19   E. Iwamoto, C-H. Chung, M. Yamamoto and Y. Yamamoto, Talanta, 33 (1986) 577.

20   P.M. Eller and J.C. Haartz, Am. Ind. Hyg. Assoc. J., 39(10) (1978) 790.

21   R.B. Denyszyn, P.M. Grohse and D.E. Wagoner, NIOSH Contrib., No. 210-76-0142, 50(8) (July 1978) 1094.

22   Y. Matsumura, Ind. Health, 26 (1988) 135.

23   P.D. Goulden and P. Brooksbank, Anal. Chem., 46 (1974) 1431.

24   E.N. Pollock and S.J. West, At. Absorpt. Newsl., 11 (1972) 104.

25   E.N. Pollock and S.J. West, At. Absorpt. Newsl., 12 (1973) 6.

26   Y. Yamamoto, T. Kumamaru and Y. Hayashi, Anal. Lett., 5 (1972) 419.

27   F.J. Fernandez, At. Absorpt. Newsl., 12 (1973) 93.

28   K.C. Thompson and D.R. Thomerson, Analyst, 99 (1974) 595.

29   F.J. Schmidt and J.L. Royer, Anal. Lett., 6 (1973) 17.

30   Anonymous, Health Lab. Sci., 14 (1977) 53.

31   M. Verlinden, J. Baart and H. Deelstra, Talanta, 27 (1980) 633.

32   F.J. Schmidt, J.L. Royer and S.M. Muir, Anal. Lett., 8 (1975) 123.

33   A.E. Smith, Analyst, 100 (1190) (1975) 300.

34   K. De Doncker, R. Dumarey, R. Dams and J. Hoste, Anal. Chim. Acta, 187 (1986) 163.

35   C.J. Peacock and S.C. Singh, Analyst, 106 (1981) 931.

36   R.G. Smith, J.C. Van Loon, J.R. Knechtel, J.L. Fraser, A.E. Pitts and A.E. Hodges,

Anal. Chim. Acta, 93 (1977) 61.

37 J.Y. Hwang, P.A. Ullucci, C.J. Mokeler and S.B. Smith Jr., Atomic Absorption Determination of Arsenic by the Arsine Gas Technique, Instrumentation Laboratory, Inc., Lexington, Mass.

38 Anonymous, Arsenic and Inorganic Compounds or Arsenic in Air, Laboratory Method Using Atomic Absorption Spectrometry, Methods for the Determination of Hazardous Substances, MDHS 41, Health and Safety Executive, London (September 1984).

39 G. Colovos, N. Hester, G.R. Ricci and L.S. Shepard, The Development of a Method for the Determination of Organoarsenicals in Air, NIOSH, Division of Physical Science and Engineering, U.S. Department of Health, Education and Welfare, Public Health Service, Center for Disease Control (September 1980).

40 J.C. Chambers and B.E. McClellan, Anal. Chem., 48 (1976) 2061.

41 Characteristics of particles and particle dispersoids, Stanford Res. Inst. J., 5 (1961) 95.

42 R.E. Kupel, R.E. Kinser and P.A. Mauer, Am. Ind. Hyg. Assoc. J., 29 (1968) 364.

43 H. Koizumi, T. Hadeishi and R. McLaughlin, Anal. Chem., 50 (1978) 1700.

44 J.C. Van Loon, Anal. Chem., 51 (1979) 1139.

45 F.E. Brinckman, W.R. Blair, K.L. Jewett and W.P. Inverson, J. Chromatogr. Sci., 15 (1977) 493.

46 T.M. Vickrey, H.E. Howell, G.V. Harrison and G.J. Ramelow, Anal. Chem., 52 (1980) 1743.

47 D.T. Coker, Anal. Chem., 47 (1975) 386.

48 R.M. Harrison and R. Perry, Atmos. Environ., 11 (1977) 847.

49 W.R.A. De Jonghe, D. Chakraborti and F.C. Adams, Anal. Chem., 52 (1980) 1974.

50 J.D. Messman and T.C. Rains, Anal. Chem., 53 (1981) 1632.

51 C.N. Hewitt, R.M. Harrison and M. Radojevic, Anal. Chim. Acta, 188 (1986) 229.

52 Anonymous, Tetra Alkyl Lead Compounds in Air, Personal Monitoring Method, Methods for the Determination of Hazardous Substances, MDHS 9, Health and Safety Executive, London (July 1981).

53 O. Royset and Y. Thomassen, Anal. Chim. Acta, 188 (1986) 247.

54 W.F. Gutknecht, P.M. Grohse, C.A. Homzak, C. Tronzo, M.B. Ranade and A. Damle, Development of a Method for the Sampling and Analysis of Organotin Compounds, Division of Physical Sciences and Engineering, NIOSH, U.S. Department of Health and Human Services, Cincinnati, Ohio, NTIS PB83-180-737, April 1982.

55 J. Savory and F.A. Sedor, in S.S. Brown (Ed.), Arsenic Poisoning, Clinical Chemistry and Chemical Toxicology of Metals, Elsevier/North-Holland Biomedical Press, Amsterdam, 1977, p. 271.

56 G. Ricci, G. Colovos, N. Hester, L.S. Shepard and J.C. Haartz, Suitability of various Filtering Media for the Collection and Determination of Organoarsenicals in Air, Chemical Hazards in the Workplace, Measurement and Control, ACS Symp. Ser., No. 149 (1981) 383.

57 L. Ebdon, S. Hill, A.P. Walton and R.W. Ward, Analyst, 113 (1988) 1159.

58 H. Mukai and Y. Ambe, Anal. Chim. Acta, 193 (1987) 219.

59 F.E. Brinckman, W.R. Blair, K.L. Jewett and W.P. Inverson, J. Chromatogr. Sci., 15 (1977) 493.

60 E.A. Woolson and N. Aharonson, J. Assoc. Off. Anal. Chem., 63 (1980) 523.

61 Shan Xiao-Quan, Ni Zhe-Ming and Zhang Li, Anal. Chim. Acta, 151 (1983) 179.

62 C.A. Bache and D.J. Liske, Anal. Chem., 43 (1971) 950.

63   R.C. Dressman, J. Chromatogr. Sci., 10 (1972) 472.
64   J.E. Longbottom, Anal. Chem., 40 (1972) 1111.
65   J.G. Gonzalez and R.T. Ross, Anal. Lett., 5 (1972) 683.
66   A. Bzezinka, J. Van Loon, D. Williams, K. Oguma, K. Fuwa and I.H. Haraguchi, Spectrochim. Acta, Part B, 38 (1983) 1339.
67   A. Paudyn and J.C. Van Loon, Fresenius Z. Anal. Chem., 325 (1986) 369.
68   J. Gui-bin, N. Zhe-ming, W. Shun-rong and H. Heng-bin, J. Anal. At. Spectrom., 4 (1989) 315.
69   S. Jiang, W. De Jonghe and F. Adams, Anal. Chim. Acta, 136 (1982) 183.
70   S. Jiang, H. Robberecht and F. Adams, Atmos. Environ., 17 (1983) 111.
71   S.G. Jiang, D. Chakraborti and F. Adams, Anal. Chim. Acta, 196 (1987) 271.
72   D. Chakraborti, D.C.J. Hillman, K.J. Irgolic and R.A. Zingaro, J. Chromatogr., 249 (1982) 81.
73   P. Tittarelli and A. Mascherpa, Anal. Chem., 53 (1981) 1466.
74   Y. Suzuki, J. Chromatogr., 415 (1987) 317.
75   Anonymous, Chromium and Inorganic Compounds of Chromium in Air. Laboratory Method using Atomic Absorption Spectrometry, Methods for the Determination of Hazardous Substances, MDHS 12, Health and Safety Executive, London (June 1981).
76   E. Brown, M.W. Skougstad and M.J. Fishman, Techniques of Water-Resources Investigations of the United States Geological Survey, Chapter A1, Methods for collection and analysis of water samples for dissolved minerals and gases, Dept. of Interior, Laboratory Analysis, Book 5, p. 77.
77   Criteria for a recommended standard... Occupational Exposure to Chromic Acid, Appendix II, Analytical Methods for Chromic Acid, DHEW (NIOSH) Publ. No. HSM 73-11021, U.S. Dept. of Health, Education and Welfare, Public Health Service, Center for Disease Control, National Institute for Occupational Safety and Health, VIII, Cincinnati, 1973, p. 70.
78   K.S. Subramanian, Anal. Chem., 60 (1988) 11.
79   R. Dutkiewicz, J. Konczalik and M. Przechera, Acta Pol. Pharm., 26 (1969) 168.
80   M.T. Abell and J.R. Carlberg, Am. Ind. Hyg. Assoc. J., 35 (1974) 229.
81   C.N. Gray, A. Goldstone, P.R. Dare and P.J. Hewitt, Am. Ind. Hyg. Assoc. J., 44 (1983) 384.
82   B. Pedersen, E. Thomsen and R.M. Stern, Ann. Occup. Hyg., 31 (1987) 325.
83   J.Y. Hwang, P.A. Ullucci and A.L. Malenfant, Can. Spectrosc., 16 (1971) 2.
84   S-L. Tong, Anal. Chem., 50 (1978) 412.
85   D.C. Wigfield and R.S. Daniels, J. Anal. Toxicol., 12 (1988) 94.
86   A.E. Moffitt Jr. and R.E. Kupel, Am. Ind. Hyg. Assoc. J., 32 (1971) 614.
87   A.O. Rathje, D.H. Marcero and D. Dattilo, Am. Ind. Hyg. Assoc. J., 35 (1974) 571.
88   A.O. Rathje and D.H. Marcero, Am. Ind. Hyg. Assoc. J., 37 (1976) 311.
89   Technical Paper, Model 401 and Dosimeters in Chlor-Alkali Industry, Jerome Instrument Corp., U.S.A., 1982.
90   SKC-West Inc., Gas Monitoring Dosimeter Badge For Mercury—Operating Instructions, SKC-West Inc., Fullerton, Calif. (no date of publication given), p. 9.
91   USDOL-OSHA, Evaluation of Mercury Solid Sorbent Passive Dosimeter—Backup Data Report, Occupational Safety and Health Administration Analytical Laboratory, Salt Lake City, Utah, 1982.
92   USDOL-OSHA, ID-145S (SCK Badges), Mercury in Workplace Atmospheres, Occupational Safety and Health Administration Analytical Laboratory, Salt Lake

City, Utah, 1987.

93   USDOL-OSHA, ID-145H (Hydrar Tubes), Mercury in Workplace Atmospheres, Occupational Safety and Health Administration Analytical Laboratory, Salt Lake City, Utah, 1987.

94   W.H. Schroeder and R. Jackson, Chemosphere, 13 (1984) 1041.

95   P.E. Trujillo and E.E. Campbell, Anal. Chem., 47 (1975) 1629.

96   NIOSH, Manual of Analytical Methods, Volume 4, 2nd ed., #757-141/1829, Method #S199, U.S. Government Printing Office, Washington, D.C.

97   NIOSH, Manual of Analytical Methods, Volume 1, 2nd ed., #017-033-00267-3, Method #PandCAM 175, U.S. Government Printing Office, Washington, D.C.

98   C.S. McCammon Jr., S.L. Edwards, R.D. Hull and W.J. Woodfin, Am. Ind. Hyg. Assoc. J., 41(7) (1980) 528–531.

99   R. Cee, J. Ku, E. Zimowski, S. Edwards and J. Septon, An Evaluation of Mercury Vapor Sampling Devices, OSHA-DOL, Occupational Safety and Health Administration Analytical Laboratory, Salt Lake City, Utah (submitted for publication).

100 W.F. Gutknecht, M.B. Ranade, P.M. Grohse, A.S. Damle and P.M. Eller, Development of a Method for Sampling and Analysis of Metal Fumes, Chemical Hazards in the Workplace, Measurement and Control, ACS Symp. Ser., 149 (1981) 95.

101 L.R. Betz and R.L. Grob, Anal. Lett., 17 (1984) 701.

102 Anonymous, Lead and Inorganic Compounds of Lead in Air, Laboratory Method using Atomic Absorption Spectrometry, Methods for the Determination of Hazardous Substances, MDHS 6, Health and Safety Executive, Bootle, Merseyside, April 1987.

103 P.M. Eller, Appl. Ind. Hyg., 1 (1986) 115.

104 J. Alary, J. Vandaele, C. Escrieut and R. Haran, Talanta, 33 (1986) 748.

105 A. Kovatsis and M. Tsougas, Bull. Environ. Contam. Toxicol., 15 (1976) 412.

Chapter 4d

Application of Atomic Absorption Spectrometry to the Analysis of Foods

T.C. RAINS

*Retired from National Institute of Standards and Technology;
Presently President of High-Purity Standards, Charleston, SC (U.S.A.)*

I. INTRODUCTION

Atomic absorption spectrometry (AAS) is widely used for the determination of trace elements in foods and food by-products. There are many approved (official, quasi-official, standardized, or recommended) methods which give detailed analytical methodologies regarding AAS for the food analyst. Since this list is quite long, only the most pertinent listing will be given in this chapter. The most recent addition of the method manual of the Association of Official Analytical Chemist (AOAC) [1] should be consulted. In addition to the AOAC methods, the Society for Analytical Chemistry (SAC) [2] publishes standardised and recommended methods of food analysis. Also, there are several monographs [3–6] published which depict step-by-step analysis procedures for AAS and related techniques.

While the technique of flame atomic absorption spectrometry (FAAS) is widely used for elemental analysis in the food industry, in recent years AAS with electrothermal atomization (graphite furnace) (ETA-AAS) is being widely used, especially for trace constituents. A new trend is the increasing number of laboratories which are using optical emission spectroscopy/inductively coupled plasma (OES-ICP) and inductively coupled plasma/mass spectrometry (ICP-MS). These last techniques have multielement capabilities; however, the initial cost of the instrumentation can be several times the cost of a good AAS unit. In the next several years, it is expected that these techniques will be widely used in the food analysis industry.

The occurrence of elements in foods is a function of the biological roles played by the elements in the structure and physiology of the food tissue, and adventitious contamination during growth, processing and preparation. Of the at least 90 naturally occurring elements, 26 are known to be essential for animal life; in addition, boron has been demonstrated essential for the higher plants. Of the 11 essential major elements (H, C, N, O, Na, Mg, P, S, Cl, K, Ca) and 16 trace elements (B, F, Si, V, Cr, Mn, Fe, Co, Cu, Zn, As, Se, Br, Mo, Sn, I), the following are usually determined or determinable directly by FAAS: Na, K, Ca, Mg, while B, Si, V, Cr, Mn, Fe, Cu, Zn, As, Se, Mo, and Sn are determined by ETA-AAS or by a hybrid technique. A large number of other elements occurs in foods as environmental contaminants, occasionally at levels toxic to the food organism or to the user of the food product. As an example, Cd and Pb are of current interest and are measurable by ETA-AAS while Hg is determined by a cold vapour AAS method.

To give the analyst some idea of the elemental content to be expected in foods, a listing of estimated typical ranges of some of the more important elements in 12 different classes of food is presented in Table 1.

II. SAMPLING AND PREPARATION OF ANALYTICAL SAMPLES

A. Sampling

Regardless of the analytical technique for a specific component, the result will be of little value if the sample analyzed does not represent the bulk sample and if the bulk sample does not reflect the original sample and the objective of the study. The analyst and food scientist must therefore be cognizant of several facets relevant to the investigation toward which the analytical information makes a contribution, when conducting bulk sampling, selecting a food component for analysis, and reducing the bulk sample to a laboratory sample on which the actual analytical determinations are to be made. Thus, if the objective of the study is to measure the average level of manganese in the fruit of an apple orchard, then all the apples collected would be reduced to one convenient homogeneous laboratory sample for analysis. If, on the other hand, one is interested in the variability of manganese among the individual fruits, analysis of the individual fruits would be indicated. Similarly, attention must be paid to other variables having a possible bearing on concentrations of the analyte of interest: food component (e.g. tissue type, edible part or as purchased), growing and processing conditions, time of sampling (e.g. diurnal, seasonal and inter-year variation in chemical composition), length and conditions of storage, etc.

Deliberations on sampling are beyond the scope of this treatment. For coverage of this topic, the reader is referred to several selected references. Basic sampling concepts and practices have been treated by Benedetti-Pechler [13], Walton and Hoffman [14] and Bicking [15].

B. Preparation of analytical samples

The bulk sample initially collected and consisting of a large, most likely heterogeneous mass of perhaps tens of kilograms, must be reduced in size prior to analysis. The preparation of an analytical sample (sub-sample) involves reduction in amount and in particle size and thorough mixing to yield a portion representing the average composition of the bulk sample from which 1 to 10 g test portion can be removed for chemical analysis. For mineral determinations, initial air- or oven-drying of materials such as those of plant origin followed by grinding can precede bulk reduction. Dry samples such as ground solids, powders, grains and pulses can be reduced to representative subsamples by quartering or mechanical reduction. Pulverization of solid and semi-solid foodstuffs can be effected by use of various mortars, food choppers, blenders, mixers and mills. Lyophilization (freeze drying) of bulk samples such as meat will facilitate pulverization, subsampling and storage. Freeze-drying is accomplished by freezing the sample and evaporating of the ice under vacuum. As analyte levels may vary greatly among the different components and size fractions of the sample, steps must be taken to include, in the analytical sample, all components, dust, etc., to ensure representative sampling. Liquids can be mixed by repeated slow inversion of the container or by stirring prior to sub-sampling. Care must be taken to prevent stratification in liquids and finely divided powders. Storage should be in air-tight containers at room or low temperature as required.

Due regard must be paid to the question of contamination of bulk and analytical samples, especially when dealing with trace analysis. Soil splash, road dust and salt spray must be removed from vegetation surfaces by rinsing with water; in the summer, perspiration contamination during collection can be avoided by the use of polyethylene gloves. The grinding of hard materials may result in some contamination form the grinders, e.g. by Fe, Mn, Mo, Co, Cr, Ni from steel, and Al, B, Ca, K and Na from ceramic construction materials. Metal-to-metal contact in food choppers can be a source of contamination. It is conceivable that contamination of liquid and semi-solid food may occur via attack of the sample on the metal blades and container. Storage may also result in positive or negative contamination from the storage container, and pseudo-contamination due to sample segregation, if attention is not paid to container suitability and proper re-mixing prior to analysis. It may be necessary to monitor contamination and gauge its magnitude by comparing analyte levels in the prepared sample with those in a stringently prepared separate portion of the bulk sample (less contamination), or with levels determined on a laboratory sample subjected to more extensive processing according to the prescribed procedure (more contamination). This area of food analysis remains incompletely explored.

Details are presented below for the preparation for elemental determinations, of analytical samples of some of the more common foodstuffs. It is assumed that bulk samples have been suitably collected; reduction to suitable size is to be by appropriate methods. Although in general these procedures are based on those advocated by official organisations for subsampling and storing samples, their suitability for all elements, particularly at low trace levels, has not necessarily been demonstrated. As some elements in certain food matrices may occur in volatile organoelemental forms, they may be lost during common drying procedures. It is in fact sometimes recommended, when analyzing biological reference materials, to conduct determinations on

TABLE 1

COMPOSITION (mg kg^{-1}) OF FOODSTUFFS IN RESPECT OF ELEMENTS DETERMINABLE BY ATOMIC SPECTROMETRY [a]

Food class	Na	K	Mg	Ca	B	Al	V
Cereal products	6– 6700	90– 5200	20 –6900	40 – 5000	0.03 – 17	0.4 – 300	0.002 –1
Dairy products	130–19000	320–21000	10 –2300	240–21000	0.002– 8	0.1 – 700	0.00005–0.1
Eggs and egg products	490–12000	830–11000	40 – 720	90 – 2800	0.01 – 1	0.2 – 2	
Meat and meat products	500– 1800	1400– 1800	70 –1000	20 – 200	0.1 – 3	0.5 – 10	0.001 –0.05
Fish and marine products	400–81000	800– 5400	90 –1000	100–60000	0.1 – 5	0.4 – 8	0.002 –0.05
Vegetables	10– 4600	950–19000	14 –2800	140– 2600	0.3 –125	0.02– 50	0.0008 –6
Fruit and fruit products	10– 340	350–16000	5 –1000	30– 1300	0.05 – 40	0.1 – 20	0.001 –0.06
Fats and oils	0– 9900	0– 230	1 – 480	0– 200	0.3 – 6	0.7 – 3	0.001 –0.6
Nuts and nut products	10– 40	4500– 7700	330 –4100	60– 2300	4 – 40		0.06 –0.2
Sugar and sugar products	10– 2300	10– 6200	2 –2600	5– 2300	0.05 – 8	0.6 – 110	
Beverages	0– 720	0–33000	0.3–6400	0– 1800	0.02 – 25	0.02–1300	0.00009–0.3
Spices and Condiments	50–10000	730–47000	100 –6900	800–21000	0.5 – 95	20 –1000	0.1 –1

Food class	Cr	Mn	Fe	Co	Ni	Cu	Zn
Cereal products	0.01 –2	0.5 –260	2 – 94	0.0001–1.2	0.08 –3.4	0.2 – 54	0.6 – 210
Dairy products	0.01 –2	0.005– 7	1 – 21	0.0005–0.07	0.004–1.4	0.03 – 15	0.2 – 105
Eggs and egg products	0.2 –0.5	0.08 – 2	0.3 – 108	0.005 –0.1	0.03 –0.4	0.05 – 2.4	0.1 – 62
Meat and meat products	0.06 –0.4	0.09 – 5	5 – 65	0.005 –0.7	0.02 –0.9	0.1 –100	1.0 – 500
Fish and marine products	0.01 –0.4	0.02 – 37	4 – 36	0.007 –2	0.02 –1.7	0.1 –140	2.0 –1600
Vegetables	0.002–0.7	0.04 –150	6 – 96	0.002 –5	0.01 –7	0.02 – 18	0.02– 280
Fruit and fruit products	0.01 –0.9	0.08 – 45	0.5 – 30	0.001 –0.3	0.06 –0.6	0.04 – 10	0.05– 22
Fats and oils		0.01 – 36	0.1	0.04 –0.6	0.01 –2.3	0.002– 7	0.02– 30
Nuts and nut products		0.4 – 42	22 – 60	0.04 –0.6	0.05 –1.3	0.2 – 70	0.5 – 145
Sugar and sugar products	0.02 –0.4	0.1 – 3	0.7 – 50	0.007 –0.1	0.03	0.1 – 14	0.2 – 8
Beverages	0.003–0.8	0.1 –900	0.09– 56	0.03 –0.9	0.01 –8	0.002– 50	0.02– 54
Spices and condiments	0.01 –10	0.5 –410	30 –1200	0.7 –1.2	0.4 –4	0.2 – 24	0.7 – 125

TABLE 1 (continued)

Food class	As	Se	Mo	Cd	Sn	Hg	Pb
Cereal products	0.01 – 0.5	0.002 – 4	0.06 –6	0.005–0.9	3 – 30	0.002 –0.13	0.05 –0.8
Dairy products	0.004– 1.5	0.003 – 0.2	0.00002–0.5	0.002–0.01		0.0006–0.3	0.001–0.8
Eggs and egg products	0.002– 0.2	0.02 – 1	0.1 –0.8	<0.01 –0.03		0.002 –0.07	0.03 –0.3
Meat and meat products	0.01 – 0.05	0.004 –12	0.07 –4	0.02 –0.5		0.0008–0.4	0.05 –1.2
Fish and marine products	0.1 –60	0.1 – 9	0.03 –0.6	0.01 –5	0.5– 5	0.002 –0.07	0.03 –0.3
Vegetables	0.000– 0.1	0.001 –12	0.004–6	0.001–0.1		0.001 –0.06	0.01 –1
Fruit and fruit products	0.02 – 0.2	0.001 – 0.2	0.01 –0.8	0.01 –0.05	8 –130	0.001 –0.04	0.01 –1
Fats and oils	0.02 – 0.3	(0.25)	0.2 –1	0.01 –0.1		0.001 –0.005	0.02 –1
Nuts and nut products	0.02 – 8	0.2 – 0.6				0.05 –0.2	
Sugars and sugar products	0.01 – 0.08	0.003 – 0.3	0.08 –0.9	0.005–0.03		0.0007–0.003	0.01 –0.8
Beverages	0.01 – 0.08	0.0006– 0.25	0.00003–0.2	0.01 –0.1	0.2– 50	<0.001 –0.5	0.01 –0.7
Spices and condiments		0.006 – 2	0.2	(0.8)			(3)

a Estimated typical total concentration ranges in the edible portion on a fresh, dry, processed or prepared weight basis; adapted from data reported in refs. [7–12]. Data in parentheses reflect limited information.

the undried material (as supplied) and adjust the results to a dry sample basis using the moisture content measured on a separate portion of sample. This procedure is advocated here. Wherever results are to be expressed on a fresh-weight basis, and preliminary drying is indicated, the sample weight prior to drying must be recorded. There is some discussion in the literature of errors in elemental determinations resulting from inhomogeneity [16–18], drying [17, 19, 20], and storage [17, 21]. As results may often have to be reported on a moisture-free or dry-matter basis, and occasionally on an ash basis.

C. Recommended procedures for preparation of analytical samples

1. Cereal products

Deliver bulk wheat flour samples to air-tight containers. Prior to opening containers for analysis, thoroughly mix contents by inverting and rolling the container. For samples of wheat, rye, oats, buckwheat, corn, barley and rice and their products, grind the sample to pass a 1 mm mesh sieve and mix. For samples too moist to grind readily, dry at ca. 100°C to remove excess moisture before grinding. When removing a sample for analysis, you should avoid extremes in temperatures and humidities. For all types of bread not containing fruit, cut into slices ca. 2 mm thick, dry at warm ambient temperature until sufficiently brittle, grind to pass a 1 mm mesh sieve, mix well and store in air-tight containers. Use a food chopper instead of a grinder for processing raisin bread. Mill-grind macaroni and similar products to produce ca. 500 g of material passing through a 1 mm mesh sieve and store in air-tight containers.

2. Dairy products

(i) Milk and cream

Immediately before withdrawing portions for analysis, bring the sample to ca. 20°C and mix thoroughly by agitation, taking care not to cause frothing, until a uniform emulsion forms. If the cream does not disperse, warm the sample slowly in a water bath to 30–40°C and mix until homogeneous.

(ii) Unsweetened evaporated milk

Heat the unopened can in a 40–60°C water bath to 2 h, shaking vigorously every 15 min. Remove from bath, allow to cool to room temperature, open, and thoroughly stir the contents.

(iii) Sweetened condensed milk

Heat the unopened can in a water bath at 30–40°C, open, remove entire contents and stir thoroughly until sample is homogeneous.

(iv) *Dried milk and nonfat dry milk*

Transfer the sample to a dry, air-tight container with a capacity of ca. twice the volume of the sample and mix by shaking and inverting. Sift the sample through a 0.85 mm mesh sieve if necessary to break up any lumps and return it to the air-tight container.

(v) *Butter and margarine*

Soften the sample in a sampling container by warming in a water bath at as low a temperature as practicable, but not exceeding 39°C. Shake at frequent intervals during softening procedure, remove from bath, shake vigorously until the sample cools to a creamy consistency and promptly remove the sample for analysis.

(vi) *Cheese*

Remove surface layer not usually consumed. Cut the sample into strips, pass several times through a food chopper and mix. Blend cream cheese and similar products in a high-speed blender for ca. 2–5 min until homogeneous.

Alternatively, mix well by intensive kneading. Store all samples in air-tight containers.

3. *Eggs and egg products*

Keep ca. 500 g of representative samples of liquid and frozen eggs frozen in air-tight containers. Warm in a 50°C bath and mix well before removing sample for analysis. For dried eggs, take ca. 500 g of a representative sample and store in an air-tight container; pass several times through a flour sifter to break up lumps before removing a sample for analysis.

4. *Meat and meat products*

(i) *Meats*

Render uniform, a representative sample free of bone, by passing at least twice through a food chopper with plate openings not larger than 4 mm, mixing thoroughly after each grinding. Remove casings from sausages before grinding. If not immediately required for analysis, dry either in vacuum at a temperature below 60°C or by evaporating on a steam bath with alcohol. Extract fat from the dried sample with petroleum ether (b.p. < 60°C). Evaporate the ether and store the fat in a cool place for analysis if required.

(ii) *Meat extracts*

Completely remove liquid and semiliquid preparations including sediment from the container, and mix thoroughly, warming if necessary. Grind in a mortar a representative quantity of products in solid form.

References pp. 224–226

5. *Fish and marine products*

(i) *Fresh and frozen fish*

Cut representative samples from fish with or without heads, skin and bones as dictated by the nature of the test. Grind the sample several times using a meat chopper, with 1.5–3.0 mm holes, each time removing unground material from chopper and mixing with the ground material. Alternatively, use a blender for soft fish. For fish packed in brine, drain the drain again for 2 min prior to proceeding with grinding. Let frozen fish thaw at room temperature before processing.

(ii) *Dried fish*

Cut samples into small pieces, mix, reduce by quartering to ca. 100 g, and grind as finely as possible.

(iii) *Canned products*

Blend solid and liquid contents of a can in a blender until homogeneous or pass several times through a meat chopper. For large cans, drain the meat, collect the liquid, determine the weight of the meat and volume of liquid, recombine each in proportionate amounts and blend until homogeneous.

(iv) *Shellfish*

Before opening, wash exterior of the shell with a brush and potable water to remove all loose silt and dirt, and drain. Separate the edible portion from shells of at least 10 specimens, place in a suitable tared container and determine the wet weight. Blend the sample (solid and liquid) using a blender with stainless steel blades and Teflon gaskets. After removing edible portions from shell oysters, shell clams and scallops, transfer the shellfish meats to a skimmer, pick out pieces of shell and drain for 2 min before homogenizing meat and liquid.

6. *Processed vegetable products*

(i) *Dried vegetables*

Grind the sample to pass a ca. 0.5 mm mesh sieve and store in an air-tight container.

(ii) *Canned products*

For products containing solid and liquid components thoroughly grind the entire contents of a can in a food chopper and mix. If only the solid portion is required for analysis, grind drained vegetables. Add, if necessary, an equal weight of water for proper operation of the blender. Transfer to an air-tight container. If the sample is to be kept for some time before analysis, dry and store in an air-tight container. For comminuted products shake the unopened container to homogenize the sample; transfer contents to a large receptacle and mix thoroughly by stirring. Dry and store in an air-tight container if the sample is to be kept for some time prior to analysis.

7. *Fruits and fruit products*

Pulp fresh fruits, dried fruits and preserves using a food chopper or suitable mixer or by grinding in a large mortar, and mixing thoroughly. For jellies and syrups, mix thoroughly to ensure homogeneity. For canned fruits, empty the contents on a 2 mm mesh sieve and leave to drain for two min. Invert all fruits having cups or cavities if they fall on the sieve with cups or cavities up, then proceed as for canned vegetable products.

8. *Fats and oils*

Melt solid fats and filter, using a heated funnel; filter oils that are not clear and make determinations on these homogeneous samples. Store in a cool place protected from light and air.

9. *Nuts and nut products*

Separate kernels from shells, grind twice at least 250 g in a food chopper equipped with 3 mm plate openings or other suitable devices that give a smooth homogeneous paste without loss of oil. Mix well and store in an air-tight container. For butters and pastes, mix, warming semi-solid products.

10. *Sugars and sugar products*

(i) *Sugars and syrups*
Grind solids, if necessary, and mix thoroughly. With semi-solids, dissolve 50 g in a minimum volume of water, wash into a 250 ml volumetric flask, dilute to volume and mix. Alternatively, weigh and dilute 50 g of sample with water to 100 g. Mix uniformly by shaking to disperse any solid material before taking aliquots for analysis. With liquids, mix thoroughly. Dissolve any sugar crystals by gently heating, taking care to avoid evaporative loss of water. Calculate all analytical results to the original sample basis.

(ii) *Molasses, molasses products, and confectioneries*
Pass dry molasses samples (after reaching room temperature) containing lumps through a ca. 2 mm mesh sieve, crush the residue with a pestle in a dry mortar, and add to the portion passing through the sieve, mix thoroughly and store in an air-tight container. Process liquids as mentioned above. Grind confectionery samples, mix thoroughly and store in an air-tight container.

References pp. 224–226

(iii) *Honey*

With liquid honey, mix by shaking and remove a portion for analysis. With granulated products, warm the closed container in a 60–65°C water bath until the contents are liquefied, shaking occasionally. Mix thoroughly, cool, and remove a sample for analysis. Foreign material may be removed by straining the warm (40°C) sample through cheesecloth in a heated funnel. With comb honey samples, separate the honey from the comb by straining through a 0.4 mm mesh sieve, straining through cheesecloth as outlined above, if comb or wax pass through the sieve.

11. *Beverages*

Remove CO_2 from beer and wine by transferring the sample to a large flask and shaking, or by pouring back and forth between large containers. Remove, by filtration, after removal of CO_2 and mixing, abnormal suspended matter or sediment present in beer, wine, fruit juices and other beverages marketed in the clear condition. Grind coffee and tea specimens to pass a 0.6 mm mesh sieve and store in air-tight bottles.

12. *Spices and other condiments*

Grind spices to pass through a 1 mm mesh sieve and mix. Immediately prior to analysis, thoroughly mix the material and remove a portion of not less that 2 g for analysis to insure uniformity. For prepared mustard, transfer the entire contents of the container to a vessel, stir well, store in an air-tight container and stir before removing a portion for analysis. Vinegars need only to be mixed and filtered.

For semi-solid and emulsified dressings, transfer a sample to a larger vessel, mix for ca. 2 min until homogeneous and without delay remove a portion for analysis. Blend separable dressings with 0.20 g of egg albumen powder as emulsifier per 100 g of sample in a high-speed blender. Mix thoroughly immediately prior to removing an aliquot for analysis. Correct analytical results for added emulsifier (include appropriate quantity of emulsifier in the blank determination).

13. *Plants*

Remove all foreign matter such as adhering soil and dust by rinsing with distilled water but avoid prolonged rinsing to minimize leaching of soluble mineral constituents. Dry at room temperature or in an oven at 35–40°C until the sample is sufficiently dry to be ground. Grind the sample to pass a ca. 0.5 mm mesh sieve and store in an air-tight container. Avoid contamination by dust during drying, and from grinders and sieves when determining elements such as Ca, Al, Fe and Cr abundant in dust and equipment.

14. *Animal feeds*

Reduce an amount of representative material to a laboratory-sized sample by the technique of quartering as follows, first breaking or cutting into smaller pieces material originally present in long form. Mix the sample on a clean non-absorbent surface by adding to the top of the heap from the bottom thus minimizing size and density bias; divide the heap into quarters, reject two diagonally opposite sections, thoroughly mix the two remaining quarters and repeat the procedure until a sample of convenient size, ca. 200–300 g, remains. Dry as mentioned above for plants or follow other instructions specific to the material and determination to be made (e.g. ref. [22]). Grind the sample to pass a 1 mm or smaller mesh sieve or otherwise reduce to as small particle size as possible, mix well and store in an air-tight container.

III. SAMPLE TREATMENT

Prior to quantification of analyte by atomic spectrometry it is usually necessary to destroy the organic matrix and bring the element into solution. Most of the multitude of decomposition procedures reported fall into one of two classes, wet digestion and dry ashing. Often many variants of each procedure provide adequate results with a variety of analytes and matrices. Several commonly used procedures of general applicability are described below; specific details are found in the sections dealing with the determination of individual elements. The reader is referred to several good sources of information on sample decomposition for fuller details and discussion [23–25]. The procedures described below result in essentially total decomposition of the sample with the exception of complete attack on silicate and refractory organic compounds [26]. Many useful extractive, partial digestion procedures reported in the literature are not treated in this section.

A. Recommended procedures of sample treatment

1. *Digestion with nitric and perchloric acids*

Exercise caution with perchloric acid [27]. Transfer to a 200 ml borosilicate Kjeldahl flask an accurately weighed amount of sample containing not more than 2 g of dry matter, add 25 ml of HNO_3 and boil gently for 30 min. Cool, add 15 ml of $HClO_4$ and boil gently (about 1 h) until the solution is colourless or nearly so and white fumes of $HClO_4$ are evolved. *Do not allow contents to go dry.* Quantitatively transfer the contents, by rinsing out with water, into a suitable volumetric flask and make to volume.

Notes

(1) This procedure, adapted from ref. [27], is applicable to a wide variety of biological materials, will result in the destruction of protein and carbohydrate, and will give a solution suitable for the determination of most of the mineral elements. Although small amounts of fat are tolerated, the method is not recommended for samples with a very high fat content. Due to the possibility of violent reaction with some highly reactive materials, this decomposition procedure should be first tested on small quantities of materials of unknown composition with due regard to possible hazards. Abide by safety precautions when working with perchloric acid [27].

(2) With such large volumes of acids, charring of sample will not usually occur. Digestions should not, however, be left unattended and more HNO_3 in 1 ml portions should be added whenever darkening of the digestion mixture occurs. This procedure has the disadvantage of requiring a higher proportion of acids to sample than may be desired. The ratio of acids to sample may be reduced with attention to appropriate operational information. Close operator attention is required; charring of the sample, indicating the onset of potentially dangerous conditions must be rectified by the addition of HNO_3 as mentioned above. The references cited above and original literature dealing with the materials and elements of interest should be consulted for details of operation and safety.

2. *Digestion with nitric, perchloric and sulphuric acids*

Transfer to a 100 ml Kjeldahl flask 2 g of an accurately weighed sample and add 5 ml of HNO_3. After vigorous initial reaction subsides, heat gently until further reaction ceases, cool, slowly add 8 ml of H_2SO_4 and heat until the liquid darkens appreciably. Add gradually a 1–2 ml portion of HNO_3, and heat until darkening occurs again, but do not heat so strongly as to cause excessive charring. Repeat the addition of HNO_3 and heating until the solution no longer darkens on heating for 5–10 min and is only pale yellow. Add 1 ml of $HClO_4$ and 2 ml of HNO_3 and heat for 15 min. Add a second 1 ml portion of $HClO_4$, heat for several minutes, cool, and add 10 ml of water. Gently boil the generally colourless solution to the evolution of white fumes of $HClO_4$, cool, add 5 ml of water and again boil to white fumes. Cool the solution, add 5 ml of water, quantitatively transfer to a suitable volumetric flask and make to volume.

3. *Dry ashing*

Accurately weigh 5–10 g of sample into a suitable silica or platinum crucible and spread thinly over the bottom. Add ashing aid if required. Dry and char the sample using an infrared lamp, hot plate or burner, taking care not to cause ignition. Place into a cold muffle furnace, raise the temperature slowly to 450–500°C and heat overnight, again not allowing the contents to ignite. If unoxidized organic matter remains, moisten the residue with water or (1 + 2) HNO_3, evaporate to dryness on a water bath and heat again in the furnace for a further period. When a suitable ash has been obtained, cool, moisten with water, carefully add 10 ml of (1 + 1) HCl and evaporate to dryness on a water bath. Dissolve the residue in (1 + 9) HCl or another suitable solvent.

4. *Microwave digestion*

Transfer 0.5–1 g sample to a 120 ml Teflon digestion vessel. Add 15 ml of HNO_3 and swirl contents in vessel. (If there is a reaction and nitrous oxide fumes are evolving, allow the sample to stand for 10 min before proceeding.) Attach cap and tighten in a capping station. Place vessel in the carousel and connect the venting tube. Place the carousel in the oven and set the power at 100% for 2 min, then 40% for 10 min. After the digestion program is complete, remove the carousel from the oven and disconnect the vent tube. Cool the digestion vessel in a cold water bath or with liquid nitrogen. Uncap the vessel with the capping station in a fume hood. Wash down with a small quantity of deionized water the cap, relief valve, and the inside walls of the vessel. Add 5 ml of 30% H_2O_2 and swirl the contents in the vessel. Place the uncapped vessel in the oven and set the power at 100% and digest for 30 min. Near the end of the digestion period, observe the volume of sample and evaporate the solution to 2–3 ml. Cool the vessel and transfer the solution to a volumetric flask and dilute to calibrated volume with deionized water.

Note

When digesting organic materials in a closed vessel, considerable pressure is produced by gaseous products. For some samples it may be necessary to remove the easily oxidizable materials by allowing the vessel to stand at ambient temperature or predigesting on a hot plate. For samples high in fat, predigest the sample in the HNO_3 for 30 min before capping and digesting. This has been found to reduce the pressure buildup in the vessel and minimize the loss of sample. Also, it is advisable to pretest a smaller sample weight for materials that have not been digested before to determine their reaction in a closed vessel [27, 28].

IV. RECOMMENDED ANALYTICAL PROCEDURES

A. General analytical protocol

1. *Apparatus*

The instrumentations used in AAS and ETA-AAS for food analysis are described in the literature. If there are any queries about the applications and operation of the instrumentation, the analyst should consult the instrument manufacturer.

2. *Reagents*

All reagents should be prepared from high-purity chemicals and checked prior to use. This is essential, especially in trace metals, when the analytical blank can be a limiting factor.

3. *Standard solutions*

Standard stock solutions should be prepared from high-purity metals or salts of known stoichiometry. High-purity materials for virtually every element are available commercially. A standard stock solution should contain at least 1000 μg/ml of the analyte in an appropriate acid to prevent hydrolysis. Working solutions should be prepared daily from this stock solution. Procedures for the preparation and storage of standard solutions have been given by Ihnat [12] and Dean and Rains [29, 30]. The analyst must remember that the accuracy of the determination can never be any better than the accuracy of the standard.

4. *Reference materials*

As a part of any laboratory's quality assurance program, a reference material should be carried through the entire analytical procedure. Table 2 lists several of the Standard Reference Materials (SRM) available from the National Institute of Standards and Technology (NIST). Table 3 gives the certified values for a few selected SRMs.

5. *Instrumental operating parameters*

As there has been so much literature published by the instrument manufacturers describing the operating parameters of their instrumentation,

TABLE 2

U.S. NATIONAL INSTITUTE OF STANDARDS AND TECHNOLOGY, BIOLOGICAL MATRIX STANDARD REFERENCE MATERIALS FOR CHEMICAL COMPOSITION

SRM 1548	Total diet	SRM 1572	Citrus leaves
SRM 1549	Non-fat milk powder	SRM 1573a	Tomato leaves
SRM 1566a	Oyster tissue	SRM 1575	Pine needles
SRM 1567a	Wheat flour	SRM 1577a	Bovine liver
SRM 1568a	Rice flour	SRM 8412	Corn stalk
SRM 1569	Brewers yeast	SRM 8413	Corn kernel
SRM 1570a	Spinach		

TABLE 3

CERTIFIED CONCENTRATIONS (mg kg⁻¹) OF CONSTITUENT ELEMENTS IN U.S. NATIONAL INSTITUTE OF STANDARDS AND TECHNOLOGY BIOLOGICAL MATRIX STANDARD REFERENCE MATERIALS[a]

Element	SRM 1567a Wheat flour	SRM 1568 Rice flour	SRM 1570 Spinach	SRM 1572 Citrus leaves	SRM 1573 Tomato leaves	SRM 1566a Oyster tissue
Al	5.7 ± 1.3		870 ± 50	92 ± 15		202.5 ± 12.5
As		0.41 ± 0.05	0.15 ± 0.05	3.1 ± 0.3	0.27 ± 0.05	14.0 ± 1.2
Ca	191 ± 4	140 ± 20	13500 ± 300	31500 ± 1000	30000 ± 300	1900 ± 190
Cd	0.026 ± 0.002	0.029 ± 0.004		0.03 ± 0.01		4.15 ± 0.38
Co		0.02 ± 0.01				0.57 ± 0.11
Cr			4.6 ± 0.3		4.5 ± 0.5	1.43 ± 0.46
Cu	2.1 ± 0.2	2.2 ± 0.3	12 ± 2	16.5 ± 1.0	11 ± 1	66.3 ± 4.3
Fe	14.1 ± 0.5	8.7 ± 0.6	550 ± 20	90 ± 10	690 ± 25	539 ± 15
Hg		0.0060 ± 0.0007	0.030 ± 0.005	0.08 ± 0.02		0.0642 ± 0.0067
K	1330 ± 30	1120 ± 20	35600 ± 300	1820 ± 60	44600 ± 300	7900 ± 470
Mg	400 ± 20			5800 ± 300		1180 ± 170
Mn	9.4 ± 0.9	20.1 ± 0.4	165 ± 6	23 ± 2	238 ± 7	12.3 ± 1.5
Na	6.1 ± 0.8	6.0 ± 1.5		160 ± 20		4170 ± 130
Ni				0.6 ± 0.3		2.25 ± 0.44
Pb			1.2 ± 0.2	13.3 ± 2.4	6.3 ± 0.3	0.371 ± 0.014
Se	1.1 ± 0.2	0.4 ± 0.1				2.21 ± 0.24
Zn	10.6 ± 0.4	19.4 ± 1.0	50 ± 2	29 ± 2	62 ± 6	830 ± 57

[a]Concentration ± uncertainty on a dry sample basis from NIST certificates [31]. The certificates also contain informational data on other elements.

it will not be described in this chapter. The analyst should consult his manufacturer for these details.

Flame AAS is widely used to determine the major analytes in the food industry. For those analytes that are below the detection limits by FAAS, the analyst can preconcentrate the analyte or resort to the more sensitive ETA-AAS. If ETA-AAS is used, the analyst should use digestion techniques that are suitable for ETA-AAS. For example, H_2SO_4 should be avoided and HNO_3 is recommended for most analytes. Also a matrix modifier is essential to prevent analyte losses or alleviate interferences by ETA-AAS. While some 40 different matrix modifiers are used in ETA-AAS procedures, only those found successful by this author are given in Section B below. This is not to say that these are the only matrix modifiers to be used. But the analyst should evaluate the different matrix modifiers and use the one which produces the best analytical results.

6. *Interferences*

Interferences are discussed at great detail in the literature [17, 32, 34]. Interferences encountered and ways to alleviate them in food analyses by FAAS are widely known. However, it is advisable to use a suitable reference material to evaluate potential interferences and the techniques used to alleviate them.

Interferences encountered in ETA-AAS are not known or understood as thoroughly as those in FAAS. Therefore, a reference material must be carried through the procedure and interferences evaluated before proceeding with the analysis.

7. *Preparation of calibration curves*

There are several ways to prepare a standard calibration curve. The simplest method is to prepare a series of standard aqueous or organic solutions of the analyte to cover the desired concentration range and to measure their absorbance at an appropriate wavelength. Failure to select the appropriate concentration range could lead to a non-linear calibration curve. Solutions of the unknown sample are then measured and their elemental concentration determined from the calibration curve. This procedure is often the least reliable because of matrix or interelement effects. The highest accuracy is always attained when the matrix of the standards and samples are matched.

When working with non-linear calibration curves, three approaches are possible. If a calibration curve is non-linear, it is advisable to increase the number of standard solutions to accurately determine the shape of the curve and closely bracket the unknown with standard solutions. This tends to reduce the error. A second approach is to dilute the sample and use lower standards that are on the linear portion of the calibration curve. A third approach is curve fitting with a microcomputer. The accuracy of a calibration curve computed for a non-linear relationship depends on the number of standards and the equation used for calibration. Regardless of the method used, it is advisable to repeat the standard solutions at a frequent interval in the analysis [32].

B. Determination of specific elements

1. *Sodium and potassium*

(i) *Summary of procedure*
Sodium and potassium are extracted from solid samples by wet or microwave digestion or dry ashing followed by acid dissolution of residue, and are determined by flame AAS using an air–acetylene flame. Liquid samples are aspirated after dilution.

(ii) *Preparation of sample solution*
Marine products. Weigh 1 g of sample into a 50 ml Pyrex beaker, add 5 ml of HNO_3 and allow to digest overnight at ambient temperature. Heat on a hot plate set at low temperature until the sample dissolves and a clear red solution results, then take to dryness. Repeat the digestion twice more, finally add 2 ml of HNO_3 and dissolve the solids by warming. Filter the solution through nitric acid-cleaned glass wool into a 100 ml volumetric flask, rinsing out the beaker with hot distilled or deionized water, cool and dilute to volume. For determination of the elements using the most sensitive wavelengths, dilute the sample solution with water by a factor of 25 and 5 for sodium and potassium, respectively. For samples with high oil content, defat the dry sample (ca. 2 h at 110°C) by extracting on a steam bath with successive 10 ml portions of petroleum ether before proceeding with the digestion. Alternatively, weigh 5 g of sample into a nitric acid-cleaned Vycor, silica or platinum crucible and char by heating under an infrared lamp or on a hot plate. Place in a cold muffle furnace, slowly bring the temperature to 500°C and ash for ca. 2 h. Dissolve the residue in 15 ml of (1 + 4) HNO_3, filter through acid-washed filter paper into a 100 ml volumetric flask and dilute to volume. For determinations using the most sensitive wavelengths, dilute the sample solution with water by factors of 100 and 25 for sodium and potassium, respectively.
Fruits and fruit products. Weigh 300 g of jelly, syrup, fresh or dried fruit or preserve into a 2 l beaker, add ca. 800 ml of water and extract by gently boiling for 1 h, replacing water lost by evaporation. Filter into a 2 l volumetric flask, cool, dilute to volume, and dilute again if necessary to bring the analyte concentration to a level suitable for FAAS.
Beverages. For wine samples dilute with water by a factor of 50–200; dilute

mineral water samples as required. Dilute liquors with 50% ethanol by a factor required to bring absorbance onto the spectrometer scale. Prepare standards in identical diluent.

Plant materials. Transfer an accurately weighed 1-g sample into a silica crucible, ash for 2 h at 500°C and let cool. Wet the ash with 10 drops of water and carefully add 3–4 ml of (1 + 1) HNO_3. Evaporate excess HNO_3 on a 100–120°C hot plate, return the crucible to the furnace and ash for 1 h at 500°C. Cool, dissolve the ash in 10 ml of (1 + 1) HCl, transfer the solution to a 50 ml volumetric flask and dilute to volume. Alternatively, weigh 1 g of sample into a 150 ml Pyrex beaker, add 10 ml of HNO_3 and allow to soak thoroughly. Add 3 ml of $HClO_4$ and heat on a hot plate, slowly at first, until frothing ceases. Heat until the HNO_3 is almost evaporated. Should charring occur, cool, add 10 ml of HNO_3 and continue heating. Heat to fumes of $HClO_4$, cool, add 10 ml to (1 + 1) HCl, transfer the solution to a 50 ml volumetric flask and dilute to volume. In both the dry ashing and wet digestion procedures allow the silica to settle and use the supernatant for determination.

(iii) *Determination*

Aspirate solutions into an air–acetylene flame (fuel-lean for sodium measurement) and measure absorbances at 589.00 nm and 766.49 nm for sodium and potassium, respectively. Use less sensitive wavelengths of 330.24–330.30 nm and 404.41 nm for sodium and potassium, respectively, when more concentrated (by a factor of about 200) solutions are used.

2. *Magnesium and calcium*

(i) *Summary of procedure*

The sample is brought into solution by wet or microwave digestion or by dry ashing followed by acid dissolution. The concentrations of magnesium and calcium are determined by flame AAS using an air–acetylene and nitrous oxide–acetylene flame for magnesium and calcium, respectively.

(ii) *Preparation of sample solution*

In general, digest sample with HNO_3–$HClO_4$ (Section III.A.1 or III.A.4), transfer the solution to a volumetric flask of suitable size and make up to volume. For plant materials, follow either the dry ashing or wet digestion procedure described in Section IV.B.1(ii).

(iii) *Determination*

Aspirate solutions into a fuel-lean air–acetylene flame to measure magnesium at 285.21 nm and into a fuel-rich nitrous oxide–acetylene flame to determine calcium at 422.67 nm.

3. *Boron*

(i) *Summary of procedure*

The sample is subjected to pressure digestion with HNO_3, or dry ashing, boron in the digest is chelated with 2-ethyl-1, 3-hexanediol, extracted with MIBK and its concentration determined by FAAS using a nitrous oxide–acetylene flame [35].

(ii) *Preparation of sample solution*

Exercise caution when digesting organic materials in a pressure digestion vessel. (Some advise against it). Transfer 1 g of an accurately weighed food sample into a Teflon pressure digestion vessel, of suitable volume, add 5.0 ml of HNO_3 and close tightly. Digest the sample by heating vessel at 150°C for 1 h in an oven. Remove from heat, allow to cool, and transfer contents with the aid of 10 ml of water to a 125 ml separatory funnel marked at 25 ml. Make the solution basic to litmus paper by slowly adding NH_4OH (ca. 5 ml is required). Acidify by the dropwise addition of (1 + 1) H_2SO_4. Add 10 ml of chelating–extracting solution consisting of 15% v/v 2-ethyl-1, 3-hexanediol in methyl isobutyl ketone (MIBK) and shake for 1 min. Let the phases separate and decant the organic layer into a 25 ml volumetric flask. Repeat the chelation–extraction with another 10 and 5 ml of chelating–extracting solution, combining organic phases into the volumetric flask. Dilute to volume with chelating–extracting solution and mix.

For plant material, weigh 10 g of sample into a platinum or porcelain crucible and moisten with ca. 20 ml of saturated $Ba(OH)_2$ solution. Dry 1 h at 150°C and ash for 10 h in a muffle furnace at 600°C. Remove from the furnace, add 10 ml of 5 N HCl and triturate with a polyethylene rod. Evaporate at 150°C to less than 6 ml, add 3 ml of extracting solution and stir vigorously. Decant into a centrifuge tube and let the phases separate.

(iii) *Determination*

Aspirate organic solutions into a nitrous oxide–acetylene flame previously optimised on boron standards in an identical organic solvent matrix and measure boron absorbance at 249.68–249.77 nm. Prepare boron standards by chelating and extracting aqueous boron solutions in the same manner as sample digests. Flush the burner with extracting solution after aspiration of each standard and sample solution.

4. *Aluminium*

(i) *Summary of procedure*

The sample is brought into solution by wet or microwave digestion and aluminum is quantified by flame AAS using a nitrous oxide–acetylene flame.

(ii) *Preparation of sample solution*

Digest the sample by one of the acid wet-digestion procedures described above (Sections III.A.1, III.A.2, III.A.4), transfer the solution to a volumetric flask of suitable small size and make up to volume.

(iii) *Determination*

Aspirate solutions into a fuel-rich nitrous oxide–acetylene flame and measure Al absorbance at 309.27–309.28 nm.

5. *Vanadium*

(i) *Summary of procedure*

The sample is brought into solution by wet or microwave digestion or dry ashing-acid dissolution, and vanadium is determined by FAAS using a nitrous oxide–acetylene flame, or by ETA-AAS with 1% HNO_3 as matrix modifier.

(ii) *Preparation of sample solution*

Digest the sample by one of the wet digestion procedures described above (Sections III.A.1, III.A.2, III.A.4), transfer solution to a small volumetric flask and dilute to volume. Alternatively, dry ash according to Section III.A.3 and the following specific instructions. Transfer 5–20 g of sample into a platinum crucible, char or evaporate to dryness on a hot plate and ash at 450°C for 16 h in a muffle furnace. Remove, add 2 ml of HNO_3, and evaporate to dryness on a hot plate. Re-ash the residue for 2 h at 450°C, dissolve in 2 ml of 4 N HNO_3, warm on a hot plate, transfer to a 5 ml volumetric flask and dilute to volume.

(iii) *Determination*

Aspirate solutions into a fuel-rich nitrous oxide–acetylene flame and measure vanadium absorbance at 318.34/318.40/318.54 nm using a spectral bandpass of 0.2 nm to pass all three emission lines. For samples with low vanadium contents, use ETA-AAS with HNO_3 as matrix modifier and background correction.

6. *Chromium*

(i) *Summary of procedure*

The sample is brought into solution by wet or microwave digestion and chromium is determined by FAAS using a nitrous oxide–acetylene flame or ETA-AAS. Vegetable oils are dissolved in MIBK prior to FAAS determination.

(ii) *Preparation of sample solution*

Digest the sample with HNO_3–$HClO_4$ according to the procedure outlined in Section III.A.1, transfer the solution to a small volumetric flask and dilute to volume. With vegetable oils, accurately weigh 5 g into a 25 ml volumetric flask, dilute to volume with MIBK and mix.

(iii) *Determination*

Aspirate standards and solutions resulting from HNO_3–$HClO_4$ digestion into a stoichiometric or fuel-lean nitrous oxide–acetylene flame and measure chromium absorbance at 357.87 nm. For vegetable oils dissolved in MIBK use the tentative AOAC procedure [36] as follows. Prepare organo-chromium stock solution containing 100 μg Cr g^{-1} by dissolving with warming 0.0516 g of dried tris(1-phenyl-1,3-butanediono)chromium(III) (NBS No. 1078a, 9.7% Cr) in 3 ml of xylene and 3 ml of 2-ethylhexanoic acid and making to 50 ml with a base oil (vegetable oil containing no more than 0.02 mg of analyte kg^{-1}). Prepare working standard solutions by diluting stock solution with base oil–MIBK to give the required concentration of chromium in the oil–MIBK matrix similar to that of sample solution. Aspirate solutions into an air–acetylene flame optimised for these organic solutions and measure absorbance at 357.87 nm. For samples containing Cr below the FAAS detection limits, determine the Cr by ETA-AAS with $(NH_4)_2HPO_4$ as matrix modifier.

7. *Manganese*

(i) *Summary of procedure*

Samples are wet- or microwave-digested or dry-ashed and manganese is determined by FAAS using an air–acetylene flame or by ETA-AAS.

(ii) *Preparation of sample solution*

In general, digest the sample with HNO_3–$HClO_4$ (Section III.A.1), transfer the solution to a small suitable volumetric flask and dilute to volume. Alternatively, dry-ash the sample according to instructions in Section III.A.3, transfer the solution to a volumetric flask and dilute to volume. For plant materials, follow either the dry-ashing or wet-digestion procedure described in Section IV.B.1(ii).

(iii) *Determination*

Aspirate solutions into a stoichiometric air–acetylene flame and measure manganese absorbance at 279.48 nm or by ETA-AAS with $(NH_4)_2HPO_4$ as a matrix modifier.

8. *Iron*

(i) *Summary of procedure*

After wet- or microwave-digestion or dry ashing, iron in samples is determined by FAAS using an air–acetylene flame or by ETA-AAS.

(ii) *Preparation of sample solution*

For a variety of cereal products, weigh 10 g into a platinum, Vycor or silica crucible and ash in a furnace at ca. 550°C until a light grey ash results. To facilitate removal of the last traces of carbon, moisten the ash with 0.5–1.0 ml of $Mg(NO_3)_2$ solution or with HNO_3. Do not add these ashing aids to products containing NaCl in a platinum crucible. Dry, carefully ignite in a furnace and cool. Add 5 ml of HCl and evaporate on a steam bath. Dissolve the residue in 2.0 ml of HCl, cover with a watch glass, and heat for 5 min on a steam bath. Transfer via a filter quantitatively into a 100 ml volumetric flask and dilute to volume. Alternatively, transfer 10 g of sample to an 800 ml Kjeldahl flask, separately add 20 ml of water, 5 ml of H_2SO_4 and 25 ml of HNO_3, mixing after the addition of each reagent. Allow to stand for several minutes, then heat gently at intervals until heavy evolution of NO_2 stops. Continue heating gently until charring begins and cautiously add several ml of HNO_3 at intervals until SO_3 is evolved and a colourless or very pale yellow liquid results. Cool, add 50 ml of water and one Pyrex glass bead, heat to SO_3 fumes, cool, add 25 ml of water, transfer via a filter to a 100 ml volumetric flask and dilute to volume.

For vegetable oils containing at least 2 mg Fe kg^{-1} accurately weigh 5 g into a 25 ml volumetric flask, dilute to volume with MIBK and mix.

For wines, dilute 20 ml of sample with 88 ml of 95% ethanol and water, to 200 ml in a volumetric flask. Analyze distilled liquor without preparation; prepare standard solutions of Fe in 43% v/v ethanol.

For beer, pipet 25 ml into a 100 ml stoppered flask. Equilibrate flask in water bath at 20°C for 30 min, add 2.0 ml of 1% APDC solution (22), mix and add 10.0 ml of MIBK. Shake vigorously for 5 min and centrifuge to separate layers. Equilibrate at 20°C for 10 min.

Process plant samples by dry ashing according to the procedure in Section IV.B.1(ii), reducing by heating the volume of (1 + 1) HCl to ca. 5 ml so that when diluted to 50 ml, the final solution will be ca. 0.5 M with respect to HCl. Pipet 20 ml or less of the digest into an extraction tube and add 5.0 ml of 1.0 M metal-free citrate buffer adjusted to pH 4.7 (bromphenol blue). Add dropwise (1 + 1) NH_4OH to the colour change, plus three drops in excess. Add 2 to 10 ml of 3-heptanone followed by water to bring the top of the solution near the constriction. Add 2.0 ml of 1% APDC and additional water to bring the solvent interface to the centre of the constriction. Stopper, and shake for 1 min.

(iii) *Determination*

Aspirate solutions into a fuel-lean air–acetylene flame and measure iron absorbance at 248.33 nm. For low-level Fe in wines and beer add 1% $(NH_4)_2HPO_4$ and determine the Fe directly by ETA-AAS. For procedures involving chelation–extraction, treat standard solutions in the same manner as samples. Aspirate organic solutions into a flame previously optimised on the organic solvent. Flush burner with extracting solution after aspiration of each standard and sample solution. Carry out determinations on beer and plant sample extracts within 30 min and 2 h, of extraction respectively. For vegetable oils in MIBK, follow the procedure described in Section IV.B.6(iii). Prepare organo-iron stock solution from 0.0486 g of dry tris(1-phenyl-1,3-butanediono)iron(III) (NBS No. 1079a, 10.30% Fe, dried 1 h at 100°C).

9. Cobalt

(i) Summary of procedure

Samples are wet- or microwave-digested, cobalt is chelated with 2-nitroso-1-napthol, solvent-extracted and determined, wether by FAAS using an air–acetylene flame, or by ETA-AAS.

(ii) Preparation of sample solution

For FAAS determination on plant materials weigh 2 g of sample into a 100 ml Kjeldahl flask and digest with 30 ml of HNO_3 + 5 ml of $HClO_4$. Continue heating for 1 h past the appearance of white fumes of $HClO_4$ to destroy refractory organic matter and to dehydrate the silica. Dilute to ca. 40 ml, heat to dissolve any precipitate of $KClO_4$ and let cool. Add 10 ml of 40% m/v sodium citrate solution and one drop each of bromthymol blue and methyl red indicator solutions. Adjust pH to within the range 5.3–5.7 (orange to golden orange colour) with 10 M NH_4OH correcting any overshoot with ca. 6 M HCl. Add 5 ml of 6% v/v H_2O_2 to reduce Fe^{3+} to Fe^{2+} and decant the solution from the precipitate into a 100 ml separatory funnel. Add 1 ml of 1% m/v ethanolic solution of 2-nitroso-1-napthol, mix and allow to stand for 1 h. Serially extract the cobalt chelate with 10, 5 and 5 ml portions of $CHCl_3$ by shaking each time for 1 min and collect the pooled extracts in a 50 ml glass-stoppered weighing bottle. Discard the aqueous phase by pouring out of the funnel and rinse the funnel with water, taking care that no aqueous phase enters the stopcock or stem of the funnel. Return the organic phase to the separatory funnel rinsing in with 5 ml of $CHCl_3$. Add 10 ml of 2 N NaOH to the extract and shake for 15 s to remove excess 2-nitroso-1-napthol. Return the organic phase to the weighing bottle and evaporate to dryness. Let cool, add 1 ml of MIBK, stopper, and carefully dissolve the residue.

For ETA-AAS determination on plant materials, digest 0.5 g of sample with 15 ml of HNO_3 + 2.5 ml of $HClO_4$ and reduce the volume of $HClO_4$ to ca. 0.5 ml. Let cool, rinse the wall of the flask with 5 ml of water and warm the solution to dissolve any $KClO_4$ precipitate. When the digests have cooled, add 1 ml of 40% m/v sodium citrate solution, one drop each of bromthymol blue and methyl red indicator solutions and adjust pH as described above with concentrated NH_4OH. Correct any overshoot with 3 M HCl or 2 M NH_4OH. Add 1 ml of 30% H_2O_2 followed by 0.3 ml of 1% m/v ethanolic solution of 2-nitroso-1-napthol and transfer to a 15 ml glass-stoppered graduated cylinder. Rinse the flask with two portions of 1.5 ml of water and mix the solution. Make the final volume of solution in the cylinder to 11–12 ml, add 1 ml of isoamyl acetate and shake vigorously for 30 s.

(iii) *Determination*

For FAAS determination, aspirate MIBK solution into an air–acetylene flame, previously optimised on cobalt standards in an identical matrix, and measure cobalt absorbance at 240.73 nm. Flush the burner with MIBK after aspiration of each standard and sample solution. For ETA-AAS determination, introduce an appropriate small volume of isoamyl acetate solution into an electrothermal atomizer, and proceed with drying, charring and atomization. For both types of determinations, prepare cobalt standard solutions by chelating and extracting aqueous cobalt standard solutions in the same manner as sample digests, beginning with the addition of the sodium citrate solution.

10. *Nickel*

(i) *Summary of procedure*

The sample is brought into solution by wet- or microwave-digestion or dry ashing. Nickel is concentrated if required by chelation–solvent extraction, and is determined by FAAS using an air–acetylene flame. Vegetable oils are dissolved in MIBK prior to FAAS measurement.

(ii) *Preparation of sample solution*

Subject the sample to one of the wet-digestion or dry-ashing procedures described in Sections III.A.1, III.A.3, or III.A.4, transfer the solution to an appropriate volumetric flask and dilute to volume. Use this solution for direct FAAS determination when the nickel content of the product (e.g. tea) is sufficiently high. For the more typical case of low nickel levels, follow for plant materials the dry-ashing APDC-3-heptanone extraction procedure outlined in Section IV.B.8(ii) using an agitation time of 20 s after the addition of APDC solution. With vegetable oils, accurately weigh 5 g into a 25 ml volumetric flask, dilute to volume with MIBK and mix.

(iii) *Determination*

Aspirate solutions into a fuel-lean air–acetylene flame and measure nickel absorbance at 232.00 nm. Aspirate organic solvent solutions into the same flame, previously optimised on nickel standards in an identical matrix. Flush the burner with 3-heptanone after aspiration of each standard and sample solution. For vegetable oils in MIBK, follow the tentative AOAC procedure [36] as described in Section IV.B.6(iii). Prepare organo-nickel stock solution from 0.0360 g of Ni cyclohexane butyrate (NBS No. 1065b, 13.89% Ni, desiccated for 48 h over P_2O_5) according to the procedure in the same section. Aspirate sample and standard solutions into an optimised air–acetylene flame.

11. *Copper*

(i) *Summary of procedure*

The sample is brought into solution by wet digestion or dry ashing and copper is determined by FAAS using an air–acetylene flame. Vegetable oils are dissolved in MIBK, whereas other liquid samples require no or minimal treatment. When necessary, copper is concentrated by chelation–solvent extraction prior to measurement by FAAS or ETA-AAS.

(ii) *Preparation of sample solution*

In general, destroy organic matter using one of the wet-digestion or dry-ashing procedures described in Sections III.A.1–III.A.3, and transfer the solution to a volumetric flask of suitable small volume. For dry ashing, use platinum or silica crucibles; avoid porcelain as copper may be extracted from the glaze. If an ashing aid is required, use $Mg(NO_3)_2$. Dissolve the residue in $(1 + 1)$ HCl or $(2 + 1 + 3)$ $HCl-HNO_3-H_2O$, provided that the latter mixture is not used in a platinum vessel.

Alternatively, for solid foodstuffs in general, proceed according to the following wet-digestion procedure. Accurately weigh a sample containing not more than 20 g solids, according to expected Cu content, and transfer to an 800 ml Kjeldahl flask. If sample contains less than 75% water, add water to obtain this dilution. Add an initial volume of HNO_3 to equal ca. twice the dry sample weight and as many ml H_2SO_4 as grams of dry sample but not less than 5 ml. Warm slightly and carefully to avoid excessive foaming. When reaction has subsided, heat cautiously and maintain oxidizing condition by cautiously adding small amounts of HNO_3 whenever the mixture darkens. Digest until organic matter is destroyed and SO_3 fumes are evolved and the solution is colourless or light straw-colored. Cool, add 25 ml of water and remove nitrosylsulphuric acid by heating to fumes of SO_3. Repeat addition of water and fuming, cool, transfer to 100 ml volumetric flask and make to volume. When the sample contains a large amount of fat, make a partial digestion with HNO_3 until only the fat remains undissolved. Cool, filter free of solid fat, wash residue with water, add H_2SO_4 to the filtrate and proceed as above.

For vegetable oils containing at least 2 mg Cu kg^{-1}, accurately weigh 5 g into a 25 ml volumetric flask, dilute to volume with MIBK and mix. Wines, distilled liquors, beer and other beverages may be analyzed directly, filtering through a membrane filter if necessary to remove suspended matter, and diluting if required to lower dissolved solids content. For plant material, follow the dry-ashing or wet-digestion procedure described in Section IV.B.1(ii).

In instances of low copper levels, treat the sample digest resulting from any of the preceding ashing or digestion procedures according to the following procedure of chelation–extraction and FAAS. Ensure that the acidity of the digest is not greater than 5 N when chelation–extraction is performed. Pipet a suitable aliquot into a narrow neck flask and dilute to the base of the neck. Add 2 ml of 1% aqueous solution of APDC, mix and allow to stand for 2 min. Add 4 ml of MIBK or 3-heptanone, shake for 1 min and let the phases separate. Use the organic layer without delay for FAAS measurement.

Alternatively, for small quantities of plant material, use the following procedure in conjunction with ETA-AAS. Weigh 0.01–0.05 g of very finely divided sample into a 15 × 150 mm Pyrex test tube and add 1 ml of (1 + 8) $HClO_4/HNO_3$. For larger quantities of sample, use larger volumes of acids and ensure that sufficient HNO_3 is present for safe digestion. Heat to fumes of $HClO_4$, dissolve and transfer the residue to a 10 ml glass-stoppered cylinder using 9 ml of purified (1 + 30) $HClO_4$. Add 1 ml of 1% aqueous APDC, mix, add 1 ml of MIBK and shake vigorously for 15 s. Allow the solution to stand for 1 h to permit phase separation prior to ETA-AAS.

(iii) *Determination*

Aspirate solutions into a fuel-lean air–acetylene flame and measure copper absorbance at 324.75 nm. For vegetable oils in MIBK, follow the procedure described in Section IV.B.6(iii). Prepare organo-copper stock solution from 0.0303 g of bis(1-phenyl-1,3-butanediono)Cu(II) (NBS No. 1080, 16.5% Cu, dried for 30 min at 110°C) according to the procedure. For distilled liquor and beer, prepare standard solutions in 50% and 3% v/v ethanol, respectively. For FAAS measurements on organic solutions, aspirate solutions into an air–acetylene flame previously optimised on copper standards in an identical matrix. Flush burner with appropriate solvent after aspiration of each standard and sample solution. For ETA-AAS determination, introduce an appropriate small volume of MIBK extract into an electrothermal atomizer and follow a previously established drying, charring and atomization program. For both chelation–solvent extraction procedures, prepare Cu standard solutions by chelating and extracting aqueous standard solutions in the same manner as sample digests.

12. *Zinc*

(i) *Summary of procedure*

The sample is brought into solution by wet or microwave digestion or dry ashing, and zinc is determined by FAAS, using an air–acetylene flame. Liquid samples require no or minimal treatment.

(ii) *Preparation of sample solution*

In general, destroy organic matter using one of the wet-digestion [37] or dry-ashing procedures described in Sections III.A.1–III.A.4, and transfer the solution to a volumetric flask of suitable volume. For dry ashing, use platinum or silica crucibles; avoid porcelain as zinc may be extracted from the glaze. Addition of 0.2 g of $CaCO_3$ to the sample before ashing facilitates ashing and dissolution of residue. Dissolve the residue in (1 + 1) HCl or (2 + 1 + 3) HCl–HNO_3–H_2O, provided that the latter mixture is not used in a platinum vessel.

Alternatively, for foodstuffs in general, proceed according to one of the following wet-digestion of dry-ashing procedures. Where glassware is called for, use only borosilicate glassware, thoroughly cleaned with hot HNO_3. Clean platinum crucibles by fusion with $KHSO_4$ followed by leaching with 10% HCl.

Wet digestion. Accurately weigh, into a 300 ml or 500 ml Kjeldahl flask, up to 10 g of sample estimated to contain 25–100 μg Zn. Evaporate liquid samples to small volume before digesting. Add ca. 5 ml of HNO_3 and cautiously heat until vigorous reaction subsides. Add 2.0 ml of H_2SO_4 and continue heating, maintaining oxidizing conditions by adding small quantities of HNO_3 until the solution is colourless and fumes of SO_3 are evolved. Cool, dilute with 20 ml of water, transfer quantitatively to a 100 ml volumetric flask and dilute to volume. Dilute further if required, with (1 + 49) H_2SO_4.

Dry ashing. Accurately weigh, into a platinum crucible, sufficient sample to contain 25–100 μg of zinc. Char under an IR lamp and ash until carbon-free in furnace with temperature slowly raised to not over 525°C. Dissolve the ash in a minimum volume of (1 + 1) HCl, add 20 ml of water and evaporate to near-dryness on a steam bath. Add 20 ml of 0.1 N HCl and heat for 5 min. Transfer quantitatively with 0.1 N HCl into a 100 ml volumetric flask and dilute to volume with 0.1 N HCl. Dilute further if required, with 0.1 N HCl.

Analyze beer, wines, other beverages and liquid samples with no or minimal treatment. Filter through a membrane filter if necessary to remove suspended matter and dilute if required to reduce dissolved solids concentration and to bring zinc concentration to scale of spectrometer. For plant material, follow the dry-ashing or wet-digestion procedure described in Section IV.B.1(ii).

(iii) *Determination*

Aspirate solutions into a fuel-lean air–acetylene flame and measure zinc absorbance at 214.86 nm. For low levels of Zn in beer and wines, add 1% $(NH_4)_2HPO_4$ and proceed with the determination by ETA-AAS.

13. *Arsenic and selenium*

(i) *Summary of procedure*

Sample organic matter is destroyed by wet digestion [37], arsenic and selenium are converted to gaseous AsH_3 and H_2Se, respectively, through reduction by sodium borohydride in a hydride generator. The volatile hydrides are transported into an argon–hydrogen-entrained air flame in which absorbances of arsenic and selenium are measured at 193.70 nm and 196.03 nm, respectively.

(ii) *Preparation of sample solution*

Accurately weigh appropriate amount of samples (e.g. 5–10 g of products containing >50% water and low in fat, starch or sugar, such as meat, fish, fruits and vegetables; 3–5 g of products containing 10–50% water; 1–3 g of products containing less than 10% water such as cereal products and dried foods; 1–2 g of products high in fat or sugar such as cheese, butteroil, syrups and jams) and transfer with three glass beads into a 100 ml Kjeldahl flask. Add 30 ml of $(5 + 1)$ HNO_3–$HClO_4$ and let the sample digest overnight at room temperature, or heat continuously until foaming subsides and the sample dissolves. Gradually increase the temperature to achieve steady vigorous boiling, reduce volume by one half, cool, and add 10 ml of $(1 + 1)$ HNO_3–H_2SO_4. Continue heating through $HClO_4$ oxidation, characterized by vigorous surface reaction and evolution of white fumes. If the yellow digest begins to darken, avoid charring by the cautious addition of 1 ml portions of HNO_3. Continue heating until the solution becomes clear and colourless and dense white fumes of SO_3 appear. Heat 5 min past this stage, cool, transfer to a 100 ml volumetric flask containing 30 ml HCl (or a volume to give final concentration appropriate to the specific hydride generator used) and 20 ml of water, and dilute to volume.

(iii) *Determination*

Transfer a 20 ml aliquot of digest (or volume required by the hydride generator used) to a hydride generator affording suitable precision, for As determination, add 2.0 ml of 10% solution of KI, let stand 2 min, and generate arsine and hydrogen selenide with $NaBH_4$, following procedures specific to the particular generator used, for this and all relevant operations. Transport the gaseous hydrides to an argon–hydrogen-entrained air flame and measure absorbances of arsenic and selenium at 193.70 nm and 196.03 nm, respectively.

14. *Molybdenum*

(i) *Summary of procedure*

The sample is wet- or microwave-digested, molybdenum is concentrated by chelation–solvent extraction, and determined by FAAS using a nitrous oxide–acetylene flame.

(ii) *Preparation of sample solution*

Digest several grams of air-dried plant sample with HNO_3–$HClO_4$–H_2SO_4 and remove all of the $HClO_4$ by boiling. Dilute the H_2SO_4 digest with water to ca. 20 ml, and filter into a 50 ml centrifuge tube. Adjust pH with ammonium hydroxide to 1.3 using a pH meter, add 5 ml of 0.5% 8-quinolinol (oxine) solution in MIBK, shake for 25 min, and let the phases separate.

(iii) *Determination*

Aspirate the MIBK solution into an optimised fuel-rich nitrous oxide–acetylene flame and measure molybdenum absorbance at 313.26 nm. Prepare standard solutions in the organic solvent by taking aqueous standards through the chelation–extraction procedure.

15. *Cadmium*

(1) *Summary of procedure*

The sample is wet-digested with HNO_3–H_2SO_4–H_2O_2, cadmium is converted to the iodocadmate ion with KI, is extracted with a solution of a liquid ion-exchange resin in MIBK and determined by FAAS in an air–acetylene flame. Alternatively, the sample is digested with HNO_3–$HClO_4$, cadmium is chelated and extracted with APDC–MIBK, back-extracted into an aqueous solution of HNO_3 and H_2O_2 and determined by ETA-AAS.

(ii) *Preparation of sample solution*

(a) Transfer to a 100 ml Kjeldahl flask 2 g or more of an accurately weighed food sample together with several glass beads and add 10 ml of H_2SO_4. To the cold mixture add 50% H_2O_2 dropwise until the reaction slows down or until the solution becomes colourless. If the reaction does not begin within two min of the addition of H_2O_2, initiate by gentle warming. Heat the solution to fumes of H_2SO_4 adding more H_2O_2 as necessary until a colourless solution is obtained. When the oxidation is complete, cool the digest, carefully add 20 ml of water and again cool the digest. Transfer the solution to a 100 ml separatory funnel with a minimum amount of water. Add 5 ml of 0.1 M KI solution, dilute to ca. 50 ml with water and mix well. Add with a pipette 10 ml of Amberlite LA-2/MIBK solution and shake vigorously for 20 s. Allow the phases to separate, run-off and discard the aqueous phase and filter the organic phase through a dry Whatman No. 541 or equivalent filter paper to remove suspended droplets of aqueous phase collecting the organic phase in a small glass-stoppered vessel.

(b) Alternatively, accurately weigh 1–15 g (dry mass) of food sample into a 250 ml Vycor beaker covered with a non-ribbed watch glass and dry overnight at 120°C if necessary. Add HNO_3 according to the formula: ml $HNO_3 = 10 + (5 \times$ dry mass of sample in grams) and allow digestion at room temperature for 2–12 h. Boil the solution until the volume is decreased to about one half of the original volume of HNO_3, and add $HClO_4$ according to the formula: ml $HClO_4 = 20 + (20 \times$ mass of fat or oil present, in grams) $+ 7 \times$ dry mass of sample in g. Heat again, adding 2 ml portions of HNO_3 in the unlikely event that evolution of gases becomes excessive or the froth above the solution darkens, and reduce the volume to 4 to 8 ml. Rinse the watch glass and walls of the beaker with ca. 40 ml of water, heat the solution to ca. 90°C and transfer to a 125 separatory funnel, rinsing the beaker with 10–15 ml of water. Adjust the acidity with NH_4OH solution to a blue-green hue of the methyl violet indicator (one drop of 0.1% aqueous solution). Carry out the adjustment rapidly as the indicator colour is stable for only a few minutes. Use a pH meter to adjust the pH to 1.6 in instances where large amounts of metal interfere with the indicator. Add 5 ml of 1% APDC solution while swirling for 10 s. Immediately after, add 70 ml of MIBK, stopper and shake for 60 s. Unstopper, let the phases separate for 20 s and drain and discard the aqueous phase, leaving behind ca. 1 ml. Pipette in stripping solution (5 ml when Cd < 0.2 μg; 10 ml when 0.2 μg $<$ Cd $<$ 0.4 μg, and 20 ml when Cd > 0.4 μg), stopper the funnel and rinse the stem with water. Shake gently for 5 min, invert to avoid seepage, stand overnight, shake a further 5 min and return to an upright position. After 1 h, drain the aqueous phase into a beaker, discarding the first 1–2 ml, and pipette 2.00 ml into a 25 ml polypropylene bottle containing 6.00 ml of modification solution.

In both sample preparation schemes, carry a procedural reagent blank through identical respective operations.

(iii) *Determination*

(a) Aspirate the MIBK solutions into an air–acetylene flame with air flow adjusted to give a flame just short of luminous while aspirating MIBK and measure cadmium absorbance at 228.80 nm. Prepare standard solutions in an identical organic matrix by taking inorganic cadmium standard solutions through the same extraction procedures as for the samples. Flush the burner with MIBK after aspiration of standard and sample solutions. Exercise care when changing solutions, as with some burner systems, such as the three-slot Boiling burner, the removal of the ketone renders the flame weak and there is a tendency to flash-back.

(b) For ETA-AAS determination, introduce an appropriate small volume of solution into an electrothermal atomizer and follow a previously established drying, charring and atomization program. Prepare standards in an aqueous dilution solution of composition identical to the bulk matrix of the sample solutions; there is no need to process standards through the complete chelation/extraction/back extraction procedure. When absorbances of sample solutions exceed those of ca. 6 ng ml^{-1} cadmium standards, reduce by diluting with dilution solution.

16. *Tin*

(i) *Summary of procedure*

The sample is wet-digested; if necessary, tin is extracted as $Sn(IV)I_4$ into toluene and back-extracted as stannate with aqueous KOH to reduce interferences in the flame, and is determined by FAAS using a nitrous oxide–acetylene or air–hydrogen flame.

(ii) *Preparation of sample solution*

Digest an accurately weighed appropriate amount of food sample according to the H_2SO_4–H_2O_2 [37] wet-digestion procedure described in Section IV.B.15(ii)(a). Transfer the digest to a graduated cylinder and add H_2SO_4 to give a volume one quarter that of a suitable calibrated flask. Transfer the solution to the flask and rinse both digestion flask and cylinder with water, combining washings into the calibrated flask. Carry out the following operations rapidly. Cool the solution, make to volume with water, pipet an aliquot containing 0 to 400 μg Sn into a separatory funnel and add 2.5 ml of 5 M KI solution to 25 ml of diluted digest. Mix, add 10 ml of toluene, shake vigorously for 2 min, allow phases to separate and discard the aqueous layer. Wash the toluene layer, without shaking, with 5 ml of a mixture containing the same proportions of KI solution and 9 N H_2SO_4 as the diluted digest (1 volume of KI solution to 10 volumes of the acid). Discard the aqueous phase; toluene layer should be pink with extracted iodine. Repeat the washing, rinsing the inner surface of the separatory funnel, and discard the wash solution. Add 5.0 ml of water and 0.50 ml of 5 M KOH, and shake the funnel for 30 s. Should the toluene layer not be colourless, add 5 M KOH drop by drop from a graduated pipette until the pink colour disappears, then 2 drops in excess. Note the total volume of KOH added and shake the funnel for 30 s. Drain the KOH phase quantitatively into a 20 ml calibrated flask, add 5.0 ml of 6 M HCl and mix. If more than 0.5 ml of KOH solution has been used for the extraction, add an equal volume of 6 M HCl in addition to the 5.0 ml. Remove the yellow iodine with 0.5 or 1 ml of 5% m/v ascorbic acid solution and dilute to volume with water.

Alternatively, digest an accurately weighed amount of sample according to any of the two wet-digestion procedures described in Sections III.A.1–III.A.2, transfer to a volumetric flask of suitable volume and dilute to volume. Use this solution directly for FAAS.

(iii) *Determination*

Aspirate solutions resulting from the first procedure, in which Sn is extracted from potential interfering elements, into an air–hydrogen flame and measure tin absorbance at 224.61 nm. Check and correct for nonatomic absorption. For solutions from the second procedure, from which tin is not isolated from interferants, aspirate solutions into a fuel-rich nitrous oxide–acetylene flame and measure tin absorbance at 235.48 nm or 286.33 nm.

17. *Mercury*

(i) *Summary of procedure*

Organic matter in the sample is destroyed by wet digestion with H_2SO_4–HNO_3–V_2O_5, with HNO_3 under pressure in a Teflon vessel, or with H_2SO_4–HNO_3–H_2O_2; the digest is reduced with hydroxylamine sulphate–$SnCl_2$ or hydroxylammonium chloride–$SnCl_2$, elemental mercury vapour is removed by aeration, and transported to a cell with quartz windows situated in an AA spectrometer where mercury is determined by AAS.

(ii) *Preparation of sample solution*

(a) Accurately weigh 5 g (fresh weight) of fish sample into a 200 ml flat bottomed digestion flask and if necessary, rinse the neck with not more than 5 ml of water. Add ca. twenty 6–8 mesh boiling stones, 10–20 mg V_2O_5 and 20 ml of (1 + 1) H_2SO_4–HNO_3. Quickly connect flask to condenser through which cold water circulates, and swirl to mix contents. Heat to produce a low initial boil (ca. 6 min) and complete the digestion with a strong boil (ca. 10 min), swirling flask intermittently during the digestion. No solid material should be apparent after ca. 4 min except for globules of fat. Remove the flask from heat and wash the condenser with 15 ml of water. Add 2 drops of 30% H_2O_2 through the condenser and wash into the flask with 15 ml of water. Cool to room temperature, disconnect the flask, rinse the ground joint with water and quantitatively transfer, with rinsing, to a 100 ml volumetric flask, diluting to volume with rinse water. Ignore solidified fat.

(b) *Exercise caution when digesting organic materials in a pressure digestion vessel (some advise against it)*. Accurately weigh 1 + 0.1 g of fish sample into a Teflon pressure digestion vessel (this procedure is specified for a 23 ml vessel available from Uni-Seal Decomposition Vessels, P.O. Box 9463, Haifa, Israel; do not change sample weight or acid volume substantially as excessive pressure may damage the vessel), add 5.0 ml of HNO_3 and close the vessel with a screw cap. Place in a preheated oven at 150°C for 30–60 min or until the solution is clear. Remove the vessel, allow to cool to room temperature, and transfer the digest with the aid of 95 ml of diluting solution (to a 1 l flask add 500 ml of water, 58 ml HNO_3 and 68 ml H_2SO_4, dilute to volume with water) to a flask suitable for mercury generation.

(c) Weigh 2.5 g of fish sample into a 200 ml Kjeldahl flask having a B24 socket, add 9 ml of H_2SO_4 and attach the upper part of the digestion apparatus to the flask. Heat on a heating mantle, and swirl vigorously until a homogeneous tarry fluid is obtained, cool the flask in ice and add 2 ml of 50% H_2O_2 through the top of the condenser. Open valve to slowly introduce the H_2O_2 into the mixture, remove the flask from the ice-bath and swirl it slowly until the reaction begins. As the reaction slows down, apply heat. Close valve, add 2 ml of HNO_3 through the top of the condenser, and allow the acid to run, through valve, slowly into the flask while the contents are still hot. After 2 min close valve, heat the flask until fumes are evolved, and run off the condensate that has collected in a beaker. Add, through the top of the condenser, 1 ml of 50% H_2O_2 and 1 ml of HNO_3 to the flask, close valve and again heat until fumes are evolved; transfer the condensate into the same beaker as before. Repeat this operation with 0.5 ml portions of both H_2O_2 and HNO_3 until the digest is a pale straw colour. Return the cool condensate from the beaker to the flask through the condenser. Cool the contents of the flask and add 6% m/v $KMnO_4$ solution until a permanent pink colour is produced. Transfer the digest to a 50 ml calibrated flask, rinse the reflux system and Kjeldahl flask with water into the calibration flask and dilute to volume. Set aside for 24 h before proceeding with the AAS determination [1].

(iii) *Determination*

(a) Pipet 25 ml of solution from digestion (a) into the original digestion flask and add 75 ml of diluting solution (300–500 ml of H_2O, 58 ml of HNO_3, and 67 ml of H_2SO_4 and dilute to 1 l with H_2O); alternatively, use entire digest from procedure (b). Follow operating instructions pertaining to the specific apparatus used [38], generate mercury with 20 ml of reducing solution (mix 50 ml of H_2SO_4 with 300 ml of H_2O; cool to ambient temperature, dissolve 15 g of NaCl and 15 g of hydroxylamine sulphate and 25 g of $SnCl_2$ in the solution and dilute to 500 ml), aerate for ca. 1 min or as required.

(b) Alternatively, pipette 10 ml of the solution resulting from digestion procedure (c) into the aeration test tube of the vapour generation apparatus [39]. Add water to bring the volume to 13 ml, then add 2.0 ml of reducing solution (mix 20 ml of 15% m/v NaCl solution with 12 ml of 21% m/v of hydroxylammonium chloride solution and dilute to 100 ml) and 0.20 ml of reducing solution. Mix well and aerate according to the pertinent operating instructions [39].

Transport the mercury vapour in the air stream to the absorption cell and measure the absorbance of the Hg vapour at 253.65 nm.

18. *Lead*

(i) *Summary of procedure*

The sample is digested with HNO_3–$HClO_4$ or H_2SO_4–H_2O_2; lead is chelated and extracted with APDC–MIBK and is determined by aspiration of the MIBK phase into an air–acetylene flame, or is back-extracted from the organic phase into an aqueous solution of HNO_3 and H_2O followed by ETA-AAS.

(ii) *Preparation of sample solution*

(a) Digest an appropriate mass of sample with H_2SO_4–H_2O_2 according to the procedure in Section IV.B.15(ii)(a) to the point prior to addition of water. Dilute the cold solution with 10 ml of water, add 2 ml of 10% sodium sulphite and boil to fumes of SO_3. Dilute the digest to 50 ml with water, cool and transfer to a 125 ml separatory funnel. Add 2 ml of 1% APDC solution, shake and let stand for 5 min. Accurately add 10 ml of MIBK and shake the mixture vigorously for 1 min. Allow the layers to separate, discard the aqueous layer and filter the organic layer through a small dry filter paper into a suitable glass-stoppered flask.

(b) Take an appropriate mass of sample through the HNO_3–$HClO_4$ digestion, and chelation–extraction, back-extraction procedures described in Section IV.B.15(ii)(b).

(iii) *Determination*

(a) Aspirate MIBK solutions into an optimised air–acetylene flame and measure lead absorbance at either 283.31 or 217.00 nm, whichever wavelength gives the greater signal-to-noise ratio. Prepare standard solutions by taking aliquots of aqueous lead standard solutions with 5 ml of H_2SO_4 diluted to 50 ml through the same chelation–extraction procedure. Flush the burner with MIBK after aspiration of standard and sample solutions. Exercise care when changing solutions, as with some burner systems, such as the three-slot Boiling burner, the removal of the ketone renders the flame weak and there is a tendency to flash-back.

(b) For ETA-AAS determination, use the solution resulting from Section IV.B.18(ii)(b) and proceed according to instructions in Section IV.B.15(iii)(b). When absorbances of sample solutions exceed those of ca. 60 ng ml^{-1} lead standards, reduce by making an appropriate dilution.

REFERENCES

1   S. Williams (Ed.), Official Methods of Analysis of the Association of Official Analytical Chemist, 14th edn., Arlington, VA, 1984.
2   N.W. Hanson (Ed.), Official, Standardised and Recommended Methods of Analysis, 2nd edn., The Society for Analytical Chemistry, London, 1973.
3   R.D. King (Ed.), Developments in Food Analysis Techniques 3, Elsevier, Amsterdam, 1985.
4   G. Charalambous, Analysis of Foods and Beverages, Modern Techniques, Academic, Orlando, Fla., 1985.
5   International Union of Pure and Applied Chemistry, Standard Methods for the Analysis of Oils, Fats and Derivatives, IUPAC, UK, 1987.

6   S.R. Koirtyohann and E.E. Pickett, in J.A. Dean and T.C. Rains (Eds.), Flame Emission and Atomic Absorption Spectrometry, Vol. 3, Marcel Dekker, New York, 1975.

7   B.K. Watt and A.L. Merrill, Composition of Foods, Agriculture Handbook No. 8, United States Dept. of Agriculture, Washington, D.C., 1975.

8   L.P. Posati and M.L. Orr, Composition of Foods, Dairy and Egg Products, Raw, Processed, Prepared, Agriculture Handbook No. 8-1, United States Dept. of Agriculture, Washington, D.C., 1976.

9   A.C. Marsh, M.K. Moss and E.W. Murphy, Composition of Foods, Spices and Herbs, Raw, Processed, Prepared, Agriculture Handbook No. 8-2, United States Dept. of Agriculture, Washington, D.C., 1977.

10 W.R. Wolf, Life Sci. Res., Rep., 33 (1986).

11 P. Valenta, GIT Fachz. Lab., 32 (1988) 312.

12 M. Ihnat, Application of Atomic Absorption Spectrometry to the Analysis of Foodstuffs, in John Edward Cantle (Ed.), Atomic Absorption Spectrometry, 1st edn., Elsevier, Amsterdam, 1982.

13 A.A. Benedetti-Pechler, Essentials of Quantitative Analysis, Ronald Press, New York, 1956.

14 W.W. Walton and J.I. Hoffman, in I.M. Kolthoff and P.J. Elving (Eds.), Treatise on Analytical Chemistry, The Interscience Encyclopedia Inc., New York, 1959.

15 C.A. Bicking, in I.M. Kolthoff and P.J. Elving (Eds.), Treatise on Analytical Chemistry, 2nd edn., Part 2, Theory and Practice, Vol. 1, John Wiley, New York, 1978.

16 M.A. Perring, J. Sci. Food Agric., 25 (1974) 247–250.

17 A.A. Brown, D.J. Halls and A. Taylor, Spectrom., 4 (1989) 47R–87R.

18 J.W. Jones and K.W. Boyer, J. Assoc. Off. Anal. Chem., 62 (1979) 122–128.

19 P. Strohal, S. Lulic and O. Jelisavcic, Analyst, 94 (1969) 678–680.

20 P.D. La Fleur, Anal. Chem., 45 (1973) 1534–1536.

21 H.L. Huffman, Jr. and J.A. Caruso, Talanta, 22 (1975) 871–875.

22 Ministry of Agriculture, Fisheries and Food, Technical Bulletin 17, The Analysis of Agriculture Materials, A Manual of the Analytical Methods used by the Agricultural Development and Advisory Service, Her Majesty's Stationery Office, London, 1973.

23 E.C. Dunlop, in I.M. Kolhoff and P.J. Ilving (Eds.), Treatise on Analytical Chemistry, Part I, Vol. 2, Interscience, New York, 1961, Ch. 25.

24 T.T. Gorsuch, The Destruction of Organic Matter, Pergamon, Toronto, 1970.

25 R. Bock, A Handbook of Decomposition Methods in Analytical Chemistry, John Wiley, New York, 1979 (English translation by I. L. Marc of 1972 German edn.).

26 G.D. Martine and A.A. Schilt, Anal. Chem., 48 (1976) 70–74.

27 N.W. Hanson (Ed.), Official, Standardised and Recommended Methods of Analysis, 2nd edn., The Society for Analytical Chemistry, London, 1973, pp. 3–23.

28 H.M. Kingston and L.B. Jassie (Eds.), Introduction to Microwave Sample Preparation, American Chemical Society, Washington, D.C., 1988.

29 J.A. Dean and T.C. Rains (Eds.), Flame Emission and Atomic Absorption Spectrometry, Vol. 3, Marcel Dekker, New York, 1975.

30 J.A. Dean and T.C. Rains (Eds.), Flame Emission and Atomic Absorption Spectrometry, Vol. 2. Marcel Dekker, New York, 1971.

31 US Department of Commerce National Institute of Standards and Technology Certificate of Analysis, Standard Reference Materials (SRM).

32 J.A. Dean and T.C. Rains (Eds.), Flame Emission and Atomic Absorption Spectrometry, Vol. 3, Marcel Dekker, New York, 1975.

226

33 J.C. Van Loon, Selected Methods of Trace Metal Analysis—Biological and Environmental Samples, John Wiley and Sons, New York, 1985.
34 G.R. Carnick and W. Slavin, Am. Laboratory, Part 1, October 1988, Part 2, February 1989.
35 J. Assoc. Off. Anal. Chem., 58 (1975) 392.
36 W.E. Link (Ed.), Official and Tentative Methods of the American Oil Chemists' Society, 3rd edn., American Oil Chemists' Society, Champaign, Ill., 1977, Method Ca 15–75.
37 R.F. Puchyr and R. Shapiro, J. Assoc. Off. Anal. Chem., 69 (1986) 868.
38 J. Assoc. Off. Anal. Chem., 60 (1977) 470–471.
39 Analytical Methods Committee, Analyst, 102 (1977) 769–776.

Chapter 4e

Applications of Atomic Absorption Spectrometry in Ferrous Metallurgy

K. OHLS and D. SOMMER

Hoesch Stahl AG, Postfach 10 50 42, 4600 Dortmund (Federal Republic of Germany)

I. INTRODUCTION

Atomic absorption spectrometry was introduced approximately 25 years ago into laboratories concerned with the analysis of materials connected with ferrous metallurgy. Thus, literature references for this area of analysis only cover this period [89, 103, 134, 135, 151, 154]. Atomic absorption apparatus have been improved significantly during the past two decades. Analysts have made demands upon manufacturers which have led to the development of sophisticated electronics and improvements in optical systems [33, 34, 52, 53, 59, 148, 149].

Atomic absorption analysis was made available to the routine laboratory analytical technique, which initially was intended to produce considerable simplification of procedures for the analysis of aqueous, acidic or basic solutions, and thereby contribute to a reduction in costs. Numerous reviews show the worldwide application of this technique [15, 40, 77, 126–129, 137]. Nevertheless, some 10 years passed before atomic absorption became part of the international standardisation of analytical methods. At present, there are many standard methods being developed on the basis of atomic absorption [35, 185, 67]. Some, dealing with the determination of metals in lubricating oils, are already in use [35, 66], although the overwhelming majority, for example those dealing with the analysis of iron ores [67], are still being developed. The first indication of standardisation of atomic absorption methods for iron and steel analysis was seen in 1973 [8]. Five EURONORM methods are published till now: No. 134 (determination of Al in unalloyed steel), No. 136 Ni (see also ISO 4940-85], No. 177 Ca, No. 181 Pb, and No. 188 Cr, all to be determined in steel and iron.

The industrial application of atomic absorption for routine analysis can be divided into four areas:

(1) The incoming inspection of all raw materials.
(2) Production testing (indirect).
(3) Final inspection of all products.
(4) Environmental analysis.

References pp. 268–274

228

A characteristic of the analytical requirement at the incoming inspection stage is the accuracy of the results, because these are frequently exchanged between the vendor and the purchaser, and become the subject of financial negotiation. Raw materials include iron ore, ferrous alloys and raw iron, but also casting aids, refractories, as well as lubricating oils, fats and fuels (see Fig. 1). Trace concentrations in raw materials are also significant for environmental reasons. Analytical methods must be capable of working over concentration ranges between 10^{-5} and 10^{-2} weight percent. Atomic absorption analysis has made a significant contribution in these areas.

For rapid analysis during the production process atomic absorption is mainly of indirect value because, due to the sequential character of the technique, it cannot be used for complete steel or slag analysis in a two- to three-minute period. The analytical requirements for the testing of rapid continuous production processes are fulfilled by the techniques of emission and X-ray spectrometry. These techniques are characterised by great speed, high precision and simultaneous multi-element analysis. Accuracy must, however, be constantly checked with a variety of special calibration samples. This requires the determination of the true concentrations of the calibration samples with chemical methods of solution analysis, whose precision is often only equal to or, when compared with X-ray spectrometry, frequently poorer. Chemical analysis is, however, the basis of all comparisons, and must be repeated frequently for the determination of the true concentrations. Atomic absorption, with its relatively good precision, has greatly simplified the analytical control of numerous elements.

An important advantage in atomic absorption spectrometry is given just now by the development of simultaneous reading methods [189, 199, 214].

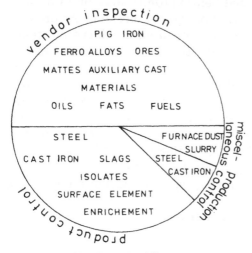

Fig. 1. Application of AAS.

If the Atomsource [163, 212], being able to generate sample atoms very constantly, becomes combined with simultaneous reading instrumentation, metallic and other conducting samples can be analysed also in a two- to three-minute period.

In slower production processes, atomic absorption can be used directly for production control of those processes for which rapid sample preparation techniques have been developed. One such example is the determination of magnesium concentration during the production of cast iron [70]. Further examples are the determination of acid soluble aluminium in steel [147] and calcium [157, 185, 216] which is becoming an important alloying element in the range below 50 microgram per gram of steel (see also Section II.A).

Considerable significance is attached to atomic absorption in production testing. The products of ferrous industries are materials with specific qualities which must be guaranteed world-wide, often regulated by law. Since finished products cannot be tested by techniques requiring mechanical destruction, correlations have been established between material qualities and chemical structure. Thus, chemical analysis has become a significant aid in the testing of materials and their qualities. Because of the need for a guaranteed product quality, the chemical analysis must be highly accurate. Because of time limitations, and therefore cost, product analyses are also done using emission and X-ray spectrometry, but here the number of check measurements with standardised methods must be increased considerably. This is one of the main applications of atomic absorption in the laboratories of the iron and steel industry.

In addition to the metallic products there are also by-products such as fertilisers, slags and recycling materials. Here too the concentration ranges between 10^{-4} and 50 weight percent describe metallic materials and oxide products.

Another application of atomic absorption is in the determination of concentrations on steel surfaces after special sample preparation, and the analysis of steel residues (purity tests) after isolation and possible selective dissolution of the iron matrix [18, 124, 139] (Fig. 1). Atomic absorption is particularly useful for environmental analysis where dust samples can be analysed in a similar manner to steel residues; water and effluents are the main examples.

Table 1 summarizes the most common materials in ferrous metallurgy, listed by element, which are analysed by atomic absorption spectrometry.

Because of the growing requirements for accurate analytical results, it is sensible to work with at least two different techniques. If the results of combined methods, such as photometry/atomic absorption, ICP spectrometry/atomic absorption, polarography/atomic absorption, etc., agree within predetermined limits in standard deviation, there is a great probability that the results are also accurate.

The future uses of atomic absorption also depend on the ability of analysts

TABLE 1

ELEMENTS DETERMINED BY AAS ROUTINE ANALYSIS IN DIFFERENT MATE-
RIALS

Element	Materials
Al	Steel, pig iron, cast iron, ferro alloys, ores, slags, refractories, furnace dust, slurry, auxiliary cast materials, isolates, fats
B	Auxiliary cast materials (¿10% B)
Ba	Ores, slags
Bi	Furnace dust, cast iron
Ca	Steel, pig iron, cast iron, ferro alloys, ores, slags, refractories, furnace dust, slurry, auxiliary cast materials, isolates, oils, fats
Cd	Furnace dust, slurry
Co	Steel, pig iron, ores, slags, furnace dust
Cr	Steel, iron, cast iron, ferro alloys, ores, slags, refractories, furnace dust, slurry
Cu	Steel, pig iron, cast iron, ferro alloys, ores, slags, furnace dust, slurry
Fe	Slurry, isolates, oils
Li	Ores, slags, auxiliary cast materials, fats
Mg	Steel, pig iron, cast iron, ferro alloys, ores, slags, refractories, furnace dust, slurry, auxiliary cast materials, oils, fats
Mn	Steel, pig iron, cast iron, ferro alloys, ores, slags, refractories, furnace dust, slurry, auxiliary cast materials, isolates
Mo	Steel, pig iron, cast iron, ores, furnace dust, slurry, isolates
Na	Oils, fats (slags, refractories)
Ni	Steel, pig iron, cast iron, ores, slags, oils
Pb	Steel, pig iron, cast iron, ores, slags, furnace dust, slurry, petrol
Sb	Steel, pig iron, cast iron, ores, slurry, oils
Si	Steel, pig iron, cast iron, ores
Sn	Steel, pig iron, cast iron, ores, slurry, slags, furnace dust
Ti	Steel, pig iron, cast iron, ferro alloys, ores, slags, refractories, furnace dust, slurry, auxiliary cast materials
V	Steel, pig iron, cast iron, slags, furnace dust, slurry, isolates, oils
W	Steel
Zn	Steel, pig iron, cast iron, ores, slags, furnace dust, slurry, auxiliary cast materials, oils, fats

to produce rapid and simple techniques for sample preparation, and to compensate for the small number of interferences that occur because of matrix elements. In routine laboratories particularly, there is an urgent need for simple methods.

In line with the real multi-element capability given only by the simultaneous reading methods and the microsampling capability which needs a minimal sample preparation, and is used mainly as solid sampling technique today [164, 212,228, 231], it seems to be totally wrong to promise a significant decrease of atomic absorption development [192] for the next decade. It is true that ICP emission spectrometry has been applied to the routine analysis of many different sample solutions, but any change of the matrix makes a new calibration necessary. In most cases the matrix is not known exact enough, and therefore, atomic absorption spectrometry is still an ideal method to check the results for accuracy. That is the main reason analysts are going to prepare sample solutions which can be used for both the methods alternately. In principle, many difficulties of ICP and other emission spectrometric methods of metal analysis caused by spectral line coincidences can be eliminated by excitation of ground-state lines mainly as usual in atomic absorption. As well as spectrophotometry is used in routine analysis today atomic absorption spectrometry will still be needed in the year 2000.

By means of the microprocessors built into recent instruments, operations such as calibration and checking of stability have been simplified. The analyst needs only to use universally accepted standard methods which can be easily applied to as many classes of materials as possible. Such methods will be described in the following sections for chemical analysis in the ferrous metallurgy industry.

II. ANALYSIS OF IRON, STEEL AND ALLOYS

Many authors [15, 21, 25, 28, 29, 31, 891, 82, 106–108, 111, 114, 119, 123, 126–129, 131, 137, 141, 142, 147, 153, 185, 190, 223] have reported studies on the analysis of common acid solutions of different steels in which 1 to a maximum of 12 elements are determined sequentially from one sample weighing. Elements commonly analysed include Al, Ca, Cd, Co, Cr, Cu, Mg, Mn, Mo, Ni, Pb, Si, Ti, V and Zn. There has also been no lack of effort to determine elements such as P [218, 219] or As [54, 101, 142, 173, 183], that are difficult to analyse using atomic absorption. Efforts to produce standard methods have also been described [5, 56, 140, 190, 223]. The closest to a universal method are the descriptions of the determination of 12 elements sequentially from two different acid solutions in all kinds of steels [190] and the one to determine 14 elements from water, coals, ashes, ores, rocks, building materials, and metals [223].

The development of analytical methods during the last decade is concerned mainly with three principles: the graphite furnace [188, 201, 243], the hydride generation methods [156, 175, 226, 237, 240] and the Smith-Hieftje [211] or Zeeman background correction techniques [170, 194, 198, 244].

The graphite furnace was used to determine the elements Al [187], As [173, 183, 196], Ca [157, 216], Cd [211], P [218], Pb [168, 220] and Si [188] in steels and alloys, mainly for trace concentrations.

References pp. 268–274

Most work has been done to develop hydride generation methods for single element determinations like As [158, 161, 184, 202, 217, 234, 236, 239], Bi [178, 209], Sb [177, 245], Se [162, 176, 235] and Te [207], or for multi-element ones like As/Se [208], As/Sn [204], As/Sb/Se [246], As/Sb/Sn [210], As/Bi/Sb [213], As/Pb/Sn [167], As/Bi/Pb/Sb/Se/Sn [193] and As/Bi/Sb/Se/Sn/Te [165, 238].

Also, combined methods were described like hydride generation/graphite furnace [180, 229] applied to the determination of As [183, 196], Se [197] and Sn [230], or hydride generation/graphite furnace/Zeeman background correction to determine Sb [233] and Bi/Pb in steel [232].

These developments are important because of the limits of detection (3 s) of As and Te, for example, obtained by flame AA analysis of steel solutions: 20 ppm As (C_2H_2/N_2O; 193.7 nm) and 30 ppm Te (C_2H_2/air; 214.3 nm). Compared to the ICP emission spectrometric data—25 ppm As and 10 ppm Te (N_2/Ar-source at the 3-kW level)—obtained from the same steel solution and using the same analytical lines, flame AA is not bad, but not good enough for real trace determination.

The often used technique of analyte addition cannot be applied to all analytical problems, especially not to those with unknown composition. Welz [241] wrote it in the best way it could be written: "The two most important basic rules are that the analyte addition technique can only correct for multipliactive but not for additative systematic errors, and that the added analyte element and the element present in the sample must behave in an analytically identical manner."

Two further applications have to be noticed, therefore. The use of chelating agents in flame AA [169, 171] mostly combined with an extraction procedure [174, 182, 186, 206, 221] is one technique to concentrate the analyte in solution. It becomes much more sensitive if the graphite furnace is used additionally [160, 173, 183]. The solid sampling technique [191] working with metal chips or residues obtained by removing the solvents, offers the most sensitive method of non-flame atomic absorption spectrometry [228, 231]. The D2-background correction [164] can be used here, as well as the Smith-Hieftje [211] or Zeeman [170] correction. In case of refractory elements, the halogen-assisted volatilization [203] becomes important. For example, the interference of chloride has been studied [225] as well as its elimination [242].

A. General solution methods

For almost all types of steel there is a similar problem in the dissolution step. Three elements, Al, Si and W, behave differently in the analysis of the total content and require special methods. These methods are simple extensions of the technique, which in the case of the determination of acid-soluble Al or Si, can simply be omitted. When, during the dissolution, an oxide residue remains, which in general consists of SiO_2 or Al_2O_3, this must be brought into solution with a fusion, and added to the rest of the sample. If the W concentration is higher than 0.5%, the dissolving acid solution should also include phosphoric acid. Thus, it is possible to produce a universal method for the elements that are contained in steel and which influence its properties.

A general method for the determination of all elements of interest in steel samples must be simple enough for routine work and able to be set out in a flow diagram (Fig. 2).

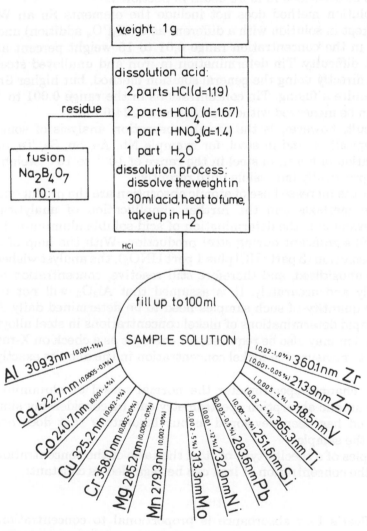

Fig. 2. General solution method.

References pp. 268–274

The acid mixture is chosen such that the majority of steels can be dissolved in it. When a residue remains, this must be fused using the well-known sodium tetraborate method. This sequential determination of 15 elements in one weighing obviously requires the previous setting up of 15 calibration curves in the appropriate concentration range (Fig. 2). This can be done, in principle, in two ways; either the calibration curves are entered using known data with the aid of a microprocessor, or appropriate standard samples are used with every analysis.

It has been demonstrated that a calibration curve set up via direct entry through a microprocessor, can be reliably recalibrated with one standard solution. This will be described in more detail in Section V.

The general solution method does not include the elements Sn an W. Tungsten can be kept in solution with a different acid (H_3PO_4 addition) and can be measured in the concentration range 0.01 to 15 weight percent at 255.1 nm without difficulty. Tin determination in iron and unalloyed steel can be performed directly using the general solution method, but higher Sn concentrations require a fusing. Tin concentrations in the range 0.001 to 1 weight percent can be measured without problems at 224.0 nm.

Far more difficult, however, is the atomic absorption analysis of some other elements typically found in steel, for example Nb, As, Sb, Se, Te, or Bi. The determination of boron in steel in the range of 10^{-3} to 10^{-5} weight percent has been practically impossible until recently.

Future goals for the increased use of atomic absorption are the production of simple general methods and the further simplification of analytical techniques. One example is the determination of acid-soluble aluminium in steel, which is still significant during steel production. With the help of a conventional acid solution (3 parts HCl plus 1 part HNO_3), the analyst wishes to determine the unoxidised, and therefore still reactive, concentration of Al in steel rapidly and accurately. It is assumed that Al_2O_3 will not be dissolved. A large quantity of such samples need to be determined daily. A large number of rapid determinations of nickel concentrations in steel alloys during the production may also be required, for example as a check on X-ray analysis, in order to measure the nickel concentration in the alloy as exactly as possible.

Using a double-channel spectrometer the normal nickel or aluminium determination by atomic absorption spectrometry can be considerably simplified as a method has been developed in our laboratory that does not require weighing the sample.

In various samples of a specific type of steel the ratio of the concentration of an element to the concentration of iron can be regarded as constant:

$$C_X/C_{Fe} = \text{const.}$$

According to Beer's Law absorbance is proportional to concentration, therefore, for the steel sample A_X/A_{Fe} is approximately constant.

TABLE 2

EFFECTS OF WEIGHT AND DILUTION RATIO (set value of standard sample 0.067 mass% Al)

Weight (g)	Volume of solution (ml)	Actual value (%)
1.8	75	0.068
1.5	75	0.066
0.9	75	0.067
2.7	100	0.069
1.5	100	0.070
0.9	100	0.067
2.1	125	0.067
1.8	125	0.067
1.2	125	0.068
2.4	150	0.070
1.2	150	0.068
0.9	150	0.068

For low-alloy steels the total iron concentration can be regarded as being 99 weight percent, so that the concentration C_X can be computed from the absorbance A_X independently, if the absorbance of iron (A_{99}) is measured simultaneously in the second channel:

$$A_X = \text{const.} \times A_{99}$$

Although the absolute values of A_X and A_{99} vary, dependent upon the amount of sample that happens to be used, the ratio always remains constant.

This requires a very insensitive iron line which produces a reproducible signal. The spectral line Fe 283.3 nm is suitable. The ratio of both values (Channel A : Channel B) is shown directly and also printed. The recorder signals show the constancy of these for iron and the dependence on concentration of those for Ni/Fe (Fig. 3). A similar ratio is true for Al, whose determination is thereby greatly simplified. A chip or chips of steel of about 1–2 g are placed into a beaker, dissolved with 20 ml aqua regia, filtered and analysed directly. Standard samples prove that no significant variations occur in results (Fig. 4). A standard sample containing 0.0067% aluminium is used to show that the method without sample weighing is independent of the sample quantity and the dilution for a certain range (Table 2). This method can be applied to different elements and different concentration ratios in low-alloy steels. It could also be useful in cases where sample weighings are difficult to make, as for example on research vessels [94].

References pp. 268–274

236

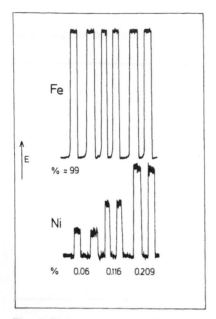

Fig. 3. Determination of Ni without sample weighing.

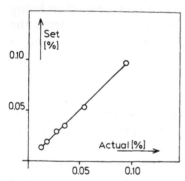

Fig. 4. Accuracy of the Al determination controlled by standard samples.

Atomic absorption is also applied to the analysis of accompanying elements in ferro alloys [113]. Methods have been published for ferro silicon, with the elements Al, Ca, Mn, Ti [32], ferro manganese [45] with silicon [46], and ferro chromium with Pb and Sn [105] and Bi [146].

In general, ferrous alloys are difficult to dissolve with acids so that a fusion, for example with sodium/potassium carbonate, is recommended. The resulting high salt concentration can produce difficulties (viscosity, nebulizer/ burner system), as is discussed in Chapter 3. Once sample solutions are available, there is no difference in analysis from the methods for iron and steel.

B. Trace-element analysis

Trace-metal analysis in pure iron and low-alloy steels as well as their high-alloy variants is still relatively undeveloped. Traditionally, this was done by means of complicated concentration procedures, for example dissolution of the iron matrix by so-called isolation or chloride techniques, and microanalysis of the residues. It is possible to determine main components in these residues with atomic absorption spectrometry [18, 124, 139]. More interesting, however, are the methods which can be used to determine trace concentrations directly in the steel sample or trace concentrations in sample sizes of the order of micrograms.

The few articles currently available regarding trace analysis without preconcentration, use in general the graphite furnace technique [10, 120, 138] with sample sizes of the order of microlitres, and deal with the elements Al [187], As [173, 183, 196], Ca [157, 216], Cd [211], Se [197], Si [188], Sb[47, 83], Pb/Bi [48–50], As/Sb/Bi/Sn/Cd/Pb [10, 57, 116] as well as Al/Cr/Sn [6, 62] and Co/Mg [104]. Alkaline earths can be determined directly with the flame method [122, 147]. Further techniques of atomic absorption by flame use concentration methods like extraction procedures, for example for the determination of small concentrations of Co [186], Sn [17, 181], Te [26], V [221], Co/Pb/Bi [104] and W [106]. From the analytical viewpoint, it is only useful to remove the iron matrix. The extraction of the elements to be determined from the matrix always carries with it the danger of losses and, therefore, results showing concentrations that are too low.

1. *Solid-sample technique*

The use of a two-channel spectrometer in combination with a furnace atomizer with a temperature program makes available an entirely new possibility for direct trace-element analysis [228].

The use of a rectangular graphite tube makes possible the placing of the solution in a sample boat (Fig. 5). The authors propose two changes: reduction of the size of the sample space, and use of a very pure graphite from emission spectroscopy (RW-O) instead of the relatively expensive pyrolytic graphite (Fig. 5). Instead of solutions, a single steel chip is introduced, heated to a minimum of 2600°C, and the resultant atomic vapour is analysed directly. When the sample is weighed (about 1 mg), at least two elements can be determined simultaneously with the two-channel instrument. Because there is no dilution, the same sensitivity produces a much lower detection limit without degrading the precision of the results. In particular, accuracy can be easily checked by standard materials. There are no contaminations due to acids or other chemicals and small amounts of ions are not adsorbed on the walls of the vessels.

Since the weighing of such small sample quantities can influence the analytical error, it is desirable to avoid the need to weigh the sample. One

Fig. 5. Graphite tube oven with standard and modified boat.

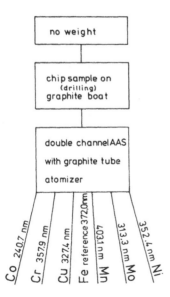

Fig. 6. Solid-sample method.

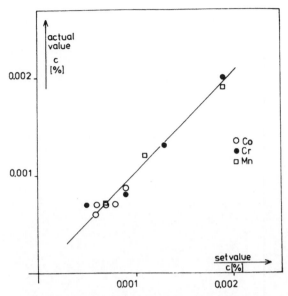

Fig. 7. Accuracy of the solid-sample methods controlled by standard samples.

channel is occupied by the reference absorption measurement, for example in the determination of pure iron or low-alloy steels, an iron hollow-cathode lamp is used and the measurement takes place at 372 nm. In this case, only one further element can be determined. The carrying out of the method is simple (Fig. 6). Tests of this compact sample technique with standard materials, of which there are at present only a few containing certified trace elements, show satisfactory agreement (Fig. 7). As regards traces, the chips of standard materials are very homogeneous.

It has been found that with repeated rapid heating, for example at 2600°C, numerous elements can be determined from one chip. Repeated heating does, however, produce increasingly smaller but determinable peaks; at present it is not yet known how many elements can indeed be determined in one chip.

Figure 8 is an example of the reduction of peak heights with increasing numbers of heating periods in the determination of copper in pure iron. The copper concentration is 0.003 weight percent. If the peak after the tenth heating is still measurable with sufficient precision, then at least 10 elements can be determined per chip.

When samples are less homogeneous, this method can be repeated frequently, since the technique is simple and the time requirement relatively small. The numerous single results permit statistical calculations and in general lead to the correct mean value. The detection limits in the compact sample method have not yet been found exactly, because the appropriate samples are not available but could lie in the range 10^{-5}–10^{-7} weight percent.

References pp. 268–274

240

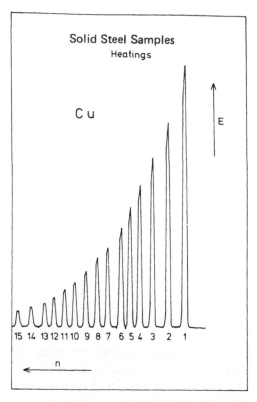

Fig. 8. Decrease of peak height by repeated heatings of the same sample.

Solid sampling AA gets a new dimension when the Atomsource [163] is applied to metal analysis. This new sputtering chamber for atomic absorption measurements, which needs a compact (conducting) sample to be put on, was designed to have maximum ground-state atoms and minimum emission, what is really the reverse situation compared to the known glow discharge configuration. Therefore, the argon gas is introduced directly at the sample surface drawing the plume of gas and atoms directly out, vertically from the sample surface. By using angled gas jets of high flows compared to the atomic diffusion rate the amount of material atomized increases and the sputtered material is directed away from the sample surface immediately. The loss of atoms due to deposition on the sample and the walls of the sputter chamber is minimized by its configuration. The deep penetration of the sample results in a more representative analysis by presenting more of the sample atoms for measurement. The appealing goals to be reached by this technique include:

(1) Analysis of elements which are usually not analysed with AA, such as those which form refractory compounds in the flame or whose resonance

lines lie in the vacuum UV.

(2) An extended dynamic range by controlling discharge parameters permitting analysis from traces to >90 mass percent.

(3) Elimination of time-consuming and error-prone dissolution steps for macro metal or powder samples (what is generally true for solid sampling methods).

(4) Reduction of chemical interference through the use of a sputtering source with optimised conditions to the sample material.

(5) Achievement of the same sensitivities in bulk material as obtained in flames.

(6) The possibility of a rapid multi-element analysis.

The conditions for steel and ferrous alloy analysis have been studied to determine high concentrations of Mn and Nb in ferrous alloys, accompanying elements in steel like Cr, Cu, Mn, Mo, Ni, Si, and low concentrations of La and Zr from steel samples [212] (see also Section VI).

2. *Preconcentration*

Flame atomic absorption can also be used for trace analysis when the iron matrix is extracted [39, 147]. When this extraction is combined with an injection technique [3, 130], trace analysis can also be performed [31, 99, 142]. By using less than 100 μl of sample solution, the improvement compared to conventional techniques is at least a factor of ten [13, 14].

In general, a simple technique (Fig. 9) which can be used with the majority of iron alloys is again available.

Through the use of a two-channel spectrometer two elements can be determined simultaneously in each case. Reduction of the sample solution to 10 μl degrades the reproducibility of the results only slightly; the RSD is in the range 5–10% for concentrations of 10^{-3} to 10^{-4} weight percent (Fig. 10). With an aspiration volume of 10 μl, a triple determination per element can be carried out, theoretically, for more than 50 elements in 1 ml.

Because standard samples contain only a limited number of elements whose concentration lies in the range of this method, we would determine, for example, 9 trace elements in 15 attempts, 3 times each, involving a consumption of 150 μl of sample solution (Table 3). The pairs Cu 324.7 nm/Cr 357.9 nm, Ni 314.5 nm/Mn 279.5 nm and Zn 213.9 nm/Co 240.7 nm were determined with the air/acetylene flame, and Al 309.3 nm/Mo 313.3 nm and Mg 285.2 nm were determined with the nitrous oxide/acetylene flame.

C. **Interference of sample matrix**

Interferences in atomic absorption occur with flame methods as well as electrothermal methods. Many articles have appeared that discuss interferences with flame methods [27, 41, 60, 84, 122, 136] as well as with the furnace

References pp. 268–274

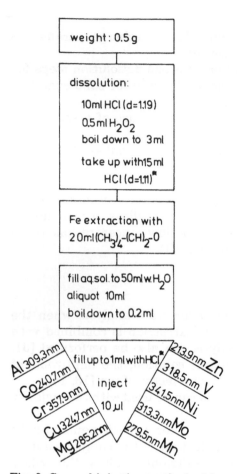

Fig. 9. General injection method with pre-concentration.

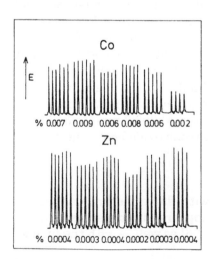

Fig. 10. Simultaneous determination of Co and Zn using the injection technique with 10 μl.

TABLE 3
CERTIFIED AND MEASURED VALUES BY THE MODIFIED INJECTION TECHNIQUE (in mass %)

		Mn	Ni	Cr	Mo	Cu	Co	Al	Zn	Mg
Fe reductum	Certified	0.0020				0.0020	0.0002			
	Measured	0.0009	0.0022	0.0015	0.0067	0.0012		0.0005	0.0004	0.0003
AKP 043-1	Certified	0.0007	0.0056	0.0052	0.0062	0.0020	0.0060	0.0014		
	SD	0.0001	0.0004	0.0009	0.0009	0.0002	0.0004	0.0001		
	Measured	0.0008	0.0055	–	0.0056	0.0020	0.0058	0.0014	0.0003	0.0004
AKP 044-1	Certified	0.0011	0.0879	0.0009	0.0070	0.0043	0.0078	(0.0001)[b]		
	SD	0.0002	0.0032	0.0001	0.0008	0.0002	0.0003			
	Measured	0.0012	–	0.0011	0.0078	0.0041	0.0083	0.0007	0.0004	0.0004
BCS 149-3[a]	Certified	0.019	0.004	0.001	0.001	0.001	0.007	(0.002)		
	Measured	0.018	0.0036	0.0007	0.0012	0.0010	0.0067	0.0009	0.0004	0.0003
BCS 260-2[a]	Certified	0.013	0.0011	(0.001)	(0.002)	0.002	0.009	(0.0012)	(0.0003)	
	Measured	0.0137	–	0.0013	0.0016	0.0020	0.0088	0.0007	0.0003	0.0003
BCS 260-4[a]	Certified	0.002	0.003	0.002	0.002	0.003	0.006	(0.001)		
	Measured	0.0016	0.0035	0.0018	0.0017	0.0029	0.0062	0.0009	0.0004	0.0003

[a] BCS standard have no certified SD.
[b] Values in parentheses are not exactly determined.

References pp. 268–274

atomizer [117, 155, 164, 194, 200, 225, 229, 242] for different techniques of background correction. With flame methods, the type of interference often depends upon the gas composition of the flame. Anionic interferences, that is, the appearance of chemical combinations [110] that cannot be broken up by the temperature of the air/acetylene flame (2400°C), can be prevented by the use of a hotter flame, e.g. nitrous oxide/acetylene (2800°C), although ion-isation interferences are particularly significant with hot flames. These can be suppressed by the addition of easily ionised elements such as potassium or cesium (ionisation buffers) [85]. Additionally, physical interferences are observed which can be shown to be due to differing viscosities of standard and sample solutions (high acid concentrations [100] and high salt concen-trations after fusion). By the addition of increased dilution steps, viscosity differences can easily be removed.

In ferrous metallurgy interference effects have a very insignificant effect, being theoretical in character. For the production of standard curves, due to the complex matrix, pure element solutions are never used. The correction of a single interference would not solve the problem. Further effects must be corrected, which leads to complicated and time-consuming standardisations.

For the analysis of iron, steel, and ferro compounds, therefore, standards are always used whose chemical composition is very similar to the sample. These standards undergo the same dissolution and dilution procedure as the sample. In special cases they are fused as for the sample. Therefore, the standard solutions differ neither in their physical characteristics nor in their concentration of acid or total salt. Any effects that occur will be the same for sample and standard measurements. They are easily eliminated by including them in the calibration.

If in special cases no appropriate standards are available, model solutions must be prepared (Section V.A). These are obtained by including in the model solution all the main components of the sample made up out of analytically pure solution. In addition, the acids and fusion materials must be added in the same concentration as is apparent in the sample. Such model solutions can be regarded as adequate standards. Interferences are thereby eliminated.

When a graphite tube is used, different interference effects appear. These can be caused by the sample as well as by the specific instrument [24]. Interferences caused by the sample occur when the element is vaporised as a molecule, the chemical compound hinders atomization, or when after vaporisation of the matrix the sample has poor contact with the graphite tube. Also, unspecific absorption can occur by simultaneous vaporisation of the matrix. Here, molecular absorption or light loss by scattering from solid particles are the most common effect.

Most of these effects can easily be removed by careful temperature programming—vaporisation of the liquid, separation of the matrix by slow ramping of the temperature and rapid atomization. Unspecific molecular absorption can be removed by the use of a continuum source, in so far as broad absorption bands occur. In practice, pipetting of the sample causes the greatest error. When the sample is pipetted with a piston pipette, the solution drops do not always strike the same place in the graphite tube. As a result, the temperature conditions could be quite different for the calibration and measurement of the sample. Thus, the reproducibility of the method becomes distinctly worse. An improvement is achieved by the use of an automatic sampling system, which always places the same size of droplet at the same angle and the same position in the graphite tube.

While chemical and physical interference possibilities can be largely removed by making the standard solution and the sample solution essentially similar, unreproducible temperature settings frequently lead to differences which cannot be corrected. Though there has been no lack of effort to achieve high temperature constancy by the construction of special feedback systems, the lifetime of these systems is very limited when the very high temperatures required with ferrous metallurgy are used. The analytical solutions, which always have corrosive acids (perchloric acid, aqua regia), required for iron and steel, limit the use of even pyrolytically coated graphite tubes to only a few times. Even after a single use the inner wall of the graphite tube has been attacked so much that reproducible temperature conditions and constant heat transfer can no longer be taken for granted. In iron metallurgical solutions the acid content must be regarded as the actual matrix and is of course present in far higher concentrations than the iron content itself.

These matrix effects can be avoided by the solid-sample system described in Section II.B.1. At temperatures around 2600°C the steel sample placed into the rectangular graphite tube melts without measurable vaporization of larger particles. Effects from the iron matrix are not observed.

The Zeeman background correction has been described to overcome many of these interferences, especially when biological or environmental sample materials are used, mainly by a direct solid-sample input [170, 194, 244].

This method was tested in our laboratory using iron matrix containing solutions, and compared with both the other techniques of background correction, Smith-Hieftje and deuterium (D_2) [198, 211]. It can be seen from the results for two different concentrations of each of the elements (Al, As, Ni, Pb, Se, Sb, and Tl) given in Table 4, that there is no significant difference between those obtained by the Zeeman or Smith-Hieftje correction, but both these methods have a significantly decreasing effect on the sensitivity of absorption signals. A good D_2-correction is still the best technique, except for some special analytical problems, such as Al in the example tested here (see Section III.B.2).

The interferences caused by the instrumentation include carbide formation, which, for example in the determination of vanadium, becomes apparent as a long tailing. The analytical results can be improved if not only

TABLE 4

DETERMINATION OF DIFFERENT ELEMENTS CONTAINED IN AN IRON MA-
TRIX USING DEUTERIUM (D$_2$), SMITH-HIEFTJE (SH) AND ZEEMAN (Z) BACK-
GROUND CORRECTION

Element ($\mu g\ g^{-1}$)	Sample I				Sample II			
	E	D$_2$	SH	Z	E	D$_2$	SH	Z
As	3.0	3.0	3.1	3.1	12.0	11.4	11.7	12.0
Sb	1.0	1.1	1.2	1.0	9.0	8.6	8.0	8.8
Se	2.0	1.0	1.3	2.2	7.0	6.2	5.5	7.0
Pb	3.5	3.6	3.8	3.4	8.0	8.0	8.3	8.1
Tl [a]	1.5	1.5	1.5	1.5	5.0	4.8	4.8	4.9

Element ($\mu g\ g^{-1}$)	Sample III				Sample IV			
	E	D$_2$	SH	Z	E	D$_2$	SH	Z
Al [b]	7.0	–	7.5	7.1	12.5	–	13.0	12.7
Ni [b]	7.0	8.0	7.8	7.3	12.5	13.0	12.3	12.6

Note: E is expected value.
Normal preparation of analyte solution: 1 g Fe sample per 100 ml dissolved in 10 ml
HCl (1.19) and 5 ml HNO$_3$ (1.4) unless otherwise stated.
[a] Dissolved in 10 ml H$_2$SO$_4$ (1 + 1).
[b] 2 g Fe sample per 100 ml dissolved as normal.

the peak height but also the peak area is measured. The determinations of
B, Nb, Ta, W and Zr are prevented completely by the formation of stable
carbides.

Influences of high DC light emission from the graphite tube are difficult
to remove. Though the DC and AC light emissions are electrically separable,
the reflection of the DC light on the cathode of the hollow-cathode lamp
causes pulsing of the light. This emission has the same frequency as the
elements' specific hollow-cathode light and cannot be electrically separated.

D. Precision and accuracy

Reproducibility is a measure of the constancy of the data from the total
procedure—dissolution of the sample, dilution, measurement, and readout
[65]. It can be determined if the procedure is repeated several times with a
single instrument. For the determination of the accuracy of analytical data a
second independent analytical procedure is always required. In the analysis
of concentrations between 0.001% and 20.0% checking of the accuracy is
readily done with photometric or gravimetric procedures. It is difficult to
check accuracy with concentrations below one microgram per millilitre.
Contamination or losses on the walls of test tubes, borders of phases, or

in the mixing chamber of the atomic absorption spectrometer, can only be checked with difficulty. Also, contaminations from laboratory glassware, laboratory air, and the solvents used must be taken into account. Very pure chemicals can only be added after determining blank solutions. The problem of the accuracy of analysis is most significant if the graphite-tube cold-vapour apparatus of special concentration methods is used. If particularly low concentrations are measured, it is desirable not only to use two analytical systems, but also different sample preparation methods, if possible. Thus, changes that occur even before the actual analysis can be noted.

The optics of all common atomic absorption spectrometers of similar price are much the same, but nebulization systems are designed according to various principles (direct injection, counter flow). If normal flame methods are used, with constant aspiration rate of the sample solution, the reproducibility of the measurements depends mainly upon the nebulizer system. There are exceptions, namely smaller atomic absorption instruments which work according to the single-beam principle, and when drifting and variability of the hollow-cathode lamp have a direct effect upon the result. Double-beam atomic absorption spectrometers compensate completely for changes in light intensity by the measurement of the light intensity in the reference beam.

As an example of the reproducibility of a modern instrument a steel sample containing 1.04 weight percent nickel was dissolved (Fig. 2) and continuously aspirated. Between two integration values (integration time one second) water was always aspirated to clean the mixing chamber.

The measured values (in absorbance) were 0.313, 0.313, 0.314, 0.318, 0.315, 0.318, 0.315, 0.314, 0.317, 0.317; mean value 0.316 mass%; SD = +0.002 mass%.

This SD is equivalent to an RSD of 0.65%. It includes all instrumental variations and can be regarded as characteristic for the concentration range between 0.0001 weight percent and 20 weight percent. If the same sample is analysed several times, the standard deviation is increased by the variations that occur because of sample preparation, to a level of about 2 rel.%.

If the total salt content of a solution is high or if only small amounts of sample are available, the injection technique may be used. If a piston burette is used for the injection, individual errors cannot be avoided, and reproducibility becomes somewhat poorer than with continuous aspiration. Also, the uncertainty of the volume measurement with small injection volumes becomes noticeable in the reproducibility.

A simultaneously measured determination of a Cu (2 mg/l) and Ni (5 mg/l) containing solution by injection of 100 microlitres, leads to a RSD of 3% for Cu and 2% for Ni. The RSD with relatively large injection volumes and pure element solutions is not significantly larger than obtained with continuous aspiration.

Trace analysis including previous element concentration procedures as well as partial matrix removal (II.B.2) produces an increase in RSD with an injection volume of only 10 μl of 7.6 relative percent in a copper concentration of 0.0021 weight percent and to 16 relative percent with a manganese content of 0.00088 weight percent. Use of an automatic injector can improve the reproducibility of the sample volume considerably [14].

When the graphite furnace is used for extreme-trace analysis, relative standard deviations are in general above 10 relative percent. The solid-sample analysis described in Section II.B.1, using a steel chip and the rectangular cuvette, leads to an RSD of 13 relative percent in the determination of a copper concentration of 0.00125 weight percent, and an RSD of 20 relative percent with a manganese concentration of 0.00071 weight percent, and lies in the range of trace analysis. The frequently mentioned assumption of poor reproducibility with solid-sample analyses in the graphite tube thus does not apply for steel chips.

III. ANALYSIS OF ORES, SLAGS AND OTHER OXIDES

Oxide materials are distinguished by the fact that the compound forms of the elements may be very different, depending on the many geological types. In general, various main components (greater than 10 weight percent) and trace concentrations (less than 0.01 weight percent) are present. The analysis must take this into account not only in the sample preparation (preparation of solutions) but also in the desired indication of results.

Basically, the literature provides two dissolution methods; sample preparation with sample weights of 0.2–1 g and large dilutions, or smaller sample weights with less dilution (Section III.B). The relatively large dilution, in general after a fusion [51], for the determination of main and lesser components, as for example in silicate analysis [2], the determination of Al, Ca, Mg, Mn and Si in slags [4], Si [55], Pb and Mn [143], and also Cd, Ca, Cu, Pb, Mg and Si in ores or iron sinter [97, 147] and Cr, Mg in refractories [93] is presently used in routine analysis.

A. General solution methods

The problem is to prepare solutions suitable for the desired analytical results. Once a homogeneous analytical solution is available, the differences between the techniques used with dissolved metals are no longer great since the same nebulizer systems can be employed. For the determination of major components it is preferable to use less sensitive spectral lines of the same elements rather than use greater dilutions.

A significant difference in iron solutions can arise from the method of sample preparation; the salt concentration can be very high, and clogging of the burner by crystallisation may be a problem.

1. *Direct dissolution*

It can be seen from above that it is desirable, wherever possible, to dissolve the oxides directly in mixed acids. Small quantities of residue can be fused and added, as described for the analysis of steel. Dissolution in pressure bombs [12, 57] is a method used to keep the salt concentration within limits. This technique is usable for single determinations in particular cases (small quantities); it has not been extensively applied in routine analysis upto now because the use of larger quantities is very difficult. The analytical procedure is rather simple (Fig. 11).

2. *Universal fusion technique*

The great majority of oxide materials can be brought into solution using a fusion technique, mixtures of tetraborates being the most widely used fusing

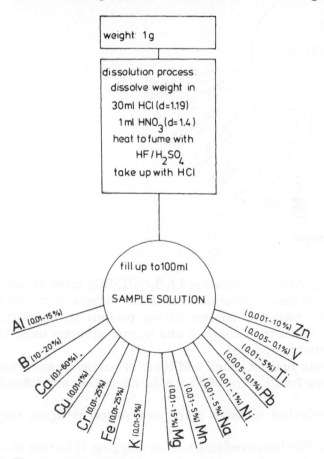

Fig. 11. General dissolution method for oxides.

References pp. 268–274

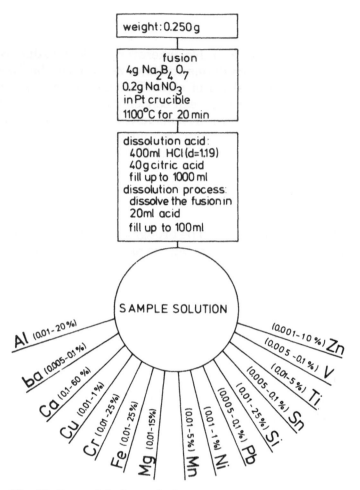

Fig. 12. General fusion technique.

materials. For measuring sodium in solution $Li_2B_4O_7/LiBO_2$ must be used instead of $Na_2B_4O_7$. Sodium can be determined with the necessary precision only with great difficulty owing to contamination present in industrial laboratories. Where it is done, the nebulizer and burner systems must be clean and air pollution must be reduced.

With sodium tetraborate fusions, small additions of oxidising materials (Na_2O_2, $NaNO_3$) have been found successful for substances that are difficult to fuse.

The procedure for analytical methods employing fusion is again very simple (Fig. 12).

There are only rarely difficulties involving burner clogging, if burner slots of 0.8 to 1 mm can be used. If it is necessary for technical reasons to work with narrower slots, or when great numbers of samples are to be analysed without cleaning the system, the use of the Hoesch injection technique [130] is recommended, for example with 100 μl injections, either manually or with the help of an automatic sampling system [14]. Different matrices lead to the consideration of different classes of material, for example based on iron (iron ores, drowses, dusts), calcium (lime, slags), aluminium (fire clay), magnesium (magnesite) or silicon (silicate rocks, sand, residues), as well as substances without a unique basis or with various major constituents (slimes, blast furnace slags). This is not really necessary from the analytical point of view in which there are only differences in the sample preparation, the dilution or the choice of spectral lines, according to differing sensitivity.

Through calibration and re-calibration techniques possible interelement effects or influences from the matrix can be largely eliminated. Because of its worldwide importance the introduction of standard methods by atomic absorption started first with iron ores [67] and refractories. The ISO/TC 102 is presently occupied with draft proposals for the atomic absorption determinations of Na/K, Ca/Mg, Zn, V, Pb and Al.

B. Trace-element analysis

In general, for trace analysis very small dilutions are preferred [54], in which case only small solution quantities are available for atomic absorption. These can be brought into the flame either with the injection technique [3, 13, 14, 31, 99, 130, 142] or with the boat technique [76], or they can be determined in the furnace atomizer [75, 90, 91, 120] so that even such elements as La in ores [109] or the rare earths generally [36, 37] as well as Pb [48, 50], Bi [49, 178], Sb [177], Sn [181, 230] and As/Sb/Sn [116] can be determined. Because of the variety of oxide materials there are a number of special methods, for example for the determination of Cr [26, 44], and Cu, Mn, V and Ni [28]; also sampling of the solid material, as in the determination of Al and Sn [6]. An attempt has been made to determine fluorine with the aid of the AlF molecule [145]. A universally applicable method is elusive.

1. *Electrothermal atomization methods (low salt content)*

The furnace atomizer is readily used for trace determination; also in the case of oxide materials it provides extremely low detection limits. Because of the inhomogeneity of the sample materials, the sample may frequently be no smaller than 1 g (or with extremely small concentrations up to 5 g), so the solution volume is limited as far as possible. Furthermore, it is useful to limit the salt content of the solution. This generally means that the furnace technique should only be used to determine trace elements soluble in acids—possible after pressure bomb decomposition. In this case the time

References pp. 268–274

of analysis becomes irrelevant. An example is the determination of Cd-traces in iron ores and related oxides. Samples are dissolved in the acid mixture by a long reaction. The solution evaporates in the furnace atomizer where it is transformed into atomic vapour and the absorption of light intensity is measured. (Addition method, see [241].)

Concentration range: 0.01–1 mg Cd kg^{-1}.

Reagents

HCl Suprapur (1.15 g/ml). HNO$_3$ Suprapur (1.4 g/ml). H$_2$SO$_4$ Suprapur (1 + 1). Cd stock solution: 1.000 g Cd dissolved in 5 ml HNO$_3$ and 25 ml H$_2$O and topped up to 1000 ml with H$_2$O (1 ml = 1 mg Cd). Cd standard solution (I): 50 ml of the stock solution diluted with H$_2$O to 500 ml (1 ml = 100 μg Cd). Cd standard solution (II): 10 ml of standard solution (I) diluted with H$_2$O to 1000 ml freshly every day (1 ml = 1 μg Cd). Cd solutions for the addition method: 10.0, 25.0 and 50 ml of standard solution (II) (10, 25 and 50 μg Cd accordingly) are topped up with H$_2$O to 1000 ml.

Operating conditions

AAS-device with graphite tube furnace and deuterium background corrector. Wavelength, 228.8 nm; EDL, Cd (6 W).

Procedure

Approximately 5 g of the sample are weighed exactly and placed in a 250 ml beaker. A second, similar beaker is used for the reference solution. To both are added, consecutively, 40 ml H$_2$O, 20 ml HCl and 20 ml HNO$_3$. To dissolve the sample it is boiled for 1–2 h. After it has cooled 50 ml of H$_2$SO$_4$ are added and evaporation encouraged until H$_2$SO$_4$ vapour escapes. After cooling, 100 ml H$_2$O are added. The solution should be left overnight on a warm hotplate. The solution in the beaker without the sample is treated in the same way. Afterwards each is filtered through an appropriate filter, washed out with diluted H$_2$SO$_4$ and hot H$_2$O, cooled, and filtered into a 250 ml flask, which is then topped up. Diluted sample solution, 40 ml, is now pipetted into each of four 50 ml graduated flasks (A, B, C, D). Then, 5 ml of the Cd solutions with 10 ng ml^{-1}, 25 ng ml^{-1} and 50 ng ml^{-1} are pipetted into the graduated flasks B, C, D, in this sequence. All graduated flasks are topped up to 50 ml and stored in polythene bottles. The reference solutions are produced by corresponding additions to 40 ml of the initial solution. At 3-minute intervals 5 μl of the test solutions at a time are injected into the furnace by a flask ejector pipette.

The sequence of measurements is important. It has proved useful to measure C, B, A, and D again in the same way after solution D has been measured 2–4 times, to recognise possible equipment changes during the series.

In addition, the 4 reference solutions are measured in the same sequence.

Evaluation

The results of the measurements are plotted on a graph against the concentration of Cd in 50 ml (addition solution) and are amended with the corresponding blank values.

The Cd concentration of the unknown sample can be read off the result of solution A after it has been amended by the blank value A. This method is also very exact in terms of the error margin (RSD = 10% at 0.1 mg Cd kg^{-1}) and is therefore a typical example for this procedure.

Analysis by AAS of (most) trace elements in bulk materials from the field of ferrous metallurgy is still in its early stages.

2. *Injection technique (high salt content)*

Clearly, trouble-free universal methods are the ideal for trace analysis of oxide materials. Numerous materials can only be dissolved via melting fusion; they then cannot be over-diluted and therefore contain a relatively high salt content. However, many laboratories are not yet equipped with furnace atomizers so the flame method must be used.

During trace analytical determinations with the flame technique a separation and therefore concentration, e.g. via ion exchange [169], can be introduced if the concentrations during normal sample preparation are likely to be below the detection limit of the instrument [92].

The direct use of the injection technique is useful for trace analysis of oxide substances after fusion [31, 99, 142]. It is possible to take up the fused substance in small quantities (max. 10 ml, usually 1 ml). It is best to use an acid mixture which contains 400 ml HCl (1.15 g ml^{-1}) and 40 g citric acid in 1000 ml. Generally, 10 μl, and in special cases 50 μl, of the analytical solution are injected. It is thus possible to determine all important ascertainable elements in 1 ml of analytical solution several times over. Salt concentrations of 100 mg ml^{-1} are no problem for the injection technique.

The following general methods can also be recommended for the analysis of almost all oxide products.

(1) Fusion of an analytical sample. The sample is ground until it passes through a 100 μm sieve. A 0.250 g quantity is weighed and mixed with 4.0 g of sodium tetraborate and 0.2 g of sodium nitrate. The fusion occurs in a platinum-crucible either at 1350°C with a coil which is heated inductively or in a muffle furnace at 1100°C. The period of fusion is at the most 5–20 min.

(2) X-ray fluorescence spectroscopy. The melt is poured into a pre-heated platinum-mould, producing a tablet of approximately 30 mm diameter. Cracking of the tablet is prevented by slow cooling. The level surface is measured directly.

(3) AAS-analysis. The same tablet or an aliquot part of it is dissolved directly in acid (400 ml conc. HCl + 40 g citric acid in 1000 ml) and topped up to 100 ml with H$_2$O. All elements which can be ascertained by AAS can

References pp. 268–274

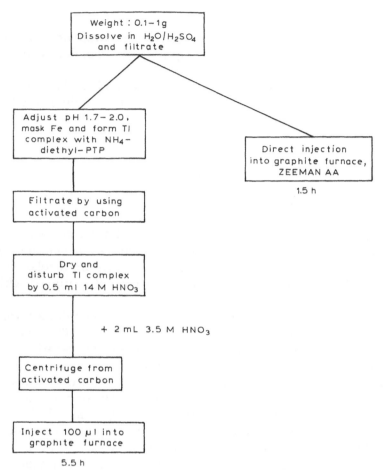

Fig. 13. Comparison of conventional non-flame and ZEEMAN AA to determine Tl in iron-containing materials.

be determined directly from this solution with the injection technique, if corresponding standard solutions have been used for calibration each time.

A successful application of the combined methods of graphite furnace/ Zeeman AA/injection technique enables the analysts to determine special elements in a much shorter period of time [198]. For example, the conventional method of non-flame AA/injection including a difficult separation to determine traces of Tl in iron containing materials (Fig. 13) can be done in about 5.5 hours. This time is reduced by the Zeeman correction to 1.5 hours because no separation is needed.

A combination of the rapid procedures of spectroscopy and corresponding control methods is presently the only analytical procedure available with the necessary precision and accuracy and at a reasonable cost. In the field discussed AAS is an essential technique for routine work.

C. Interference of sample matrix

The chemical make-up of oxide substances is often very complex and the matrix frequently contains two or three compounds of almost equal concentration. The determination of individual interference effects is very complicated and time consuming, because the composition of these substances can be subject to wide variations. As in the case of steel, pig iron and cast iron the interference effects are generally not determined.

By using suitable calibration systems the interferences can be compensated for without being determined in detail. For this purpose calibration using model solutions or the use of the addition method have proved useful.

In the case of the addition method an analytical-grade solution of the element to be determined is added to the sample solution in graded concentration steps. The concentrations of the additional solutions are measured so that only a few millilitres have to be added. Thus the change in the sample matrix is so insignificant that normally neither physical nor chemical interference effects are altered. The prepared solutions are measured and the determined absorption values are plotted against the concentration of the added solution. The values are always made up of two components, one from the sample and the other from the added analytical-grade element solution. If the straight line obtained is extended beyond the ordinate, it will intersect the abscissa. This value always lies to the left of the origin and is thus a negative value. The absolute value at this point corresponds with the concentration of element in the sample. The addition method can also be applied if an atomic absorption spectrometer plus microprocessor is being used. In this case the sample solution is aspirated and the concentration adjusted to zero. Then the sample with the highest addition is aspirated and the microprocessor is fed the concentration of the element solution. If a dummy solution containing all acids and solvents is then used, the microprocessor will print out the required concentration of the sample with a "minus" sign.

This method of internal calibration may only be used for linear calibration lines. For a curved calibration plot a linear extrapolation against zero would be very inaccurate.

D. Precision and accuracy

Statistical data on the analysis of oxide products should be based on the same criteria as for iron and steel analysis. Reproducibility limits of about 1–2 rel.% are valid in the case of direct analysis with AAS. The use of the injection technique leads to reproducibilities of 2–5 rel.% for high salt content and 10–20 rel.% for the graphite tube technique (low salt content) used as a trace method.

While analysis in an iron matrix ranges from the trace range to 20%, element concentrations of upto 60% are determined in oxide substances.

By using an electrically controlled "diluter" reproducibility of 1–3 rel.% can be achieved even for concentrations not normally suitable for determination by AAS.

Since classical techniques such as gravimetry and photometry produce similar variations of reproducibility in these ranges of concentration, the use of AAS for these determinations is justified. The gain in time compared with traditional methods is considerable. However, the accuracy of the analysis of oxide products has to be checked by the reference method (XRF, classical chemistry).

IV. SPECIAL AAS METHODS

There have been attempts to develop special methods for some elements which cause problems with solvent analysis when using the flame or furnace atomizer techniques.

Two basic trends can again be recognised, i.e. variations in the equipment, or in the associated chemical reactions. For example, to be able to determine as low P contents in steel as necessary today, an indirect method was applied [219]. After formation of the bismuthophosphomolybdate complex and its extraction by methyl isobutyl ketone the Bi can be determined in the extract and correlated with the P content.

The direct introduction of atomic vapour into the AAS flame can also be seen as an equipment variation [148, 152]. Here, the evaporation of the steel samples can be carried out with the help of the glow-discharge lamp [148] or an aerosol generator with a low current d.c.-arc discharge [152]. In line with these developments the Atomsource [163, 212] can play a very important role in routine analysis of metals and mixed oxide/metal powder samples in the near future.

Another example is the combination of gas chromatography and AAS [92], where AAS is used as an element detector.

A. Development of methods for gases

Chemical reactions whose product is gaseous and contains the element to be determined, have proved useful. The determination of Hg as metal vapour [79, 150, 213] and the determination of elements which form volatile hydrides, for example As [30, 43, 158, 161, 165, 167, 183, 184, 193, 196, 202, 204, 208, 210, 213, 217, 234, 236, 238, 239, 246] and Sb [165, 177, 193, 213, 233, 238, 246], Ge [19, 132], Se [23, 98, 162, 165, 176, 193, 197, 208, 235, 238, 246] and Te [165, 207, 238], Bi [165, 178, 193, 209, 213, 232,

238], Pb [167, 193, 232] and Sn [42, 165, 167, 193, 204, 210, 230, 238] have been described. The Hg determination and those of As and Sb relevant to ferrous metallurgy, are now part of the routine analysis of materials. The combination of EDL and the evolution of gas method has proved especially useful for this. For reasons of environmental control the determination of Hg in bulk materials has become very important, for example in iron ores; the procedure for this is now given as an example.

1. *Determination of mercury in iron ores, dust and slurry*

In this analysis samples are dissolved by pressure fusion, are then reduced in a wash bottle with $SnCl_2$ and analysed flameless as cold vapour.

Concentration range: 0.005–1 mg Hg kg^{-1}.

Reagents

HCl Suprapur (1.15 g ml^{-1}); H_2SO_4 Suprapur (1 + 1); HNO_3 Suprapur (1.4 ml^{-1}); HF, 40 mass% p.a.

$K_2Cr_2O_7$ solution, 10 mass% p.a. NH_2OH solution, 5 mass% p.a. $SnCl_2$ solution, 100 g $SnCl_2 \cdot 2H_2O$ in 50 ml HCl to be topped up to 1000 ml with $H_2O \cdot N_2$ to be passed through for 30 min.

Hg stock solution: 108.0 mg HgO are dissolved in 5 ml HNO_3 and topped up to 1000 ml with H_2O; 1 ml = 0.1 mg Hg. Hg standard solution I: 5 ml of $K_2Cr_2O_7$ solution and 10 ml HNO_3 are added to 10 ml of stock solution and topped up to 1000 ml with H_2O; 1 ml = 1 μg Hg. Hg standard solution II: 5 ml of $K_2Cr_2O_7$ solution and 10 ml HNO_3 are added daily to 10 ml of standard solution I and topped up to 1000 ml with H_2O; 1 ml = 10 ng Hg.

Operating conditions

AAS device with background corrector, absorption cuvette open (15 cm length, 15 mm diameter) and closed system with crystal windows circulating pump, 2.5 l min^{-1} pressure fusion vessel. Wavelength, 253.7 nm; EDL, Hg (5 W); deuterium compensator.

Procedure

According to the Hg concentration, 0.2–1.0 g of sample are weighed into the pressure fusion vessel, mixed with 4–10 ml H_2SO_4, 1–5 ml HNO_3 and 0.2 ml HF, close and heat for 45 min at 140°C stirring constantly with a magnetic mixer. After cooling, transfer to a gas wash bottle (250 ml), add 5 ml of $K_2Cr_2O_7$ solution and top up to 100 ml with H_2O. The excess oxidiser is reduced with NH_2OH solution (decolorising). Then 5 ml of $SnCl_2$ solution is added, the wash bottle is closed and Hg-free air is passed through the reagent vessel with the help of the circulation pump and through the absorption cuvette, until the maximum value appears (closed system). In the

258

case of the open system wait for 15 s after the reduction with SnCl₂ solution before air is passed through with the pump.

Calibration

With the standard solutions a calibration curve is produced by following the same sequence; this curve should be linear for the closed system for the concentration range 10–250 ng Hg. A linear calibration curve is produced for the concentration range 200–1000 ng for the open system. In the case of samples of unknown composition the addition method is recommended for best results. In this case certain quantities of the standard solutions can be added after pressure fusion.

Today, metal hydrides can be synthesised in commercial equipment. Principally a solution which contains AsCl₃ or SbCl₃ is pre-reduced, with for example KI and ascorbic acid, and the reduced with sodium borohydride to arsine or stibine. This is done in a closed system and the method is simple (Fig. 14). The determination of As via arsine in iron ores is now available

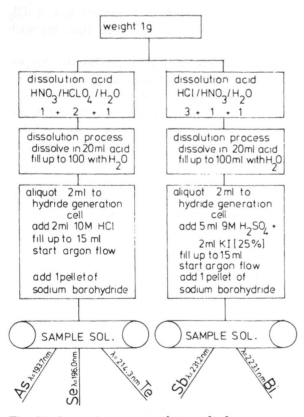

Fig. 14. General gas-generation method.

as ISO/TC 102 [185]. The method described for the concentration range of 0.0001–0.05 mass% is as follows.

Decompose the test portion by sintering with sodium peroxide and leaching with water and HCl. Transfer the solution to a distillation flask and evaporate a portion of the solution. Reduce arsenic to arsenic(III) by treatment with potassium bromide and hydrazine sulphate. Adjust the acidity and distill $AsCl_3$, collecting the distillate in water. Reduce an aliquot portion of the solution with potassium iodide and ascorbic acid, and treat with sodium borohydride to reduce arsenic ions to the volatile hydride, arsine. Rapidly sweep the arsine into a hydrogen–nitrogen (or argon)-entrained air flame and measure the transient peak absorbance at 197.2 nm in an atomic absorption spectrometer.

The determination of As in iron and steel for the concentration range 0.002–0.01 mass% is possible with the hydride generation method with an RSD of approximately 2–5%.

Many problems arise with the conventional procedure caused mainly by local over-concentrated agents during addition. Therefore, the continuous flow-method helps very much to minimize these effects [213].

B. Metal determination in organic solvents

A concentration procedure with the help of extraction [171, 174, 182, 186, 206] or chromatographic separation [169] of the elements to be determined from the matrix is often available using the organic phase.

Disregarding possible losses during extraction or elution it is a common procedure to aspirate organic solutions with metal ions into the AAS flame since this generally improves the measuring sensitivity [78, 92, 151, 206].

In industry, analysis for metals in lubricating oils, greases or fuels is important for checking characteristics on delivery or for testing the quality of used lubricants for the purpose of preventing maintenance. In the first case one talks of the determination of metal ions in oils among others, in the second of wear metals, such as Cd, Cr, Cu, Fe, Mn, Ni, Sb, Sn and Ti. Metals which may be present in lubricants or fuels, are Al, Ba, Ca, K, Li, Mg, Mo, Na, P, Pb, S, Si, V and Zn [74]. This analytical area is very important because considerable financial considerations are related to the use of the correct lubricants, to the possibility of harmful substances developing in fuel oils (S or V) and also to maintenance.

This is also shown by the existence of a large and increasing number (Table 5) of standardised AAS-methods [7, 35, 66] relating to metal ions in lubricating oils or greases (Al, Ba, Ca and Zn). In the literature, methods can be found which check additive contents (Ba, Ca, Zn) with the help of AAS [16, 20, 61, 63, 112, 133].

There are also methods for V in fuel oil [88], as well as Hg [80, 144] and Cd [188] and the AAS determinations of Ti [121], Pb [61, 95] and Mn [96] in

References pp. 268–274

TABLE 5

STANDARDISATION ACTIVITIES

Standard	Lubricants, fuels	Elements	AAS method Fuel/oxidant	
ASTM D 2788-69 T [7]	Gas turbine fuels	Ca, P, Mg, Na, K V	C_2H_2/air N_2O/C_2H_2	Dilution with organic solvent Direct nebulisation
IP 308/74 [66]	Lubricating oils	Ba, Ca, Mg, Zn	N_2O/C_2H_2	Dilution with organic solvent Direct nebulisation
DIN 51 391, Teil 1	Oils	Ba, Ca, Zn	C_2H_2/air	Dilution with petrol Direct nebulisation
DIN 51 393, Teil 1	Lubricants	Pb	C_2H_2/air	Dilution with petrol Direct nebulisation
DIN 51 395, Teil 1	Oils	Mo	C_2H_2/air	After incineration
DIN 51 395, Teil 2	Oils	Mo		Direct nebulisation
DIN 51 397, Teil 1	Oils	Fe	Non-flame	After incineration dissolution in HCl
DIN 51 401 [35]	–	–	General working principles	
DIN 51 431	Oils	Mg	C_2H_2/air	Dilution with petrol Direct nebulisation
DIN 51 769, Teil 7	Petrol	Pb	C_2H_2/air	Dilution with isooctane Direct nebulisation
DIN 51 769, Teil 8	Petrol	Pb (5–25 mg l^{-1})	C_2H_2/air	Direct nebulisation
DIN 51 790, Teil 3	Fuels, oils	Na	C_2H_2/air	After incineration dissolution in HCl
	Oils, fuels	V	Non-flame	
DIN 51 793, Teil 3	Fuels, oils	Na	C_2H_2/air	After incineration dissolution in H_2O
DIN 51 797, Teil 3	Oils	Na	C_2H_2/air	After incineration
DIN 51 797, Teil 4	Oils	Na	C_2H_2/air	Dilution with ethanol/xylene direct nebulisation
DIN 51 815	Greases	Li, Na, Ca	C_2H_2/air	Dilution with cyclohexene, HCl, H_2O-extraction

TABLE 6

DETECTION LIMITS AND LIMITS OF DETERMINATION OF DIRECT AAS ANALYSIS OF WEAR METALS FROM ORGANIC SOLUTION [74]

Element	AAS		NFAAS	
	Detection limit (mg kg^{-1})	Limit of determination (mg kg^{-1})	Detection limit (mg kg^{-1})	Limit of determination (mg kg^{-1})
Ag	0.1	0.4	0.01	0.15
Al	3	10	0.2	0.6
Co	0.2	0.6	0.02	0.06
Cr	0.2	0.6	0.02	0.06
Cu	0.2	0.6	0.01	0.03
Fe	0.2	0.6	0.01	0.03
Mg	0.008	0.025	0.0002	0.0006
Mn	0.1	0.3	0.002	0.01
Ni	0.2	0.6	0.02	0.06
Pb	0.8	2.5	0.005	0.02
Si	3	10	0.05	0.2
Sn	5	15	0.01	0.03
Ti	0.4[a]	2[a]	–	–
W	1[a]	20[a]	–	–
Zn	0.01	0.05	0.001	0.005

[a] Only from water solution after incineration.

petrol. The use of the graphite tube furnace in the determination of metal ions in oils and greases, including wear metals, has been described [9, 22, 69, 86].

Other reports deal with individual elements, such as Ni [1, 86, 87] or Fe [11, 84]. The efficiency [71–73] of flame methods (AAS) has been compared with non-flame techniques (NFAAS) (Table 6). Because of their significance there have been attempts to determine the elements P [38] and S [78] directly with AAS. This, however, requires a device which can measure ultraviolet lines (ca. 180 nm) with sufficient sensitivity. Good results can also be achieved by gas chromatographic separation and successive AAS determination [92], and simultaneous multi-element analysis with a Vidicondetector has been tried [68] because the speed with which the information is gained can be very important in practice. Some work [39, 53] reports on the problem of molecular bands which can appear when working with organic solutions. This demands specially selected lines or background correction.

Owing to the difficulty in extracting a homogeneous sample from used oils, or to the uncertain distribution, types and sizes of particles, an early incineration is generally undertaken, which should ideally be without waste. The treatment of greases is similar. The analysis of the ash is basically similar to the analysis of oxide substances in an aqueous medium.

References pp. 268–274

Fresh oil and petrol are normally analysed directly by AAS; oils of differing viscosity should be diluted with organic solvents such as MIBK, xylene etc. until the rate of atomization is no longer influenced by the viscosity.

Two basic methods of sample preparation can be distinguished. The direct method includes all bound or completely dissolved ions. No agreement has been reached concerning particles or their diameters determined in the process. The indirect method includes all metal or non-metal ions which are not volatilised during incineration.

To measure the increase or the decrease of a certain type of ion which corresponds to contamination, wear or thinning (consumption of additives), it is necessary to determine even slight variations of concentration with high precision [125]. The efficiency and precision of the AAS method is such that slight variations can be detected even for concentrations around 1 mg kg^{-1}. In practice, far-reaching decisions often depend on this. For example, the checking of accuracy and construction of calibration curves is now done world-wide, with oil-soluble standards available for all possible types of ions (e.g. CONOSTAN or MERCK). Due to the increasing importance of preventive maintenance this analytical field will expand considerably.

V. CALIBRATION TECHNIQUES

There are various calibration systems for AAS which differ mainly in the time they need.

For standardised methods the calibration curve is always produced with a pure-element solution. Hence, interferences have to be considered and eliminated, leading to time-consuming checks of variable parameters such as matrix effects, anion or cation influences or interferences caused by acids.

In the analytical procedures of ferrous metallurgy calibration methods have quickly proved useful, interferences being compensated for from the outset although they are not known individually.

A. Calibration using standard solutions

Atomic absorption spectrometers fitted with microprocessors allow calibration points to be stored. The necessary data can be made available to the microprocessor by aspirating suitable calibration solutions whose concentrations should be such that they cover the entire concentration range to be analysed, at appropriate intervals. After storing the calibration points the microprocessor constructs the calibration curve, inspection of which shows whether the plot is linear or curved.

With the help of microprocessors, recalibrations have become so easy and so quick, that a new calibration curve can be plotted using different parameters (a different analytical line, slit width, position of burner or dilution of sample solution).

When using a dual-channel spectrometer the calibration is easier since one concentration range can be covered by channel A and another by channel B. Standards for all categories of substances which are being used routinely in the laboratories of steel works can be purchased.

For calibration, a series of standard solutions is prepared which differ in their concentrations of all other elements. When calibrating with a pure-element solution a stock solution is used which contains a high concentration of the element to be determined and which only has to be diluted in certain ratios. However, in the case of an analytical sample, interferences have to be expected. It is recommended that a model solution is prepared and adjusted to the necessary concentration by diluting with a separately prepared solution which contains all matrix elements and accompanying elements but not the element to be analysed. Thus, standard solutions with different concentrations of the element to be analysed but with the same matrix are produced. For steel, pig iron and cast iron it is mostly sufficient to prepare a stock solution. Dilution of the stock solution for calibration to various concentration is done with an analytical grade iron solution.

The concentration of iron in the standard has to correspond with that of the sample solution. In the case of alloy steel of varying iron concentration, pure-iron solution is added to the stock solution and to the sample solution in excess. Thus, the difference in the concentration of iron in the sample compared with that in the standard is negligible. Interferences due to the matrix then remain constant during calibration and analysis. Calibration with model solutions is carried out by following the same pattern as with standard solutions.

B. Recalibration procedures

The standardised sequence for measuring calibration solution and solutions to be analysed is: (1) measuring of the calibration solution in order of increasing concentration; (2) measuring of sample solution, twice; (3) measuring the calibration solutions in order of decreasing concentration.

Measuring the calibration solution in order of decreasing concentration, after the solution to be analysed has been measured, has no influence on the results for the sample solutions. It is better to calibrate in order of increasing concentration before measuring the sample solutions and to check the calibration values by measuring in order of decreasing concentration; memory effects can then be recognised when calibration points deviate.

Once the data for the calibration solution have been stored, it is recommended to measure a recalibration solution after every fifth sample solution. This solution must be prepared as for a calibration solution but may not be used for calibration. The concentration of the element in the sample solution to be determined, should lie in about the middle of the concentration range of the calibration solutions and must be known. If the calibration curve does not change while measuring the sample solution, then the recalibration value must correspond with the known nominal value. Should deviations occur, the calibration curve has to be checked with the calibration solutions and, if necessary, the calibration data fed in again. This should not, however, be done without having checked for causes such as blocking of the burner slit, twisting of the burner port or blocking of the atomizer.

If a measurement has to be interrupted because of time, or if the same problem occurs again after a few hours or the next day (same element in the same matrix, e.g. Al in steel), then the calibration curve may be stored. With the spectrometer switched to standby, data are preserved in the microprocessor while other functions, such as the fuel gas supply or the chopper motor, are switched off. The lamp current is adjusted to a low setting so that the hollow-cathode lamps are kept prewarmed and ready to operate. To balance slight alterations in equipment parameters the lowest concentration calibration sample is aspirated and the new value fed into the microprocessor by passing on the nominal concentration. The remaining calibration points of the curve are then related to the new recalibration value. Thus the complete curve is recalibrated using one calibration sample. After approximately two minutes the spectrometer will be operational again.

When analysing low concentrations, in the range of a few $\mu g\ ml^{-1}$ or even less, it has to be remembered that solutions of very low concentrations are not very stable. It is essential that fresh solutions are prepared for calibration as well as for recalibration. Thus, it is recommended to prepare a solution of a higher concentration, such as a stock solution, since it will remain stable longer. Just before measuring the calibration solutions are diluted to their nominal concentrations with the appropriate solutions.

C. Background correction (conventional)

Background correction is carried out with a continuum source, e.g. a hydrogen hollow-cathode lamp or a deuterium arc lamp.

If the background absorption is wide-banded as in the case of dissociation continua of alkali halides, non-specific loss of light can be compensated for using the background compensator. However, in the case of finely structured molecular bands, e.g. electron stimulation spectra, use of background correction leads to overcompensation and consequent inaccurate measurements. Therefore, the background correctors, now part of nearly all modern atomic absorption spectrometers, should only be used if a thorough knowledge of the spectral background has been gained. The operational range of the background corrector lies between 190 and 350 nm.

In ferrous metallurgy background correction is mainly used after fusion, when light absorption occurs on salt particles because of high total salt concentrations. By dissolving the fusion substance in inorganic acids, such as HCl, alkali halides develop in high concentration (NaCl, KCl with soda potassium fusions; LiCl with lithium tetraborate fusions etc.). These result in the molecular absorption already described.

Even if the non-specific light losses do not lead to analytical mistakes when balancing the calibration solution and the sample solution, the signal specific to the element can be increased by using the background absorption adjustment and thus the determination limit is lowered.

When analysing solid substances using graphite tube absorption, signals arising from formation of smoke from the matrix have often been observed. The solid-sample method, with steel chips as described in Section II.B.1, shows neither smoke or fog phenomena, nor other unspecified absorptions.

The direct analysis of oxide products such as iron sinter, slag or refractories, leads to considerably non-specific loss of light due to the large silicate matrix. It has yet to be investigated whether the use of background correction will make direct oxide analysis possible.

First experiments indicate that a direct determination of trace elements in oxide substances is possible but the determination of matrix elements, because of their high concentrations, should be left to other methods of analysis, such as solution analysis with AAS or direct X-ray fluorescence spectrometry.

D. Calculations and data handling

The increasing use of advanced computing equipment can be applied to the processing of AAS data. Some modern atomic absorption spectrometers allow calibration curves to be stored. The adaptation of the equipment parameters to a base calibration curve can then be carried out with just one recalibration solution. This, however, is only valid for the calibration curve of one element in a specific matrix.

The microprocessor can store calibration curve data, compute statistical data such as the mean value, standard deviation and relative standard deviation, and compensate for equipment parameters. In addition it is easily possible to couple it to a large computer via an interface. Thus, the functions of the equipment as well as calibration values and analytical values can be controlled on line and evaluated. Direct coupling has been in use for years for simultaneously registering emission and X-ray fluorescence spectrometers. On-line coupling has proved useful when similar analytical problems with a fixed number of elements to be analysed, occur. It leads to fewer errors in operation and to quicker analyses. However, the value of on-line coupling for sequential analytical methods such as polarography or AAS is questionable. Personnel have to be present continuously to operate the equipment (changing lamps, feed samples), so that no economy of effect can be achieved. When using automatic sample feeders, direct coupling to a computer is only feasible if the same elements in the same matrix are always determined.

References pp. 268–274

In ferrous metallurgy analysis, where AAS is used as a special procedure for special analyses, direct coupling is not profitable [58]. Installing a spectrometer controlled by a computer would require analysis of a far larger number of samples or possible problems. The ratio of computer location to frequency of use would be very unfavourable and would increase the cost of the individual analysis. It is more useful to feed the analytical data, issued by the microprocessor and printed by an automatic printer, to the laboratory computer via an off-line system, which would then include the data together with those from on-line analytical systems.

If the data are stored on floppy discs, they are always available for further statistical investigations concerned with reproducibility, ring experiments or survey statistics, without taking up storage space.

VI. SIMULTANEOUS READING ATOMIC ABSORPTION SPECTROMETRY

In metallurgical analysis it is very important to produce large amounts of analytical data because of statistical reasons to calibrate big analysis systems, and to be able to use two or more different analytical methods in reference to guarantee for accuracy.

The application of the Atomsource [163] initiated a strong development of a simultaneous reading detection in atomic absorption. But also, any multi-element information of the vaporized and atomized sample is contained in both of the other sources, the flame and the graphite furnace. Therefore, there is a great demand for a real AA multi-element technique as it would be helpful to solve more difficult analytical problems.

Two ways of development have been reported the last years: the emitted light of three 5-element hollow-cathode lamps is reflected into the optical pathway of radiation and detected by a polychromator instrument [199], and fiber optics are used to transport the light of some hollow-cathode lamps through an atom reservoir into a polychromator detector [189]. Another way has just been completed and is briefly described here.

The GRIMM-type glow discharge (GDS) can be used as primary light source in atomic absorption spectrometry [214], whereby the number of element radiations is given by the composition of the sample (target) which is sputtered. Its composition can be adapted to the analytical problem to be solved, and it is easy to change the target.

The GDS generated at relatively low power (10–24 W) is burning stably for >20 minutes on the same sputtering spot.

The cathode sputtering process occurring in the discharge area of the GDS is responsible for the sample atom introduction into the plasma. Depending on the sample material this process becomes stable after a short time (about 10 s), resulting in a stable emission of spectral lines of all elements contained

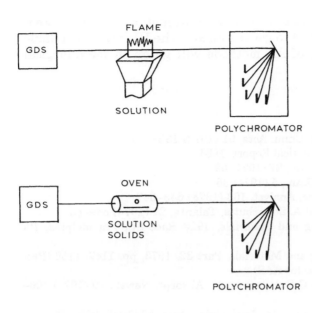

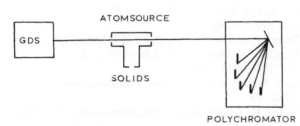

Fig. 15. Schematic of glow discharge system for simultaneous mult-elemental detection, illustrating its use with three types of atom cell.

in the target. Compared to hollow-cathode lamps replaced by the GDS it offers much more spectral lines of much higher intensity, especially below 280 nm, which can be used well for analytical purposes. The GDS can be optimized in relation to the actual analytical problem as well.

The spectrometer used was constructed by Spectruma, Munich, and enables the application of all the three sources (Fig. 15) which generate the sample atoms. The main problem of this development is to modulate the GDS in the same way as hollow-cathode lamps are modulated normally. The first results do show an improved determination of elements which have the best analytical lines emitted in the near UV region of the spectrum [214].

The combination of GDS/Atomsource/AA polychromator and possibly the closed graphite furnace unit offers the possibility to install an AA spectrometer working under vacuum or flushed by inert gas. One has to

268

combine only the single units with a vacuum polychromator by a tube system which can be evacuated or flushed to get the lines of the UV region available for analysis. The development of AA methods will go on in metallurgical analysis.

REFERENCES

1   J.F. Alder and T.S. West, Anal. Chim. Acta, 61 (1972) 132–135.
2   E. Althaus, Perkin-Elmer Analytical Report, 1964.
3   A.D. Ambrose, Proc. Chem. Conf., 27 (1974) 59.
4   E. Ametrano and B. Grasso. Scan., 5 (1974) 35.
5   Analytical Methods Committee, Analyst, 103 (1978) 643–647.
6   M.A. Ashy, J.B. Headridge and A. Sowerbutts, Talanta, 21 (1974) 649–652.
7   American Society for Testing and Materials, 1916 Race St., Philadelphia, PA 19103.
8   American Society for Testing and Materials, Part 32, 1973, pp. 1147–1159 (Proposed Recommended Practices for AAS).
9   W.B. Barnett, H.L. Kahn and G.E. Peterson, At. Absorpt. Newsl., 10 (1971) 106–110.
10 W.B. Barnet and E.A. McLaughlin, Jr., Anal. Chim. Acta, 80 (1975) 285–296.
11 T.T. Bartels and M.P. Slater, At. Absorpt. Newsl., 9 (1970) 75–77.
12 B. Bernas, Anal. Chem., 40 (1968) 1682–1685.
13 H. Berndt and K. Jackwerth, Spectrochim. Acta, Part B, 30 (1975) 169–177.
14 H. Berndt and E. Jackwerth, At. Absorpt. Newsl., 15 (1976) 109–113.
15 P.L. Bertolaccini, Metall. Ital., 61 (1969) 396–403.
16 H. Binding and H. Gawlick, Dtsch. Forschungsber., 1969 (TIB Hannover).
17 H. Boesch, Mikrochim. Acta, (1976) 49.
18 H. Bosch, E. Buchel, H. Grygiel and K. Lohau, Arch. Eisenhüttenwes., 45 (1974) 699–704.
19 R.S. Braman and M.A. Tompkins, Anal. Chem., 50 (1978) 1088–1093).
20 M.P. Bratzel, Jr. and C.L. Chakrabarti, Anal. Chim. Acta, 61 (1972) 25–32.
21 F. Brivot, I. Cohort, G. Legrand, J. Louvrier and I.A. Voinovitch, Analusis, 2 (1973) 570–576.
22 K.G. Brodie and J.P. Matousek, Anal. Chem., 43 (1971) 1557–1560.
23 K.G. Brodie, Int. Lab., (1977) 65–74.
24 W.M.G.T. van den Broek and L. de Galan, Anal. Chem., 49 (1977) 2176–2186.
25 W.D. Cobb, W.W. Foster and T.S. Harrison, Lab. Pract., 24 (1975) 143.
26 W.D. Cobb, W.W. Foster and T.S. Harrison, Analyst, 101 (1976) 39–43, 255–259.
27 W.D. Cobb, W.W. Foster and T.S. Harrison, Anal. Chim. Acta, 78 (1975) 293–298.
28 J. Collin, J. Sire and I.A. Voinovitch, Bull. Liaison Lab. Pons Chaussees, 79 (1975) 78.
29 A. Condylis and B. Mejean, Analusis, 3 (1975) 94–109.
30 E.A. Crecelius, Anal. Chem., 50 (1978) 826–827.
31 M.S. Cresser, Anal. Chim. Acta, 80 (1975) 170–175.
32 M. Damiani, M.G. Del Monte Tamba and F. Bianchi, Analyst, 100 (1975) 643–647.
33 M.B. Denton and H.U. Malmstadt, Anal. Chem. 44 (1972) 241–246.
34 H.A. von Derschau and H. Prugger, Fresenius Z. Anal. Chem., 247 (1969) 8–12.
35 Deutsches Institut für Normung (DIN), P.O. Box 1107, D-1000 Berlin 30.

36 K. Dittrich, E. John and I. Rohde, Anal. Chim. Acta, 94 (1977) 75–81.
37 K. Dittrich and K. Borzym, Anal. Chim. Acta, 94 (1977) 83–90.
38 D.J. Driscoll, D.A. Clay, C.H. Rogers, R.H. Jungers and F.E. Butler, Anal. Chem., 50 (1978) 767–769.
39 Y. Endo and Y. Nakahara, Tetsu To Hagane, 59 (1973) 800–807.
40 Y. Endo and Y. Nakahara, Trans. Iron Steel Inst. Jpn., 16 (1976) 396–403.
41 T.A. Eroshevich and D.F. Makarov, Zavod. Lab., 41 (1975) 186.
42 H.D. Fleming and R.G. Ide, Anal. Chim. Acta, 83 (1976) 67–82.
43 D.E. Fleming and G.A. Taylor, Analyst, 103 (1978) 101–105.
44 A.G. Fogg, S. Soleymanloo and D.T. Burns, Talanta, 22 (1975) 541–543.
45 P. Foster, R. Molins and H. Bozon, Analusis, 1 (1972) 434–438.
46 P. Foster and J. Garden, Analusis, 2 (1974) 675.
47 W. Frech, Talanta, 21 (1974) 565–571.
48 W. Frech, Anal. Chim. Acta, 77 (1975) 43–52.
49 W. Frech, Fesenius Z. Anal. Chem., 275 (1975) 353–357.
50 W. Frech and A. Cedergren, Anal. Chim. Acta, 82 (1976) 83–102.
51 T.W. Freudiger and C.T. Kenner, Appl. Spectrosc., 26 (1972) 302–305.
52 L. de Galan and G.F. Samaey, Anal. Chim. Acta, 50 (1970) 39–50.
53 L. de Galan and E.J. Benes, Metall. Ital., 64 (1972) 343–348.
54 C.A. Gomez and L.M.T. Dorado, Rev. Metal. (Madrid), 10 (1974) 355.
55 R.J. Guest and D.R. McPherson, Anal. Chim. Acta, 71 (1974) 233–253.
56 R.K. Hansen and R.H. Hall, Anal. Chim. Acta, 92 (1977) 307–320.
57 J.B. Headridge and A. Sowerbutts, Analyst, 98 (1973) 57–64.
58 W. Heinemann and W. Prinz, Fresenius Z. Anal. Chem., 289 (1978) 17–23.
59 R. Herrmann, Z. Instrum., 75 (1967) 101–111.
60 R. Hohn and F. Umland, Fresenius Z. Anal. Chem., 258 (1972) 100–106.
61 H.E. Hoffmann, H. Nathansen and F. Altmann, Jenaer Rundsch., 6 (1976) 302–304.
62 M.E. Hofton, British Steel Corp., Report 1976 (GS/TECH/558/2/76/C).
63 S.T. Holding and P.H.D. Matthews, Analyst, 97 (1972) 189–194.
64 S.T. Holding and J.J. Rowson, Analyst, 100 (1975) 465–470.
65 I.D. Ingle, Jr., Anal. Chem., 46 (1974) 2161–2171.
66 IP Standards for Petroleum and its Products, Part I: Methods for Analysis and Testing, Sect. 2, IP Method 308 (1975), Applied Science Publishers, Barking.
67 International Organization for Standardisation (ISO), TC 102/SC 2 (Iron Ore/ Chemical Analysis).
68 K.W. Jackson, K.M. Aldous and D.G. Mitchell, Appl. Spectrosc., 28 (1974) 569–573.
69 E. Jantzen, Schmiertech. Int. Tribol., 22 (1975) 31–37.
70 A.H. Jones and W.D. France, Jr., Anal. Chem., 44 (1972) 1884–1886.
71 S.H. Kägler, Erdoel/Kohle, 27 (1974) 514–517.
72 S.H. Kägler, Erdoel/Kohle, 28 (1975) 232–237.
73 S.H. Kägler, Erdoel/Kohle, 29 (1976) 362.
74 S.H. Kägler, Schmiertech. Tribol, 25 (1978) 84–88.
75 H.L. Kahn and S. Slavin, At. Absorpt. Newsl., 10 (1971) 125.
76 H.L. Kahn, G.E. Peterson and J.E. Schallis, Perkin-Elmer Analytical Report No. 17, 1969.
77 I.P. Kharlamov and G.V. Eremina, Zavod. Lab., 40 (1974) 385–391.
78 G.F. Kirkbright, M. Marshall and T.S. West, Anal. Chem., 44 (1972) 2379–2382.
79 G.F. Kirkbright and P.J. Wilson, Anal. Chem., 46 (1974) 1414–1418.

270

80 H.E. Knauer and G.E. Millman, Anal. Chem., 47 (1975) 1263–1268.
81 K.H. Koch and K.Ohls, Arch. Eisenhüttenwes., 39 (1968) 925–928.
82 P. Konig, K.H. Schmitz and E. Thiemann, Arch. Eisenhüttenwes., 40 (1969) 553–556.
83 J. Kozusnikova and A. Kolarova, Hutn. Listy, 32 (1977) 810–812.
84 J. Komarek, J. Jambor and L. Sommer, Fresenius Z. Anal. Chem., 262 (1972) 91–94.
85 G.R. Kornblum and L. de Galan, Spectrochim. Acta, Part B, 28 (1973) 139–147.
86 J.J. Labrecque, J. Galobardes and M.E. Cohen, Appl. Spectrosc., 31 (1977) 207–210.
87 I. Lang, G. Sebor, V. Sychra, D. Kolihova and O. Weisser, Anal. Chim. Acta, 84 (1976) 299–305.
88 I. Lang, G. Sebor, O. Weisser and V. Sychra, Anal. Chim. Acta, 88 (1977) 313–318.
89 L.L. Lewis, Anal. Chem., 40 (1968) 28A–47A.
90 D. Littlejohn and J.M. Ottaway, Analyst, 103 (1978) 595–606.
91 D. Littlejohn and J.M. Ottaway, Anal. Chim. Acta, 98 (1978) 279–290.
92 J.C. van Loon, B. Radziuk, N. Kahn, J. Lichwa, F.J. Fernandez and J.D. Kerber, At. Absorpt. Newsl., 16 (1977) 79–83.
93 R.P. Lucas and B.C. Ruprecht, Anal. Chem., 43 (1971) 1013–1016.
94 H.M. Luschow, D. Lindenberger and G. Kraft, Fresenius Z. Anal. Chem., 279 (1976) 347–349.
95 R.J. Lukasiewicz, P.H. Berens and B.E. Buell, Anal. Chem., 47 (1975) 1045–1049.
96 R.J. Lukasiewicz and B.E. Buell, Appl. Spectrosc., 31 (1977) 541–547.
97 O. Luzar and V. Sliva, Hutn. Listy, 30 (1975) 55.
98 D.C. Manning, At. Absorpt. Newsl., 10 (1971) 123–124.
99 D.C. Manning, At. Absorpt. Newsl., 14 (1975) 99–102.
100 T. Maruta, M. Suzuki and T. Takeuchi, Anal. Chem. Acta, 51 (1970) 381–385.
101 H. Massmann, Fresenius Z. Anal. Chem., 225 (1967) 203–213.
102 H. Massmann, Spectrochim. Acta, Part B, 23 (1968) 215–226.
103 H. Massmann, Angew. Chem., 86 (1974) 542–552.
104 G.L. McPherson, At. Absorpt. Newsl., 4 (1965) 180–191.
105 J. Musil, Hutn. Listy, 30 (1975) 292.
106 J. Musil and J. Dolezal, Anal. Chim. Acta, 92 (1977) 301–305.
107 G. Nonnenmacher and Fr.H. Schleser, Fresenius Z. Anal. Chem., 209 (1965) 284–293.
108 N. Oddo, Metall. Ital., 64 (1972) 359–362.
109 W. Ooghe and F. Verbeek, Anal. Chim. Acta, 73 (1974) 87–95.
110 W. Ooghe and F. Verbeek, Anal. Chim. Acta, 79 (1975) 285–291.
111 B.D. Pederson and R.W. Taylor, Electr. Furnace Conf. Proc., 1974, 32 (1975) 168.
112 G.E. Peterson and H.L. Kahn, At. Absorpt. Newsl., 9 (1970) 71–74.
113 G.E. Peterson and J.D. Kerber, At. Absorpt. Newsl., 15 (1976) 134–143.
114 A.A. Petrov and G.V. Skvortsova, Zh. Prikl. Spektrosk., 22 (1975) 991.
115 W.J. Price and P.J. Whiteside, Analyst, 103 (1978) 643–647.
116 D.B. Ratcliffe, C.S. Byford and P.B. Osman, Anal. Chim. Acta, 75 (1975) 457–459.
117 R.D. Reeves, B.M. Patel, C.J. Molnar and J.D. Winefordner, Anal. Chem., 45 (1973) 246–249.
118 W.K. Robins and H.H. Walker, Anal. Chem., 47 (1975) 1269–1275.
119 R.C. Rooney and C.G. Pratt, Analyst, 97 (1972) 400–404.
120 C.J. Rowe and M.W. Routh, Res. Dev. (Nov. 1977) 34–30.

121   C.S. Saba and K.J. Eisentrant, Anal. Chem., 49 (1977) 454–457.
122   Z. Samsoni, Mikrochim. Acta, Suppl. II (1978) 177–190.
123   A. Sato and T. Ito, Tetsu To Hagane, 56 (1970) 144.
124   K.H. Sauer and M. Nitsche, Arch. Eisenhüttenwes., 40 (1969) 891–893.
125   W. Schmidt, F. Dietl, G. Schadow and D. Ade, Motortech. Z., 37 (1976) 307–310.
126   P.H. Scholes, Proc. Soc. Anal. Chem., 5 (1968) 114–115.
127   P.H. Scholes, Proc. BISRA Conf. 1968 (MG/D/Conf. Proc./115/73).
128   P.H. Scholes, British Steel Corp., Report 1972 (MG/CC/588/72).
129   P.H. Scholes, British Steel Corp., Report 1973 (CAC/Conf. Proc./115/73).
130   E. Sebastiani, K. Ohls and G. Riemer, Fresenius Z. Anal. Chem., 264 (1973) 105–
      109.
131   I.L. Shresta and T.S. West, Bull. Soc. Chim. Belg., 84 (1975) 549.
132   R.K. Skogerboe and A.P. Bejmuk, Anal. Chim. Acta, 94 (1977) 297–305.
133   S. Skujins, Techtron Appl. Notes, No. 4, 1970.
134   W. Slavin, Appl. Spectrosc., 20 (1966) 281–288.
135   W. Slavin and S. Slavin, Appl. Spectrosc., 23 (1969) 421–433.
136   B.G. Stephens and H.L. Felkel, Jr., Anal. Chem., 47 (1975) 1676.
137   K.A. Stewart, Proc. Soc. Anal. Chem., 5 (1968) 116–117.
138   R.E. Sturgeon, Anal. Chem., 49 (1977) 1255A–1267A.
139   D.H. Svedung, Jernkontorets. Ann., 155 (1971) 295–297.
140   D.R. Thomerson and W.J. Price, Analyst, 96 (1971) 825–834.
141   W. Thomich, Arch. Eisenhüttenwes., 42 (1971) 779–781.
142   K.C. Thompson and R.G. Godden, Analyst, 101 (1976) 96–102.
143   M. Tomljanovic and Z. Grobenski, At. Absorpt. Newsl., 14 (1975) 52.
144   W.Ch. Tsai and L.J. Shiau, Anal. Chem., 49 (1977) 1641–1644.
145   K.J. Tsunoda, K. Fujiwara and K. Fuwa, Anal. Chem., 49 (1977) 2035–2039.
146   G.L. Vassilaros, Talanta, 21 (1974) 803–808.
147   Verein Deutscher Eisenhüttenleute, Handbuch für das Eisenhüttenlaboratorium,
      Vol. 5, Verlag Stahleisen, Düsseldorf, 1977.
148   A. Walsh, Appl. Spectrosc., 27 (1973) 335–341.
149   H.C. Wagenaar and L. de Galan, Spectrochim. Acta, Part B, 28 (1973) 175–
      177.
150   R.J. Watling, Anal. Chim. Acta, 94 (1977) 181–186.
151   B. Welz, CZ Chem. Tech., 1 (1972) 373–380, 455–460.
152   R.K. Winge, V.A. Fassel and R.N. Kniseley, Appl. Spectrosc., 25 (1971) 636–641.
153   R. Whitman, Analyst, 100 (1975) 555–562.
154   J.D. Winefordner, J.J. Fitzgerald and N. Omenetto, Appl. Spectrosc., 29 (1975)
      369–383.
155   R. Woodriff, M. Marinkovic, R.A. Howald and I. Eliezer, Anal. Chem., 49 (1977)
      2008–2012.
156   J. Agterdenbos and D. Bax, Fresenius Z. Anal. Chem., 323 (1986) 783–787.
157   J. Alvarado, F. Campos and J.M. Ottaway, Talanta, 33 (1986) 61–65.
158   H.M. Arbab-Zavar and A.G. Howard, Analyst, 105 (1980) 744–750
159   S. Arpadjan and V. Krivan, Fresenius Z. Anal. Chem., 329 (1988) 745–749; or Anal.
      Chem., 58 (1988) 2611–2614
160   P. Battistoni, P. Bruni, L. Cardellini, G. Fava and G. Gobbi, Talanta, 27 (1980)
      623–626.
161   D. Bax, J.T. van Elsteren and J.Agterdenbos, Spectrochim. Acta, 41B (1986) 1007–
      1013.

272

162 D. Bax, F.F. Peters, J.P.M. van Noort and J. Agterdenbos, Spectrochim. Acta, 41B (1986) 275–282.

163 A.E. Bernhard, Spectroscopy, 2 (1987) 38–43.

164 A. Besse, A. Rosopulo, C. Busche and G. Küllmer, LaborPraxis-Special, (1987) 102–107.

165 H. Bombach, B. Luft, E. Weinhold and F. Mohr, Neue Hütte, 29 (1984) 233–236.

166 C. Brescianini, A. Mazzucotelli, R. Frache and G. Scarponi, Fresenius Z. Anal. Chem., 332 (1988) 34–36.

167 K.G. Brodie and J.J. Rowland, Eur. Spectrosc. News, 36 (1981) 41–44; or Int. Lab., (1979) 40–47.

168 W.G. Brumbaugh and S.R. Koirtyohann, Anal. Chem., 60 (1988) 1051–1055.

169 P. Burba and P.G. Willmer, Fresenius Z. Anal. Chem., 329 (1987) 539–545

170 G.R. Carnrick, W. Barnett and W. Slavin, Spectrochim. Acta, 41B (1986) 991–997.

171 J.R. Castillo, J.M. Mir, C. Bendicho and F. Laborda, Fresenius Z. Anal. Chem., 332 (1988) 37–40.

172 J.R. Castillo, J.M. Mir and C. Bendicho, Spectrochim. Acta, 43B (1988) 263–271.

173 D. Chakraborti, W. de Jonghe and F. Adams, Anal. Chim. Acta, 120 (1980) 121–127.

174 M. de la Guardia and M.T. Vidal, Talanta, 31 (1984) 799–803.

175 J. Dedina, Fresenius Z. Anal. Chem., 323 (1986) 771–782.

176 J. Dedina, Anal. Chem., 54 (1982) 2097–2102.

177 K. de Doncker, R. Dumarey, R. Dams and J. Hoste, Anal. Chim. Acta, 153 (1983) 33–40.

178 K. de Doncker, R. Dumarey, R. Dams and J. Hoste, Anal. Chim. Acta, 161 (1984) 365–368.

179 K. Dittrich and R. Mandry, Analyst, 111 (1986) 269–280.

180 K. Dittrich, R. Mandry, C.Udelnow and A. Udelnow, Fresenius Z. Anal. Chem., 323 (1986) 793–799.

181 E.M. Donaldson, Talanta, 27 (1980) 499–505.

182 E.M. Donaldson and M. Wang, Talanta, 33 (1986) 233–242.

183 G. Drasch, L. v. Meyer and G. Kauert, Fresenius Z. Anal. Chem., 304 (1980) 141–142.

184 A. Dornemann and H. Kleist, Fresenius Z. Anal. Chem., 305 (1981) 379–381.

185 EURONORM, ECISS/TC 20, Postboks 77, DK-2900 Hellerup; or ECISS, 200 rue de la Loi, B-1049 Bruxelles.

186 C.J. Eskell and M.E. Pick, Anal. Chim. Acta, 117 (1980) 275–283.

187 J. Fazakas and M. Hoenig, Talanta, 35 (1988) 403–405.

188 W. Frech and A. Cedergren, Anal. Chim. Acta, 113 (1980) 227–235.

189 A. Golloch, A. Brockmann and H.-M. Kuß, Pittsburgh Conf. on Analytical Chemistry, Atlantic City, 1987, Paper No. 640.

190 Handbuch für das Eisenhüttenlaboratorium, Bd.2 (1987) 10.7, Verlag Stahleisen, Düsseldorf.

191 J.B. Headridge and I.M. Riddington, Mikrochim. Acta (Wien), II (1982) 457–467.

192 G.M. Hieftje, J. Anal. At. Spectrom., 4 (1989) 117–122.

193 P.K. Hon, O.W. Lau, W.C. Cheung and M.C. Wong, Anal. Chim. Acta, 115 (1980) 355–359.

194 H. Hocquaux, P. Bonnet and J.B. da Costa, C.C.E. Rapport (1985), UNIREC, B.P. 34, F-42701 Firminy, France.

195 M. Ikeda, Anal. Chim. Acta, 167 (1985) 289–297.

196 T. Inui, S. Terada and H. Tamura, Fresenius Z. Anal. Chem., 305 (1981) 189–192.

197 T. Inui, S. Terada, H. Tamura and N. Ichinose, Fresenius Z. Anal. Chem., 311 (1982) 492–495.

198 K.H. Koch, D. Sommer and D. Grunenberg, GIT Fachz. Lab., 32 (1988) 766–771.

199 W. Lautenschläger, R. Wagner and A.E. Bernhard, LaborPraxis, 10 (1986) 1004–1013.

200 R.P. Liddell, N. Athanasopoulos, R.G. Grey and M.W. Routh, Int. Lab., (1987) 82–87; or LaborPraxis, 11 (1987) 34–40.

201 B. Magyar, B. Wampfler and J. Zihlmann, GIT Fachz. Lab., 28 (1984) 301–308.

202 W.A. Maher, Talanta, 29 (1982) 532–534.

203 J.P. Matousek and H.K.J. Powell, Spectrochim. Acta, 41B (1986) 1347–1355.

204 K. Matsubara, N. Ota, M. Taniguchi and K. Narita, 99th ISIJ Meeting, 1980, Paper No. S395 (Wakinohama-cho, Fukiaki, Kobe 651).

205 E.B. Milosavljevic, L. Solujic, J.H. Nelson and J.L. Hendrix, Mikrochim. Acta (Wien), III (1985) 353–360.

206 J. Mohay and Zs. Vegh, Fresenius Z. Anal. Chem., 329 (1988) 856–860.

207 F. Mohr and B. Luft, Neue Hütte, 26 (1981) 431–433.

208 H. Narasaki and M. Ikeda, Anal. Chem., 56 (1984) 2059–2063.

209 H. Narasaki, Fresenius Z. Anal. Chem., 321 (1985) 464–466.

210 E.A. Norman, E.S. Orlova and A.I. Sestakova, Zavodskaja Lab., 46 (1980) 1108–1109.

211 K. Ohls and K.H. Koch, LaborPraxis, 9 (1985) 1336–1343; or Spectrochim. Acta, 39B (1984) 9–11.

212 K. Ohls, Fresenius Z. Anal. Chem., 327 (1987) 111–118.

213 K. Ohls, LaborPraxis, 11 (1987) 668–673.

214 K. Ohls, J. Flock and H. Loepp, Fresenius Z. Anal. Chem., 332 (1988) 456–463.

215 J.M. Ottaway and F. Shaw, Anal. Chim. Acta, 99 (1978) 217–223.

216 J.M. Ottaway, B. Fu, J. Marshall and D. Littlejohn, Anal. Chim. Acta, 161 (1984) 265–273.

217 C.J. Peacock and S.C. Singh, Analyst, 106 (1981) 931–938.

218 J.-A. Persson and W. Frech, Anal. Chim. Acta, 119 (1980) 75–89.

219 R. Ramchandran and P.K. Gupta, Talanta, 35 (1988) 653–654.

220 T.M. Rettberg and J.A. Holcombe, Anal. Chem., 60 (1988) 600–605.

221 N.K. Roy, D.K. De and A.K. Das, At. Spectrosc., 5 (1984) 126–128.

222 G. Schlemmer and B. Welz, Fresenius Z. Anal. Chem., 328 (1987) 405–409.

223 H. Schinkel, Fresenius Z. Anal. Chem., 317 (1984) 10–26.

224 J.G. Sen Gupta, Talanta, 31 (1984) 1045–1057.

225 J.M. Shekiro, Jr., R.K. Skogerboe and H.E. Taylor, Anal. Chem., 60 (1988) 2578–2582.

226 A.E. Smith, Analyst, 100 (1975) 300–306.

227 D. Sommer and K. Ohls, GIT Fachz. Lab., 26 (1982) 1015–1020.

228 D. Sommer and K. Ohls, Fresenius Z. Anal. Chem., 298 (1979) 123–127.

229 R.E. Sturgeon, S.N. Willie and S.S. Berman, Fresenius Z. Anal. Chem., 323 (1986) 788–792.

230 R.E. Sturgeon, S.N. Willie and S.S. Berman, Anal. Chem., 59 (1987) 2441–2444.

231 T. Takada and K. Hirokawa, Talanta, 29 (1982) 849–855.

232 B. Vanloo, R. Dams and J. Hoste, Anal. Chim. Acta, 151 (1983) 391–400.

233 B. Vanloo, R. Dams and J. Hoste, Anal. Chim. Acta, 160 (1984) 329–333.

234 B. Vanloo, R. Dams and J. Hoste, Anal. Chim. Acta, 175 (1985) 325–328.

235 M. Verlinden, J. Baart and H. Deelstra, Talanta, 27 (1980) 633–639.

274

236   W.J. Wang, S. Hanamura and D. Winefordner, Anal. Chim. Acta, 184 (1986) 213–218.
237   B. Welz, Z. Grobenski and M. Melcher, Perkin-Elmer Report, 17 (1979) 2–14.
238   B. Welz and M. Melcher, Spectrochim. Acta, 36B (1981) 439–462.
239   B. Welz and M. Melcher, Anal. Chim. Acta, 131 (1981) 17–25.
240   B. Welz and M. Schubert-Jacobs, Fresenius Z. Anal. Chem., 324 (1986) 832–838.
241   B. Welz, Fresenius Z. Anal. Chem., 325 (1986) 95–101.
242   B. Welz, G. Schlemmer and J.R. Mudakavi, Anal. Chem., 60 (1988) 2567–2572.
243   W. Wendl and G. Müller-Vogt, GIT Fachz. Lab., 28 (1984) 271–276.
244   G. Wibetoe and F.J. Langmyhr, Anal. Chim. Acta, 176 (1985) 33–40.
245   M. Yamamoto, T. Shohji, T. Kumamaru and Y. Yamamoto, Fresenius Z. Anal. Chem., 305 (1981) 11–14.
246   M. Yamamoto and T. Kumamaru, Fresenius Z. Anal. Chem., 281 (1976) 353–359.

Chapter 4f

The Analysis of Non-Ferrous Metals by Atomic Absorption Spectrometry

M.R. NORTH

Willan Metals Limited, Poplar Way, Rotherham, South Yorkshire (Great Britain)

I. INTRODUCTION

Atomic absorption spectroscopy (AAS) has been shown to be capable of determining almost 70 elements and finds wide application in the analysis of non-ferrous metals and alloys. This review gives a cross-section of the techniques used, which vary from conventional flame AAS for the determination of major and minor elements, to the use of more sensitive techniques such as electrothermal atomization, for the determination of elements at sub-ppm levels.

II. DETERMINATION OF IMPURITIES IN ALLOYS USING FLAME AAS APPLICATIONS

A. Aluminium alloys

One of the earliest applications for the analysis of aluminium alloys was that of Mansell et al. [1], who determined copper, calcium, manganese, zinc and magnesium after a caustic soda dissolution. The most widely used method for aluminium analysis is however based on the work of Bell [2], who used dissolution in hydrochloric acid.

Aluminium alloys may be dissolved using the following procedure. Weigh out 0.50 g of sample into a 250 ml beaker, add 30 ml of 50% hydrochloric acid in 5 ml aliquots. When the reaction subsides add 3 ml of 20 vol hydrogen peroxide, evaporate to approximately 15 ml and transfer to a 100 ml graduated flask, make up to 100 ml using deionised water. Standards should be prepared to contain 0 to 25 mg l^{-1} of the elements of interest in a 0.5% solution of high-purity aluminium in 15% hydrochloric acid. Copper, iron, magnesium, manganese, nickel and zinc can all be successfully determined using this method, though for magnesium determinations the sample and standard solutions should contain 0.1% lanthanum chloride as releasing agent to prevent aluminium interference.

If silicon is to be determined the method can be adapted in the following way. Weigh out 0.50 g of sample into a PTFE beaker, add 10 ml of deionised

References p. 287

TABLE 1

INSTRUMENTAL CONDITION FOR THE ANALYSIS OF ALUMINIUM-BASED AL-LOYS

Analyte	Wavelength (nm)	Flame conditions	Recommended range [a] (%)
Aluminium [b]	394.4	N$_2$O/C$_2$H$_2$, rich	50.0–100.0
Magnesium [c, d]	285.2	N$_2$O/C$_2$H$_2$, stoichiometric	0.0– 1.0
Copper	324.7	Air/C$_2$H$_2$,	0.0– 2.0
	222.6	lean	2.0– 10.0
Iron	372.0	Air/C$_2$H$_2$, lean	0.0– 2.0
Tin	235.5	N$_2$O/C$_2$H$_2$, rich	0.0– 10.0
Zinc [c]	213.9 [e]	Air/C$_2$H$_2$,	0.0– 0.5
	307.6	stoichiometric	0.5– 10.0
Silicon [b]	251.6	N$_2$O/C$_2$H$_2$, rich	0.0– 3.0
Manganese	403.1	N$_2$O/C$_2$H$_2$, stoichiometric	0.0– 2.0
Nickel	232.0	Air/C$_2$H$_2$,	0.0– 0.1
	346.2	lean	0.1– 2.0

[a] Reduced sample weight or higher dilution may be required.
[b] Ionization suppressor required.
[c] Burner rotation may be required.
[d] Releasing agent required.
[e] Not recommended for high iron levels.

water and 8 ml of conc. hydrochloric acid in 2 ml aliquots. When the reaction has subsided add 15 ml of 50 vol hydrogen peroxide in small aliquots whilst heating the solutions. After cooling add 5 ml of hydrofluoric acid and dilute to 100 ml. Standard solution of 0 to 250 mg l^{-1} silicon in 0.5% high-purity aluminium should also be prepared, taking care to match the sample acid concentration.

A summary of the typical instrumental conditions used for aluminium alloy analysis by flame AAS are given in Table 1.

B. Copper alloys

A general method for the analysis of copper alloys is made difficult by the tendency of tin to precipitate as metastannic acid when nitric acid is used and also by the insolubility of lead chloride when hydrochloric acid is used. A hydrochloric acid/nitric acid mixture is however recommended, providing the final solutions are not allowed to stand for long periods and the hydrochloric acid strength is maintained when the solutions are diluted [3].

The following method has been found to be satisfactory for dissolution of copper alloys with up to 5% of tin. Weigh 1.00 g of sample into a 250 ml beaker, add 20 ml of acid mixture (25% conc. nitric/25% conc. hydrochloric acids in deionised water). After dissolution, cool and transfer to a graduated flask and dilute with deionised water to 100 ml. Further dilutions from this solution can be made providing the hydrochloric acid strength is maintained by adding 50% hydrochloric acid per 100 ml of final solution.

As copper has an enhancement effect on the determination of aluminium and tin, calibration standards should contain copper concentrations similar to the test solution.

TABLE 2

INSTRUMENTAL CONDITIONS FOR THE ANALYSIS OF COPPER-BASED ALLOYS

Analyte	Wavelength (nm)	Flame conditions	Recommended range [a] (%)
Aluminium [b]	309.3	N_2O/C_2H_2, rich	0.0 – 10.0
Antimony	206.8	Air/C_2H_2, stoichiometric	0.0 – 4.0
Arsenic	193.7	Air/C_2H_2, just luminous	0.0 – 0.2
Copper	249.2	Air/C_2H_2, lean	30.00–100.0
Iron	372.0	Air/C_2H_2, lean	0.0 – 5.0
Lead	217.0	Air/C_2H_2,	0.0 – 1.0
	261.4	lean	1.0 – 6.0
Manganese	403.1	N_2O/C_2H_2, stoichiometric	0.0 – 2.0
Nickel	232.0	Air/C_2H_2,	0.0 – 0.1
	346.2	lean	0.1 – 2.0
	352.5		2.0 – 40.0
Silicon [b]	251.6	N_2O/C_2H_2, rich	0.0 – 3.0
Tin	235.5	N_2O/C_2H_2, rich	0.0 – 10.0
Zinc [c]	213.9 [d]	Air/C_2H_2,	0.0 – 0.5
	307.6	stoichiometric	0.5 – 50.0

[a] Reduced sample weight or higher dilution may be required.
[b] Ionization suppressor required.
[c] Burner rotation may be required.
[d] Not recommended for high iron levels.

References p. 287

Aluminium, iron, silicon, lead, copper, nickel, manganese, antimony, arsenic, zinc and tin can all be successfully determined using this dissolution method; suitable instrumental conditions are given in Table 2.

C. Nickel and cobalt alloys

The use of nickel and cobalt alloys has greatly increased in the last twenty years, mainly in the aerospace industry where there is a demand for high-strength and corrosion-resistant alloys capable of maintaining specific metallurgical properties at high temperatures. The corrosion resistance of these alloys makes formulation of a general analysis method difficult and involves the use of hydrofluoric acid mixtures. The majority of such methods is based on the works of Welcher and Kriege [4, 5].

The following method can be used for dissolution of the majority of nickel- and cobalt-based alloys, though for the analysis of alloys containing high levels of tungsten excessive heating with high nitric acid concentrations can lead to the precipitation of tungstic acid.

Weigh 1.00 g of sample into a PTFE beaker and add 20 ml of concentrated hydrochloric acid, 5 ml of concentrated nitric and 5 ml of deionised water. Heat the mixture gently for approximately 15 minutes, whilst adding in small aliquots a mixture of 5 ml hydrochloric acid and 1 ml nitric acid. Cool the sample, add 10 ml of conc. hydrofluoric acid and fume to near dryness. After cooling add to the mixture 20 ml of hydrochloric acid, 1 ml of nitric acid and 5 ml of water, heat to boiling, cool the solution, transfer to a 250 ml polythene graduated flask and make up to volume using deionised water.

The base elements, nickel and cobalt can be determined by diluting the sample, but care should be taken to maintain the acid strength between samples and standards. The conditions suitable for flame AAS analysis are given in Table 3.

D. Lead alloys

The dissolution of lead-based alloys for subsequent determination by flame AAS presents a number of difficulties due to the tendency of tin, when present in high levels, to precipitate from solution if nitric acid is used, or silver to precipitate as silver chloride if hydrochloric acid is used. Trace levels of tin in lead alloys can be successfully determined by dissolution in nitric acid, but any antimony present needs to be complexed by addition of ammonium fluoride [6].

The most widely adopted dissolution method is based on the work of Price [7], who used a mixture of hydrobromic acid and bromine; the method is however unpleasant and can present problems on dilution due to precipitation.

TABLE 3

INSTRUMENT CONDITIONS FOR THE ANALYSIS OF NICKEL- AND COBALT-BASED ALLOYS

Analyte	Wavelength (nm)	Flame conditions	Recommended range[a] (%)
Cobalt	240.7	Air/C_2H_2,	0.0– 1.0
	352.7	lean	1.0– 50.0
Chromium	357.9	N_2O/C_2H_2,	0.0– 0.1
	427.5	rich	0.1– 2.0
Iron	248.3	Air/C_2H_2,	0.0– 0.1
	372.0	lean	0.1– 5.0
	344.1		5.0– 20.0
Manganese	403.1	N_2O/C_2H_2,	0.0– 2.0
	222.2	stoichiometric	2.0– 5.0
Nickel	232.0	Air/C_2H_2,	0.0– 0.1
	346.2	lean	0.1– 2.0
	352.5		2.0– 40.0
	339.1		40.0–100.0
Molybdenum	313.3	N_2O/C_2H_2,	0.0– 0.5
	315.8	rich	0.5– 10.0
Copper	324.7	Air/C_2H_2, lean	0.0– 2.0
Vanadium	318.4	N_2O/C_2H_2, stoichiometric	0.0– 2.0
Niobium	334.9	N_2O/C_2H_2, rich	0.0– 2.0
Tin	235.5	N_2O/C_2H_2, rich	0.0– 1.0
Aluminium[b]	309.3	N_2O/C_2H_2, rich	0.0– 10.0
Lead	217.0	Air/C_2H_2, lean	0.0– 1.0
Tungsten	255.1	N_2O/C_2H_2, rich	0.0– 10.0
Titanium	365.4	N_2O/C_2H_2, rich	0.0– 2.0

[a] Reduced sample weight or higher dilution may be required.
[b] Ionization suppressor required.

An alternative method is that first described by Hwang and Sandonato [8], who used a solvent acid consisting of nitric acid, fluoroboric acid and water in a ratio of 3 : 2 : 5 by volume. Preparation and dilution of samples and standards should always be made to maintain this acid combination. The following method is successful for the majority of lead/tin solders and white metals.

Weigh 1.000 g of sample into a 250 ml polythene beaker and add 50 ml of the solvent acid; agitate the mixture at room temperature as heating

References p. 287

280

can lead to precipitation of metastannic acid. When dissolution is complete, dilute to 100 ml using further solvent acid.

When the determination of tin is not required, an alternative solvent mixture of nitric acid and perchloric acid (1:5) can be used [9]. In this case add 20 ml of acid mixture to 10.0 g of sample. When dissolution is complete, evaporate the solution to fumes of perchloric acid. Cool the solution and add 50 ml of dilute hydrochloric acid (50%), heat to boiling and cool to room temperature. Finally dilute to 100 ml in a graduated flask. Standards should be similarly prepared from high-purity metals ensuring that the acid concentration is matched between samples and standards.

This method can be successfully used for the determination of copper, bismuth, silver and zinc, and Table 4 shows suitable instrumental conditions for the analysis of lead-based alloys.

TABLE 4

INSTRUMENTAL CONDITIONS FOR THE ANALYSIS OF LEAD-BASED ALLOYS

Analyte	Wavelength (nm)	Flame conditions	Recommended range [a] (%)
Copper	324.7	Air/C_2H_2, lean	0.0– 2.0
Bismuth	223.1	Air/C_2H_2, lean	0.0– 2.0
Silver	328.1	Air/C_2H_2, lean	0.0– 0.2
Zinc [b]	213.9 [c]	Air/C_2H_2, stoichiometric	0.0– 0.5
Aluminium [d]	309.3	N_2O/C_2H_2, rich	0.0–10.0
Antimony	206.8	Air/C_2H_2, stoichiometric	0.0– 4.0
Arsenic	193.7	Air/C_2H_2, just luminous	0.0– 0.2
Cadmium	228.8	Air/C_2H_2, lean	0.0– 2.0
Iron	372.0	Air/C_2H_2, lean	0.0– 2.0
Nickel	232.0 346.2	Air/C_2H_2, lean	0.0– 0.1 0.1– 2.0
Tin	235.5	N_2O/C_2H_2, rich	0.0–10.0

[a] Reduced sample weight or higher dilution may be required.
[b] Burner rotation may be required.
[c] Not recommended for high iron levels.
[d] Ionization suppressor needed.

TABLE 5

INSTRUMENTAL CONDITIONS FOR THE ANALYSIS OF ZINC-BASED ALLOYS

Analyte	Wavelength (nm)	Flame conditions	Recommended range[a] (%)
Lead	217.0	Air/C_2H_2, lean	0.0– 1.0
Iron	372.0	Air/C_2H_2, lean	0.0– 2.0
Magnesium[b, c]	285.2	N_2O/C_2H_2, stoichiometric	0.0– 1.0
Aluminium[d]	309.3	N_2O/C_2H_2, rich	0.0–10.0
Cadmium	228.8	Air/C_2H_2, lean	0.0– 2.0
Copper	324.7	Air/C_2H_2, lean	0.0– 2.0

[a] Reduced sample weight or higher dilution may be required.
[b] Burner rotation may be required.
[c] Releasing agent required.
[d] Ionization suppressor required.

E. Zinc alloys

Zinc-based alloys present no particular dissolution problems and can be prepared by the following method. Weigh 1.00 g of sample into a 250 ml beaker and add 10 ml of hydrochloric acid. Care should be taken when adding the acid if the sample is in a finely divided form. When dissolution is complete, warm the sample, add dropwise 2 ml of 20 vol hydrogen peroxide and boil the solution to remove excess hydrogen peroxide. Cool the solution and transfer to a 100 ml graduated flask. Dilute to 100 ml with deionised water.

An alternative internal standard method first described by Smith et al. [10] uses nitric acid in place of hydrogen peroxide, the method giving a higher level of precision.

Table 5 shows suitable instrumental conditions for the analysis of zinc-based alloys.

III. DETERMINATION OF IMPURITIES IN ALLOYS USING ENHANCED FLAME TECHNIQUES

Two types of method have been introduced in recent years that have improved the sensitivity of flame AAS determinations.

The first of these is the "atom-trapping" technique, first described by Watling [11, 12], which uses a quartz tube to prolong the residence time

References p. 287

TABLE 6

FLAME SENSITIVITY IMPROVEMENTS USING A TUBE ATOM TRAP [13]

Element	Flame type	Conventional flame AAS [a]	Atom trap AAS
Lead	Air/Acetylene	0.1	0.03
Cadmium	Air/Acetylene	0.014	0.004
Arsenic	Argon/Hydrogen	0.3	0.06
Selenium	Argon/Hydrogen	0.26	0.08
Copper	Air/Acetylene	0.04	0.015
Zinc	Air/Acetylene	0.01	0.004
Tin	Air/Hydrogen	0.35	0.1
Thallium	Air/Acetylene	0.28	0.1
Silver	Air/Acetylene	0.03	0.01
Gold	Air/Acetylene	0.12	0.05
Mercury	Air/Acetylene	2.7	0.85
Platinum	Air/Acetylene	1.2	0.9
Bismuth	Air/Acetylene	0.28	0.08
Antimony	Air/Acetylene	0.36	0.12
Tellurium	Air/Acetylene	0.20	0.08

[a] The concentration of analyte required to give an absorbance of 0.0044 A.

TABLE 7

TYPICAL DETECTION LIMITS OBTAINABLE USING HYDRIDE GENERATION [16]

Element	Wavelength (nm)	Reduction [a]	Characteristic concentration ($\mu g\ ml^{-1}$)	Detection limit ($\mu g\ ml^{-1}$)
As	193.7	$NaBH_4$	0.00052	0.0008
As	193.7	ZnHCl	0.001	0.0015
Bi	223.1	$NaBH_4$	0.00043	0.0002
Ge	265.1	$NaBH_4$	1.0	0.5
Pb	283.3	$NaBH_4$	0.08	0.1
Sb	217.6	$NaBH_4$	0.00061	0.0005
Se	196.1	$NaBH_4$	0.0021	0.0018
Sn	224.6	$NaBH_4$	0.00044	0.0005
Te	214.3	$NaBH_4$	0.002	0.0015

[a] Silica tube in air/acetylene flame–nitrogen flushed.

of the analyte atom population within the flame, the tube being mounted above the burner slots just below the optical path. Typical improvements in sensitivity given by this type of system are shown in Table 6.

The second method, based on the work of Manning [14], involves the formation of gaseous analyte hydrides, which are passed to a heated silica tube mounted above the burner in the optical path. This hydride generation technique typically involves the premixing of an acidic sample solution with sodium borohydride, though there are numerous variations on the basic technique [15]. Sodium borohydride is preferred as the hydrides are formed more rapidly and a collection reservoir is not needed. The hydrides of antimony, arsenic, bismuth, germanium, lead, selenium, tellurium and tin, can all be successfully generated using this technique, and as nebulization has been replaced by this more efficient gaseous sample transport system, an improvement in detection limit for all these elements, except lead, is achieved. The method is prone however to interelement interferences, particularly from copper, nickel, silver and gold, an effect that seriously limits application of the technique for analysis of these metals and their alloys.

Typical detection limits and reduction techniques are shown in Table 7.

IV. DETERMINATION OF IMPURITIES IN ALLOYS USING ELECTROTHERMAL ATOMIZATION–AAS

The methods described in the previous sections are adequate for the determination of major and minor element concentrations, but with the development of high-performance alloys the need has arisen to determine some elements at levels unattainable by conventional flame AAS. The introduction of electrothermal atomization methods by L'vov [17] and Massmann [18] have shown to give sensitivities which allow determinations in the part per billion region.

Two alloy types that have a need for this type of technique are copper alloys and nickel/cobalt alloys. In each case variations in methodology are needed to suit the type of equipment used. A powerful background correction is usually required, coupled with precise control of furnace conditions to overcome the often significant chemical interferences that occur.

The following generalised dissolution methods have been shown to be suitable.

A. High-purity nickel

Dissolve 2.00 g of sample in 20 ml of dilute nitric acid and dilute to 100 ml in a graduated flask. Suitable aliquots of the solution can be pipetted directly into the graphite furnace without further pretreatment, standards being similarly prepared from high-purity metals or by standard addition to previously characterised samples. Results obtained for bismuth, lead and selenium in three NBS standards are shown in Table 8, the values obtained showing good agreement with certified values [19].

TABLE 8

RESULTS OBTAINED FOR THE ANALYSIS OF THREE NBS STANDARDS [19]

Standard	Bismuth (μg ml^{-1})		Lead (μg ml^{-1})		Selenium (μg ml^{-1})	
	Certified	Found	Certified	Found	Certified	Found
NBS 671	0.07	0.07	16	17	2.0	1.8
NBS 672	0.3	0.26	38	40	0.40	0.4
NBS 673	0.06	0.06	3.5	2.9	0.2	0.2

B. High-performance nickel base alloys

The dissolution method for this type of material needs a mixed acid solution of hydrochloric, nitric and hydrofluoric acids as previously described. Some types of nickel alloys may not dissolve easily using this method, though modification of the relative proportions of hydrochloric acid to nitric acid will often prove successful. It is particularly important to prepare calibration standards using the same acid combination, with the use of high-purity reagents being essential. Typical operating parameters are shown in Table 9, though these should be optimised for the equipment in use.

C. High-purity copper

Weigh 1.00 gm of sample into a 25 ml beaker, add 3 ml of deionised water and 5 ml of concentrated nitric acid. When dissolution is complete, boil the solution to remove nitrous fumes, cool to room temperature, transfer to a 25 ml graduated flask and dilute to 25 ml using deionised water.

Calibration standards should be prepared daily from a copper stock solution prepared in the same way, to which suitable aliquots of 1000 mg l^{-1} standard element solutions have been added. A reagent blank should be carried through the entire procedure, with high-purity reagents used for sample and standard preparation.

In most cases aliquots of the sample or standard solutions can be pipetted directly into the graphite furnace, though addition of matrix modifiers may also be required.

An example of the results that can be obtained using electrothermal atomization are shown by the work of Sentimenti and Mazzetto [21], who determined fourteen trace elements in high-purity copper. The operating conditions used to achieve the results are summarised in Table 10, though it should be noted that the dry, char, atomization times and temperatures vary between different instruments, and should be optimised for the particular instrument used. A summary of the results obtained by Sentimenti and Mazzetto for two copper standards is shown in Table 11.

TABLE 9

TYPICAL OPERATING PARAMETERS FOR THE DETERMINATION OF TRACE ELEMENTS IN NICKEL-BASED ALLOYS

	Pb	Bi	Te	Se	Tl	Sb	As	Cd	Ag
Wavelength (nm)	283.3	223.1	214.3	196.0	276.9	217.6	193.7	228.4	328.11
Band width (nm)	0.7	0.2	0.7	0.7	0.7	0.7	0.7	0.7	0.7
Sample volume (µl)	50	20	50	20	50	20	20	20	20
Drying temp. (°C)	150	150	150	150	150	200	150	150	150
Drying time (s)	20	20	20	30	20	40	30	30	30
Char temp. (°C)	400	800	600	1000	500	600	900	500	800
Char time (s)	60	45	60	20	60	20	20	20	20
Atomization temp. (°C)	2000	2200	2200	2800	2000	2800	2800	2800	2800
Atomization time (s)	5	5	5	5	5	8	5	5	5
Reference	[20]	[20]	[20]	[19]	[20]	[19]	[19]	[19]	[19]

References p. 287

TABLE 10

TYPICAL ELECTROTHERMAL ATOMIZATION PARAMETERS FOR THE DETER-
MINATION OF TRACE ELEMENTS IN HIGH-PURITY COPPER [21]

	Ag	As	Bi	Cd	Co	Cr	Fe
Wavelength (nm)	328.1	193.7	233.0	228.8	240.7	357.9	248.3
Bandwidth (nm)	0.7	0.7	0.2	0.2	0.2	0.7	0.2
Sample volume (μl)	10	10	10	10	10	10	10
Matrix modifier	none	none	none	[a]	[a]	none	none
Drying temp. (°C)	150	150	150	150	150	150	150
Char temp. (°C)	500	1400	900	800	1400	1650	1100
Atomization temp. (°C)	1900	2600	2000	1800	2700	2500	2600
Purge temp. (°C)	2750	2750	2750	2750	2750	2750	2750

	Mn	Ni	Pb	Sb	Se	Sn	Te
Wavelength (nm)	279.5	232.0	283.3	217.6	196.0	224.6	214.3
Bandwidth (nm)	0.2	0.2	0.7	0.2	2.0	0.7	0.2
Sample volume (μl)	10	10	10	10	10	10	10
Matrix modifier	[b]	[a]	none	none	none	[a]	none
Drying temp. (°C)	150	150	150	150	150	150	150
Char temp. (°C)	1000	1300	600	1100	900	900	1100
Atomization temp. (°C)	2600	2700	200	2400	2600	2500	2500
Purge temp. (°C)	2750	2750	2750	2750	2750	2750	2750

[a] Matrix modifier magnesium nitrate.
[b] Matrix modifier ammonium hydroxide.

TABLE 11

COMPARISON VALUES FOR TWO HIGH-PURITY COPPER STANDARDS [21]

Element	NBS 395 (μg ml^{-1})		NBS 398 (μg ml^{-1})	
	Certified	Found	Certified	Found
Se	0.60 ± 0.05	0.63 ± 0.05	14 ± 3	13 ± 3
Te	0.32 ± 0.03	0.30 ± 0.05	11 ± 1	10 ± 1
Bi	0.5 ± 0.10	0.5 ± 0.08	2 ± 0.3	2.1 ± 0.2
Mn	5.3 ± 0.8	6.0 ± 0.5		
Sb	8.0 ± 0.5	7.6 ± 0.4	7.5 ± 0.1	7.2 ± 0.2
Cd	~0.4	0.3 ± 0.05		
As	1.6 ± 0.3	1.4 ± 0.4	25 ± 3	
Pb	3.25 ± 0.02	3.3 ± 0.05	9.9 ± 0.6	10 ± 0.5
Sn	1.5 ± 0.2	1.1 ± 0.3	4.8 ± 0.6	5 ± 0.5
Ni	5.4 ± 0.1	5.2 ± 0.2	7.0 ± 0.1	6.8 ± 0.2
Fe	96 ± 3	90 ± 4	11.5 ± 0.5	10.8 ± 0.5
Co	0.3 ± 0.1	0.3 ± 0.1	2.8 ± 0.1	2.4 ± 0.1
Ag	12.2 ± 0.5	12.2 ± 0.2	20.0 ± 0.2	20.5 ± 0.2
Cr	6.0 ± 0.5	1.5 ± 0.4		

REFERENCES

1 R.E. Mansell, H.W. Emmel and E.L. McLaughlin, Appl. Spectrosc., 20 (1966) 231.
2 G.F. Bell, At. Absorpt. Newsl., 5 (1966) 73.
3 P. Johns and W.J. Price, Metallurgia, 81 (1970) 75.
4 G.G. Welcher and O.H. Kriege, At. Absorpt. Newsl., 9 (1970) 61.
5 G.G. Welcher and O.H. Kriege, At. Absorpt. Newsl., 8 (1969) 97.
6 B. Perry, Spectrovision, 25 (1971) 8.
7 W.J. Price, Analytical Atomic Absorption Spectroscopy, Heyden, London, 1972.
8 J.Y Hwang and L.M. Sandonato, Anal. Chem., 42 (1970) 744.
9 A.S.T.M. Standard No. E37, 1978.
10 S.B. Smith, J.A. Blasi and F.J. Feldman, Anal. Chem., 40 (1968) 1525.
11 R.J. Watling, Anal. Chim. Acta, 94 (1977) 181.
12 R.J. Watling, Anal. Chim. Acta, 97 (1978) 395.
13 M. Wassall, J. Inst. Met., 4 (1988) 618.
14 D.C. Manning, At. Absorpt. Newsl., 10 (1971) 86.
15 T. Nakahara, Proc. Anal. At. Spectrosc., 6 (1983) 163.
16 A.J. Thomson and D.R. Thomerson, Analyst, 99 (1974) 595.
17 B.V. L'vov, Inzh. Fiz. Zh., 2 (1959) 44.
18 H. Massmann, Spectrochim. Acta, 23B (1968) 215.
19 J.E. Forrester, V. Lehecka, J.R. Johnston and W.L. Ott, At. Absorpt. Newsl., 18 (1979) 73.
20 G.G. Welcher, O.H Kriege and J.Y. Marks, Anal. Chem., 46 (1974) 1227.
21 E. Sentimenti and G. Mazzetto, At. Spectrosc., 7 (1986) 181.

Chapter 4g

Atomic Absorption Methods in Applied Geochemistry

M. THOMPSON and E.K. BANERJEE

Department of Chemistry, Birkbeck College, Gordon House, 29 Gordon Square, London WC1H 0PP (Great Britain)
Imperial College, London SW7 2BP (Great Britain)

I. INTRODUCTION

A. Analytical requirements in applied geochemistry

Applied geochemistry consists of two aspects, namely the scientific prospecting for mineral reserves, and environmental studies. Both aspects employ the wide-scale surveys of the concentrations of various elements in samples taken at or near the surface of the earth. The primary purpose is to identify geochemical "anomalies", that is, areas which have unusually high (or unusually low) concentrations of one or more elements. Such surveys may cover many thousands of square kilometres and produce comparably large numbers of samples. As a second stage the anomalies found in the primary survey are investigated in detail, with much greater sampling densities. The purpose of the follow-up is to determine whether the anomaly is due to mineralisation, to naturally occurring concentration processes, metalliferous but unmineralised rock strata, or to pollution. Sampling media most frequently used in geochemical surveys are (in decreasing order) soil, stream and lake sediment, rock, water and herbage.

Geochemical surveys impose upon the analytical chemist certain special constraints [1]. The over-riding factor is the cost-effectiveness of the analytical method, as can be gauged from the fact that, for about 95% of all analyses in this field, the sole outcome is to eliminate localities from further consideration. Large numbers of samples have to be analysed with an optimally low cost and a fast turn-around time, sometimes at a temporary camp remote from normal services.

A reasonable standard of trueness and precision must be consistently obtained. The analytical precision required depends on the geochemical contrast, which is the ratio of the anomalous concentration to the background level, but also on the sampling precision which is usually the limiting factor. As a general rule a relative standard deviation (RSD) of 5% is completely acceptable. Somewhat higher RSD's are often usable, and sometimes have to be tolerated, for example, when the only economic analytical method is being used near the detection limit.

References pp. 319–320

B. Atomic absorption spectrometry in applied geochemistry

The larger part of this analytical requirement is met by atomic absorption spectrometry (AAS) methods: at the time of the last survey in 1970 about 70% of all geochemical samples were analysed by conventional flame AAS methods, usually for several elements. At that time the emphasis was on the base metals copper, lead, zinc and nickel. Flame methods are completely satisfactory for the determination of these elements in soil and sediments, and of several others: cadmium, iron, manganese, cobalt, calcium and magnesium. Many other elements require more sensitive methods for their determination. This can be achieved by preconcentration, gas injection (for mercury and elements that form volatile hydrides) or electrothermal atomization (ETA) for a range of elements. However, these more sensitive alternatives cannot match the flame technique in speed.

Preconcentration combined with a flame technique can often compete with ETA-AAS determination, but the two methods have complementary capability. Preconcentration largely eliminates the matrix effects that plague ETA-AAS. However, the experimental simplicity and small volumes used in ETA make it less prone to contamination. Neither technique makes a major contribution to prospecting analysis, except possibly ETA for gold [2]. A rapid multi-element preconcentration has also been described [3]. In environmental studies, however, both methods are widely used.

Gas injection methods are particularly useful for both exploration and environmental geochemistry. Mercury, arsenic, antimony, bismuth, selenium and tellurium are valuable "pathfinders" for precious metal mineralisation, while tin is an important industrial raw material. Environmentally, mercury and some of the other elements are notorious pollutants, while selenium is an essential element that is in many areas at deficiency levels in both livestock and man.

The application of AAS in geochemistry has declined somewhat in recent years, a trend due in part to the reduction in the scale of exploration for base metals, in part to the development of inductively coupled plasma atomic emission spectrometry. ICP-AES offers better detection limits than flame AAS for a range of elements (refractory elements, lanthanides, non-metals) and can provide simultaneous determination 20–30 elements from a single decomposition.

In a geochemical laboratory the output from a flame AAS instrument can be as high as 800–1000 determinations in an eight-hour period. To meet this capability and minimise costs, special sample preparation methods are adopted. The main emphasis is on simplifying the analytical procedure so that large batches of samples can be handled by a single person, typically with 50–300 samples per batch. As an example of the efficiency obtainable, under appropriate conditions, during the preparation of a geochemical atlas of England and Wales [4] in the authors' laboratory, a two-person team analysed stream sediments for zinc and cadmium at an average rate of 300 samples per working day over a ten-month period. They carried out every aspect of the task themselves, including weighing, sample digestion, dilution, instrumental analysis, washing up, and preliminary stages of data quality control.

This rate of working is achieved by using, as far as possible, a number of special techniques, including: (i) carrying out the whole procedure in a single vessel, which avoids time-consuming processes such as quantitative liquid transfer or filtration; (ii) the use of accurate liquid dispensers to provide a known final volume, rather than volumetric flasks; (iii) use of a fixed sample weight so that a single dilution factor applies to a whole batch; (iv) use of partial sample digestion procedures which may extract only about 90% of the analyte from the sample; (v) use of optimised glassware cleaning procedures; and (vi) use of a rapid automatic diluter for out-of-range samples. Space is conserved where possible by using test tubes as digestion vessels rather than the wider beakers or flasks. Test tubes, in addition, provide a good depth of liquid for a given volume, which facilitates nebulisation for several elements from a small volume, and they can be conveniently heated in large numbers in aluminium block baths or air baths.

C. Instrumental requirements

Almost any standard AAS instrument can be used for flame atomization applied geochemistry, but there are a number of features which can improve the scope and speed of analysis. Facilities such as autozero, automatic linearisation of the calibration curve, digital readout in concentration units, and a printer are all essential for cost-effective geochemical analysis, because the cost of the operator and the instrument can be distributed among a disproportionately larger number of samples. However, it has yet to be demonstrated that, over a long period, an automatic sample introduction system is as quick or reliable as a motivated human operator in flame AAS. Fully automatic systems, however, are now regarded as essential for satisfactory operation of the ETA-AAS technique.

The correction of background interference (see below) is especially important in geochemical analysis because of the prevalence of calcium as a major constituent and the dramatic effect this element can have on the determination of elements such as cadmium.

D. Interference effects

In the context of applied geochemical analysis by flame AAS, i.e. the determination of traces of heavy metals in the presence of large concentrations of major constituents (Na, K, Ca, Mg, Fe and Al), numerous instances of

References pp. 319–320

interference can be found. These interferences have either to be rendered insignificant directly (e.g. by the addition of a further reagent to the sample solution) or instrumentally, or ignored. If they are too large to be ignored alternative analytical methods are used. Chemical separation techniques are often inadmissible on cost grounds.

Two distinct types of interference can be recognised, namely background (translational) effects and enhancement/suppression (rotational) effects, sometimes occurring together in the same interferant/analyte combination. Background effects can be largely overcome by the use of one of the several standard methods available in modern instruments. These procedures are widely used in geochemical analysis, especially in the determination of Cd, Co, Ni and Pb in the presence of calcium [5]. Rotational effects can, in principle, be overcome by the method of standard additions, but this would not be practicable in applied geochemistry because of the additional cost of this labour-intensive technique. Consequently, rotational effects are often ignored.

In a comprehensive survey [6] most of the important interference effects relevant to the analysis of stream sediments by flame AAS have been identified. A large number of significant effects was detected, but only those effects judged likely to adversely affect interpretation of the data were distinguished as important. Important translational (background) effects were those of calcium on Cd, Co, Ni and Pb, and aluminium on Pb. Rotational effects were found to be important only in the instances of calcium on Li and aluminium on Co, Li and Ni. Interference effects are even more numerous and severe in the ETA-AAS technique, and background correction is universally required. However, ETA-AAS methods are not covered in this chapter, so no attempt to summarise these problems will be made.

E. Sample decomposition methods

Applied geochemistry seldom requires the complete chemical attack of the sample. In soil and sediment, for example, the trace elements incorporated in the crystal lattice of quartz particles are usually of little interest. Consequently, sample attacks which liberate trace elements only from less resistant minerals such as clays, hydrolysates (Fe and Mn oxides) or organic matter are usually completely satisfactory. Sometimes methods designed to attack a specific constituent are used: examples are cold reducing buffers for extracting trace metals from precipitated manganese(IV) oxide, or cold ammonium acetate to extract metal ions bound to ion-exchange sites on clay minerals. Results of such methods need to be interpreted with caution. At the other extreme are methods designed to bring into solution metals which are present in discrete resistant minerals such as cassiterite, which are not amenable to attack with the usual mineral acids.

Fusion methods are avoided where possible in applied geochemistry, especially in atomic absorption work. Fusions are not convenient for rapid large batch methods because of the manipulative difficulties. The alkali metal salt introduced into the sample solution is invariably a drawback in atomic absorption methods because the high solid content of the solution can cause blockages of the nebuliser and burner slot, as well as giving poorer detection limits caused by background absorption and noisy signals. Fluxes tend to be expensive because of the high degree of purity required, as they need to be used in at least a five-fold excess over the sample. Mineral acids, by contrast, are easy to purify, quick to dispense, and, apart from sulphuric acid, do not produce large matrix effects in AAS.

II. GENERAL ASPECTS OF SAMPLE PREPARATION METHODS

A. Digestion vessels

Many of the decomposition procedures used in applied geochemistry can be undertaken in glassware. Except where special equipment is required or constant shaking is necessary, all such attacks on rock, soil or sediment samples can be carried out in test tubes, the most convenient type being medium-wall rimless borosilicate tubes, size 19×250 mm. These can be handled in batches of 50 in nylon-coated wire racks. Cold extractions are best carried out in small polypropylene bottles (50 ml or 100 ml). Hot attacks involving hydrofluoric acid can be carried out in large batches in open 50 ml PTFE beakers on PTFE-coated or "Sindanyo" topped hot plates, but preferably in PTFE test-tubes.

Cleaning of glass test-tubes in regular use is straightforward. After any unused liquid has been poured away, the solid residue is washed out with a jet of tap water. The tube is then rinsed once with 1% nitric acid in demineralized water and dried in an inverted position in a wire rack in an oven at about 50°C. This procedure is effective, and no measurable contamination of a sample occurs even when the sample previously attacked in the tube was 1000 times more concentrated in trace elements. This is true of both new and of heavily used tubes. Wire stemmed test tube brushes must not be used, because of the danger of contamination from zinc and other metals. PTFE beakers can be rendered free from trace elements derived from previous samples by thorough rinsing in dilute nitric acid or by boiling in a dilute solution of a laboratory glassware detergent.

B. Heating equipment

The most convenient and effective method of heating large batches of test tubes is undoubtedly by means of heating blocks. A simple heating block (see Fig. 1) can be cheaply made by obtaining a 2-inch thick block of aluminium

References pp. 319–320

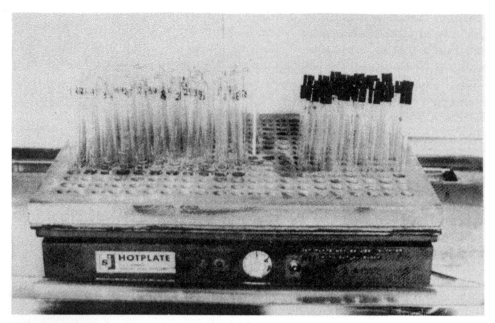

Fig. 1. Shallow aluminium block for heating many test tubes.

or Dural, of the same size as a standard hot plate. This is drilled at about 2-mm intervals to a depth of about 30 mm with holes about 1 mm larger than the test tubes to be accommodated. By use of the normal hot-plate controls, a temperature steady to within ±5°C can be obtained at any point over the temperature range 50–200°C. These shallow blocks can be used for a variety of purposes where heating below the boiling point or gentle refluxing is required.

Deeper blocks can be obtained commercially (see Fig. 2) and can accommodate up to two hundred 150-mm test tubes to a depth of 130 mm. As the temperature control is very fine, deep blocks are ideal for critical work and suitable for the gentle evaporation of acids such as perchloric acid without excessively heating the residue. Tops of aluminium blocks can be made somewhat resistant to corrosion by acid fumes if they are given a heavy coating of PTFE from an aerosol spray. Alternatively the top of the block can be protected with a sheet of aluminium cooking foil, which is disposed of after each use.

A temperature programmer with several heating stages is an invaluable addition to a heating block. By this means a decomposition of a large batch of samples can be handled with no attention at all from the analyst over a 24-hour period. This enables decompositions to be optimised and closely standardised. Several of the methods described in this chapter depend on such decompositions.

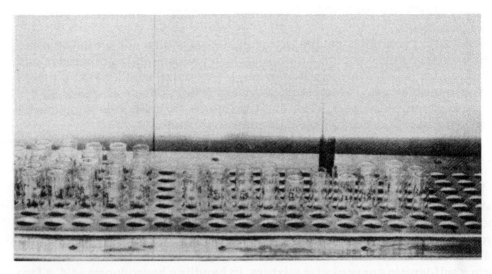

Fig. 2. Top surface of deep aluminium block bath for evaporation to dryness in test tubes.

C. Dispensers and diluters

The repetitive measuring of small (<20 cm^3) volumes of liquid is best undertaken by means of equipment such as the Oxford all-glass dispenser (The Bohringer Corporation Ltd., Bell Lane, Lewes, Sussex BN7 1LG, U.K.). Even concentrated nitric acid or perchloric acid can be permanently kept in such a device. The volume dispensed is remarkably precise, and the action very rapid. Other manufacturers' equipment may be suitable but those with metal parts (such as springs) must be avoided because of the inevitable corrosion caused by acid fumes and the possibility of contamination. Recently satisfactory dispensers for hydrofluoric acid have become available.

In geochemical analysis the concentration range of an analyte often spans several orders of magnitude, and dilutions are frequently required to bring the sample solution within the calibration range of flame AAS. Many manufacturers produce suitable models of automatic diluters, but corrosion problems are almost inevitable with the acidic solutions which are regularly used. Small amounts of acid can escape between the piston and the barrel, and cause corrosion, especially where metal parts are enclosed.

D. Reagents and calibrators

Analytical grade reagents (i.e. "Analar" or equivalent) are suitable and should be used for most geochemical analyses, but for very low concentrations of analytes, as may be encountered in water analysis, high purity reagents (e.g. "Aristar") are preferable. Demineralised water is sufficiently pure for most purposes.

Calibration solutions (calibrators) should always be made up in the same medium as the final sample solution, and can be conveniently prepared in one dilution from stock solutions of 1000 mg l^{-1} by using small piston pipettes with disposable tips. The pipettes provide a relative error of less than 1% when maintained properly. The stock solutions of standards supplied by chemical houses are cost-effective in commercial analysis.

E. Safety

Most of the procedures involving attacks with strong acids must be carried out in adequately constructed fume cupboards. Great care is required in the design and choice of constructional materials for cupboards used for hydrofluoric acid or perchloric acid. In particular, perchloric acid vapours or condensate must not be allowed to come into contact with wood, which can form a self-igniting or explosive mixture. In handling hydrofluoric acid, a face mask, rubber gloves and a plastic apron are essential, because irreversible tissue damage can be caused very quickly by this acid.

Perchloric acid mixtures can be used with complete safety for attacks on virtually all rocks, soils and sediments, as for methods described in Section III, B and C. Over half a million such samples have been thus treated in the authors' laboratory without incident. Even peaty soils can be safely treated by these methods. The only explosive hazard that may be encountered is with rocks containing appreciable quantities of oil or bitumen. These types of sample are, however, easily recognised during sample preparation and weighing. Perchloric acid can also be used with complete confidence for herbage analysis so long as the procedures described (Section III, H and I) are followed closely and no attempt is made to treat fatty or oily samples with this acid. The Society for Analytical Chemistry's guide to the use of perchloric acid should be studied before attempting any attacks with perchloric acid [7]. Operations involving perchloric acid should be undertaken only under the supervision of a professionally qualified analytical chemist.

F. Mechanical sample preparation

1. Rocks

Rock samples are reduced to 1–2 mm fragments by a jaw crusher or percussion mortar. They are then reduced to a fine powder to pass a 200 mesh (63 μm aperture) sieve by means of a swing-mill such as the one produced by Tema Machinery Ltd., Banbury, Oxon, U.K. Agate mortars for this device are expensive but do not contaminate the sample with trace metals. A hard steel mortar is suitable if contamination with traces of chromium can be tolerated. Small hammer mills are suitable for the fine grinding, but as the hammers are commonly stainless steel, they contaminate the sample with nickel and chromium. Non-metallic sieves can be cheaply constructed from perspex tubing (10–15 cm diameter) and monofilament polyester bolting cloth, which is supplied in a wide range of mesh sizes by Henry Simon Ltd., PO Box 31, Stockport, Chesire, SK3 0RT, U.K.

2. Soils

The dried soil is gently disaggregated (not crushed) and the particles greater than 2 mm sieved out. The <2 mm fraction is then crushed in a swing mill or hammer mill as described for rocks.

3. Sediments

Sediments are usually prepared by simple sieving of the natural material after drying and disaggregation. A number of size fractions is used for special purposes, but the most widely used is the −80 mesh (ca. 200 μm aperture) fraction, which is analysed without further preparation. Coarser fractions may need to be crushed before analysis.

4. Herbage

Herbage samples must be milled to pass a 1 mm aperture before analysis. This can be achieved by means of a beater mill with a carbon steel construction (as produced by Christie and Norris Ltd., Broomfield Road, Chelmsford, Essex, U.K.). This device introduces no detectable concentrations of foreign metals into the sample. A small hammer mill or knife mill is also satisfactory, but may contaminate the sample lightly with traces of nickel and chromium.

G. Analytical quality control

A prominent feature of applied geochemical analysis is the regular use of quality control methods to ensure that a satisfactory level of accuracy and precision is maintained from batch to batch of samples. The methods are based on the planned use of house reference materials and/or duplication, as much as 5–10% of the analyses being devoted to the control system. For example, in a batch of 300 samples, it would be usual to insert 5 each of 2 reference materials (representing low and medium levels of the analyte), 10 samples (selected at random) in duplicate and 10 reagent blanks. The entire batch of test materials, reference materials and blanks would then be processed in randomised order. While the statistical principles behind the control methods are elementary, their application needs a considerable amount of close attention to ensure that realistic (rather than optimistically biased) results are obtained [8].

References pp. 319–320

III. SAMPLE DIGESTION METHODS

A. Nitric acid digestion for soil or stream sediment

Reagents
(a) Nitric acid (70%).

Equipment
(a) Test tubes, wire test tube racks (plastic coated) and stainless steel test tube racks.
(b) Liquid dispensers (2).
(c) Shallow aluminium heating block sited in a suitable fume cupboard.
(d) Centrifuge tubes, polystyrene, disposable.
(e) Vortex tube mixer.
(f) Centrifuge.

Procedure
(a) Weigh each sample (0.250 g, −80 mesh) into a clean, dry, numbered test tube.
(b) Weigh standard and duplicate samples and leave empty test tubes at random intervals for blank determinations.
(c) Add nitric acid (1.0 ml) to each tube.
(d) Place test tubes into the aluminium heating block at 105° ± 5°C and leave for one hour.
(e) Transfer test tubes to a stainless steel rack and allow to cool.
(f) Add water (9.0 ml) to each tube and mix, using a vortex mixer.
(g) Allow solid residue to settle (at least 4 hours) and nebulise solution directly into a flame AAS from the test tube. Alternatively transfer solutions to centrifuge tubes and centrifuge solution for 2 min at 2000 rpm.

Remarks
(a) This method is designed for the digestion of clay minerals, usually the major constituent of the −80 mesh fraction of soils and sediments. It is completely effective (i.e. >90%) in solubilising trace elements such as Cu, Pb, Zn, Cd, Mn, Fe, Co, Ni, Hg and Ag. For other rock forming minerals the attack may be virtually complete (feldspar, olivine), partial (e.g. about 50% for pyroxine, biotite, amphibole), or negligible (quartz). Lateritic soils are only partially attacked.
(b) Samples with a high organic matter content (e.g. peaty soils) may react vigorously and char. Such samples should be left standing overnight in nitric acid (procedure step c) prior to hot digestion.
(c) The dilution factor for this method of digestion is 40.

B. Nitric acid–perchloric acid digestion for rock, soil or stream sediment

CAUTION: This method must not be attempted on samples containing oil or bitumen.

Reagents
(a) Nitric acid (70%).
(b) Perchloric acid (60%).
(c) Hydrochloric acid (5 M). Dilute hydrochloric acid (430 ml, 36%) to 1 l with water.

Equipment
(a) Test tubes, wire test tube racks (plastic coated) and stainless steel test tube racks.
(b) Liquid dispensers (4).
(c) Deep aluminium heating block sited in a suitable fume cupboard.
(d) Vortex tube mixer.
(e) Centrifuge tubes, polystyrene, disposable.
(f) Centrifuge.

Procedure
(a) Weigh each sample (0.250 g) into a clean, dry, numbered test tube.
(b) Weigh standard and duplicate samples and leave empty test tubes at random intervals for blank determinations.
(c) Add nitric acid (4.0 ml) to each tube.
(d) Add perchloric acid (1.0 ml) to each tube.
(e) Place the test tubes into the cold aluminium heating block and raise the temperature to $150° \pm 5°C$ over 2–3 h. Leave at this temperature until the copious evolution of fumes ceases.
(f) Increase temperature of the heating block to $185° \pm 5°C$ and when the residue is dry, transfer the test tubes to stainless steel test tube racks. Allow to cool.
(g) Add hydrochloric acid (2.0 ml, 5 M) to each tube.
(h) Place tubes in a shallow heating block (60°C) and leave for one hour.
(i) Transfer the tubes to wire racks and allow to cool.
(j) Add water (8.0 ml) to each tube and mix.
(k) Allow residue to settle (at least 4 h) and nebulise the supernatant liquid directly into a flame AAS from the test tube. Alternatively, decant solutions into disposable centrifuge tubes and centrifuge for 2 min at 2000 rpm.

Remarks
(a) This is a more powerful attack than nitric acid alone. In addition to the minerals completely attacked by method III.A, pyroxenes, biotite,

limonite and some amphiboles are almost completely attacked. Trace element extraction from lateritic soil is almost complete. However, some common minerals containing important metals are attacked only to a negligible extent (e.g. rutile, chromite, cassiterite, zircon, beryl), or to a minor degree (barite).

(b) Samples with a high organic matter content may react vigorously with nitric and perchloric acids. Such samples should be kept overnight in the heating block at 50°C after the addition of the acids at stage (d) above. The normal procedure can be resumed at stage (e).

(c) The dilution factor for this method of digestion is 40.

C. Hydrofluoric, nitric and perchloric acid digestion for rock, soil or sediment

Reagents
(a) Hydrofluoric acid (40%).
(b) Perchloric acid (60%).
(c) Nitric acid (70%).
(d) Hydrochloric acid (6 M). Dilute hydrochloric acid (516 ml, 36%) to 1 l with water.

Equipment
(a) PTFE beakers (50 ml).
(b) Graduated flasks (25 ml) or graduated test tubes (10 ml).
(c) Hotplate sited in suitable fume cupboard.
(d) Polythene measuring cylinder and plastic tray for dispensing hydrofluoric acid.
(e) Liquid dispensers (4).

Procedure
(a) Weigh each sample (0.250 g, −200 mesh) into a clean, dry, numbered PTFE beaker.
(b) Weigh standard and duplicate samples and leave empty beakers at random intervals for blank determinations.
(c) Add nitric acid (3.0 ml) followed by perchloric acid (3.0 ml) to each beaker.
(d) Add hydrofluoric acid (10 ml) to each beaker.
(e) Heat the beakers on a hotplate until dense white fumes are seen (1–1.5 h).
(f) Heat for a further 2 min and then allow the beakers to cool.
(g) Add further hydrofluoric acid (2.0 ml) to each beaker.
(h) Heat the beakers on the hotplate until the solution is gently evaporated to dryness (about 4 h). Allow the beakers to cool.
(i) Add further perchloric acid (2.0 ml) to each beaker.

(j) Heat gently, evaporate to dryness and allow the beakers to cool.

(k) Add hydrochloric acid (2.0 ml if the final volume is 10 ml, 5.0 ml if the final volume is 25 ml) to each beaker and warm gently.

(l) Transfer the solutions from the beakers to either graduated flasks (25 ml for a dilution factor of 100) or to graduated test tubes (10 ml for a dilution factor of 40) and dilute to volume with water.

Remarks

(a) This method will completely digest most constituents of rocks, soils and sediments. A few minerals will partly or completely resist attack, e.g. barite chromite, cassiterite, tourmaline, kyanite, some spinels and magnetites, rutile, zircon and wolframite.

(b) When using PTFE beakers on a hotplate, care should be taken not to exceed the temperature at which PTFE becomes plastic (240°C).

(c) A shortened form of this method can be used for less resistant samples by omitting steps (f), (g), (i) and (j) from the above method. The double fuming with perchloric acid is necessary for calcareous samples (i.e. >10% Ca) to destroy the insoluble calcium fluoride residue.

(d) This method must not be attempted on samples containing oil or bitumen.

D. Hydrofluoric, nitric and perchloric acid digestion in test tubes for rock, soil or sediment

Reagents

(a) Nitric acid (70%).

(b) Perchloric acid (60%).

(c) Hydrofluoric acid (40%).

(d) Hydrochloric acid (4 M). Dilute hydrochloric acid (344 ml, 36%) to 1 l with water.

(e) Hydrochloric acid (0.3 M). Dilute hydrochloric acid (26 ml, 36%), to 1 l with water.

Equipment

(a) PTFE test tubes (19 mm × 98 mm), wire test tube racks (plastic coated) and stainless steel test tube racks.

(b) PTFE liquid dispenser (or polythene measuring cylinder) and (plastic tray for dispensing hydrofluoric acid.

(c) Liquid dispensers (5).

(d) PTFE rod.

(e) Vortex tube mixer.

(f) Heating block sited in a suitable fume cupboard.

(g) Centrifuge tubes (polystyrene, disposable).

References pp. 319–320

Procedure

(a) Weigh each sample (0.1000 g, −200 mesh) into a clean, dry, numbered PTFE test tube.

(b) Weigh standard and duplicate samples and leave empty test tubes at random intervals for blank determinations.

(c) Add nitric acid (2.0 ml) to each tube.

(d) Add perchloric acid (1.0 ml) to each tube.

(e) Add hydrofluoric acid (5.0 ml) to each tube, using a PTFE dispenser or a plastic measuring cylinder.

(f) Place the tubes into the cold heating block and raise the temperature to 90°C. Leave for 3 h.

(g) Raise temperature to 140°C and leave for 3 h.

(h) Raise temperature to 190°C and leave for 10 h (or until the solution has evaporated).

(i) Transfer the PTFE tubes to a stainless steel test tube rack and allow to cool. Place the tubes into a wire rack.

(j) Add hydrochloric acid (4M, 2.0 ml) to each tube.

(k) Place tubes in the shallow heating block and leave for 1 h at 70°C.

(l) Transfer tubes to wire racks and allow to cool.

(m) Add hydrochloric acid (8.0 ml, 0.3 M) to each tube and mix.

(n) Decant into disposable centrifuge tubes.

Remarks

(a) This method will completely digest most constituents of rocks, soils and sediments. A few minerals will partly or completely resist attack, e.g. barite, chromite, cassiterite, tourmaline, kyanite, some spinels and magnetites, rutile, zircon and wolframite.

(b) When using PTFE test tubes in a heating block, care should be taken not to exceed the temperature at which PTFE becomes plastic (240°C).

(c) This method must not be attempted on samples containing oil or bitumen.

(d) The dilution factor for this method is 100.

(e) The final drying stage of the procedure (h) can be extended to 15 h (overnight) without detriment.

E. Hydrofluoric acid–boric acid digestion for silicon determination in rock, soil or sediment

This method is useful for the determination of silicon, but the solution can also be used for other major constituents. It is based on decomposition of the sample with hydrofluoric acid in a polypropylene bottle. Boric acid solution is added to dissolve precipitated fluorides [9, 10].

Reagents

(a) Hydrofluoric acid (40%).

(b)   Hydrochloric acid (36%).
(c)   Saturated boric acid solution. Weigh out 200 ± 5 g boric acid into a beaker, add 1000 ± 50 ml water. Cover the beaker and heat until the acid has dissolved. Cool to $40° \pm 10°C$ and decant into a bottle.

Equipment
(a)   Polypropylene bottles with screw caps (125 ml).
(b)   Water bath or oven.
(c)   PTFE dispenser or plastic measuring cylinder and plastic tray for hydrofluoric acid.
(d)   Liquid dispenser.
(e)   Measuring cylinder, 50 ml.

Procedure
(a)   Weigh each sample (0.100 g, −80 mesh) into a dry, numbered polypropylene bottle.
(b)   Weigh standard and duplicate samples, and leave empty bottles at random intervals for blank determinations.
(c)   Add hydrochloric acid (1.0 ml) to each bottle, wetting the sample thoroughly.
(d)   Add hydrofluoric acid (5.0 ml) to each bottle and close firmly.
(e)   Place the bottles in an air oven, or in a water bath at $95° \pm 5°C$ and leave for one hour. Allow the bottles to cool. (f) Add boric acid solution (50 ml) to each bottle, close firmly and replace it in the air oven for a further hour. Allow the bottles to cool.
(g)   Add water (44.0 ml) to each bottle and mix thoroughly.
(h)   Use this solution for the determination of silicon.

Remarks
   (a) Ensure that the bottles used are of polypropylene or other plastic material that will withstand temperatures up to about 130°C. If the screw caps do not give a tight seal, this can be improved by using "washers" cut from thin plastic film.
   (b) This method is suitable for the same range of minerals as method III.C.
   (c) The hydrofluoric–boric acid solutions should not be left in contact with glass apparatus for more than two hours to avoid etching the glassware and contaminating the sample solutions with silicon.
   (d) Other elements can be determined on the same solution. Make the calibrators for aluminium with the same concentration of hydrofluoric–boric acid as the sample solution. Determine magnesium and calcium by using the nitrous oxide–acetylene flame, making a dilution of the sample solution to contain 1000 μg ml^{-1} of potassium as an ionisation suppressant. Determine sodium and potassium by using air–acetylene flame making an appropriate

dilution of the sample to contain 1000 μg ml^{-1} of caesium as an ionisation suppressant.

(e) For samples with low (<5%) silicon content, use higher sample weights (up to 0.5 g).

(f) The dilution factor for this method is 1000.

F. Lithium metaborate fusion treatment for the determination of silicon and other major elements in rock, soil or sediment

This method of decomposition is useful when silicon determinations are required on samples that are not decomposed by hydrofluoric–boric acids at 95°C. The finely ground sample is fused with lithium metaborate in a graphite crucible and the melt is dissolved in dilute nitric acid [11].

Reagents
(a) Lithium metaborate high purity flux.
(b) Nitric acid (4% v/v).

Equipment
(a) Graphite crucibles. Supplied by Heydon & Son Ltd., 24 Ninian Avenue, Hendon, London NW4 3XP.
(b) Porcelain crucibles.
(c) Magnetic stirrer with plastic-coated stirrer bar.
(d) Plastic bottles (125 ml wide mouth) with screw caps.
(e) Graduated flasks (100 ml).
(f) Beakers (100 ml, polypropylene).
(g) Muffle furnace.

Procedure
(a) Pre-ignite graphite crucibles for 30 min at 950°C and then cool, taking care not to disturb the powdery inside surface.
(b) Mix the sample (0.2000 g, −200 mesh) with lithium metaborate (1.0 g) in a porcelain crucible, transfer to a graphite crucible and heat in a muffle furnace at 900°C for 15 min.
(c) Add nitric acid (50 ml) to a polypropylene beaker.
(d) Remove the crucible from the furnace and immediately pour the melt into the nitric acid. Introduce a plastic-coated stirrer bar into the solution and stir to dissolve the melt (about 10 to 15 min).
(e) Transfer the solution into a graduated flask (100 ml) and dilute to volume with nitric acid (4% v/v).
(f) Immediately transfer the solution to a clean, dry, numbered plastic bottle, and use to determine silicon.

Remarks

(a) Samples with an appreciable organic matter content should be ignited prior to fusion.

(b) Aluminium can also be determined on the solution directly, by using the nitrous oxide–acetylene flame.

(c) Calcium and magnesium are usually determined on dilutions of the original sample solution. The diluted solutions and calibrators are prepared to contain potassium (1000 μg ml^{-1}) as ionisation suppressant and the determinations carried out in the nitrous oxide–acetylene flame.

(d) Sodium and potassium are determined in the air–acetylene flame.

(e) All calibration solutions must contain appropriate concentrations of lithium metaborate and nitric acid.

(f) The dilution factor for this method is 500.

G. Chelation–solvent extraction method for determining trace metals in water samples

This method is for the determination of cadmium, cobalt, copper, iron, manganese, nickel, lead and zinc, which are solvent extracted and concentrated as their diethyldithiocarbamate chelates. After destruction of the organic complexes, dissolution of the residue in dilute acid gives a solution suitable for atomic absorption analysis [12].

Reagents

(a) Acetic acid (100%).
(b) Ammonia solution, isothermally distilled from the low-lead reagent S.G. = 0.880. Allow the reagent to equilibrate in a desiccator at room temperature, with an equal volume of water.
(c) Chloroform ("Aristar" grade).
(d) Deionised water.
(e) Hydrochloric acid (1 M). Dilute hydrochloric acid (89 ml, 36%, "Aristar") to 1 l.
(f) Nitric acid (70% "Aristar").
(g) Nitric acid (2 M). Dilute nitric acid (125 ml, 70%) to 1 l.
(h) Sodium acetate trihydrate.
(i) Sodium diethyldithiocarbamate (SDDC).
(j) SDDC/buffer solution. Dissolve sodium acetate trihydrate (250 g) in water (500 ml), add acetic acid (6 ml) to the solution and mix well. Add SDDC (50 g) to this solution and mix. Dilute the solution to 1 l with water. If necessary adjust the pH of the solution to between 8.0 and 9.0. Extract any SDDC–metal complexes by successive treatments in a separating funnel with 30 ml aliquots of chloroform, until extract is colourless.

References pp. 319–320

Equipment
(a) pH narrow range test paper.
(b) Separating funnels (1 l).
(c) "Quickfit" conical flasks (100 ml).
(d) Hotplate sited in suitable fume cupboard.
(e) Membrane filters (0.45 μm) and a suitable filtering assembly.
(f) Sample bottles, high density, polythene 1 l capacity.
(g) Glass rods.
(h) Liquid dispensers (2).
(i) Measuring cylinder (1 l).

Procedure

Sampling procedure:
(a) Fill sampling bottles with nitric acid (2 M) and leave for 24 h.
(b) Rinse out each bottle three times with water.
(c) Rinse out each bottle with the water being sampled, and take a 1-l sample.

Preliminary treatment—"soluble metal fraction":
(a) Wash a membrane filter with hydrochloric acid and rinse with water.
(b) Filter the sample through the membrane as soon as possible after collection.
(c) Add nitric acid (2 ml) to filtrate, mix well and store in a polyethylene bottle.

Alternative preliminary treatment—"total metal fraction":
(a) Add nitric acid (2 ml) to the sample in the sample bottle and mix well.
(b) Set the sample aside for at least four days, shaking the bottle each day.
(c) Wash a membrane filter with hydrochloric acid and rinse with water. Filter the sample through this washed membrane and store in a sample bottle.

Extraction procedure:
(a) Place the pretreated sample (500 ml) in a separating funnel (1 l) and adjust the pH to approximately 7 with ammonia solution.
(b) Add SDDC/buffer solution (20 ml) and if necessary, adjust the pH to between 5.8 to 6.1 by dropwise addition of either ammonia solution or nitric acid.
(c) Shake the funnel for 5 min and then add chloroform (30 ml).
(d) Shake the funnel for 5 min then allow the phases to separate. Run the chloroform layer into a conical flask.
(e) Add chloroform (20 ml) to the funnel and repeat the extraction. Add the separated chloroform phase to the conical flask.
(f) Add nitric acid (1 ml) to the combined extracts and place the flasks on a hotplate.
(g) Gently evaporate the chloroform extracts to dryness. Repeat the addition of nitric acid until a white or pale yellow residue remains.

(h) Add hydrochloric acid (5.00 ml of 1 M) to the residue, stopper the flask and leave to dissolve.

(i) After mixing determine the metal concentrations by atomic absorption analysis.

(j) Carry out blank determinations by repeating steps (a) to (i) on deionised water (500 ml), which has already been stripped of any trace metals by the extraction procedure.

Remarks

(a) Cleanliness of apparatus and precautions against contamination at all stages are essential. Soak glassware in nitric acid (2 M) followed by a solution of "Decon" (5% v/v) solution in deionised water. Wash out five times in deionised water before use. Equipment required immediately can be cleaned by shaking with the SDDC/buffer solution and rinsing five times with deionised water. Separating funnels should be cleaned in this way before beginning a series of extractions.

(b) Where high trace-metal concentrations are indicated by the formation of an immediate deep colour or a precipitate on the addition of the SDDC/buffer solution, further extractions with chloroform will be necessary.

(c) The SDDC solution is unstable under acid conditions and decomposes rapidly. At a pH of 9, the SDDC solution is stable for at least one month.

(d) Acidification of the sample as described will stabilise the concentration of extractable trace elements for at least 35 days.

(e) A duplicate and a blank determination should be carried out after every tenth sample. Spiked samples should be analysed periodically, particularly at the beginning and end of each batch of SDDC/buffer solution.

(f) In this method "soluble" refers to those metal species capable of passing through a 0.45 μm membrane whilst "total" refers to that metal fraction of the particulate solubilised by dilute nitric acid and capable of passing through a 0.45 μm membrane, plus the soluble fraction. A genuine "total" fraction would include metals still not leached from the particulate matter.

(g) The concentration factor for this method is 100.

H. Nitric acid–perchloric acid digestion for herbage samples

CAUTION: Fatty or oily samples must not be digested by this procedure.

Reagents
(a) Nitric acid (70%).
(b) Perchloric acid (60%).
(c) Hydrochloric acid (6 M). Dilute hydrochloric acid (516 ml, 36%), to 1 l with water.

References pp. 319–320

Equipment
(a) Conical flasks (250 ml) or Philips beakers each with a watch glass cover.
(b) Hotplate sited in a suitable fume cupboard.
(c) Graduated flasks (25 ml) or graduated test tubes (10 ml), depending on the dilution factor required.
(d) Liquid dispensers.

Procedure
(a) Weigh each sample (2.0 g, dried at 105°C and ground) into a clean, dry, numbered conical flask.
(b) Weigh standard and duplicate samples and leave empty flasks at random intervals for blank determinations.
(c) Add nitric acid (40 ml) to each flask, cover with watch glass and set aside in a fume cupboard overnight.
(d) Place covered flasks on a hotplate and warm gently until frothing ceases.
(e) Allow the flasks to cool and add perchloric acid (3 ml).
(f) Replace the flasks on the hotplate, remove the covers and heat cautiously just to dryness.
(g) Allow the flasks to cool, add hydrochloric acid (2.0 ml) and deionised water (2–3 ml) to each flask. Warm gently to dissolve the residue.
(h) Transfer the cooled solutions either to graduated flasks or to graduated test tubes.
(i) Dilute to volume with water and use the solutions for atomic absorption analysis.

Remarks
   (a) On completion of the attack, most samples exhibit a residue to insoluble silica.
   (b) This method is suitable for a large number of metals.
   (c) The dilution factor for this attack is 12.5 if 25 ml flasks are used, or 5.0 for 10 ml test tubes.

I. Nitric and perchloric acid digestion of herbage samples in large test tubes

CAUTION: Fatty or oily samples must not be digested by this procedure.

Reagents
(a) Fuming nitric acid (95%).
(b) Perchloric acid (60%).
(c) Hydrochloric acid (5 M). Dilute hydrochloric acid (430 ml, 36%) to 1 l with water.
(d) Decon solution (0.1%) in high purity water.

Equipment
(a) Test tubes (28 mm × 140 mm, borosilicate) with air condensers (Quickfit, size 24/25), test tube racks.
(b) Deep heating block sited in a suitable fume cupboard.
(c) Liquid dispensers (3).
(d) Autopipette.
(e) Centrifuge tubes 18 mm × 110 mm (polystyrene, disposable).
(f) Vortex tube mixer.

Procedure
(a) Weigh each sample (2.0 g dried at 105°C and ground) into a clean, dry and numbered test tube.
(b) Weigh standard and duplicate samples and leave empty tubes at random intervals for blank determinations.
(c) Dampen samples with Decon solution (2.0 ml, 0.1%) using an autopipette. Leave for 2–3 h.
(d) Attach a condenser to each tube.
(e) Add 5.0 ml (5 × 1 ml) fuming nitric acid to each tube, but wait for frothing to cease before each 1 ml portion is added.
(f) Add further fuming nitric acid (3 × 5.0 ml) to each tube. If frothing persists allow 30 min intervals between each 5 ml.
(g) Place tubes in the cold block and raise temperature to 50°C. Leave at this temperature for 3 h.
(h) Increase the temperature to 100°C and leave for 3 h.
(i) Raise temperature to 150°C and leave for 10 h.
(j) Allow the tubes to cool. Wash down condensers with a fine jet of deionised water (~1 ml), then remove condensers.
(k) Add perchloric acid (3.0 ml) to each tube and place tubes in the block.
(l) Leave at 50°C for 15 min, then raise temperature to 100°C and leave for a further 15 min.
(m) Increase temperature to 150°C and leave for 18 h or until just dry.
(n) Transfer the tubes to wire racks and allow to cool. Add hydrochloric acid (2.0 ml, 5 M) to each tube and replace clean condensers.
(o) Place tubes in heating block and leave at 70°C for 1 h. Transfer the tubes to the wire racks and allow to cool.
(p) Add water (8.0 ml) to each tube, remove condensers and mix.
(q) Decant into centrifuge tubes.

Remarks
(a) On completion of the attack most samples exhibit a residue of silica.
(b) This method is suitable for a large number of metals.
(c) The dilution factor for this attack is 5.0.
(d) This method is not suitable for the determination of potassium, because of the precipitation of potassium perchlorate.

References pp. 319–320

J. Extraction of mercury from rocks, soils and sediments

Reagents
(a) Nitric acid (70%).
(b) Tin(II) chloride solution; dissolve 25 g of $SnCl_2 \cdot 2H_2O$ in 100 ml diluted (1 + 1) hydrochloric acid. Pass oxygen-free nitrogen through the solution for 3 hours or until the mercury blank is negligible.
(c) Magnesium perchlorate (anhydrous).
(d) Soda asbestos.
(e) Nitrogen.

Equipment
(a) Test tubes (100 × 12 mm) 2 for each sample; test tube racks.
(b) Liquid dispensers (2).
(c) Shallow aluminium heating block sited in a suitable fume cupboard.
(d) Piston pipettes.
(e) Bubbler and atomic absorption cell as shown in Fig. 3.

Procedure
(a) Weigh each sample (0.300 g) into a clean, dry, numbered test tube.
(b) Weigh standard and duplicate samples and leave empty test tubes at random intervals for blank determinations.
(c) Add nitric acid (1.0 ml) to each tube and heat at 80°C for 1 h in the heating block.
(d) Cool, add water (2.00 ml), mix, and allow the solid residue to settle.
(e) Add water (2.00 ml) and the test solution (1.00 ml) to a second tube, and then tin chloride solution (200 μl). Use piston pipettes for the sample

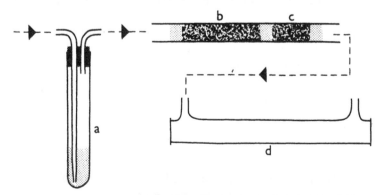

Fig. 3. Schematic diagram of equipment for the determination of mercury, showing: (a) tube for reduction with $SnCl_2$ and outgassing of mercury vapour; (b) magnesium perchlorate desiccant; (c) soda-asbestos; and (d) silica-windowed cell to fit in light path of instrument.

and tin chloride solutions.

(f) Pre-set the nitrogen carrier gas flow rate at $0.5 \, l \, min^{-1}$.

(g) Attach the second tube to the carrier gas line and record the peak absorption on a chart recorder.

(h) If the absorption is greater than the maximum of the calibration range, immediately repeat the determination with a smaller test portion of sample solution at stage (e).

(i) Calibrate the system, starting at stage (e), by adding 1.00 ml of a solution of mercury(II) chloride in diluted $(1 + 3)$ nitric acid. Suitable calibration solutions are made at concentrations of 0, 10, 20, 40, 60, 80 and 100 ng ml^{-1}, and should be prepared fresh daily. Check the calibration and blank after every 10 samples.

Remarks

(a) The detection limit is about 0.5 ng Hg, depending on the internal volume of the apparatus and the stability of the spectrometer. The upper limit of the useful calibration range is about 100 ng Hg. For the weights and volumes given the working range is 5–1000 ppb. A lower range can be obtained by using a 0.600 g sample and by using the whole of the solution at stage (d) in the procedure above.

(b) Special care is required in pre-handling of samples to avoid loss of mercury. Rocks should be subjected to a minimum of grinding: a particle size of $-200 \, \mu m$ is suitable. Soil samples should be air-dried only, the water content being determined on a separate sub-sample. Sediments should be deep-frozen until just before analysis and weighed wet, to avoid loss of the volatile organo-mercury compounds which may account for a substantial proportion of the mercury present in sediments.

(c) The internal dimensions of the bubbler, cell and tubing are kept as small as possible for good sensitivity.

(d) The absorption cell should be as narrow as possible without occluding the light beam, and as long as can be accommodated in the instrument.

(e) Samples can be analysed at the rate of about one per minute.

(f) Water used in the sample solutions should be demineralised, and outgassed to remove traces of mercury metal.

(g) Background correction should not be necessary.

K. Extraction of tin from rocks, soils and sediments

Cassiterite-bearing samples are attacked by volatilisation with ammonium iodide. The sublimate containing the tin is dissolved in dilute tartaric acid. The solution can be used for normal nebulisation into an air–acetylene flame for high levels, or for hydride formation for low levels of tin. Interfering elements are not volatilised [13].

References pp. 319–320

Reagents
(a) Ammonium iodide, ground to −200 μm.
(b) Tartaric acid solution (1% m/v). Dissolve 10 g of tartaric acid in 1 l of water. (c) Sodium tetrahydroborate solution. Dissolve 10 g of reagent in 1 l of 0.1 M sodium hydroxide solution.

Equipment
(a) Modified 3-slot burner for argon-hydrogen diffusion flame as shown in Fig. 4.
(b) Volatilisation tubes as shown in Fig. 5.
(c) Heating block (600°C).
(d) Hydride generator as shown in Fig. 6.
(e) Peristaltic pump, Watson Marlow MHRE200 or equivalent with silicone rubber tubing (0.8 mm i.d. for reagent, 0.5 mm i.d. for sample solution).

Procedure A (nebulisation)
(a) Grind the sample to pass 200 mesh (63 μm) sieve.
(b) Weigh the sample (0.200 g) and the ammonium iodide (0.30 g) into a heating tube and mix.
(c) Heat the tube at 600°C for 10 min in the aluminium block, with the condenser and glass sphere in place, the condenser being cooled by a forced draught (e.g. from a vacuum cleaner) or by a cooling jacket. Lay the tubes horizontally to cool, with the condensers still in place.
(d) When cool, detach the condenser and place it in a test tube. Add tartaric acid solution (20.0 ml) and heat at 50°C for 20 min.

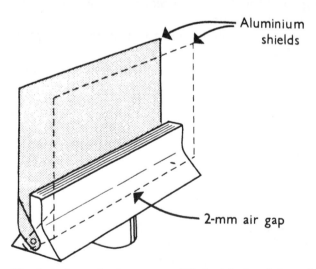

Fig. 4. A three-slot burner modified with aluminium shields for tin determination in an entrained air–hydrogen flame.

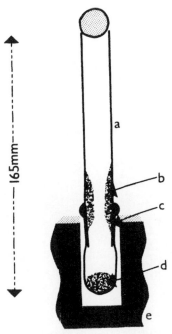

Fig. 5. Apparatus for the volatilisation of tin with ammonium iodide showing: (a) 14/23 cone used as an air condenser; (b) sublimate above hot zone; (c) 14/23 socket adapted to form a short tube; (d) sample residue; and (e) part of a multi-hole hot-block.

(e) When cool, nebulise the solution into the air–acetylene flame, and record the absorption.

(f) Prepare calibration solutions in the range 2 to 250 μg ml^{-1} of tin(II) chloride in a solution containing tartaric acid (1% m/v) and ammonium iodide (1.5% m/v). The calibration range is linear up to 230 μg ml^{-1} and serviceable up to at least 250 μg ml^{-1}.

Procedure B (hydride generation)

(a) Follow procedure A up to stage (d).

(b) Block the nebuliser capillary and attach the hydride generator to the auxiliary oxidant inlet. Light the argon–hydrogen diffusion flame supported on the shielded 3-slot burner (argon flow 13 l min^{-1}, hydrogen flow 8 l min^{-1}).

(c) Operate the hydride generator with the following flow rates: carrier gas (argon) 1 l min^{-1}; reagent 4.5 ml min^{-1}; sample solution 9.2 ml min^{-1}.

(d) After about 5 s for signal stabilisation, integrate for 10 s and record the absorption.

(e) Prepare calibration standards in the range 0.02 to 2.0 μg ml^{-1}, as before, containing tartaric acid and ammonium iodide.

References pp. 319–320

314

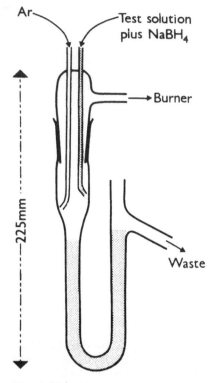

Ar

Test solution
plus NaBH₄

Burner

225mm

Waste

Fig. 6. Hydride generator for the determination of tin.

Remarks

(a) Fine grinding of the sample is necessary to ensure complete attack and homogeneous distribution of the cassiterite throughout the sample.

(b) The heating can be carried out with a bunsen burner if a 600°C heating block is not available.

(c) No interfering elements are volatilised with the tin.

(d) The equipment can be set up so that nebulisation or hydride generation can be employed in quick succession depending on the concentrations of tin encountered in the samples. The appropriate ranges are: nebulisation from about 200 μg g^{-1} (detection limit) to 25000 μg g^{-1}; hydride generation from about 0.80 μg g^{-1} (detection limit) to 200 μg g^{-1}. Higher ranges could be accommodated by dilution of the solution obtained at stage (d) in procedure A.

(e) Dilute tin standards in tartaric acid solution are stable for at least one week.

(f) The dilution factor for this method is 100.

L. Nitric and perchloric acid digestion with lanthanum precipitation for selenium by hydride generation in rocks, soils or sediments

CAUTION: Samples containing oil or bitumen must not be digested by this procedure.

Reagents
(a) Nitric acid (70%).
(b) Perchloric acid (60%).
(c) Ammonia solution (35%).
(d) Hydrochloric acid (36%).
(e) Potassium bromide solution (4%) in hydrochloric acid (5 M). Dissolve potassium bromide (4 g) in water, add hydrochloric acid (430 ml, 36%) and make up to 1 l with water.
(f) Lanthanum nitrate solution (5%). Dissolve lanthanum nitrate (10 g, La(NO$_3$)$_3$·6H$_2$O) in water (200 ml).
(g) Sodium tetrahydroborate solution (1%). Dissolve sodium tetrahydroborate (10 g) in water (1 l).

Equipment
(a) Test tubes (18 mm × 180 mm) and test tube racks (plastic coated).
(b) Glass beads (~3 mm diameter).
(c) Liquid dispensers (3).
(d) Dropper pipette.
(e) Deep aluminium heating block sited in a suitable fume cupboard.
(f) Water bath.
(g) Vortex tube mixer.
(h) Centrifuge tubes polystyrene, disposable.
(i) Centrifuge.
(j) Hydride generator as shown in Fig. 6, or similar arrangement.
(k) Quartz atomization tube or other system.
(l) Peristaltic pump, Watson Marlow MHRE200 or equivalent with silicone rubber tubing, 0.8 mm i.d. for sample, 0.5 mm i.d. for reagent.
(m) Argon cylinder.

Procedure A (decomposition)
(a) Weigh each sample (0.500 g) into a clean, dry, numbered test tube.
(b) Weigh standard and duplicate samples and leave empty test tubes at random intervals for blank determinations.
(c) Add nitric acid (2.0 ml) and a glass ball to each tube. Mix by a vortex mixer.
(d) Place test tubes into the cold aluminium heating block and leave overnight at 50°C.
(e) Transfer tubes to a wire rack and add perchloric acid (1.0 ml). Return test tubes into the heating block and leave for 1 h at 100°C.

References pp. 319–320

(f) Raise temperature to 150°C and leave for 2 h.

(g) Raise temperature to 170°C and leave until solid residues are bleached (*not dry*).

(h) Transfer test tubes to a stainless steel rack and allow to cool.

(i) Add hydrochloric acid (4 drops, 36%) and mix.

(j) Add water (5.0 ml) and mix again.

(k) Transfer solutions to centrifuge tubes and make up with water to 10.0 ml.

(l) Centrifuge solutions for 2 min at 2000 rpm.

(m) Decant supernatant solutions into clean centrifuge tubes.

Procedure B (precipitation)

(n) Add lanthanum nitrate (0.5 ml) to each centrifuge tube.

(o) Add ammonia (2.0 ml) to each tube and mix.

(p) Centrifuge solutions for 2 min at 2000 rpm. Decant and discard the supernatant liquid.

(q) Add potassium bromide solution (~6 ml at ~50°C) to each tube containing residue and shake tube by hand to dissolve precipitate. (r) Make up solutions to 10.0 ml with potassium bromide (4% at 50°C), transfer tubes into a water bath. Leave for 1 hour at 50°C.

(s) Transfer tubes to test tube racks and allow to cool.

(t) Make up volumes to 10.0 ml with potassium bromide (4%, room temperature) and mix.

Procedure C (determination)

The selenium in the solution produced at stage (t) in procedure B above can be determined by continuous hydride generation with equipment equivalent to that shown in Fig. 6, with test solution flow rate of 9 ml min^{-1}, reagent (NaBH$_4$) solution flow rate of 4.5 ml min^{-1}, and carrier gas (argon) flow rate of 500 ml min^{-1}. Atomization is achieved in a commercial heated quartz tube atomizer.

Remarks

(a) Overheating *must* be avoided at stage (g) in procedure A above to avoid loss of selenium.

(b) Interferences from base metals are removed by the lanthanum precipitation.

(c) Very peaty samples should be allowed to stand overnight after the addition of nitric acid (procedure step (c) above). (d) Dilution factor for this attack is 20.

(e) This procedure was originally devised for an ICP determination [14], but the atomic absorption finish with a heated quartz tube atomizer provides almost equivalent performance, with detection limit around 10–20 ng g^{-1} selenium in the solid.

M. Hydrobromic acid and bromine digestion for gold in rocks and sediments

Reagents
(a) Hydrobromic acid (49%, Analar).
(b) Bromine (99%, Analar).
(c) Hydrobromic acid 0.1N. Dilute hydrobromic acid (23 ml, 49%) to 2 l with water.
(d) Hydrobromic acid/bromine. Add bromine (10 ml) to hydrobromic acid (49%, 2 l).
(e) 4-methyl pentan-2-one (MIBK) (Analar).

Equipment
(a) Porcelain crucibles.
(b) Culture tubes (borosilicate, 18 mm × 170 mm, screw capped with Teflon cap lining).
(c) Autopipette.
(d) Measuring cylinders (25 ml, 2 l).
(e) Shaking machine.
(f) Pasteur pipettes.
(g) Centrifuge.
(h) Muffle furnace.

Procedure
Samples containing organic material or sulphides should be ignited first at 650°C overnight.
(a) Weigh each sample (8.00 g) into a clean, dry culture tube.
(b) Weigh standard and duplicate samples and leave empty tubes at random intervals for blank determinations. (c) Add HBr/Br solution (16.0 ml) to each tube and allow to stand for 10–15 min. (d) Cap tubes firmly and shake on shaking machine for 15 min.
(e) Add water (16.0 ml) and MIBK (10 ml).
(f) Shake for 2 min on shaker.
(g) Centrifuge each tube at 600 rpm for 5 min.
(h) Add hydrobromic acid (25 ml, 0.1 N) to clean culture tubes.
(i) Transfer MIBK layer from each sample tube quantitatively into a tube containing hydrobromic acid (0.1 N), using a pasteur pipette.
(j) Shake tubes by hand for ~15s.
(k) Transfer MIBK layer into centrifuge tubes and centrifuge at 600 rpm if necessary. (Quantitative transfer not necessary here.)
(l) Nebulise MIBK layer directly into a flame AAS, against matching standards.

Calibration standards
(a) Prepare Au standard solution (10μg ml^{-1}) in hydrobromic acid (1.5 N).

References pp. 319–320

(b) Place hydrobromic acid (1N, 30.0 ml) into three 100 ml volumetric flasks.

(c) Add standard gold solution: (i) 3.00 ml and (ii) 12.00 ml using an autopipette, into two of the volumetric flasks containing the hydrobromic acid.

(d) Add MIBK (30.00 ml) to each flask, using an autopipette and shake flasks vigorously for 1 min.

(e) Add water to bring the MIBK phase into the neck of the flasks.

Remarks

(a) Care must be taken when adding the hydrobromic acid–bromine reagent to any sample with high carbonate content, because it may effervesce from the tube.

(b) Any gold minerals completely enclosed in insoluble particles such as quartz will not be brought into solution and results will be low.

(c) If the sample is ground too fine it is difficult to get a separation of the ketone layer. Fine grinding may also result in loss of gold by plating on the grinding surfaces.

(d) Iron is coextracted into MIBK and it is presumed to interfere by causing light scattering. Iron is removed by washing the extract with hydrobromic acid at stages (i) and (j) in the procedure above.

(e) The dilution factor for this method is 12.5.

(f) This method is based on a U.S. Geological Survey method [15].

N. Ashing with magnesium nitrate for arsenic solubilisation and subsequent determination by hydride generation in rocks, soils and sediments

Reagents

(a) Magnesium nitrate solution (saturated). Dissolve magnesium nitrate $((MgNO_3)_2 \cdot 6H_2O$, 100 g, Analar) in water (100 ml).

(b) Hydrochloric acid (36%).

(c) Potassium iodide solution (0.2%, freshly prepared). Dissolve potassium iodide (1.0 g) in water (500 ml).

(d) Sodium tetrahydroborate solution (1%). Dissolve sodium tetrahydroborate (10 g) and sodium hydroxide (0.4 g) in water (1 l).

Equipment

(a) Beakers, 50 ml, borosilicate.

(b) Liquid dispensers (3).

(c) Centrifuge tubes (polystyrene, disposable) and test tube racks.

(d) Clingfilm (PVC kitchen film).

(e) Rocking machine.

(f) Muffle furnace.

(g) Hydride generator as shown in Fig. 6, or similar arrangement.

(h) Quartz atomization tube or other system.

(i) Peristaltic pump, Watson Marlow MHRE200 or equivalent with silicone rubber tubing, 0.8 mm i.d. for sample and 0.5 mm i.d. for reagent.

(j) Argon cylinder.

Procedure A (decomposition)

(a) Weigh each sample (0.250 g) into a clean, dry, numbered beaker.

(b) Weigh standard and duplicate samples and leave empty beakers at random intervals for blank determinations.

(c) Add magnesium nitrate solution (1.0 ml) to each beaker.

(d) Place beakers into the cool furnace, raise temperature to 450°C and leave for 6 h.

(e) Allow to cool, remove beakers from the furnace. (f) Add hydrochloric acid (5.0 ml, 36%) to each beaker. Allow to stand for 10 min, then cover beakers with clingfilm.

(g) Place beakers on a rocking machine and leave on overnight.

(h) Remove beakers from rocking machine and take off clingfilm covers.

(i) Add potassium iodide solution (5.0 ml) to each beaker and cover again with clingfilm. Swirl beakers gently by hand and allow to stand for 30 min.

(j) Decant solutions into disposable centrifuge tubes.

Procedure B (hydride generation)

The arsenic in the solution produced at stage (j) above can be determined by continuous hydride generation by using equipment equivalent to that shown in Fig. 6, with flow rates of 9.0 ml min^{-1} for the test solution and 4.5 ml min^{-1} for the reagent (NaBH$_4$) solution. Atomization is achieved in a commercial heated quartz tube atomizer.

Remarks

(a) After step (i) of the procedure, samples should be analysed within 24 h.

(b) The dilution factor for this method is 40.

(c) This method was originally developed with an ICP determination, but provides almost equivalent performance with an AAS finish with a quartz tube atomizer, with detection limits of about 100 ng g^{-1} of arsenic in solid.

REFERENCES

1   J.S. Webb and M. Thompson, Pure Appl. Chem., 49 (1977) 2085.
2   A.L. Meier, J. Geochem. Explor., 13 (1980) 77–85.
3   J.G. Viets, Anal. Chem., 50 (1978) 1097.
4   J.S. Webb, I. Thornton, M. Thompson, R.J. Howarth and P. Lowenstein, The Wolfson Geochemical Atlas of England and Wales, Oxford University Press, 1978, 66 pp.

5 J.R. Foster, Can. Min. Metall. Bull., 60 (1973) 85.

6 M. Thompson, S.J. Walton and S.J. Wood, Analyst, 104 (1979) 229.

7 Analytical Methods Committee, Analyst, 84 (1959) 214.

8 M. Thompson, Control Procedure in Geochemical Analysis, in R.J. Howarth (Ed.), Statistics and Data Analysis in Geochemical Prospecting, Elsevier, Amsterdam, 1982, pp. 39–58.

9 F.J. Langmyhr and P.E. Paus, Anal. Chim. Acta, 43 (1968) 397.

10 B. Bernas, Anal. Chem., 40 (1968) 1682.

11 J.C. Van Loon and C.M. Parisis, Analyst, 94 (1969) 1057.

12 H. Watling, D.I.C. Thesis, Imperial College, London, 1974.

13 D. Gladwell, M. Thompson and S.J. Wood, J. Geochem. Explor., 16 (1981) 41.

14 B. Pahlavanpour, J.H. Pullen and M. Thompson, Analyst, 105 (1980) 274–278.

15 C.E. Thompson, H.M. Nagakawa and G.H. Van Sickle, Geological Survey Research 1968, Denver, Colo.

Chapter 4h

Applications of Atomic Absorption Spectrometry in the Petroleum Industry

J. MARSHALL

ICI plc, Wilton Materials Research Centre, P.O. Box 90, Wilton, Middlesborough, TS6 8JE (Great Britain)

I. INTRODUCTION

There have been a number of significant advances made in the field of trace element analysis in the last decade. Although atomic absorption spectrometry (AAS) is still widely used in the petroleum industry, other newer techniques such as inductively coupled plasma atomic emission and mass spectrometries, and energy dispersive X-ray fluorescence spectrometry also now play an important role in solving analytical problems. Consequently it is important to recognise the advantages and limitations of AAS in this context. The more general application of atomic spectrometric techniques to the analysis of petroleum and related products has recently been reviewed [1].

The virtues of AAS remain those of spectral simplicity, high sensitivity, relatively wide elemental coverage and low capital and running cost. In recent years, the advances in computer technology have been incorporated into instrument design resulting in highly automated analytical systems. The processes occurring in AAS are now well understood, and the technique has entered a phase of relative maturity. Consequently, methodologies well established, and advances are now application-driven rather than technique-led.

The petroleum and related industries have wide ranging requirements for trace element analysis. Many of these needs are process- or product-related and specific methods are subsets of a more general approach to analysis in organic media. In this chapter, an overview will be presented of the applications of AAS to such problems, highlighting the main topics of recent research.

A. Flame atomization

Most laboratories involved in trace metal analysis are equipped with flame AAS instrumentation. Descriptions of this type of apparatus can be found in Chapter 2. The flame is still the predominant atom cell used in AAS, for reasons of convenience and speed of analysis. Although there has

References pp. 339–340

been little change, in terms of performance, in conventional air–acetylene and nitrous oxide–acetylene flame systems in the last decade, aspects of operational safety (such as the fully automatic control of gas supplies) have been significantly improved. This is obviously an important issue in cases where organic solutions are to be aspirated directly into the flame. Some manufacturers now offer solvent-resistant sample introduction systems to overcome problems of swelling of o-ring seals and polymeric components on exposure to solvents.

Two technique developments are worthy of mention. Sensitivity can be improved for volatile elements by use of atom-trapping procedures, in which a silica tube is inserted into the flame to either improve residence time [2] or to preconcentrate analyte by condensation on the water cooled substrate [3]. The application of flow injection (FI) procedures for the introduction of discrete samples for flame AAS is now well established [4]. This has relevance in the present context opposite the aspiration of solvents which destabilise the flame, or where the amount of sample is limited.

B. Electrothermal atomization

The electrically heated tubular graphite furnace is now firmly established as the alternative to the flame as an atomizer for AAS. Modern instruments, such as those described in Chapter 2 provide the user with the means to control the temperature heating rate, gas flow and timing parameters, and consequently obtain more reproducible results. In addition, the majority of commercially available furnaces are now broadly similar, in terms of tube size, and the general operational facilities offered. It is becoming easier to transfer methodologies from one instrument type to another, although care must be exercised in assuming that spectrometer background correction facilities will be adequate in all cases.

The principal reasons for using electrothermal atomization in petroleum analysis apply equally in other application areas. The furnace offers sensitivity enhancements of up to a factor of 100 over conventional flame atomization, and can be used to analyse samples which are too small to allow nebulization. There is some potential for the direct analysis of solids using the furnace technique, although precision is generally rather poor, and standardisation may be problematic. Liquid samples are introduced to the furnace in aliquots of between 5 and 100 μl. In the past, variable spreading of the sample along the tube walls either during injection or on drying, as a result of low surface tension, often gave rise to poor reproducibility. This was compounded by the necessity to manually inject samples using a micropipette. Developments in autosampler technology have greatly reduced the magnitude of such problems. It is now possible to vary sample volume as required, and to automatically inject the sample into a preheated furnace, typically set at 80°C for organic solvents. Thus the solvent immediately evaporates on contact with the tube, thus minimising spreading effects and significantly improving the reproducibility of analyte positioning and hence overall precision.

Much has been made of the concept of "platform atomization" as a means of reducing chemical interference effects which are sometimes encountered when using electrothermal atomization [5]. Essentially, a graphite platform is placed in the centre of the tube, beneath the injection port. The sample is deposited on the platform, either by manual injection, or using an autosampler, and the atomizer put through the heating cycle in the usual way. Since the platform is in poor thermal contact with the tube, being heated only by its edges, there is a delay in the time taken for the sample to heat. Consequently, when the sample is atomized, the vapour temperature experienced by the analyte is increased relative to atomization from the tube wall, resulting in improved molecular dissociation, and a reduction in interference effects. This approach is most effective in the determination of volatile elements such as Pb, but in application to the detection of refractory elements such as V or Mo, the slower heating rate of the platform may be a disadvantage. There is a secondary advantage of platform atomization over the conventional procedure when handling organic liquids. The platform has the effect of constraining the sample droplet, thus minimising the extent of sample spreading, and resulting in improved precision. Many manufacturers now supply platforms as furnace accessories, and the approach should be considered in situations where interferences are encountered.

II. SAMPLING

Reference should be made to ASTM Methods D4057 and D4177 [6, 7] for details of appropriate procedures in the sampling of petroleum products. It is not widely appreciated that sampling for the purposes of trace element analysis requires a certain amount of planning and foresight. It is essential that the sample material is representative of the whole, and the analytical chemist should provide guidance on this subject in order that the AAS result itself is meaningful. In this section, a brief assessment of the most important aspects of sampling is presented.

A. Sample contamination

Contamination can arise from a variety of sources, and its significance will depend on a combination of limits of specification and detection limits. Thus sampling procedures found to be adequate when used in conjunction with flame AAS, may give rise to substantial blanks when electrothermal atomization is employed. It is not possible to define sampling conditions and storage media for all situations. Nevertheless, some general guidelines can be provided. Exposure of the sample to metallic surfaces should, where possible,

References pp. 339–340

be avoided. For example, solder seams in metal containers often used for sampling can give rise to significant contamination. Coating containers with an inert polymeric lining may reduce contamination from this source. However, some polymers may contain significant levels of catalyst residue left from their manufacture, and care must be taken to avoid extraction into organic media in such cases. Plastic bottles may exhibit porosity to certain solvents, to a lesser or greater degree, and in such circumstances glass vessels may be preferred. In all cases, checks should be made for blanks and carry-over contamination, both from the sampling process itself and in all the analytical steps through to the actual measurement itself. In many industrial environments, the presence of dust and fumes in the atmosphere can cause problems, usually indicated by high and erratic blanks. In some circumstances, it may be necessary to identify and eliminate the source of the contamination before meaningful analytical results can be obtained. This might ultimately require the installation of "clean" or filtered air facilities in severe cases of airborne contamination where the high sensitivity of electrothermal AAS is required.

B. Reagent impurities

In the analysis for metals in petroleum and petrochemical products, one of the most common sample preparation procedures is the dilution of the sample with an organic solvent such as white spirit, methyl isobutyl ketone (MIBK) or xylene. It is clearly of importance therefore, that the solvent system chosen is as free as possible from metallic contamination. Elements such as sodium and zinc are commonly found in many organic solvents. Redistillation of solvents or extraction with mineral acids may be employed to further purify reagents where such problems are encountered. Ashing or wet oxidation methods are also widely employed, and this usually necessitates the use of acidic media which itself may be subject to contamination. Thus blank levels should be monitored in all cases.

C. Sample storage

Although it is well known that solutions of metals at trace levels may not be stable with respect to time, it is not possible to specify general storage conditions for all combinations of analytes and samples. A variety of processes such as precipitation, volatilisation, adsorption on or diffusion through container walls, may result in losses of metallic analytes. Sample stability will also depend on the solvent or sample matrix, the metal of interest and its form of association, the nature of the sample container, and the conditions of storage. In general, glass bottles and containers made from polyethylene derived from the high-pressure free radical process are suitable. In particular, the use of rubber stoppers should be avoided as these are subject to attack by hydrocarbons.

Clearly, it is desirable that the storage container should not contribute to sample contamination. A strong solution of metal-free detergent can be used for cleaning purposes. Containers should be soaked in detergent for 24 hours, and rinsed repeatedly with deionised or distilled water prior to use. It is prudent to dry containers in an oven at 50°C in preference to air-jet drying after application of a volatile solvent such as acetone, since this procedure carries the additional risk of contamination.

D. Sample preparation

In most cases, some form of sample preparation is necessary prior to analysis by AAS. It is often physically impossible to introduce samples to the atomizer directly, and in such circumstances, the inclusion of a preparation stage allows conversion of the matrix to a more convenient form for analysis. Alternatively, sample preparation may be used for preconcentration purposes to improve the sensitivity of the determination. The principal methods of sample preparation employed are dry ashing, reagent-aided dry ashing, wet oxidation using acids, and simple dilution with an organic solvent.

Provided that the analyte of interest is present in the sample as a soluble complex, it is often sufficient to use an appropriate dilution with solvents such as xylene, white spirit MIBK or n-heptane. The optimum working concentration level is usually about 50 times the instrument manufacturer's quoted sensitivity for a given element, so it is prudent to use a dilution factor which provides a response in the range of 0.2 to 0.8 absorbance. A standard procedure will also help to reduce sample viscosity effects, which may have a significant effect on accuracy and precision [8]. It should be noted that although some solvents can be readily aspirated, albeit resulting in modified conditions, others such as n-hexane and benzene give smoky, unstable flames and their use should be avoided. Samples too, will vary widely in behaviour, and consequently, dilution with a well characterised solvent is an advisable precaution if direct analysis is contemplated.

Petroleum oils often contain suspended or colloidal inorganic materials. In the specific case of used lubricating oils, small metallic particles are present. In some instances, these suspended solids are of sufficiently small particle size that efficient atomization still occurs in the flame and in such cases a dilution procedure will be appropriate. Since this problem is a function of atomizer temperature, there may be something to be gained in the use of the nitrous oxide flame, or electrothermal atomization. However, it may be necessary to resort to further sample preparation prior to analysis, partly to remove interferences of this type, and also to ensure sample homogeneity.

Both wet digestion and ashing procedures may be used to overcome difficulties of the type mentioned above. The advantage of these procedures is that the sample matrix is converted from an organic to a dilute mineral acid medium, which is less hazardous to analyse by flame AAS, and which invokes fewer problems of standardisation. Ashing procedures are widely employed in the industry [9, 10, 11]. A general dry-ashing procedure, suitable for use in conjunction with AAS, is as follows.

Weigh 5–50 g of sample into a platinum or silica crucible. The sample is subjected to temperatures of 150–200°C using an infra-red heater to remove low-boiling components, and then ignited in a muffle furnace at 500–550°C. The ash is usually dissolved in nitric or hydrochloric acids, and diluted to a suitable volume with distilled water.

The main limitation of this procedure is that volatile elements such as Zn, Cd and Pb may be lost in the ashing phase. Also elements present in the form of volatile compounds (e.g. Ni and V in porphyrins) may be lost. This particular problem may be minimised by employing ashing aids such as benzene sulphonic acid or p-toluene sulphonic acid to stabilise volatile analytes, an anorganic variant of the more general sulphated ash procedure. It is worth noting that hydrochloric acid is preferable for flame applications whereas nitric acid is favoured in electrothermal AAS measurements.

Until recently, wet digestion procedures such as those employed in biological and food analysis, were considered too slow and labour-intensive for routine use in petroleum analysis. This situation is rapidly changing with the commercial introduction of microwave digestion systems [12, 13] which are substantially faster, as a result of improved efficiency of energy transfer, and can be fully or partially automated. Such methods offer particular advantages in retention of volatile elements, which might be lost on ashing. Matching of the acid concentration (usually nitric or sulphuric) in samples and standards is critical to achieving accurate and reproducible results. Using electrothermal atomization, it is advisable to restrict the maximum acid level to 5% or less, to maintain reproducibility and to protect the tube surface.

III. STANDARDS FOR PETROLEUM ANALYSIS

Standardisation is an important issue in the analysis of petroleum and related materials. Unlike the more general AAS situation, standards in an organic medium are not widely available and where these can be obtained, the elemental range is rather restricted. A list of organometallic compounds suitable for use as primary standards in AAS is provided in Table 1. The procedure for preparing such a standard from the solid organometallic compound is as follows.

Calculate the weight of material containing 50 mg of the element of interest. Dry the material in a low-temperature oven and then weigh the necessary amount. Transfer the weighed material to a weighed 250 ml volumetric flask. Add 5 ml of 2-ethyl hexanoic acid, 4 ml of 6-methyl heptane-2,4-dione and 2 ml of xylene. Gently heat on a hot plate and swirl

TABLE 1

ORGANOMETALLIC STANDARDS FOR AAS

Element	Compound
Al	Aluminium 4-cyclohexanebutyrate
Ag	Silver 2-ethylhexanoate
Ba	Barium 4-cyclohexanebutyrate
Ca	Calcium 2-ethylhexanoate
Cd	Cadmium 4-cyclohexanebutyrate
Co	Cobalt 4-cyclohexanebutyrate
Cr	Tri (1-phenyl-1,3,-butanedione) chromium(III)
Cu	Copper 4-cyclohexanebutyrate
Fe	Iron 4-cyclohexanebutyrate
Hg	Mercuric 4-cyclohexanebutyrate
K	Potassium 4-cyclohexanebutyrate
Li	Lithium 4-cyclohexanebutyrate
Mg	Magnesium 4-cyclohexanebutyrate
Mn	Manganese 4-cyclohexanebutyrate
Na	Sodium 4-cyclohexanebutyrate
Ni	Nickel 4-cyclohexanebutyrate
Pb	Lead 4-cyclohexanebutyrate
Si	Octaphenylcyclotetrasiloxane
Sn	Dibutyltin bis (2-ethylhexanoate)
V	Bis (1-phenyl-1,3-butanedione)-oxovanadium(IV)
Zn	Zinc 4-cyclohexanebutyrate

The above materials may be obtained from NBS (now NIST), Washington D.C., or from local agencies.

until a clear gel forms, then add 2 ml of 2-ethyl hexylamine. Heat and swirl until a clear solution is obtained. Immediately add approximately 80 ml of metal-free lubricating oil, mix and allow to cool to room temperature. Make the contents of the flask up to 100 g with lubricating oil, stopper and shake.

Calibration standards in the chosen solvent may be prepared by subsequent dilution of the above standard. As an alternative to lubricating oil, solvents such as MIBK, xylene or white spirit may be used directly to prepare the master standard. It is not always necessary to follow the above procedure rigidly, as many of the organometallic compounds listed dissolve relatively easily in a range of solvents. Dissolution of the salt in 2-ethyl hexylamine in xylene, followed by dilution with the required solvent, is often sufficient. When in doubt, consult the manufacturer of the salt being used. It should be noted that it is possible to obtain different AAS responses for organometallic standards and the analyte of interest, present in the sample in a different form [14]. Consequently, the validity of this approach must be verified either by comparison with an alternative sample preparation procedure, or with another technique, for a given sample type.

References pp. 339–340

A range of single and blended organic sulphonate standards in oil bases is available. These have been found to provide satisfactory stability, and may be diluted with paraffinic and aromatic hydrocarbons as well as ketones. These may be obtained from Conostan Division, Continental Oil Company, P.O. Box 1267, Ponca City, Oklahoma 74601, USA. Many suppliers of aqueous standard solutions for use in AAS now provide a limited range of single metal standards in widely used solvents such as xylene. Consequently it is worth consulting your local supplier in some cases.

Obviously the problems of standardisation can be obviated to a large extent if ashing or wet digestion procedures can be employed, since aqueous standards are readily available. One option is to use samples previously analysed by this route as secondary standards. Alternatively it may be possible to use inorganic salt standards in the form of emulsions [15, 16] or in mixed solvent systems [14, 17].

Increasingly, however, quality issues are being addressed throughout the industry. International standards on manufacturing quality have forced a re-examination of traceability of standards both for instrument calibration and analysis. It is highly recommended that the sources of standards used for calibration are certified, and this may involve identifying suppliers accredited for manufacture to a national (e.g. BS5750) or international standard.

IV. APPLICATIONS

A. Crude and residual fuel oils

Atomic absorption spectrometry is widely used for the determination of trace elements in crude oils. The trace metal "fingerprint" of a crude oil may be of use in identifying its origin, since there is a marked variation in the concentration levels encountered in oil from different fields. Also, some crude oils become contaminated during transport from the well to the refinery, for example, by pipeline material or seawater. A number of metals can cause undesirable process effects, and consequently the concentration of these have to be monitored. Thus the levels of metals such as Ni, V and Na, must be controlled in order to minimise production problems arising from plant corrosion or catalyst poisoning. Environmental issues are becoming increasingly important, and an indication of the outfall of heavy metals arising from burning oil fuels may be obtained by analysing the original material.

Residual fuel oils, consisting mainly of the residue from distillation of the more asphaltic crudes, are largely used as fuels for ships, locomotives and various heating purposes. Thus many of the inorganic components originally present in the crude oil will still be present in the residual fuel oil. Deposits from the oil may reduce boiler efficiency. In particular, vanadium in the deposits may catalyse the conversion of SO_2 to SO_3 resulting in sulphuric acid production in cooler regions, giving rise to corrosion.

Several methods have been proposed for the determination of metals in crude and residual fuel oils. The sample itself presents handling difficulties, as it may be in the form of an extremely viscous liquid. In such cases, the oil should be placed in an oven set at 60°C for at least one hour prior to sampling. All samples must be stirred or vigorously agitated to ensure homogenisation.

1. *The determination of metals in crude and residual fuel oils by solvent dilution using flame and electrothermal AAS*

This method has been used for the determination of Ca, Fe, V, Ba, Ni, Na and Mg in crude and fuel oils. It may be suitable for conversion to the determination of other elements. The sample is dissolved in an appropriate solvent in a dilution ratio dependent on the sensitivity required. Solvents which have been found useful include MIBK, xylene [19], and a mixture of 10% isopropanol and 90% white spirit [21]. As a general rule, a 1 : 10 dilution is most useful and further dilution may be made if required.

Weigh accurately approximately 10 g of the oil sample and dissolve with shaking in the chosen solvent. Make up to 100 ml and stopper. Dilute further if necessary to match calibration standards. Since alkali and alkaline earth elements are easily ionised in the flame, it is necessary to add an ionisation suppressant to samples and standards when analysing for these elements. Addition of the potassium salt of naphthasulphonic acid (or alternative) should be added to give a final potassium content of about 1000 μg ml^{-1}.

Either commercially available metal in oil, or stock organometallic standards, prepared as described in Section III, can be used for instrument calibration. These stock solutions should be diluted with the same solvent as that used for the oil samples to cover the linear response range of concentration for each analyte. This will be instrument-specific and details for flame and electrothermal operation will be given in the manufacturer's manual. At least four standards and a blank should be employed. Most standards will be in the range 0–10 μg ml^{-1} analyte. Such solutions have limited long-term stability and should be prepared freshly as required.

Adjust the instrumental parameters (e.g. wavelength, slit width, photomultiplier gain, lamp current) in accordance with the manufacturer's recommended conditions for each element. The flame conditions should be established whilst aspirating the appropriate solvent blank. It is recommended that a fuel-lean nitrous oxide–acetylene flame is used in the measurement of Mg, Ca and Ba, and is made fuel-rich for the determination of V. A fuel lean air–acetylene flame should be used for the determination of Fe, Ni and Na. It will normally be necessary to decrease the acetylene flow rate to accommodate the fuel contribution of the solvent. It may be helpful to increase the oxidant flow and/or adjust the nebulizer uptake rate in some cases. Background correction may be required to compensate for non-specific absorption or scatter effects. These can be checked for by the use of a nearby non-absorbing line.

Many modern instruments have preset "optimum" conditions for the determination of a given element. It may be necessary to alter these parameters for work with organic solvents because of dramatic changes in flame characteristics. Some instruments have built-in maximum and minimum flow rates, or ratios, for fuel and oxidant gases for safety reasons. It is advisable to consult the manufacturer for advice on the operation of such systems with organic solvents.

The simple solvent dilution approach is not universally recommended, as undissolved solids may remain, which can give rise to poor accuracy and precision using flame AAS. Viscosity differences in oil samples may also give rise to small changes in uptake rate with respect to both other samples and standards. This problem may be overcome by the use of the standard addition technique, at the penalty of a considerable increase in analysis time. Mixed solvent systems have been used with flame AAS to allow standardisation with inorganic salt standards for the determination of trace metals in crude oils [14]. Samples were dissolved in toluene and diluted with acetic acid. Inorganic salt standards were dissolved in distilled water and diluted to volume with ethanol. The standard additions method was adopted to minimise matrix effects. It was found that the toluene/acetic acid (1 + 4) solvent system provided accurate results for Ni, Cu, Zn, Na, Pb, Cd and Fe in petroleum crudes.

However, there have also been reports of success using the solvent dilution procedure in conjunction with electrothermal AAS. The levels of Ni in heavy crude oils were also estimated using electrothermal AAS [18]. The sample (5 g) was completely dissolved in 25 ml of tetrahydrofuran stabilised with 0.1% hydroquinone. The solution thus obtained was further diluted with tetrahydrofuran as required, and analysed using electrothermal AAS, using standards in the range 0.1 to 1 μg ml^{-1}. Precision was found to improve when a heated injection protocol was used.

Vanadium was determined in crude oil by dilution of samples in the ratio 1:15 to 1:30 depending on concentration, followed by direct AAS measurement at 318.5 nm using a CRA63 atomizer [19]. An organic vanadium compound (bis 1-phenyl-1,3-butadione oxovanadium) prepared in xylene was used for calibration purposes. In addition to the need for total levels of trace metals in petroleum and related products, there is now an interest in obtaining information on the form in which the metal is present. Thus the molecular characterisation and identification of vanadyl compounds in heavy crude petroleums has been the subject of study [20]. High-performance liquid chromatography coupled with electrothermal AAS was employed to speciate vanadyl porphyrin and non-porphyrin compounds. The study indicated that a considerable proportion of non-porphyrin vanadyl compounds were complexed to the large molecular weight asphaltenic component of heavy crude petroleums. It is likely that AAS will be increasingly employed in diagnostic studies of this type, as the importance of speciation information is recognised.

2. The determination of metals in crude and residual fuel oil by ashing and flame AAS

The use of solvent dilution procedures as given above may be subject to error, and relatively poor precision, if sample dissolution is incomplete. Ashing overcomes this problem, but is rather more time-consuming. The method given below has been applied in the determination of Ba, Ca, Fe, Mg, Na, Ni and V, and may be applicable to other elements. Similar alternative methods for the determination of Al [25] and Na [22] may be of interest. Care should be taken to adequately control the temperature to which the sample is exposed to prevent the loss of volatile analytes.

Accurately weigh approximately 20 g of the oil sample (preheated in an oven at 60°C if viscous) into a silica crucible and add 4 ml of a 20% solution of benzenesulphonic acid in butanol. An empty crucible should also be put through the procedure as a blank. Place the crucible under an infra-red heater, or similar apparatus, at approximately 150–200°C and allow slow charring of the sample to occur. Introduce the crucible, slowly, in stages, into a muffle furnace set at about 550°C until all the carbon has been removed. Allow the crucible to cool and add 2 ml of 50% reagent grade hydrochloric acid to dissolve the ash. Quantitatively transfer the contents to a 25 ml volumetric flask and make up to the mark with distilled or deionised water.

Adjust the instrumental parameters (i.e. wavelength, slit width, lamp current, photomultiplier gain) in accordance with the manufacturer's recommended conditions. A fuel-lean nitrous oxide–acetylene flame should be used for the determination of Ba, Ca and Mg, and fuel-rich conditions should be used for measurement of V. In the determination of Fe, Ni and Na, a fuel-lean air–acetylene flame is suitable. Background correction maybe required to compensate for non-specific absorption or scatter effects. These can be checked for by the use of a nearby non-absorbing line.

Aqueous standard stock solutions should be used to prepare calibration standards for the AAS instrument. Four standards and a blank should be prepared to cover the linear concentration range for each analyte. These should be prepared in 4% v/v hydrochloric acid in order to match the acid concentration in the samples. An ionisation suppressant, usually 1000 μg ml^{-1} potassium should be present in both sample and standard solutions where alkali and alkaline earth elements are to be determined. The presence of 1% w/v lanthanum or high levels of aluminium or titanium may improve flame AAS sensitivity for vanadium [24]. In some cases, vanadium is measured by the integrated absorbance of the triplet at 318 nm [23]. More generally, where higher elemental sensitivity is required, electrothermal AAS may be employed. However, in this case nitric acid is preferred in the dissolution of the ash.

B. Fuel and gas oils

The presence of elements such as lead, copper and zinc in fuel and gas oils may accelerate the oxidative deterioration of refined products or otherwise reduce their stability during storage [14]. Vanadium in boiler firing oils is known to produce corrosion and may give rise to toxic emissions. Clearly, therefore, the determination of these elements is of some importance. Ashing procedures are viable for the AAS determination of elements such as Al [25, 26] and Si [26] in fuel oils, but are unsuitable for the measurement of relatively volatile analytes, because of the potential for losses. The use of solvent dilution sample preparation protocols is more appropriate in such cases. The method given below has been used to determine Cu, Na, Ni, Pb, V and Zn in fuel and gas oils, but may be suitable for application to other elements. A similar method has been described elsewhere [27].

1. *The determination of metals in fuel and gas oils by flame AAS*

Accurately weigh approximately 5 g of the oil into a 50 ml volumetric flask and dilute with MIBK or xylene. Shake to dissolve and mix. Use either commercially available oil base standards or organometallic standards prepared as described in Section III. Dilute these with the same solvent as used for the samples to prepare the calibration standards. At least three standards and a blank should be prepared freshly as required. In the determination of sodium, ionisation suppressant must be added to samples and standards in the manner described in Section IV.A.

Adjust the instrumental parameters (i.e. wavelength, slit width, lamp current, photomultiplier gain) in accordance with the manufacturer's recommended conditions. A fuel-rich nitrous oxide–acetylene flame should be used for the determination of V and Mg, and may be useful for other elements if interferences are suspected. The air–acetylene flame is suitable for the determination of Na, Ni, Pb and Zn. Consult the manufacturer's handbook if other elements are to be determined. Establish suitable flame conditions while aspirating blank solvent. In some cases the use of auxiliary air in the air–acetylene flame may prove advantageous. Background correction may be required to compensate for non-specific absorption or scatter effects. These can be checked for by the use of a nearby non-absorbing line. Where higher sensitivity is required, the electrothermal AAS methodology for analysing organic solutions given below will be appropriate.

2. *The determination of low levels of nickel and vanadium in fuel using electrothermal AAS*

This method is suitable for the determination of low levels of Ni and V (down to approximately 0.01 $\mu g\ g^{-1}$) in fuel oils.

Accurately weigh the fuel oil sample in a volumetric flask and dilute to the mark with xylene. The weight taken will depend on the concentrations of Ni and V in the fuel oil. A minimum sample/solvent dilution ratio of 1:2 is necessary. Shake the flask well to homogenise the solution. Standards for the analysis may be prepared from organometallic standards, analysed samples, or the NBS (now NIST) GM-5 Heavy Oil Standard. The most satisfactory results are likely to be obtained from the latter options. Most modern electrothermal atomizers give relatively similar performance with respect to analyte sensitivity. However, tube materials vary greatly in quality between batches, and this has significant implications for the determination of relatively involatile elements such as V, which form stable carbides. Tubes coated with a layer of pyrolytic graphite are available from most manufacturers and are preferred to uncoated cuvettes. There may be some analytical utility in the use of totally pyrolytic graphite for the determination of involatile elements [28]. The size of injection aliquot selected will determine sensitivity levels, and hence the range of analyte standards required. Refer to the manufacturer's information on sensitivity and linear range and prepare standard solutions accordingly. Always prepare (freshly) a minimum of 3 standards and a blank solution for calibration.

Adjust the instrument parameters (i.e. wavelength, slit width, lamp current and photomultiplier gain) according to the manufacturer's recommended conditions for Ni and V, respectively. In almost all cases it will be necessary to employ background correction for the determination. It has been found effective to use a relatively slow ramp rate in both drying and ashing phases to allow a controlled removal of volatiles prior to atomization. The drying time can be minimised, however, if the sample volume is restricted to 5–10 μl, and if a hot injection facility is available. Ashing temperatures in the range 800–1000°C are suitable for the determination of Ni and V, but this setting may have to be lowered for the measurement of more volatile elements, or alternatively a matrix modifier such as Pd may be used to stabilise the analyte [29]. A compromise atomization temperature of 2700°C is suitable for the determination of Ni and V. Higher temperature settings may result in improved sensitivity for V but at the cost of shortened tube lifetime. The temperature programme settings may have to be substantially modified for the determination of other less volatile elements, or if platform atomization is used. The choice of measurement mode, i.e. peak height or peak area, will depend on the instrument facilities available. It has been reported that the use of peak area measurement in conjunction with a rapid heating (up to 2000 K s^{-1}) furnace and platform

References pp. 339–340

atomization provides an interference-free measurement [5]. Nevertheless, many methods have been satisfactorily developed based on the peak height mode of operation. All measurements should be made in duplicate and an average value taken. Further replicates may be required where large signal variations are encountered. Standards must be run frequently to monitor tube sensitivity changes.

C. Unused lubricating oils

Lubricants contain additives which are used for the purpose of controlling the physical and chemical properties of the product. The concentration of certain additives in lubricating oils may conveniently be determined by trace element analysis. Problems associated with metallic particulate matter are not normally encountered when analysing unused lubricating oils. The most frequently measured elements are Ba, Ca, Mg and Zn. A number of standard methods is available for the determination of these metals based on ashing [30–32], and dilution with an organic solvent [33, 34]. The application of mixed solvent systems has also been shown to be a viable approach [17, 36, 37]. The concentrations of these analytes are usually sufficiently high to allow analysis by flame AAS. However, the detection of compounds containing P is becoming more common, and electrothermal AAS is preferred in such cases for sensitivity reasons [38]. It may also be necessary to determine elemental levels in lubricating oil additive concentrates [24]. Generally, for unused lubricating oils, methods involving dilution with white spirit followed by determination by AAS using a nitrous oxide flame provide good accuracy. Such a method is described below.

1. *The determination of barium, calcium, magnesium and zinc in unused lubricating oils by dilution and flame AAS*

The method has been found suitable for the determination of the above metals in unused lubricating oils, but may also be applicable to the determination of other elements. Where known additives are to be determined, it may be appropriate to check their solubilities in the solvent used as diluent.

Accurately weigh the lubricating oil in a volumetric flask and dilute to the mark using high-quality white spirit. Anionisation suppressant such as K (e.g. as salt of naphthasulphonic acid) should be present in the final solution at a concentration of 1000 μg ml^{-1} if Ca, Mg or Ba are to be determined. Organometallic standards (see Section III) diluted in white spirit are used to calibrate the AAS instrument. Ionisation suppressant should also be added to the standards as above, if appropriate. Prepare a minimum of three calibration standards, plus the blank, at concentration levels covering the full linear range of the instrument.

Set the instrumental parameters for the determination of each element according to the manufacturer's recommended conditions. It is preferable to use the nitrous oxide–acetylene flame for the determination of all four elements, to reduce potential interference effects which may be observed when using the air–acetylene flame [17, 36]. Establish flame conditions while aspirating a white spirit solvent blank. It may be necessary to use background correction, particularly for the determination of Mg and Zn. The level of non-specific absorption or scatter of source radiation can be determined by using a continuum lamp or a nearby non-absorbing line.

The determination of Ba at 553.6 nm in the presence of high concentrations of Ca is known to be subject to interference from spectral band emission [39]. This results in a significant degradation in signal-to-noise ratio as a consequence of increased radiation flux at the detector, leading to erratic absorbance measurements. Increasing the spectrometer slit width will only increase the problem. Instead, the detector gain should be reduced to a minimum, and the Ba lamp current increased to its highest recommended setting. This procedure will discriminate against the spectral interference and minimise its effect. There may be some advantage in selecting an alternative solvent such as xylene, which gives improved sensitivity for Ba [40], although there is some evidence that poor precision may be obtained in this case [17]. Isobutyric acid has been claimed to provide improved stability when used instead of xylene as a diluent [41]. Solvents such as toluene and MIBK have been found to give relatively poor sensitivity for Ba additives. The general purpose solvent (10% isopropanol/90% white spirit) may also be used in this analysis.

D. Used lubricating oils

The analysis of used lubricating oils is accepted as an effective and practical means of monitoring engine wear. Examination of the concentration of various metals in used lubricating oils often gives an early indication of component failure. It is not possible to give an exhaustive list of the elements of importance in this area, as this very much depends on the engine construction materials used. For example in turbine engines, wear metals such as Al, Ag, Cr, Cu, Fe, Mg, Mo, Ni, Pb, Si, Sn and Ti have been found [42]. Thus analysis can be used as a relatively sophisticated diagnostic aid in determining preventative maintenance schedules for aeroplanes, railways, truck fleets, and ship engines.

Since the lubricating oil is used to reduce frictional wear between metal parts, it is possible that particulate matter will be present, in suspension, in the samples. This can give rise to significant atomization interferences and wide variations in instrumental response. This is particularly true where simple dilution is used for sample preparation [43–45]. However, it is not the absolute concentration of an element which is important, but rather the

References pp. 339–340

change in concentration with respect to time. Thus a modification of the dilution procedure (e.g. using MIBK as solvent) given for unused lubricating oils may still be valid in such cases. There is a number of such methods developed for the determination of wear metals in lubricating oils in the literature [35, 41, 46–50]. These may be suitable where the particle size is known to be very small, and as a consequence the flame chemistry is similar to that of a true solution. Extraction procedures specific for certain metals may be of some interest [51, 52].

A study has been made of the efficiency of flame AAS sample introduction for the transport of suspended wear metal particles [43]. It was reported that different observation heights were obtained for oil standards compared to nebulized Fe particulates. Significant losses were observed in transport of particles larger than 7 μm diameter. In such circumstances, the analytical result will be extremely dependent on the nature and size of the metal particulates in the lubricating oil.

It has been reported that the addition of a hydrofluoric acid/aqua regia mixture prior to dilution with MIBK/isopropyl alcohol removes the influence of particle size on the analytical result [44]. The elements studied included Al, Cu, Fe, Mg, Mo, Ni, Sn and Ti, and recoveries were found to be in the range 97 to 103% for these wear metals in aircraft lubricating oils. An allied approach, using a hydrofluoric/nitric acid pretreatment was shown to provide quantitative analysis for Mo in lubricating oils which had not been possible using a direct dilution procedure because of particulate interference [45]. It has been shown that this pretreatment can also be used as part of an emulsion-based method for the determination of iron in used lubricating oils [53].

Electrothermal AAS offers significant potential as an alternative atomizer to the flame for the analysis of lubricating oils containing particulate matter. Jet engine oil, characterised as part of the US Air Force Spectrographic Oil Analysis Program, was analysed directly by graphite filament AAS, and good agreement was obtained using aqueous and organometallic standards [54]. A study of the effect of particle size on the electrothermal AAS determination of Al, Cu, Fe, and Mg was carried out using a more conventional tubular graphite furnace atomizer [42]. The sample was diluted 1 + 4 with kerosene prior to analysis, using diluted Conostan oil standards. Iron particle sizes in the range 1–30 μm were studied, and it was found that the furnace method was least susceptible to interference effects from this source. This method was preferred to flame AAS procedures as it provided better precision for Fe and Cu.

E. Gasoline

Organometallic additives are used in gasoline as anti-knock agents. Perhaps the best known of these are the alkyl Pb compounds tetramethyl and tetraethyl lead. In addition to the requirement for analytical methods for control of formulations, there has been increasing concern regarding potential adverse effects on health arising from environmental pollution. Consequently many AAS methods have been proposed for the determination of the lead content of gasolines [56].

It has been found that the atomic absorption response for Pb is affected by the associated alkyl species present [57]. Thus unless it is known that only one Pb alkyl is present in a gasoline sample, it is not possible to employ a simple dilution procedure. A general method which has been employed to overcome this difficulty involves the addition of iodine to the sample to form di-alkyl lead iodides which gave more uniform response [58]. Small differences in response are still obtained using this approach, and it has been modified in standard methods by including stabilisation with a liquid anion exchanger [58, 59]. A comparison of various standard methods for this analysis has been published [56]. More recently, new methods based on hydride [60] and vapour generation [61], extraction [62], and emulsion calibration [63, 64] have been reported. Alternatively the technique of simple dilution followed by furnace Zeeman-effect AAS may be appropriate [65], particularly for low concentrations (e.g. in the case of lead-free gasoline). The identification and quantification of tetra-alkyl lead compounds is increasingly of interest, and gas chromatography-AAS [66] and liquid chromatography-AAS [67] approaches have been used effectively in this field.

1. *The determination of lead in leaded gasoline*

This method is suitable for the determination of Pb in gasoline in the range $1-1000$ μg ml^{-1}. Pipette a suitable volume of gasoline (depending on sensitivity required) into a 100 ml volumetric flask in duplicate. Add approximately 50 ml of MIBK and 0.2 ml of a 3% w/v solution of iodine in toluene and swirl the mixture, allowing it to stand for two minutes. Add 5 ml of a 1% v/v solution of Aliquat 336 (tricapryl methyl ammonium chloride, see ref. 59), swirl to react and dilute to the mark with MIBK.

A stock Pb standard must also be prepared. Place reagent grade Pb chloride in an oven set at 100°C for two hours. Accurately weigh 0.3355 g of this salt and dissolve it in 200 ml of a 10% v/v Aliquat 336 in a 250 ml volumetric flask. Dilute to the mark with 10% v/v Aliquat 336 solution. This standard contains 1000 μg ml^{-1} Pb. Store in a darkened bottle, tightly stoppered. Prepare from this stock solution appropriate Pb standards by dilution with MIBK, adding iodine and Aliquat 336 as with the sample above. Prepare a blank in the same manner from Pb-free iso-octane. These standards should be prepared freshly as required and should not be stored.

Set the instrumental parameters according to the manufacturer's recommended conditions for Pb, using the 283.3 nm line. Use an air/acetylene flame under fuel-lean conditions. This is most easily set up by adjusting the fuel flow while aspirating a solvent blank.

2. *The determination of lead in lead-free gasoline using electrothermal AAS*

In unleaded gasoline, Pb is present in a variety of organometallic forms, which may give rise to different AAS responses. The addition of iodine may be used to overcome this problem. Electrothermal methods generally allow the determination of Pb in unleaded gasoline at sub-ppm levels.

Dilute the Pb-free gasoline with MIBK to bring the concentration of Pb within the linear range of the instrument. Add iodine from the stock solution to give a final solution concentration of 1000 μg ml^{-1} as iodine. Use Pb in oil standards, diluted appropriately with MIBK, and matched with the sample for iodine concentration. Prepare at least four calibration standards to cover the linear range, including a blank.

Adjust the instrument parameters as recommended by the manufacturer for the determination of Pb. Background correction facilities [65] will be necessary. The Pb 283.3 nm line is preferred to that at 217.0 nm because of lower background and hence improved signal-to-noise ratio. Sample volumes between 5 and 100 μl may be used depending on the instrument employed. The drying time will be volume-dependent, at a temperature setting of 100°C. An ashing temperature of 500°C may be employed, but care must be taken to ensure that Pb is not prematurely volatilised. Atomization under gas stop conditions at a temperature of 1800–2200°C is appropriate. Peak area or peak height measurement may be employed. Solutions should be analysed in duplicate at minimum.

3. *The determination of manganese in gasoline*

The use of methylcyclopentadienyl manganese tricarbonyl (MMT) as an anti-knock agent for gasoline has become increasingly widespread in recent years. This is partly due to the decline in use of Pb alkyl compounds. A standard method is available based on reaction of the sample with bromine, followed by dilution with MIBK and flame AAS determination [68]. A vapour generation procedure has also been reported [61].

Dilute the sample with MIBK so as to fall within the chosen calibration range. Add 1 ml of a 50% bromine/carbon tetrachloride mixture to the sample prior to making up to volume and shake well. Prepare Mn standards from oil-based stock standards by dilution with MIBK. Add 1 ml of bromine solution as for the sample. The calibration range should be approximately 0–4 μg ml^{-1}. Prepare three standards and a blank to cover this range.

Adjust the instrumental parameters in accordance with the manufacturer's recommendations for the determination of Mn. It has been found that a lean air/acetylene flame is most suitable for this analysis. Adjust the acetylene fuel flow to achieve this condition while aspirating an MIBK blank solution.

This above method is suitable for the determination of Mn in gasoline in the range 0.5–40 $\mu g\ ml^{-1}$.

REFERENCES

1   F. Buckley, M.H. Ramsey, J.M. Rooke, H. Hughes and P. Norman, J. Anal. At. Spectrom., 3 (1988) 203R.
2   R.J. Watling, Anal. Chim. Acta, 94 (1977) 181.
3   Annual Reports on Analytical Atomic Spectroscopy, Royal Society of Chemistry, London, 12 (1982) 31.
4   J.F. Tyson, Analyst, 110 (1985) 419.
5   W. Slavin and D.C. Manning, Spectrochim. Acta, 37B (1982) 955.
6   Manual Sampling of Petroleum and Petroleum Products, Annual Book of ASTM Standards, D 4057-81.
7   Automatic Sampling of Petroleum and Petroleum Products, Annual Book of ASTM Standards, D 4177-82.
8   R.J. Ibrahim and S. Sabbah, At. Absorpt. Newsl., 14 (1975) 131.
9   Ash from Petroleum Products, Institute of Petroleum, Wiley, London, 1988, IP 4/81.
10 Ash from Petroleum Products Containing Mineral Matter, Institute of Petroleum, Wiley, London, 1988, IP 223/68.
11 Sulphated Ash from Lubricating Oils and Additives, Institute of Petroleum, Wiley, London, 1988, IP 163/88.
12 R. Blust, A. van der Linden and W. Decleir, At. Spectrosc., 6 (1985) 163.
13 L.A. Fernando, W.D. Heavner and C.C. Gabrielli, Anal. Chem., 58 (1986) 511.
14 O. Osibanjo, S.E. Kakulu and S.O. Ajayi, Analyst, 109 (1984) 127.
15 M. De La Guardia and M.J. Lizondo, At. Spectrosc., 4 (1983) 208
16 M. De La Guardia and M.J. Sanchez, At. Spectrosc., 3 (1982) 37.
17 S.T. Holding and J.J. Rowston, Analyst, 100 (1975) 465.
18 J.J. Labreque, J. Galobardes and M. E. Cohen, Appl. Spectrosc., 31 (1977) 207.
19 M.M. Barbooti and F. Jasim, Talanta, 29 (1982) 107.
20 R.H. Fish and J.J. Komlenic, Anal. Chem., 56 (1984) 510.
21 Sodium, Nickel and Vanadium in Fuel Oils and Crude Oils by AAS, Institute of Petroleum, Wiley, London, 1988, IP 288/74.
22 Sodium in Residual Fuel Oil, Annual Book of ASTM Standards, D1318-88 (1991).
23 W.J. Price, Spectrochemical Analysis by Atomic Absorption, Heyden, London, 1979, p. 353.
24 Barium in Lubricating Oil Additive Concentrates, Institute of Petroleum, Wiley, London, 1988, IP 271/70.
25 Aluminium in Fuel Oils by Ashing and AAS, Institute of Petroleum, Wiley, London, 1988, IP 363/83.
26 Aluminium and Silicon in Fuel Oils by Ashing, Fusion and AAS, Institute of Petroleum, Wiley, London, 1988, IP 377/88.
27 L. Capacho-Delgado and D.C. Manning, At. Absorpt. Newsl., 5 (1966) 1.
28 D. Littlejohn, I. Duncan, J. Marshall and J.M. Ottaway, Anal. Chim. Acta, 157 (1984) 291.

340

29   L.M. Voth-Beach and D.E. Schrader, J. Anal. At. Spectrom., 2 (1987) 45.
30   Zinc in Lubricating Oil, Institute of Petroleum, Wiley, London, 1988, IP 117/82.
31   Calcium in Lubricating Oil, Institute of Petroleum, Wiley, London, 1988, IP 111/82.
32   Barium in Lubricating Oil, Institute of Petroleum, Wiley, London, 1988, IP 110/82.
33   Barium, Calcium, Magnesium and Zinc in Lubricating Oils by AAS, Institute of Petroleum, Wiley, London, 1988, IP 308/85.
34   Analysis of Barium, Calcium, Magnesium, and Zinc in Unused Lubricating Oils by AAS, Annual Book of ASTM Standards, D4628-86 (1991).
35   J.E. Schallis and H.L Kahn, At. Absorpt. Newsl., 7 (1968) 84.
36   S.T. Holding and P.D.H. Matthews, Analyst, 97 (1972) 189.
37   Z. Wittman, Analyst, 104 (1979) 156.
38   P. Tittarelli and A. Mascherpa, Anal. Chem., 53 (1981) 1466.
39   L. Capacho-Delgado and S. Sprague, At. Absorpt. Newsl., 4 (1965) 363.
40   G.E. Petersen and H.L. Kahn, At. Absorpt. Newsl., 9 (1970) 71.
41   P.K. Hon, O.W. Lau and C.S. Monk, Analyst, 105 (1980) 919.
42   C.S. Saba, W.E. Rhine and K.J. Eisentraut, Appl. Spectrosc., 39 (1985) 689.
43   C.S. Saba, W.E. Rhine and K.J. Eisentraut, Anal. Chem., 53 (1981) 1099.
44   J.R. Brown, C.S. Saba, W.E. Rhine and K.J. Eisentraut, Anal. Chem., 52 (1980) 2365.
45   C.S. Saba and K.J. Eisentraut, Anal. Chem., 51 (1979) 1927.
46   S. Sprague and W. Slavin, At. Absorpt. Newsl., 4 (1965) 367.
47   S. Slavin and W. Slavin, At. Absorpt. Newsl., 5 (1966) 106.
48   W.B. Barnett, H.L Kahn and G.E. Peterson, At. Absorpt. Newsl., 10 (1971) 106.
49   J.M. Palmer and M.W. Rush, Analyst, 107 (1982) 994.
50   K.W. Jackson, K.M. Aldous and D.G. Mitchell, Appl. Spectrosc., 28 (1974) 569.
51   M. Ejaz, S. Zuha, W. Dil, A. Akhtar and S.A. Chaudhri, Talanta, 28 (1981) 441.
52   S. Bajo and A. Wyttenbach, Anal. Chem., 51 (1979) 376.
53   A. Salvador, M. De La Guardia and V. Berenguer, Talanta, 1983, 30, 986.
54   F.S. Chuang and J.D. Winefordner, Appl. Spectrosc., 28 (1974) 215.
55   R.L. Boeckx, Anal. Chem., 58 (1986) 275A.
56   S.T. Holding and J.M. Palmer, Analyst, 109 (1984) 507.
57   M. Kashiki, S. Yamazoe and S Oshima, Anal. Chim. Acta, 53 (1971) 95.
58   Lead in Gasoline by AAS, Annual Book of ASTM Standards, D3237-90 (1991).
59   Total Lead in Gasoline by AAS, Institute of Petroleum, Wiley, London, 1988, IP 362/83.
60   J. Aznarez, J.C. Vidal and R. Carnicer, J. Anal. At. Spectrom., 2 (1987) 55.
61   M. de la Guardia, A.R. Mauri and C. Mongay, J. Anal. At. Spectrom., 3 (1988) 1035.
62   S. Banerjee, Talanta, 33 (1986) 358.
63   E. Cardarelli, M. Cifani, M. Mecozzi and G. Sechi, Talanta, 33 (1986) 279.
64   L. Polo-Diez, J. Hernandez-Mendez and F. Pedraz-Penalva, Analyst, 105 (1980) 37.
65   D.R. Scott, L.E. Holboke and T. Hadeishi, Anal. Chem., 55 (1983) 2006.
66   D.T. Coker, Anal. Chem., 47 (1975) 386.
67   J.D. Messman and T.C. Rains, Anal. Chem., 53 (1981) 1632.
68   Manganese in Gasoline by AAS, Annual Book of ASTM Standards, D3831-90 (1991).

Chapter 4i

Methods for the Analysis of Glasses and Ceramics by Atomic Absorption Spectrometry

W.M. WISE, R.A. BURDO and D.E. GOFORTH

Research and Development Laboratories, Corning Glass Works, Corning, New York 14831 (U.S.A.)

I. INTRODUCTION

The implementation of modern spectrometers with stable excitation sources and electronics has engendered atomic emission (AES) and atomic absorption spectrometry (AAS) methods for determining trace to major amounts of selected elements in glasses, ceramics and other similar materials. Since analyte detection is accomplished using elementally characteristic and unique spectral lines, separations are usually unnecessary, leading to quick results. In most cases AES and AAS signals are sufficiently stable, for even major components, to produce final results which are as accurate and precise as those acquired by classical gravimetric and titrimetric procedures. It is fortunate that many simple AES and AAS methods are available to replace those classical procedures which are so arduous that past results for some elements were often obtained by difference.

Recent AES and AAS reviews have appeared on the analysis of glasses and related materials [1–3]. It is the purpose of this chapter to present to the laboratory technician with experience in AES and AAS some selected methods for determining the elements common to such substances. Since AES and AAS methods are routinely based upon relative comparisons with standards, it is desirable to have reasonable estimates of the elemental concentrations in a sample before beginning an analysis. Sometimes this information is available from batch compositions. If the material is an unknown, then qualitative and semiquantitative estimates of the composition can be obtained by optical emission spectroscopy, for which a procedure has been previously published [2].

II. APPARATUS

A. Spectrometers and sources

There are many quality AAS and AES analysis systems available from such manufacturers as Perkin-Elmer Corp., Varian Associates, Jarrell-Ash Corp., Applied Research Laboratories (ARL), Instruments S.A., Baird Corp., and others. Quality work in atomic absorption can be accomplished with both single- and double-beam spectrometers. It is recommended that a combination flame emission and atomic absorption unit be considered as alkalis can be determined well by either flame AES or AAS. Capabilities for both air–acetylene and nitrous oxide–acetylene flames are important.

Plasma analysis systems include both DCP (direct current plasma) and ICP (inductively coupled plasma) sources. DCP systems were originally manufactured by Spectrametrics, Inc. and included a high-resolution echelle spectrometer. These systems are now manufactured by ARL. DCP sources are exceptionally tolerant of hydrofluoric acid and high salt matrices, but tend to otherwise have higher matrix effects compared to an ICP source. The ICP source tends to have better long-term precision. Better overall precision is expected of ICP spectrometers which integrate the analyte signal by sitting or standing on the spectral line rather than stepping or scanning across the line profile.

B. Platinum utensils

In general, standard platinum ware as supplied by Johnson Mathey or Englehard Industries is acceptable. The most useful size for fusion crucibles and evaporation dishes is 30 ml. Fusion crucibles can be expected to last for over 500 fusions with proper care and provided that fusions are completed in a muffle furnace at 1000°C as opposed to the use of a blast burner.

III. CHEMICALS

Only established analytical reagent grade and high-purity chemicals should be employed for the preparation of standard and reagent solutions and for the treatment of samples. All water employed should be distilled and preferably deionized. Unless otherwise specified, all dilutions to volume are done with water. Reagents for preparing stock solutions are as follows:

Anhydrous Li_2CO_3, $LiBO_2$ (−200 mesh), $BaCO_3$, $CaCO_3$, Na_2CO_3 (−200 mesh), K_2CO_3, $SrCO_3$, $Na_2B_4O_7$ (−200 mesh) and SiO_2 (99.9+%, −100 mesh). Metals (99.99%) of Fe, Zn, Cd, Al and Mg. $Pb(NO_3)_2$ (99.5+%), H_3BO_3 (99.99%), $(NH_4)_6Mo_7O_{24} \cdot 4H_2O$, NaCl (<0.001%K), KCl (<0.002% Na), $LaCl_3 \cdot 6H_2O$ (<0.001% Mg + Ca + Sr + Ba), EDTA (99.6% free acid), HF (29 M), $HClO_4$ (12 M), HCl (12 M), HNO_3 (16 M), H_3PO_4 (15 M), NH_4OH (15 M, filtered).

IV. REAGENT SOLUTIONS

(1) *KCl flame buffer* (2500 μg KCl ml^{-1}). Dissolve 2.5 g of KCl in water and dilute to 1000 ml.

(2) *NaCl flame buffer* (25 000 μg NaCl ml^{-1}). Dissolve 25 g of NaCl in water and dilute to 1000 ml.

(3) *LaCl$_3$ flame buffer* (25 000 μg LaCl$_3$ ml^{-1} + 0.3 M HClO$_4$). Dissolve 36.0 g of LaCl$_3\cdot$6H$_2$O in water, add 25 ml of 12 M HClO$_4$ and dilute to 1000 ml.

(4) *Methyl orange indicator* (0.1%). Dissolve 0.1 g of methyl orange in 100 ml of ethyl alcohol.

(5) *LaCl$_3$ + EDTA flame buffer* (25 000 μg LaCl$_3$ ml^{-1} + 0.25 M EDTA). Add 73.1 g of EDTA and 10 drops of methyl orange indicator to 800 ml of water containing 67 ml of 12 M HClO$_4$. Stir and add 15 M NH$_4$OH until all of the EDTA is dissolved. Introduce 36.0 g of LaCl$_3\cdot$6H$_2$O and continue stirring and adding NH$_4$OH until the solution is clear and yellow. Cool and dilute to 1000 ml.

(6) *Ca(ClO$_4$)$_2$ + HClO$_4$ flame buffer* (25 000 μg Ca ml^{-1} + 1.8 M HClO$_4$). Introduce 31.25 g of calcium carbonate into a 500-ml volumetric flask which is half-filled with water. With slow stirring add 125 ml of 12 M HClO$_4$. After the effervescence ceases, heat the solution to expel excess CO$_2$, cool to room temperature and dilute to volume.

(7) *NaClO$_4$ flame buffer* (10 000 μg Na ml^{-1}). Dissolve 11.54 g of Na$_2$CO$_3$ with 100 ml of water and 20 ml of 12 M HClO$_4$. Heat to boiling to expel the CO$_2$, cool to room temperature and dilute to 500 ml.

(8) *Molybdate complexing solution* (39 000 μg (NH$_4$)$_6$Mo$_7$O$_{24}\cdot$4H$_2$O ml^{-1}). Dissolve 78 g of the molybdate hydrate salt in 700 ml of water and dilute to 2000 ml after filtering through a Whatman No. 41 filter paper.

(9) *Molybdate diluting solution*. Dissolve 0.5 g of Na$_2$CO$_3$ and 0.5 g of Na$_2$B$_4$O$_7$ in 100 ml of 0.39 M HNO$_3$. Add 100 ml of molybdate complexing solution and dilute to 500 ml with 0.126 M HNO$_3$. This solution is stable only for a few days after which a scale of hydrated molybdenum oxide develops on the walls of the container. The scale can be removed with ammonium hydroxide solution. Fresh solutions should always be employed.

(10) *Silicon plasma buffer solution* (10 000 μg Si ml^{-1} + 5.8 M HF). Place 21.4 g of SiO$_2$ powder into a 1000-ml plastic jar. Add 300 ml of cold water and then slowly add 250 ml of chilled 29 M HF. Allow the solution to slowly approach room temperature in order to dissipate the heat of reaction safely and without the loss of silicon as its volatile fluoride. Let the jar stand at room temperature until dissolution is complete. Dilute to 1000 ml in a plastic volumetric flask and store in plasticware.

(11) *Silicon plasma diluting solution* (300 μg Si ml^{-1} + 1.45 M HF). To a 1000-ml plastic volumetric flask, add 30 ml of silicon buffer solution and 42.5 ml of 29 M HF. Dilute to volume with water. Store in plasticware.

References p. 357

V. STANDARDS

1. *Lithium, sodium, potassium, strontium and barium stock solutions*

To obtain 1000 μg ml^{-1}, place the following quantities of each carbonate salt into a 1000-ml volumetric flask, and mix with 100 ml of water (5.324 g Li$_2$CO$_3$, 2.3042 g Na$_2$CO$_3$, 1.7674 g K$_2$CO$_3$, 2.4973 g CaCO$_3$, 1.685 g SrCO$_3$, 1.4371 g BaCO$_3$). Slowly add 15 ml of 12 M HClO$_4$. After the bubbling has ceased, heat to boiling to expel the excess CO$_2$, cool to room temperature and dilute to volume.

Prepare 50 μg ml^{-1} stock solutions by diluting 50 ml of each of the above solutions to 1000 ml.

Lithium, sodium and potassium AES flame calibration standards

Pipet portions of 5, 10, 20, 30, 40, 60 and 80 ml of the 50 μg ml^{-1} stock solutions into 1000-ml volumetric flasks and add 2.5 ml of 12 M HClO$_4$. Dilute the lithium calibration standards to volume. For the sodium calibration standards, add 40 ml of the KCl flame buffer. For the potassium calibration standards, add 40 ml of the NaCl flame buffer. Dilute the sodium and potassium standards to volume. The lithium, sodium, or potassium standards now have concentrations of 0.25, 0.50, 1.00, 1.50, 2.00, 3.00, and 4.00 μg ml^{-1}.

Calcium AES and AAS, strontium AES, and barium AES flame calibration standards

Pipet portions of 5, 10, 15, 20, 25, 30, 40 and 60 ml of the 50 μg ml^{-1} stock solutions into 1000-ml volumetric flasks. Add 500 ml of water, 10 drops of methyl orange solution and 40 ml of the LaCl$_3$ + EDTA buffer solution. Add (dropwise) sufficient 15 M NH$_4$OH to each flask such that the color of the solution becomes yellow. For the barium standards, add 4 ml of the 2500 μg KCl ml^{-1} solution. Dilute all solutions to volume to yield calcium strontium, or barium concentrations of 0.25, 0.50, 1.00, 1.50, 2.00, 3.00, and 4.00 μg ml^{-1}.

Barium AAS flame calibration standards

Pipet portions of 5, 10, 15, 20, and 25 ml of the 1000 μg ml^{-1} barium stock solution into 500-ml volumetric flasks. Add 250 ml of water, 5 drops of methyl orange solution, 20 ml of LaCl$_3$ + EDTA flame buffer and sufficient 15 M NH$_4$OH (dropwise) to turn the solution color to yellow. Dilute to volume to yield barium concentrations of 10, 20, 30, 40 and 50 μg ml^{-1}.

2. *Magnesium stock solution*

Place 1.000 g of clean magnesium metal into a 1000-ml volumetric flask. Add 100 ml of water and 20 ml of 12 M HClO$_4$. After all of the metal is dissolved (heating may be necessary), cool the solution to room temperature and dilute to volume. Prepare a stock solution containing 20 μg Mg ml^{-1} by diluting 20 ml of the 1000 μg ml^{-1} solution in a one-liter volumetric flask.

Magnesium AAS flame calibration standards

Pipet aliquots of 5, 10, 20, 30, 40, 50 and 60 ml of the stock solution into 1000-ml volumetric flasks. Add 40 ml of LaCl$_3$ flame buffer to each flask and dilute to volume to yield magnesium concentrations of 0.10, 0.20, 0.40, 0.60, 0.80, 1.00 and 1.20 μg ml^{-1}.

3. Aluminum stock solution

Prepare a 5000 μg Al ml^{-1} solution by dissolving aluminum metal in 100 ml of water containing 50 ml of 12 M HCl and 10 ml of 16 M HNO$_3$. Cool the solution to room temperature and dilute to volume in a 1000-ml volumetric flask. Make a 500 μg Al ml^{-1} stock solution by diluting a 100-ml aliquot to volume in a 1000-ml volumetric flask.

Aluminum AAS flame calibration standards

Pipet aliquots of 5, 10, 15, 20, 25, 30, 35 and 40 ml of the 500 μg ml^{-1} stock solution into 1000-ml volumetric flasks. Add 40 ml of the Ca(ClO$_4$)$_2$ + HClO$_4$ flame buffer to each flask and dilute to volume to yield aluminum concentrations of 2.5, 5.0, 7.5, 10.0, 12.5, 15.0, 17.5 and 20.0 μg ml^{-1}.

4. Iron stock solution

Dissolve 1.000 g of clean iron wire in 100 ml of 6 M HClO$_4$ and dilute to volume in a 1000-ml volumetric flask. Prepare a 50 μg ml^{-1} stock solution by diluting a 25-ml aliquot to 500 ml.

Iron AAS flame calibration standards

Pipet aliquots of 5, 10, 20, 30, 40 and 50 ml of the 50 μg ml^{-1} stock solution into 250-ml volumetric flasks. Add 10 ml of 1.25 M H$_3$PO$_4$ and dilute to volume to yield iron concentrations of 1.0, 2.0, 4.0, 6.0, 8.0 and 10.0 μg ml^{-1}. These standards should be prepared fresh on the day that they are needed.

5. Zinc stock solution

Prepare a 1000 μg Zn ml^{-1} solution by dissolving 1.000 g of clean zinc metal in 100 ml of 3 M HCl and diluting to volume in a 1000-ml volumetric flask. Dilute a 20-ml aliquot to 1000 ml to obtain a 20 μg Zn ml^{-1} stock solution.

References p. 357

Zinc AAS flame calibration standards
Pipet aliquots of 5, 10, 20, 30, 40 and 50 ml of the 20 μg ml^{-1} stock solution into 1000-ml volumetric flasks. Add 2.5 ml of 12 M HClO$_4$ and dilute to volume to yield zinc concentrations of 0.10, 0.20, 0.40, 0.60, 0.80 and 1.00 μg ml^{-1}.

6. Cadmium stock solution

Dissolve 1.000 g of mossy cadmium metal in 20 ml of 5 M HNO$_3$ and dilute to volume in a 1000-ml volumetric flask. Dilute a 10-ml aliquot to 500 ml to prepare a 40 μg Cd ml^{-1} stock solution.

Cadmium AAS flame calibration standards
Pipet aliquots of 5, 10, 20, 30, 40 and 50 ml of the 40 μg ml^{-1} stock solution into 1000-ml volumetric flasks. Add 2.5 ml of 12 M HClO$_4$ to each flask and dilute to volume to yield cadmium concentrations of 0.20, 0.40, 0.80, 1.20, 1.60 and 2.00 μg ml^{-1}.

7. Lead stock solution

Dissolve 1.5985 g of Pb(NO$_3$)$_2$ in 200 ml of water. Add 0.5 ml of 16 M HNO$_3$ and dilute to 1000 ml in a volumetric flask. Prepare a 100 μg Pb ml^{-1} stock solution by diluting 50 ml to 500 ml in a volumetric flask.

Lead AAS flame calibration standards
Pipet aliquots of 5, 10, 20, 30, 40 and 50 ml of the 100 μg ml^{-1} stock solution into 500-ml volumetric flasks. Add 1.25 ml of 12 M HClO$_4$ and 5 ml of the NaClO$_4$ buffer. Dilute to volume to yield lead concentrations of 1.0, 2.0, 4.0, 6.0, 8.0 and 10.00 μg ml^{-1}.

8. Silicon flame AAS stock solution

Thoroughly mix 0.1070 g of SiO$_2$ powder with 0.5 g of Na$_2$CO$_3$ and 0.5 g of Na$_2$B$_4$O$_7$ in a 30-ml platinum crucible. Cover and place in a muffle furnace at 1000°C for 30 minutes. Insert a small Teflon-coated magnetic stirring bar into the cooled crucible. Immerse the crucible into a 250-ml Teflon beaker containing 100 ml of molybdate complexing solution and 100 ml of 0.39 M HNO$_3$. After the melt dissolves (20–30 min), dilute to 500 ml with 0.126 M HNO$_3$ to yield 100 μg Si ml^{-1}.

Silicon flame calibration standards
Pipet aliquots of 10, 20, 30, 50, 60 and 80 ml of the stock solution into 100-ml plastic volumetric flasks and dilute to volume with the molybdate diluting solution to yield silicon concentrations of 10, 20, 30, 50, 60 and 80

μg ml^{-1}. Standards below 40 μg ml^{-1} are stable for up to ten days until a scale of hydrated molybdenum oxide forms on the container walls. Standards of higher concentration are stable for several months.

9. *Boron plasma stock solutions*

Dissolve 5.7194 g of H_3BO_3 in 500 ml of water and dilute to 1000 ml, yielding 1000 μg B ml^{-1}. Add 100 ml of this solution to a 1000-ml volumetric flask, and dilute to volume with water to yield 100 μg B ml^{-1}.

Boron plasma calibration standards

Place aliquots of 10, 20, 30, 40, 50, 60, 80 and 100 ml of the 100 μg ml^{-1} stock solution into 1000-ml plastic volumetric flasks. Add 30 ml of silicon buffer solution and 42.5 ml of 29 M HF to each flask. Dilute to volume to yield 1.0, 2.0, 3.0, 4.0, 5.0, 6.0, 8.0 and 10.0 μg B ml^{-1}. Higher concentrations of boron standards can be made, but it is preferable to dilute the samples, if required, to achieve a boron concentration of 5 to 10 μg ml^{-1}. Plasma linearity extends well above 100 μg ml^{-1}. Plasma sensitivity for boron extends well below 1 μg ml^{-1}.

VI. SAMPLE PREPARATION

Most commercial glasses and ceramics are prepared under conditions designed to promote homogeneity, e.g. pre-mixing of batch materials and/or stirring during the melting process. Consequently, it is not usually difficult to acquire a representative sample for reduction to a suitable particle size by crushing and grinding operations. It is important to first clean the sample, particularly if it was obtained as a result of some previous cutting or grinding operation which has the potential of contaminating the sample surface. After cleaning of the sample, crushing and grinding can be accomplished using several types of mortars and pestles, depending on the type of contamination that can be tolerated. Steel utensils are adequate where trace iron contamination is of little concern. Any iron chips introduced by grinding can be removed from the sample by passing a magnet over the sample powder. Sintered sapphire mortars and pestles are preferred over corundum as they are harder and introduce much less alumina contamination. Agate or steel utensils are preferred when low levels of alumina are to be analysed. It is good practice to pre-clean mortars and pestles by thoroughly grinding a portion of the sample which is then discarded.

Samples should not be ground any finer than is necessary for subsequent decomposition. Most acid and fusion decompositions require comminution to -100 mesh (150 μm), but some fusions of extremely refractory materials may require -200 mesh powders. Overgrinding not only leads to greater contamination from the grinding vessels but also increases the risk of adsorption of airborne water and/or CO_2 by the sample powder. Ground samples should be stored in a desiccator containing BaO. Samples which are suspected to have adsorbed water or CO_2 should be dried at 120°C, or ignited at 1000°C provided that the sample does not melt or decompose at this higher temperature. Extreme care must be taken with samples that are naturally hygroscopic such as high-phosphate or high-borate materials. These should be stored in a desiccator before grinding and efforts should be made to dissolve larger sample particles or chips by an acid decomposition method.

There are particular types of samples that are often associated with heterogeneous characteristics. These include small batch melts and opal or opaque glasses that contain discrete phases of variable composition and hardness. Because of the small sample sizes (0.1–0.2 g) employed by AES and AAS, homogenization procedures become important. For samples of doubtful homogeneity, the entire portion selected for grinding must be comminuted fine enough to pass the sieve, followed by thorough mixing of the powder.

VII. DETERMINATION OF LITHIUM, SODIUM, POTASSIUM, MAGNESIUM, CALCIUM, STRONTIUM, BARIUM, ALUMINIUM, IRON, ZINC, CADMIUM AND LEAD

A. Procedure for acidic decomposition

This procedure is recommended for materials that are decomposed by attack with HF and $HClO_4$. When using either of these two acids appropriate precautions should be taken. Perchloric acid in particular should not be allowed to come into contact with easily oxidisable organic matter or strongly reducing substances, and should not be allowed to become hot or anhydrous. Failure to observe these precautions can lead to a serious explosion. If a fume cupboard is used for perchloric acid digestion, acid may condense in the upper parts of the cupboard and its ducts, which can also lead to eventual explosions. Perchloric acid and hydrofluoric acids are, however, both very useful digesting acids, when used with extreme care, and analysts should consult the appropriate literature if they intend to use such materials.

Weigh, to the nearest 0.1 mg, a 0.1 to 0.2 g portion of the ground sample (about −100 mesh) directly into a 30-ml or 50-ml platinum dish. Add several ml of water and 4 ml of 6 M $HClO_4$. Gently swirl the mixture to avoid the formation of lumps and add 5 ml of 29 M HF. After all visible reaction has ceased, place the dish on a steam bath and evaporate the solution until just $HClO_4$ remains. Place the dish on a hotplate and heat until fumes of $HClO_4$ are evolved. (At this point, insoluble fluorides may begin to decompose and effervesce causing loss of solution. In this unlikely event about 40 mg of boric acid can be added to effect non-effervescent fluoride decomposition. Gentle swirling of the dish and contents until all bubbling ceases can also help to prevent any loss.) Wash down the sides of the dish with a stream of water. Evaporate the solution again to $HClO_4$ fumes, and continue with hard fuming to incipient dryness. Repeat the wash down of the sides of the dish and fuming two more times, adding small amounts of $HClO_4$ when required to prevent evaporation to dryness. Moisten the residue with 10 drops of $HClO_4$ and 25 ml of water (15 ml for a 30-ml dish). If the solids do not dissolve with gentle heating on a hotplate, cover the dish with a platinum lid and heat to near boiling on a hotplate for up to an hour to effect complete dissolution. It may be necessary to add water periodically to maintain the original volume. (Note that some samples containing high levels of zirconia, titania and alpha alumina may produce residues of these oxides which will not dissolve.) Transfer the solution to a 100-or 200-ml volumetric flask. Add about 20 ml of water and 0.5 ml of 6 M HCl to the dish. Omit the HCl if lead is present. Heat gently for several minutes and quantitatively transfer the liquid to the volumetric flask. Cool the flask to room temperature and dilute to volume. Dilute appropriate aliquots of sample solution and flame buffers (where required) so as to bring the concentrations of the analytes in the optimum ranges of the calibration curves and the concentrations of flame buffers to the same levels as are present in the calibration standards.

B. Procedures for fusions

Fusions are performed for materials that are resistant to attack by acids (refractories, abrasives etc.), for the analysis of silicon which is lost in $HF/HClO_4$ evaporation procedures, and for the analysis of elements such as zirconium and titanium, which may not be rendered totally soluble after acid attack.

Of the many types of fluxes, $LiBO_2$ or a mixture of alkali carbonate with borate prove to be the most useful. An excellent flux for glasses and ceramics is an equal weight mixture of Na_2CO_3 and $Na_2B_4O_7$. $LiBO_2$ is useful when sodium is to be determined. Na_2CO_3 is useful when lithium is to be determined. In general, the $LiBO_2$ flux weight should be about five times the sample weight. Alkali carbonate–borate flux weights should be about ten times the sample weight. The actual ratio of flux to sample may be varied somewhat for the purposes of accelerating acid dissolution of the cooled melt, reducing flux blank levels, or reducing high salt concentrations which can clog burners and nebulizers. The temperature of fusion should be at least 100°C above melting point of the flux. A common fusion temperature is 1000°C when using a muffle furnace. A common muffle fusion time is 20 minutes. It is more difficult to control fusion temperature when blast burners are used as heat sources, but higher temperatures can be reached and total fusion time can be reduced when only a few samples are to be fused.

Standard 30-ml platinum crucibles are adequate for carbonate and borate fusions. Normally about 1.0 g of flux is necessary to evenly coat the crucible bottom. However, as little as 0.5 g of flux can be successfully employed in some cases. Samples for fusion should be at least -100 mesh, but -200 mesh may be necessary for resistant samples or to reduce the fusion time.

Whereas volatile fluorides (B, Si, As) can be lost in $HF/HClO_4$ acid decomposition, certain other elements can be lost during fusion by volatilization or by alloying with the platinum after the oxide has been chemically reduced to metal. For example, lead and cadmium can be volatilized after prolonged fusion times. Iron can be reduced to metal by small residual amounts of CO which result from the minor decomposition of carbonate. Both volatilization and reduction can be decreased or eliminated by adding an oxidizing agent such as $NaNO_3$ to the flux. The amount of oxidizing agent added is usually in the range of 0.05 to 0.10 times the flux weight.

Since silicon may interfere with certain AAS procedures for the determination of some elements such as alkaline earths and aluminum, it is sometimes removed by volatilizing it as a fluoride before fusion. A common procedure is to weigh, to the nearest 0.1 mg, 0.1 g of sample powder into a 30-ml platinum crucible. Then, 5 ml of water and 5 ml of 29 M HF are added and the crucible is heated on a hotplate at low temperature until dryness is attained and HF fumes cease to evolve.

A general fusion procedure begins by weighing the sample into a platinum crucible as already described, volatilizing the silicon if necessary, adding 1.00 g of flux, and mixing the flux with the sample powder (or residue from the HF evaporation) using a platinum stirring rod for the mixing. The crucible is then placed into a muffle furnace at 1000°C for 20 minutes (or longer if necessary), or the crucible is placed over an oxidizing blast burner flame until decomposition is complete. The use of platinum crucible covers during fusion is recommended, but may not be necessary in some cases as determined by empirical analysis. Acid dissolution of the cooled melt is then accomplished by a variety of acid types and procedures depending on the flux employed, the elements to be determined, and the types of contaminations that can be tolerated. The melt can be dissolved by adding acid solution directly into the crucible, or by immersing the crucible into a larger beaker containing the acid solution. When alkali contamination is of concern, the acid solution should be added directly into the crucible (e.g. 25 ml of hot 1.2 M $HClO_4$ or HNO_3 or HCl), followed by magnetic stirring with further heating if necessary. The prior removal of silicon before fusion is advantageous because it may precipitate under these dissolution conditions or at a later time after cooling of the solution for dilution to an appropriate volume. If Na_2CO_3 flux is employed, there is a danger of losing solution through effervescence of CO_2 which forms as carbonate is neutralized by the dissolving acid. If such is the case, dissolution of the melt should begin with cold acid to reduce the reaction rate. The normal dilution volume is 100 ml, with further dilutions as required for a proper match of analyte levels to the optimum calibration ranges.

The alternate procedure of immersing the crucible into a beaker containing the acid solution is employed in this lab for $Na_2CO_3/Na_2B_4O_7$ fusions which are useful for plasma determinations of silicon and many other elements including zirconium, titanium and alkaline earths. A magnetic stirring bar is inserted into the crucible, which is then immersed into a 100-ml Teflon beaker containing 55 to 60 ml of water, 15 ml of 3.2 M HNO_3, and 5 ml of 1.86 M HF. The dissolution occurs over a period of 15 to 30 minutes at room temperature with magnetic stirring. The dissolved melt is then diluted to an appropriate volume (250, 500, or 1000 ml) with the concomitant addition of sufficient 3.2 M HNO_3 to adjust the final HNO_3 concentration to 0.21 M. The use of HCl as the dissolution acid is also appropriate for most sample types.

Regardless of the type of fusion employed, it is good practice to match the concentrations of flux salts and acids in the sample and calibration standard solutions. It is also good practice to prepare a similar matched flux blank solution, especially for alkali determinations. The flux blank should be subjected to the fusion process as are the samples.

C. Measurement of analytes

Manufacturers' handbooks contain valuable information for selecting instrumental operating conditions. However, parameters such as flame stoichiometry and burner position should be optimized for each analyte's signal before each group of standard and sample measurements are acquired. Optimizations and measurements should be completed only after a suitable system and burner warm-up time.

The recommended flame AES and AAS operating conditions for various elements are given in Table 1. For flame work, the measurement sequence begins by acquiring a reading on each calibration standard and sample solution using a signal integration time of 8–10 s. This process is repeated at least two or three more times to acquire an average of three or four readings for each solution.

Calibration curves for flame AES and AAS work should be based on at least five, and preferably six, standards. The curve is graphed either manually or by a curve-fitting computer program residing in an external computer or in the instrument itself. A typical formula for calculating the percent oxide in a sample is as follows:

$$\% \text{ oxide} = \frac{(C)(V)(DF)(GF)(0.1)}{W}$$

where C is the concentration (μg ml^{-1}) of the element as determined from the calibration curve, V is the original dilution volume (ml), DF is the

References p. 357

TABLE 1

RECOMMENDED CONDITIONS FOR FLAME AES OR AAS MEASUREMENTS FOR
Li, Na, K, Mg, Ca, Ba, Al, Fe, Zn, Cd, AND Pb

Element	Mode	Wave-length (nm)	Oxidant + fuel	Final buffer concentration
Li	AES	670.8	air + C_2H_2 [a]	0.03 M $HClO_4$
Na	AES	589.0	air + C_2H_2 [b]	100 μg KCl ml^{-1} + 0.03 M $HClO_4$
K	AES	766.5	air + C_2H_2 [c]	1000 μg NaCl ml^{-1} + 0.03 M $HClO_4$
Mg	AAS	285.2	N_2O + C_2H_2 [c]	1000 μg $LaCl_3$ ml^{-1} + 0.01 M $HClO_4$, pH = 4.5
Ca	AES, AAS	422.7	N_2O + C_2H_2 [c]	1000 μg $LaCl_3$ ml^{-1} + 0.01 M EDTA, pH = 4.5
Sr	AES	460.7	N_2O + C_2H_2 [c]	1000 μg $LaCl_3$ ml^{-1} + 0.01 M EDTA, pH = 4.5
Ba	AES, AAS	553.5	N_2O + C_2H_2 [c]	1000 μg $LaCl_3$ ml^{-1} + 0.01 M EDTA + 100 μg KCl ml^{-1}, pH = 4.5
Al	AAS	309.3	N_2O + C_2H_2 [c]	1000 μg Ca ml^{-1} + 0.07 M $HClO_4$
Fe	AAS	248.3	air + C_2H_2 [b]	0.05 M H_3PO_4
Zn	AAS	213.8	air + C_2H_2 [b]	0.03 M $HClO_4$
Cd	AAS	228.8	air + C_2H_2 [b]	0.03 M $HClO_4$
Pb	AAS	216.9	air + C_2H_2 [b]	100 μg Na ml^{-1} + 0.03 M $HClO_4$
Si	AAS	251.6	N_2O + C_2H_2 [d]	7.8 g ammonium molybdate l^{-1} + 2 g flux l^{-1} + 0.16 M HNO_3

[a] Stoichiometric flame (burner slot = 10 cm).
[b] Stoichiometric to oxidizing flame (burner slot = 10 cm).
[c] Oxidizing flame (burner slot = 6 cm).
[d] Stoichiometric to slightly reducing flame (burner slot = 6 cm).

dilution factor, GF is the gravimetric factor for the oxide, and W is the
sample weight (mg).

Calibration curves and measurement sequences for plasma AES work can
be similar to those for flame work. However, because of the extended dynamic
range and linearity of plasma sources for most elements, it is possible to
operate with a two-point calibration curve (blank and one standard) provided
that good linearity has been verified and that the analyte concentration in
the standard is within 10% of the sample analyte concentration. In this
measurement mode, the blank is measured and then measurements are
alternated between the standard and the sample so that four to five readings
are obtained for each solution. The blank is re-checked and its intensity
value is subtracted from all sample and standard intensity values. The
average sample intensity is then divided by the average standard intensity
and multiplied by the standard concentration to arrive at the concentration
of the analyte in the sample.

D. Discussion

1. *Alkalis*

Alkali oxides in glasses function as network modifiers, lowering the glass liquidus temperature. The sensitivities and stabilities of alkali resonance-line response with the air + C_2H_2 flame are about equal using either the AES or AAS mode. Consequently, to obviate the hollow cathode as an unnecessary potential variable, alkalis are best measured in the AES mode. Although the sensitivities of alkalis are better using a cooler air + propane flame (compared to air + C_2H_2), the stabilities of alkali response in the air + propane flame are inferior and not as adequate for determining minor and major alkali levels. Alkalis are better determined by flame than plasma AES due to reduced sensitivities and stabilities in the much hotter plasma source.

An ionization suppressor is not required for the determination of lithium with a stoichiometric air + C_2H_2 flame. However, the strontium hydroxide band at 670 nm may interfere with the lithium resonance line at 671 nm, requiring wavelength scanning. Ionization suppressors are required for the determination of sodium and potassium with a stoichiometric to oxidizing flame. Sodium and potassium determinations require a buffer of 100 μg KCl ml^{-1} and 1000 μg NaCl ml^{-1}, respectively. The use of a Corning 2408 red filter to remove radiant energy below 625 nm results in more stable lithium and potassium signals.

2. *Alkaline earths*

Major and minor amounts of alkaline earths can also function as glass modifiers. In addition, they act as stabilizers to improve chemical durability, mechanical strength and to produce desired electrical properties. The alkaline earths are determined with an oxidizing N_2O + C_2H_2 flame. In the AAS mode, 1000 μg $LaCl_3$ ml^{-1} is employed as an ionization suppressor for magnesium. Calcium can be determined by AES or AAS and requires 0.01 M EDTA (pH = 4.5) as a releasing agent in addition to the $LaCl_3$ ionization buffer. Strontium and barium are determined in the AES mode and require the same $LaCl_3$ + EDTA additions as calcium. However, the AES determination of barium also requires the addition of 100 μg KCl ml^{-1} to obtain more stable readings. Barium is determined in the AAS mode when it is present as a major constituent and the more concentrated standards can be employed. If an appreciable level of calcium is present in the analyte solution, then the calcium hydroxide band at 554 nm interferes with the barium AAS resonance line and the determination must be done by AES wavelength scanning.

All alkaline earths are extremely sensitive by plasma AES and can be determined well with an ICP at trace to major levels. For the most part, standard acid decomposition or the plasma fusion method mentioned in the sample decomposition section can be used to prepare the sample. ICP analyses of alkaline earths normally do not require special buffers, but do require proper matching of acid and fusion salt levels in the standards. Because of the high plasma sensitivity for alkaline earths, large dilutions of the sample can be made to suppress potential physical or chemical interferences. Standards containing low levels of alkaline earths, especially strontium and barium, should be prepared fresh for each use. The most sensitive plasma lines for alkaline earths are 455.4 nm for barium, 393.4 nm for calcium, and 279.6 nm for magnesium. Weaker plasma lines are sometimes chosen for major levels of alkaline earths.

3. *Aluminum*

Aluminum oxide is also a glass stabilizer and additionally improves resistance to thermal shock. Because of the refractory nature of aluminum, an oxidizing $N_2O + C_2H_2$ flame is used for analysis in the AAS mode. A satisfactory buffer is the 1000 μg Ca ml^{-1} + 0.07 M HClO$_4$ solution. Aluminum can also be determined by ICP-AES by methods similar to those mentioned for the plasma determination of alkaline earths. For ICP analyses, the aluminum line at 309.3 nm is more stable than the more sensitive line at 396.2 nm and does not suffer from the spectral background problems caused by calcium line broadening which occurs near the more sensitive line.

4. *Iron*

Usually, iron oxide is added to glasses as a coloring agent; or it can be present as an undesirable impurity in batch components. If the analyte solution is 0.05 M in H_3PO_4, iron produces a stable and reproducible AAS signal with a stoichiometric to oxidizing air + C_2H_2 flame. It is preferable to prepare samples for iron determinations by acid decomposition methods, as fusions generally introduce an iron blank or incur the problem of iron reduction and alloying with the platinum crucible. Iron can also be determined by ICP-AES using the line at 259.9 nm which is relatively free of spectral interferences.

5. *Zinc and cadmium*

Zinc and cadmium oxides are included in glass compositions as stabilizers and to improve chemical durabilities. With a stoichiometric to oxidizing air + C_2H_2 flame the zinc and cadmium signals are quite satisfactory when 0.03 M HClO$_4$ is present in the analyte solution. Background correction may be necessary at the wavelengths given in Table 1 if the analyte is high in salt concentration. In general, there is no particular advantage of plasma over flame determinations for these elements.

6. Lead

The oxide of lead acts as a glass stabilizer, lowers the melting temperature of the batch and produces glasses with the high indices of refraction necessary for some optical applications. The 100 μg Na ml^{-1} + 0.03 M HClO$_4$ flame buffer acts as an ionization suppressor to enhance the AAS lead signal using the stoichiometric to oxidizing air + C$_2$H$_2$ flame. If light scattering becomes a problem, background correction may be required. Plasma sensitivity for lead is not always as good as flame sensitivity, so that lead is just as well done by flame.

7. Silicon

The most commonly employed glass-former is silicon. Silicon has poor flame sensitivity, but can be determined at high minor to major levels by flame AAS and at trace to major levels by plasma (DCP and ICP) AES methods which have much better relative sensitivity for silicon.

The flame AAS method for silicon has been described in the literature by Burdo and Wise [4]. It is a fusion method in which a molybdate complexing agent is used to prevent polymerization of silica during and after dissolution of the melt. Basically, 0.1 g of sample is fused with 1.0 g of a homogeneous flux consisting of 0.5 g of Na$_2$CO$_3$ plus 0.5 g of Na$_2$B$_4$O$_7$ (see sample decomposition section). The fusion occurs in a muffle furnace at 1000°C for 20 to 30 minutes. The cooled crucible is immersed into a 250-ml Teflon beaker containing 100 ml of ammonium molybdate solution (39 g l^{-1}) and 100 ml of 0.394 M HNO$_3$. Magnetic stirring dissolves the melt in 10 to 30 minutes, and the solution is diluted to 500 ml with 0.126 M HNO$_3$. A high silica standard is prepared by treating 0.1 g of SiO$_2$ powder in the same manner as a sample. Lower standards are made by diluting the high standard with a molybdate diluting solution containing 7.8 g l^{-1} ammonium molybdate + 2 g l^{-1} flux + 0.16 M HNO$_3$. Samples are measured against standards using a slightly reducing N$_2$O + C$_2$H$_2$ flame in the AAS mode at 251.6 nm. The measurement sequence is essentially the two-point calibration technique (see measurement section), except that blank measurements are included in the alternation sequence. Because of the instability of the AAS signal, attributed to burner drift and carbon deposition, the alternation cycle is repeated at least five or six times. Excellent results can be obtained, but readings are tedious and not all laboratories may be capable of reproducing the method, although it has yielded quality results in this laboratory for many years.

An ICP method for silicon in glasses employs the general sodium carbonate–borate fusion mentioned for plasma analysis in the sample decomposition section, except that the sample weight is 50 mg and the final dilution volume is to one liter in a plastic volumetric flask. This yields a solution which contains 0.1% flux + 0.21 M HNO_3 + 0.0093 M HF. A high standard is prepared by fusing 30 mg of SiO_2 powder in the same manner as the sample. Lower standards are made by diluting the high standard with a diluting solution having the same acid and salt concentrations as the sample and standard solutions. Analyte concentration is determined by the two-point calibration method (see sample measurement section). The best silicon line for plasma methods is 251.6 nm.

The DCP-AES technique is not as accurate as the ICP technique for the determination of silicon in samples prepared by fusion. However, the DCP can be accurately employed for samples which are totally soluble in HF solution. The decomposition method has been previously published by Burdo [5] and relies on the dissolution of 0.1 g of powdered sample in a 30-ml sealed plastic jar containing 5 ml of cold water and 5 ml of cold HF, with subsequent attack at room temperature. The DCP silicon method has been found to be accurate for optical waveguide powders and glasses which are totally soluble in HF and contain other HF-soluble elements such as boron, germanium and phosphorus (all of which can be concomitantly analysed by the DCP for these material types). Samples which contain fluoride-precipitating elements, or high amounts of alkali, or are resistant to acid attack generally do not produce consistent results for silicon by the DCP acid-dissolution method.

8. *Boron*

Boron oxide is also a glass-former, but its main function in a glass is to lower the thermal expansion and thereby increase resistance to thermal shock. Boron has very poor sensitivity by flame techniques and excellent sensitivity by plasma AES. A DCP plasma method for boron in glasses has been published by Burdo and Snyder [6]. The method relies on the dissolution of 0.1 g of powdered sample in a sealed 30-ml plastic jar containing 5 ml of cold water and 5 ml of cold HF, with subsequent decomposition at room temperature for 1 to 24 hours depending on the sample type. Dissolution can be hastened by mild heating as long as a viable seal can be maintained to prevent the loss of boron as its fluoride. The formation of insoluble fluorides and the presence of other undissolved oxides does not normally prevent the complete dissolution and solubility of boron. The room-temperature solution is diluted to 100 ml in a plastic volumetric flask after decomposition is complete. When the concentration of boron is above 10 μg ml^{-1} in the initial 100-ml sample solution, the solution is diluted to achieve a boron concentration of about 5 to 10 μg ml^{-1}. Dilutions are accomplished using a diluting solution containing 1.45 M HF and 300 μg Si ml^{-1} as a buffer. The advantage of diluting is to reduce the concentrations of matrix elements, especially alkalis, which could affect the response of the DCP. The most useful boron line is at 249.7 nm and is free from most spectral interferences (including iron) when a high-resolution echelle spectrometer is employed.

The DCP is exceptionally tolerant of HF. The same boron method can be executed with an ICP having an HF-resistant torch. However, most ICP's do not have spectrometers with the high resolution of an echelle and in some cases a less sensitive boron line (e.g. 208.89 or 208.96 nm) must be used to avoid spectral interferences. Signal integration times are generally in the order of 10 to 15 seconds for precise DCP and ICP determinations of most elements, including boron.

REFERENCES

1   W.M. Wise and J.P. Williams, in I.L. Simmons and G.W. Ewing (Eds.), Progress in Analytical Chemistry, Vol. VIII, Plenum, New York, N.Y., 1976, p. 29.
2   W.M. Wise, R.A. Burdo and J.S. Sterlace, Prog. Anal. Atom. Spectrosc., 1 (1978) 201.
3   C.B. Belcher, Prog. Atom. Spectrosc., 1 (1979) 299.
4   R.A. Burdo and W.M. Wise, Anal. Chem., 47 (1975) 2360.
5   R.A. Burdo, J. Non-Cryst. Solids, 38 and 39 (1980) 171.
6   R.A. Burdo and M.L. Snyder, Anal. Chem., 51 (1979) 1502.

Chapter 4j

Clinical Applications of Flame Techniques

A. TAYLOR

Trace Element Laboratory, Robens Institute, University of Surrey, Guildford, Surrey GU2 5XH, and Department of Clinical Biochemistry, St. Lukes Hospital, Guildford, Surrey GU1 3NT (Great Britain)

I. INTRODUCTION

Measurement of calcium and magnesium concentrations in serum were among the first published applications of AAS following the development of the technique by Willis [1, 2]. Clinical and biological laboratories are one of the largest group of users, providing much of the published literature [3]. Although most of the current papers are concerned with electrothermal atomic absorption spectrometry (ETA-AAS) most practical measurements in hospitals, etc., involve flame atomic absorption spectrometry (FAAS). The great increase in interest in the importance of metals to human biochemical activity extends into occupational health, toxicology, nutrition, in addition to clinical situations, and is a direct development of the wide availability of this important technique.

This chapter will describe procedures for sample preparation as applied to FAAS, including those required for hydride generation and mercury (although the latter is not a "flame procedure", it is conveniently included in this section), give representative methods for the determination of elements of clinical interest and indicate how quality control can be applied to trace element analyses. Specimens relevant to clinical situations are blood, serum and/or plasma, urine and occasionally tissue or other samples. Remarks are appropriate, therefore, to analytical tasks with similar samples from related disciplines, such as veterinary and food sciences etc.

The practical advantages of FAAS such as speed, simplicity, precision, cost and ease of operation are ideal and necessary for the many situations where large numbers of accurate results are required for diagnosis and management of patients' clinical conditions. Although the relative lack of sensitivity of FAAS restricts its use to a limited number of clinical applications, these tend to be those where urgency is of importance and where large numbers of specimens have to be analysed. Electrothermal atomization techniques are appropriate for a wider range of elements at lower concentrations and where a result is not required so quickly. For several reasons, this more demanding type of work is best carried out at specialist laboratories.

References p. 379

II. SAMPLE PREPARATION

A. Introduction

The primary objectives of preparative procedures are: (a) the removal of components in the sample likely to interfere with the analysis (see Chapter 2), e.g. protein, other ions etc.; (b) to increase the concentration of analyte in the solution delivered to the flame, to improve precision and detection limit.

This section describes procedures relevant to the preparation of clinical samples for FAAS. Many will be appropriate to ETA-AAS and for other sample types also. The term "sample preparation" is used loosely to allow discussion of accessories and procedures which can simplify and/or improve sensitivity of flame analyses.

With all these methods it is necessary to be continually conscious of possible contamination which can easily invalidate an otherwise carefully performed measurement. It is very important, therefore, to use the highest quality reagents and water, and to take precautions to ensure the cleanliness of equipment. These measures relate to all stages of sample collection, storage, handling and preparation. Methods refer to distilled water, but deionised or reverse osmosis purified water could also be used.

B. Dilution

Elements present at concentrations of μg ml^{-1}–mg ml^{-1} (10 μmol l^{-1}–mmol l^{-1}) can be diluted with water, lanthanum or some other appropriate solution. The design of the nebulizer, premix chamber and burner may not allow complete removal of the matrix (viscosity) effect caused by protein in serum following low (2–10 fold) dilution [4] and compensating factors may have to be introduced. These include the addition of Triton X-100, butanol or a similar agent to the diluent, preparation of calibration standards in a glycerol solution or the use of the standards addition technique. Such procedures are intended to provide identical viscosities, and hence flow rates through the nebulizer, of the unknown and standard solutions.

C. Protein precipitation

An alternative approach to overcome the matrix effect is to eliminate protein from the measurement solution. For many years protein precipitation was a fundamental step in most analytical methods carried out in clinical chemistry laboratories. A number of different reagents have been developed to remove the protein while the analyte of interest remained in solution [5]. Of these, few have been used for protein precipitation preparatory to flame (or other) atomic absorption measurements. Trichloroacetic acid (5–10% w/v) is the precipitant that has been generally employed although a number of limitations and disadvantages have been shown with this reagent:

(a) Heating at 90°C for ten minutes was recommended to release copper (and possibly other elements) from the precipitate [6].

(b) A small but significant change of volume was measured by Boyde and Wu [7] and this was sufficient to suggest a 10–20% inaccuracy for the determination of zinc in serum. Inaccuracies in the measurement of zinc after protein precipitation were also shown by Taylor and Bryant [4], who noted that there was no such effect on the levels of copper when similarly prepared. These authors suggested therefore that the positive bias of 10.9% found for measurements of zinc was caused by contamination derived from the steps of solution transfer, centrifugation etc.

Precipitation of blood cells with release of lead was included in the methods developed by Taylor and Brown [8] and by Nygren et al. [9], that utilise the increased sensitivity afforded by recent developments mentioned below.

D. Chelation/solvent extraction

Chelation extraction techniques are based on the properties of compounds to form complexes with metals and which can then be extracted into an organic solvent. These procedures have two particular advantages: (a) the analyte can be taken from a difficult and complicated matrix such as whole blood, into a clean, simple solvent with greater nebulization efficiency than obtained with aqueous solutions; (b) by extraction from a larger volume of sample into a smaller volume of solvent an effective increase in concentration can be achieved. Although many chelating agents have been prepared which can be taken for this purpose, the dithiocarbamates, sodium diethyl and ammonium pyrollidine dithiocarbamate (NaDDC and APDC) are general purpose compounds which can be used for most metals. Similarly, while a variety of organic solvents can be employed, methyl isobutyl ketone (MIBK) is very widely used. Combustion in the flame causes fewer problems with MIBK than is encountered with other organic solvents. If required, selective extraction can be performed by adjustment of the sample pH and the use of more specific chelation agent–organic solvent combinations. In this way, a particular metal which may cause subsequent interference in the flame analysis, can be removed or the analyte under consideration can be isolated from the residual matrix. A refinement to this approach was developed which allowed FAAS measurements after the sequential extraction of 11 metals from 1 ml blood [10]. Chelation-extraction was a widely used procedure for the determination of lead in blood prior to the development of reliable ETA-AAS methods, and a few laboratories still continue to use it [11]. The usual applications now are for concentration of low levels of metals from specimens of urine and from tissues after the destruction of the organic material.

References p. 379

E. Destruction of organic matrix

Conversion of an organic matrix to a relatively simple inorganic form in solution is generally necessary for the introduction of tissue specimens to the flame. A digestion step is also necessary for the preparation of blood and urine specimens for an analysis requiring the determination of mercury and the hydride-forming elements. There are two main processes for tissue destruction: wet digestion and dry ashing. These techniques have been extensively investigated by Gorsuch [12] and his observations relating to losses by volatilisation, adsorption and precipitation remain fundamental to this work. The contribution of contamination and an increase in blank levels from the acids and other reagents has also to be noted.

Wet digestion methods include destruction of the matrix by concentrated acids and alkaline solubilisation with reagents such as tetra methylene ammonium hydroxide or other quaternary ammonium hydroxides. The reactions can be carried out in open vessels or tubes on a hotplate or in an aluminium heating block, or in closed vessels at increased pressure (bomb digestion). Recently, digestion or dissolution procedures have been found to be accelerated by using microwave heating.

The simplest technique for preparation of tissue material is to ash specimens in a conventional muffle furnace. Acceleration of the reaction can be accomplished for small masses of sample by heating at less than 150°C within an atmosphere of radio frequency-generated ionised oxygen. The relatively low temperature minimises loss of volatile elements.

The preparation of specimens for the determination of mercury requires special precautions. The carbon–mercury bonds of methyl or other alkyl–mercury compounds have to be destroyed in order that the cold vapour-generation technique can function.

This has to be carried out without loss of the mercury. Acid digestion is therefore preferable and can be carried out either with acid-permanganate at room temperature (for urine) or in systems protected by reflux condensers (blood, tissues).

In recent work with dried tissues the effectiveness of slurry atomization has become very apparent (Chapter 2). Measurements carried out in this way have shown that it is not always necessary to achieve complete destruction of organic material for flame analyses and that short digestion times with only partial elimination of the matrix will suffice to provide reproducible and accurate results.

F. Hydride generation

Most of the hydride forming elements are of clinical relevance. Many are very toxic and increased exposure to arsenic, selenium etc., can be associated with harmful effects or even death. Occupational exposure and absorption has to be carefully monitored and other accidental or intentional exposures can be detected by analysis of body tissues or fluids. Therapeutic administration of bismuth compounds must be carried out with caution to avoid encephalopathy and suspected overdosage needs to be diagnosed by measurement in blood or urine. Selenium is, in addition, an essential trace element and deficiency can be determined by measurement in appropriate samples.

Conventional FAAS offers very poor sensitivity for the determination of these elements for a variety of reasons (see Chapter 4k) and there are severe interferences in the ETA-AAS procedures with many biological specimens. Measurement via the hydride generation reaction circumvents many of the problems and provides a very useful method for analysis. Factors of general relevance to hydride formation include: (a) steps to ensure reduction, initially from higher to lower valency states (e.g. As(V) to As(III)) with subsequent formation of the fully reduced hydride; and (b) elimination of interferences caused by transition metals, residual nitric acid etc. These problems are common to all applications. Two further features, digestion and speciation, may also be of importance when carrying out measurements of biological specimens.

As discussed in the previous section, destruction of the organic material to provide a simple inorganic matrix is necessary for many assays. Whilst dry ashing has been employed by some workers, most published methods describe acid digestion using nitric, perchloric and sulphuric acids. Residual nitric acid, however, can interfere with analyte reduction [13] and must therefore be boiled off. Bunker and Delves [14] were able to successfully digest biological tissues with nitric and sulphuric acids and achieve accurate results for selenium. Thus digestion can be completed in such cases by the use of large volumes of nitric acid, without risk of acid interference. Loss of analyte during digestion is more of a potential problem and recovery studies indicate that it is crucial to ensure that the digestion vessel has a sufficiently long neck (Kjeldahl tube or tall boiling tube) to act as a condenser and to prevent excessive heating [15, 16] or to use a closed, pressure digestion vessel. Arsenic may be present in biological specimens in several organic and inorganic forms depending on the compounds absorbed. The various species of the other hydride elements have not been so well examined or documented. There are occasions when the measurement of total concentrations are not as informative as determination of the individual species and separation of arsenic components in urine has been reported by using HPLC [17] or by selective reduction [18].

G. Atom trapping and preconcentration; devices to improve sensitivity

Sensitivity of FAAS is limited by inefficiency of pneumatic nebulizers and the dispersion and transience of atom populations in the light path. One of the earliest attempts to increase sensitivity for a clinical measurement was the cup accessory (Chapter 4k) of Delves [19] which was developed for the determination of lead in blood. Although most laboratories now elect to use electrothermal atomization [11], some with large workloads continue with cup analysis to utilise the faster rate of measurement. Improved sensitivity is gained by direct introduction of all, i.e. 100%, of the sample into the flame and by the effect of the tube held in the light path to retard dispersion of the atoms. The cup technique has been shown to be suitable for determination of lead, cadmium, copper and silver in blood and other biological specimens.

Atom dispersion is also inhibited by the slotted quartz tube, constructed by Watling [20]. Three- to ten-fold sensitivity gains for volatile elements were exploited for the measurement of lead, cadmium, copper and zinc in clinical fluids by Brown and Taylor [21]. The use of such a tube permits flame measurements of these metals with very little sample preparation and with small sample volumes. The true flame atom-trapping devices (Chapter 2) can give up to 80-fold increases of sensitivity [22], but as yet these have not found any clinical applications.

Devices for the preconcentration of metals onto alumina or chelating resins with a sudden release of the accumulated analyte, have been described and applied to measurements in water by inductively coupled plasma atomic emission spectrometry and FAAS. Conditions for the determination of cadmium in urine using an on-line column of basic alumina and the slotted quartz tube were investigated and the levels in normal urine samples were successfully measured [23].

With a recently developed highly efficient nebulizer interface, lead in blood at normal concentrations has been determined after dilution with an acid precipitation reagent. A twelve-fold reduction in the limit of detection was reported with this system [9].

From this discussion it is apparent that almost the entire range of techniques that may be applied to sample preparation, are appropriate to clinical samples.

III. METHODS FOR ANALYSIS

A. Introduction

It is important to emphasize the remarks made above, that the appearance of interferences associated with viscosity or with ionising elements are influenced by nebulizer and burner design [4]. Not every method given in this chapter will, therefore, necessarily transfer directly to all atomic absorption instruments. Readers are cautioned to check themselves whether these interferences will take place and need to be removed as described. The methods given here refer to specimens generally examined. However, a laboratory may be requested to carry out an unusual measurement, e.g. lithium in CSF or gold in breast milk, and will need to consider which procedures may be most appropriate.

B. Sodium and potassium

These elements are placed together because they are usually measured by flame atomic emission, whereas most other elements considered in this chapter are determined by atomic absorption spectrometry. Sodium and potassium are the major physiological cations and are present in biological samples at relatively high concentrations (mmol l^{-1}). The emission instruments available for clinical measurements of these metals are designed for these high concentrations and do not provide very low detection limits. The instruments are intended to be stable so that rapid, precise readings can be obtained with simultaneous measurement of both metals. An air–propane flame is generally used. Atomic absorption/emission instruments are much more sensitive and because of the large dilutions that are then necessary, are less suitable for measurement of sodium and potassium. The concentration ranges at which these elements are present in biological samples, are very narrow and any dilution errors therefore will have a major effect on the reported result. Ionisation effects are likely to be relevant to the determination of these elements. Standards should be prepared to match the specimens or a suppressor should be included in the diluent.

Manufacturers of "clinical" flame emission instruments will have a recommended procedure which is appropriate to the design of the equipment. For measurement of sodium and potassium in plasma or urine these procedures should be followed. For other instruments the following methods are suggested but different dilutions may be found to be optimal. The use of alternative (less sensitive) wavelengths and rotation of the burner may also be considered for these elements.

1. *Determination of sodium and potassium in plasma*

Reagents
(1) Standard solution: 140 mmol l^{-1} Na, 5.0 mmol l^{-1} K.
(2) Diluent: distilled water.

Procedure
(1) Dilute the standard, samples and quality control specimens 50-fold with the distilled water. Mix well and take for measurement of potassium.

References p. 379

(2) Dilute the standard, samples and quality control specimens 250-fold with the distilled water. Mix well and take for measurement of sodium.

2. *Determination of sodium and potassium in urine*

Reagents
(1) Standard solutions: 20, 50, 100, 150, 200, 250 mmol l^{-1} Na, K combined standards.
(2) Diluent: distilled water.

Procedure
(1) Dilute the standards, samples and quality control specimens 500-fold with the distilled water. Mix well and take for measurement of potassium.
(2) Dilute the standard, samples and quality control specimens 50-fold with the distilled water. Mix well and take for measurement of sodium.
Note: Haemolysis or failure to separate plasma from cells within about 4 hours causes release of potassium from erythrocytes and results will be erroneously high. Unreliable results can be obtained for potassium with serum, and plasma is preferred for this determination.

Reference ranges
Potassium: plasma 3.5–5.2 mmol l^{-1}, urine 40–120 mmol/24 h. Sodium: plasma 135–147 mmol l^{-1}, urine 100–250 mmol/24 h.

C. Lithium

Lithium is measured to monitor treatment of patients with depressive illness where the plasma concentration should be maintained within a narrow therapeutic range. Accumulation to too high levels can cause toxicity [24]. Lithium can be measured by flame atomic emission or by FAAS and many of the flame photometers designed for the atomic emission measurement of sodium and potassium will also measure lithium. The methods described by the manufacturers should be used. The following procedure is suggested for FAAS.

1. *Determination of lithium in serum*

Reagents
(1) Standard solution: 1.0 mmol l^{-1} Li.
(2) Diluent: 140 mmol l^{-1} Na, 5.0 mmol l^{-1} K.

Procedure
(1) Dilute the standard, samples and quality control specimens 10-fold with the diluent. Mix well and take for measurement of lithium.

Note: Lithium heparin is a widely used anticoagulant and the plasma from blood samples will be grossly contaminated with lithium. Serum should be used for the determination.

Therapeutic range
  0.5–1.5 mmol l^{-1}

D. Calcium and magnesium

Aspects of the clinical chemistry of calcium and magnesium are related and the two metals are often measured in the same specimens. Increased or reduced levels can be found in sick patients and their concentrations have to be measured to a high level of accuracy and reproducibility. Calcium forms a refractory complex with phosphate that is not completely dissociated at the temperature of the air–acetylene flame. Although this interference can be eliminated by use of a nitrous oxide–acetylene flame the more usual approach is to include a releasing agent in the diluent. Sodium EDTA will complex the calcium, or metals such as strontium or lanthanum, which preferentially bind the phosphate, can be added to free the calcium. Most laboratories use a lanthanum diluent and although not necessary for measurement of magnesium it allows the determination of both metals in the same solution.

1. *Determination of calcium and magnesium in serum, plasma and urine*

Reagents
(1) Stock calcium solution: 25 mmol l^{-1} Ca.
(2) Working calcium standards: 1.5, 2.0, 2.5, 3.0, 3.5, 4.0, 5.0 mmol l^{-1} Ca. Prepared by dilution of the stock solution with distilled water.
(3) Stock magnesium solution: 41.1 mmol l^{-1} Mg.
(4) Intermediate magnesium solution: 5 mmol l^{-1} Mg. Working magnesium standards: 0.25, 0.50, 0.75, 1.0, 1.25 mmol l^{-1} Mg. Prepared by dilution of the stock solution with distilled water.
(5) Stock lanthanum solution: 10% w/v La (La Cl$_3$·7H$_2$O).
(6) Diluent: 0.1% w/v La in 1% v/v HCl.

Procedure
(1) Dilute the standards, samples and quality control specimens 50-fold with the lanthanum diluent. Mix well and take for measurement of calcium and/or magnesium.
Note: The working standards are stable for several weeks. The calcium standards at 4.0 and 5.0 mmol l^{-1} are necessary only for measurement of urine.
   Further dilutions may be necessary for measurement of calcium and magnesium in urine from individuals who are hyperexcretors of these metals.

References p. 379

As with potassium, loss of magnesium from red cells can give false high results.

    Calcium: serum/plasma 2.2–2.6 mmol l^{-1}, urine 2.5–10.0 mmol/24 h.
    Magnesium: serum/plasma 0.65–1.0 mmol l^{-1}, urine 1.0–7.0 mmol/24 h.

E. Copper and zinc

Copper and zinc are two of the essential trace elements, and deficiencies cause profound clinical consequences. Assessment of body status is best achieved by measurement of the concentrations in serum although there are some situations in which measurement of urinary excretion is valuable.

It is suggested that zinc in leucocytes may be more representative of tissue levels than is the serum concentration, but the evidence for this is ambiguous [25]. Preparation of specimens for leucocyte zinc measurements has to be specially arranged so that the cells can be harvested within about one hour of collection. There is one report that zinc concentrations in serum are greater than the corresponding plasma specimens by an average of 16%. This statement is quoted in several reviews but subsequent attempts by other workers have failed to confirm this observation. As with potassium and magnesium, haemolysis can give increased results.

Taylor and Bryant [4] examined the many variations to the methods in use for the measurement of these elements in serum. The procedures suggested here have proved to be accurate and precise for many years in the authors' laboratory, but other analysts may prefer to adopt one of the alternative methods.

1. *Determination of copper in serum, plasma or blood*

Reagents
(1) Stock copper solution: 5 mmol l^{-1} Cu.
(2) Intermediate copper solution: 100 μmol l^{-1}.
(3) Working copper standards: 10, 20, 30, 40 μmol l^{-1}.
(4) Distilled water.

Procedure
(1) Dilute the standards, samples and quality control specimens 5-fold with distilled water. Mix well and take for the measurement of copper.

1. *Determination of zinc in serum*

Reagents
(1) Stock zinc solution: 5 mmol l^{-1} Zn.

(2) Intermediate zinc solution: 100 μmol l^{-1}.
(3) Working zinc standards: 10, 20, 30, 40 μmol l^{-1}.
(4) Distilled water.

Procedure
(1) Dilute the standards, samples and quality control specimens 5-fold with
distilled water. Mix well and take for the measurement of zinc.
Note: Zinc in whole blood can be measured by this procedure after an initial
5-fold dilution of the specimens. The procedures for copper and zinc can
easily be combined by preparation of a common intermediate solution from
the stock solutions.

2. *Determination of copper in urine*

Reagents
(1) Stock copper solution: 5 mmol l^{-1} Cu.
(2) Working copper solution: 100 μmol l^{-1} Cu.

Procedure
The copper is measured by a standards additions procedure using undiluted urine.
(1) Pipette 5 ml urine into a series of 5 tubes.
(2) Add working copper solution as shown in Table 1 .
(3) Mix well and take for the measurement of copper.
Note: Patients with Wilson's disease can have very high copper excretion
and the urine samples will need to be diluted before the analysis is set up.

3. *Determination of zinc in urine*

Reagents
(1) Stock zinc solution: 5 mmol l^{-1} Zn.

TABLE 1

PREPARATION OF SAMPLES FOR THE MEASUREMENT OF COPPER IN URINE
BY A STANDARDS ADDITION PROCEDURE

Tube number	Urine volume (ml)	Volume of working solution (μl)	Added copper concentration (μmol l^{-1})
1	5.0	0	0
2	5.0	10	0.2
3	5.0	30	0.6
4	5.0	70	1.7
5	5.0	150	3.0

References p. 379

TABLE 2

PREPARATION OF SAMPLES FOR THE MEASUREMENT OF ZINC IN URINE BY A STANDARDS ADDITION PROCEDURE

Tube number	Urine volume (ml)	Volume of working solution (μl)	Added zinc concentration (μmol l^{-1})
1	5.0	0	0
2	5.0	20	4
3	5.0	30	6
4	5.0	40	8
5	5.0	50	10
6	5.0	60	12

(2) Working zinc solution: 1 mmol l^{-1} Zn.
(3) Distilled water.

Procedure

The zinc is measured by a standards addition procedure.
(1) Pipette 5 ml urine into a series of 6 tubes.
(2) Add working zinc solution as shown in Table 2.
(3) Mix well and take for measurement of zinc.
Note: If concentrations are high the additions curve will not be linear. The assay will have to be repeated after dilution of the urine.

Reference ranges

Copper in serum: 3–11 μmol l^{-1} neonates; 11–20 μmol l^{-1} children over 6 months and adults; 27–40 mol l^{-1} pregnancy; 6 weeks to term.

Zinc in serum: 11–24 μmol l^{-1}. Copper in urine: less than 0.8 μmol/24 h. Zinc in urine: 4.5–9.0 μmol/24 h. Sensitivity for the determination of copper and zinc is increased 300–500% when the slotted quartz tube is used and this permits larger dilutions of serum to be made without loss of performance. Thus, specimens of 20 μl can be diluted 20-fold and will provide sufficient volume for measurement.

F. Iron

The measurement of this metal and of the total iron binding capacity (TIBC) are of some help in the diagnosis of iron-deficiency anaemia. These parameters are of greater value for the assessment of iron overload which can occur following multiple blood transfusions or in childhood poisoning. Iron toxicity is treated by chelation therapy and measurement of urinary excretion is important to monitor progress and prevent excessive loss. Many colorimetric procedures are available for the determination of iron in serum and these are usually preferred in clinical situations. However, iron in urine cannot be easily measured spectrophotometrically and atomic absorption is then the technique of choice.

1. *Measurement of iron in serum*

Reagents
(1) Stock iron solution: 20 mmol l^{-1} Fe.
(2) Intermediate iron solution: 100 mol l^{-1} Fe in Na-K diluent.
(3) Working iron standards: 10, 20, 30, 40 μmol l^{-1} in Na-K diluent.
(4) Protein precipitation solution: 10% w/v trichloroacetic acid and 3% v/v thioglycolic acid in 16.6% v/v hydrochloric acid.
(5) Na-K diluent: 140 mmol l^{-1} Na; 5.0 mmol l^{-1} K.

Procedure
(1) Dilute the standards, samples and quality control specimens 1 + 1 with the protein precipitation solution.
(2) Mix well and stand at room temperature for 5 min.
(3) Centrifuge at 3000 rpm for 5 min to give a clear supernatant. Take for the measurement of iron.

2. *Measurement of total iron binding capacity*

Reagents
(1) Stock ferric chloride: 1 mmol l^{-1} in 0.5% v/v HCl.
(2) Saturating iron solution. Dilute the stock ferric chloride 7-fold with distilled water.
(3) Magnesium carbonate powder (chromatographic grade).
(4) Other solutions for measurement of iron.

Procedure
(1) Place 0.5 ml serum into a 10 ml plastic tube.
(2) Add 0.5 ml distilled water and 0.5 ml saturating iron solution. Mix and stand at room temperature for 15 min.
(3) Add 0.20 g magnesium carbonate, stopper and mix vigorously for 10–15 s. Mix for 20 min by repeated inversion on a rotating mixer. (If inversion mixer is not available stand for 30 min but remix by vigorous shaking 4–5 times during this interval).
(4) Centrifuge at 3000 rpm for 15 min.
(5) Take 1 ml supernatant and proceed exactly as with the serum iron procedure.
(6) Multiply results by 3 to give the total iron binding capacity.

References p. 379

TABLE 3

PREPARATION OF SAMPLES FOR THE MEASUREMENT OF IRON IN URINE BY
A STANDARDS ADDITION PROCEDURE

Tube number	Diluted urine volume (ml)	Volume of working solution (μl)	Added iron concentration (μmol l^{-1})
1	5.0	0	0
2	5.0	10	20
3	5.0	25	50
4	5.0	35	70
5	5.0	50	100
6	5.0	60	120

3. *Determination of iron in urine following chelation therapy*

Reagents
(1) Stock iron solution: 20 mmol l^{-1} Fe.
(2) Working iron standard: 10 μmol/ml Fe.

Procedure
   The iron is measured by a standards addition procedure using diluted
urine.
(1) Dilute the urine ten-fold.
(2) Set up a series of tubes as shown in Table 3.
(3) Mix well and take for measurement of iron.
   The measured concentration should be multiplied by 10 to show the
concentration in the undiluted urine specimen.
Note: If concentrations are very high the additions curve will not be linear.
The assay will have to be repeated after further dilution of the urine.

Reference ranges
   Serum iron: male 14.3–25.6; female 11.6–24.2 μmol l^{-1}.
   TIBC: male 45–72; female 36–77 μmol l^{-1}.

G. Lead

   The most appropriate specimen for assessment of exposure to inorganic
lead is whole blood. More than 95% is contained within red cells and measurements are made with specimens collected using K_2EDTA anticoagulant.
Exposure to organic lead compounds is shown by the concentration of the
metal in urine. Three quite distinct FAAS procedures can be used to measure
lead in clinical samples. Blood samples should be mixed for at least 1 hour
before starting.

1. *Measurement of lead using the slotted quartz tube*

Reagents
(1) Stock lead solution: 1 mmol l^{-1}.
(2) Working lead standards: 0.5, 1.0, 2.0, 3.0, 4.0 μmol l^{-1}. Prepared by addition of small volumes of lead to aliquots from a pool of blood with a low endogenous concentration. These standards can be kept for several weeks.
(3) Trichloroacetic acid: 10% w/v.

Procedure
(1) Place 1 ml trichloroacetic acid into a small stoppered test tube.
(2) Hold tube against vortex mixer and add, dropwise, 1.0 ml standard, sample or quality control specimen.
(3) Stopper and mix vigorously to ensure that all the cells and protein have precipitated. Stand for 15 min.
(4) Centrifuge at 3000 rpm for 10 min.
(5) Take clear supernatant for measurement of lead with slotted quartz tube in position on burner in the light path.
Note: The slotted quartz tube can be used to measure lead in urine but to retard chloride-induced devitrification of the quartz aspirate a solution containing lanthanum 10% w/v for 15 min before any samples.

2. *Measurement of lead by chelation-solvent extraction*

Reagents
(1) Standards: prepared as for slotted quartz tube.
(2) Ammonium pyrollidine dithiocarbamate, APDC (ammonium tetra methylene dithiocarbamate) 2% w/v in 5% w/v Triton X-100. Prepare fresh.
(3) Water-saturated methyl isobutyl ketone, MIBK (4-methylpentan-2-one). Place 500 ml into a glass, stoppered container and add about 10 ml distilled water. Stopper and mix well.

Procedure
(1) Pipette 5 ml blood standards, samples or quality control specimens into a series of glass stoppered tubes.
(2) Add 1 ml APDC solution.
(3) Add 5 ml water-saturated MIBK. Stopper and mix by repeated inversion on a rotating mixer for 30 min. Stand upright to allow the organic layer to separate. (If inversion mixer is not available, shake tubes vigorously by hand for 2 min. It may then be necessary to centrifuge gently to remove the emulsion from the organic layer).
(4) Aspirate organic layer for the measurement of lead.

References p. 379

Note: The Triton X-100 haemolyses the red cells to release lead. Combustion of organic solvents causes a very fuel-rich flame and to achieve a stable base line the acetylene flow has to be reduced. This requires careful adjustment to ensure that the flame does not extinguish when air is aspirated between samples. Concentrations of lead in urine are usually low and a larger volume, e.g. 20 ml, should be taken. The volumes of APDC and MIBK remain the same as for blood.

3. *Measurement of lead by Delves' microcup technique*

Reagents
(1) Standards: prepared as for slotted quartz tube.
(2) Hydrogen peroxide: 30% w/v.

Procedure
(1) Set up instrument with the Delves' accessory in position as described by the manufacturer. Use background correction.
(2) Turn on flame to heat the absorption tube to working temperature.
(3) Clean the nickel cups by burning off contamination in the flame.
(4) Prepare each standard, sample and quality control specimen in triplicate. Pipette 10 μl of blood into a nickel cup and place on hotplate at 140°C to dry.
(5) Add 20 μl hydrogen peroxide, allow the initial oxidation reaction to subside and then replace on the hotplate. Remove when the reaction is complete.
(6) Place each cup in turn into the sampler loop, introduce into the flame and record the lead absorbance.
Note: Cups should be matched to ensure that equivalent responses are given.

Reference ranges
Blood lead: children less than 1.2 μmol l^{-1}; adults less than 1.4 μmol l^{-1}.
Urine lead: less than 400 nmol/24 h.

H. Cadmium

Cadmium levels in blood and urine together provide information relevant to recent exposure. Cadmium accumulates in the kidney, but if renal damage occurs, it is released to the urine in large amounts. Normal concentrations are low, but it is possible to measure cadmium by following the same procedures described for lead, particularly if there has been increased exposure. The appropriate range of standard concentrations are: blood 0–100 nmol l^{-1}, urine 0–50 nmol l^{-1}.

Reference ranges

Blood cadmium: non-smokers less than 27 nmol l^{-1}; smokers less than 54 nmol l^{-1}.

Urine cadmium: less than 10 nmol l^{-1}.

I. Gold

Gold has been used for more than 60 years to successfully treat patients with rheumatoid arthritis. There is no therapeutic range and therefore little value in regular measurements of concentrations in serum or urine. In certain situations, however, special investigations may be appropriate and the concentrations found during loading or maintenance therapy are high enough to be measured by FAAS after two-fold dilution. To overcome the matrix effects evident at this dilution, a detergent is included in the diluent and calibration is achieved by standard additions.

1. *Measurement of gold in serum and urine*

Reagents
(1) Stock gold solution: 1 mg/ml Au.
(2) Working gold standard: 5 μg/ml Au.
(3) Triton X-100: 5% w/v.
(4) Distilled water.

Procedure
(1) Set up two tubes (A and B) for each specimen and quality control specimens as shown in Table 4.
(2) Mix well and take for the measurement of gold.
(3) Use the absorbance readings to calculate the concentrations of gold as follows:

TABLE 4

PREPARATION OF SAMPLES FOR THE MEASUREMENT OF GOLD IN SERUM OR URINE (in ml)

	Serum		Urine	
	A	B	A	B
Sample	1.0	1.0	1.5	1.5
Water	0.5	–	0.5	–
Standard	–	0.5	–	0.5
Triton X-100	0.5	0.5	–	–

References p. 379

$$\text{Serum Au} \ (\mu g \ l^{-1}) = \frac{A \times 2500}{B - A}$$

$$\text{Urine Au} \ (\mu g \ l^{-1}) = \frac{A \times 4 \times 1250}{(B - A) \times 3}$$

J. Hydride forming elements

Preparation of specimens for the formation of hydrides should ensure that digestion is completed, nitric acid is removed, and loss by volatilization does not occur. The following general procedure is suitable for many types of sample but should be evaluated in the user's laboratory to determine whether any variations or modifications have important effects.

Reagents
(1) Stock standard solutions: 1 mg ml^{-1}.
(2) Intermediate standard solutions: 100 $\mu g \ l^{-1}$.
(3) Working standard solutions: 10–100 $\mu g \ l^{-1}$ in 1.5% v/v HCl.
(4) Nitric acid.
(5) Sulphuric acid.
(6) Perchloric acid.
(7) Hydrochloric acid (Aristar grade).
(8) Hydrochloric acid: 1.5% v/v.
(9) Sodium borohydride: 3% w/v in 1% w/v NaOH.

Procedure
(1) Place sample into a 100 ml digestion tube (31 × 2.0 cm i.d.).
(2) Add 2.5 ml HNO$_3$ and place in an aluminium heating block (maximum depth of hole to be no more than 7 cm).
(3) Heat to 140°C in 15 min and hold for 25 min. Allow to cool to room temperature.
(4) Add 0.5 ml H$_2$SO$_4$ and 0.2 ml HClO$_4$.
(5) Heat to 140°C in 15 min, hold for 5 min; heat to 200°C in 10 minutes, hold for 15 min; heat to 250°C in 10 min, hold for 15 min; heat to 310°C in 10 min, hold for 20 min, allow to cool.
(6) Add 5 ml 5 M HCl. Heat to 90°C in 10 min, hold for 20 min.
(7) Allow to cool and add 5 ml H$_2$O.
(8) Take for hydride generation analysis with NaBH$_4$ reducing agent.
Note: The hydride response is similar for standards in 1.5% v/v HCl and for samples in 2.5 M HCl. The amount of sample taken for digestion and the volume of digestate taken for hydride generation have to be determined by the analyst depending on the particular application. If large amounts

TABLE 5

SELECTED REFERENCE RANGES

Element	Serum/plasma (μg l^{-1})	Blood (μg l^{-1})	Urine (μg/24 h)
Arsenic	–	<40	<50
Selenium	87 –150	89–157	25 –50
Antimony	–	–	<1
Bismuth	0.1– 3.5	–	0.3–46

of organic material are necessary, the digestion process may have to be repeated.

The selected reference ranges are given in Table 5

K. Tissues

The concentrations of many elements are greater in tissue specimens than in body fluids and it is possible to determine metals such as chromium, cobalt etc., by FAAS. The range of procedures for destruction of organic material were described above and it is not suggested that the example given here should always be used.

1. *Acid digestion using a hotplate and conical flasks*

Reagents
(1) Stock standard solutions: 1 mg ml^{-1}.
(2) Nitric acid.
(3) Perchloric acid.

Procedure
(1) Accurately weigh about 1 g tissue into a 100 ml conical flask.
(2) Add 2.5 ml HNO$_3$ and 0.5 HClO$_4$. Stand at room temperature for 1 h.
(3) Place on a hotplate and heat to about 100°C. Maintain the temperature until brown fumes have been evolved.
(4) Raise temperature to 150–200°C and heat until white fumes of perchloric acid are evolved. The residue should be white or pale yellow.
(5) Dark residues indicate that digestion is incomplete. Cool, add further HNO$_3$ and HClO$_4$ and repeat the procedure.
(6) Dissolve the residue in a known volume of 1% v/v HNO$_3$ and take for the required measurements.
Note: Digestion should be carried out in washdown perchloric acid fume cupboards. The initial heating should be cautious to prevent rapid boiling off of the nitric acid with an increased risk of perchloric reaction.

References p. 379

IV. QUALITY CONTROL

Laboratories must demonstrate the accuracy of procedures by reference to certified materials and participation in external quality assessment programmes, and maintain performance by establishment of internal quality control protocols. Quality control is widely practised in general clinical analysis and in many areas of specialist analysis. Laboratories with an interest in trace elements are in a favoured position since there are many reference materials with certified values for inorganic constituents. A system for use of internal quality control specimens according to a strictly defined protocol has been developed by a group of specialist laboratories. The protocol specifies inclusion of the quality control specimens at regular intervals throughout analytical batches. It has been convincingly demonstrated that such approaches can be extremely effective in achieving accurate results [26, 27]. External quality assessment programmes for trace elements in clinical samples are well established [28] and all laboratories should be encouraged to participate. With attention to this topic it is hoped that there will no longer be the publication of results that are clearly improbable and inaccurate [28].

V. SUMMARY AND CONCLUSIONS

Flame atomic absorption spectrometry will rarely provide the detection limits that are achieved with ETA-AAS. However, a laboratory may be faced with a request to provide a result within minutes in order that a life-threatening medical crisis can be resolved, or with hundreds of specimens collected from a hospital in-patient population or from an environmental survey. At such times ETA-AAS would be inappropriate, whereas flame analysis is ideal for these situations. Thus, FAAS will allow the measurement of calcium and magnesium in no more than 50 μl serum from a sick neonate, to demonstrate if the nutritional support needs to be supplemented with these elements. In a young child admitted to hospital with coma, lead poisoning can be diagnosed within minutes using the slotted quartz tube method. On other occasions speed of response may be less important than the facility to obtain exceptional batch-to-batch precision. Confident detection of an increase in concentration of an analyte by less than 5% can be sufficient to diagnose a tumour and require the patient to undergo a serious surgical operation. Only rarely could the reproducibility provided by FAAS be achieved by electrothermal methods. Examples such as these demonstrate the rapidity and reliability of FAAS which is ideal for the clinical laboratory. Whether for non-urgent investigations or to allow crucial decisions to be made with an immediate effect on diagnosis or treatment of seriously ill patients, FAAS has extensive and challenging applications.

REFERENCES

1 J.B. Willis, Spectrochim. Acta, 16 (1960) 259–272.
2 J.B. Willis, Spectrochim. Acta, 16 (1960) 273–278.
3 A.A. Brown, D.J. Halls and A.J. Taylor, Anal. At. Spectrosc., 4 (1989) 47R–110R.
4 A. Taylor and T.N. Bryant, Clin. Chim. Acta, 110 (1981) 83–90.
5 R.J. Henry, D.C. Cannon and J.W. Winkelman, Clinical Chemistry Principles and Technics, Harper and Row, Hagerstown, 1974.
6 A.D. Olson and W.B. Hamlin, At. Absorpt. Newsl. 7 (1968) 69–71.
7 T.R.C. Boyde and S.W.N. Wu, Clin. Chim. Acta, 88 (1978) 49–56.
8 A. Taylor and A.A. Brown, Analyst, 108 (1983) 1159–1161.
9 O. Nygren, E.A. Nilsson and A. Gustavsson, Analyst, 113 (1988) 591–594.
10 H.T. Delves, G. Shepherd and P. Vinter, Analyst, 96 (1971) 260–273.
11 D.G. Bullock, N.J. Smith and T.P. Whitehead, Clin. Chem., 32 (1986) 1884–1889.
12 T.T. Gorsuch, Analyst, 84 (1959) 135–173.
13 R.R. Brooks, D.E. Ryan and H. Zhang, Anal. Chim. Acta, 131 (1981) 1–16.
14 V.W. Bunker and H.T. Delves, Anal. Chim. Acta, 201 (1987) 331–334.
15 M. Ihnat and B.K. Thompson, J. Assoc. Off. Anal. Chem., 63 (1980) 814–839.
16 B. Welz, M.S. Wolynetz and M. Verlinden, Pure Appl. Chem., 59 (1987) 927–936.
17 B.S. Chana and N.J. Smith, Anal. Chim. Acta, 197 (1987) 177–186.
18 J. Aggett and A.C. Aspell, Analyst, 101 (1976) 341–347.
19 H.T. Delves, Analyst, 95 (1970) 431–438.
20 R.J. Watling, Anal. Chim. Acta, 94 (1977) 181–186.
21 A.A. Brown and A. Taylor, Analyst, 110 (1985) 579–582.
22 J. Khalighie, A.M. Ure and T.S. West, Anal. Chim. Acta, 131 (1981) 27–36.
23 A. Karakaya and A. Taylor, J. Anal. At. Spectrosc., 4 (1989) 261–263.
24 A. Taylor, Clin. Endocrin. Metab., 14 (1985) 703–724.
25 H. T. Delves, Clin. Endocrin. Metab., 14 (1985) 725–760.
26 R.A. Braithwaite and A.J. Girling, Fresenius Z. Anal. Chem., 332 (1988) 704–709.
27 A. Taylor, Fresenius Z. Anal. Chem., 332 (1988) 732–735.
28 A. Taylor and R.J. Briggs, J. Anal. At. Spectrosc., 1 (1986) 391–394.

Chapter 4k

Elemental Analysis of Body Fluids and Tissues by Electrothermal Atomization and Atomic Absorption Spectrometry

H.T. DELVES and I.L. SHUTTLER

Trace Element Unit, University Clinical Biochemistry, University of Southampton, Southampton General Hospital, Level D, South Laboratory and Pathology Block, Tremona Road, Southampton, SO9 4XY (United Kingdom)
Bodenseewerk, Perkin-Elmer GmbH, Postfach 10 11 64, D-7770 Uberlingen (Federal Republic of Germany)

I. INTRODUCTION

Elemental analysis of body tissues and fluids by electrothermal atomization–atomic absorption spectrometry has continued to advance the understanding of the roles of trace elements in clinical biochemistry. All of those aspects of metabolic processes that are affected by changes in the concentrations of accessible trace elements have been studied. These include: deficiencies of essential trace elements as a result of inherited or acquired metabolic disorders, or from nutritional inadequacy; and excesses of trace elements producing toxicity states as a result of inherited metabolic disorders involving essential trace elements or from the inappropriate exposure to, or ingestion of, non-essential trace elements.

Three general headings are used in the discussion of the analytical measurements: essential trace elements; non-essential trace elements used therapeutically; and non-essential, toxic trace elements. Since the publication of a similar review chapter to this present one [1], there have been many improvements in instrumentation, and an increased awareness of the stringent precautions necessary to avoid contamination during sample collection, preparation and analysis [2, 3, 4, 5, 6]. It is now clear that many early studies were compromised by contamination, so that current reference ranges for healthy unexposed subjects have fallen dramatically [2, 7]. The normal concentrations of many essential elements are at or near the detection limit for most current ETA-AAS instruments, so that measurement of levels associated with deficiency is impossible.

Rather than provide a complete literature review of methods, which has been admirably completed by Subramanian [8, 9, 10], we have attempted a selective, critical review of those, which from our own practical experience appear to be applicable in a routine clinical laboratory. The elemental analyses discussed here are intended to reflect the recent, current and probable future interest in the application of ETA-AAS.

II. ELECTROTHERMAL ATOMIZATION–ATOMIC ABSORPTION SPECTROMETRY

The past decade has witnessed a transition in the role of ETA-AAS from research applications in specialist clinical laboratories to routine assays done in many pathology laboratories of District General Hospitals. Other techniques restricted to the specialised laboratories include: isotope dilution mass spectrometry, neutron activation analysis, total reflection X-ray fluorescence, and inductively coupled plasma–mass spectrometry. Although these are more sensitive, they are also far more expensive and less amenable to routine analysis in a general clinical laboratory. Those electrochemical techniques which are less expensive, require significant sample pretreatment, to present the analyte in a suitable form for measurement and to reduce interferences and offer a restricted range of elemental analytes.

The instrumental sophistication of ETA-AAS has advanced tremendously over this past decade. Fast digital electronics allied with computer control, allow the immediate and accurate display of atomic and background signals which is essential for method development and monitoring routine analyses. Many ETA-AAS interference effects observed using older equipment were compounded by electronically slow signal processing of instruments originally designed to measure steady state flame signals, and not suitable for the transient signals obtained from electrothermal atomizers. It is now clear that the design considerations for flame and electrothermal atomizer instruments are totally different.

Fundamental studies within electrothermal atomizers of condensed- and vapour-phase chemical and physical reactions and of the temporal, radial and lateral temperature distributions have led to new designs of atomizers: L'vov platforms [11], probe atomizers [12] and the constant temperature furnace [13]. These have been fashioned from a variety of newer materials, e.g. pyrolytic (solid and coated) graphite and glassy carbon. These newer atomizers and associated instrumentation, together with improved chemical modifiers, both solution and gaseous, have made ETA-AAS a very sensitive yet reliable routine analytical technique.

A. Interferences in ETA-AAS

Atomic absorption measurements suffer from physical, chemical and spectral interferences which can significantly affect accuracy if they are not adequately overcome. Although these interferences affect both FAAS and ETA-AAS they are generally more severe with the latter, and are summarised in Table 1 [14].

TABLE 1

INTEFERENCE WITH ETA-AAS ANALYSIS OF BODY TISSUES/FLUIDS

Electrothermal treatment of the sample	Physico-chemical variable	Effect	Classification of interference
Liquid sample (10–20 µl) 100°C	Differences in surface tension and viscosity of sample, diluent, standard.	(I) Varying degrees of diffusion of sample along and into graphite occlusion of analyte within sample matrix and charred residues: leading to losses of analyte; variable rate of transfer of analyte into vapour phase.	Condensed phase interference
Dried residue 500–1100°C	Volatility of analyte and matrix affected by endogenous Na, Ca, Mg, PO_4^{3-}, SO_4^{2-}, Cl and added reagents; HCl, HNO$_3$, etc.		
Charred residue	Carbonaceous residue, e.g. from blood. Chemisorbed O$_2$ onto graphite. Graphite surface C, CO, O$_2$, SO$_2$, HCl, Cl$_2$, NaCl.	(II) Varying rates of formation of atoms variable dissociation equilibria in the vapour phase. $M_{o(s)} + C_{(s)} \rightleftharpoons M_{o(g)} + CO_{(g)} \rightleftharpoons MCl_{2(g)} + Cl_2$	Vapour phase interference
Atomic and molecular vapour from sample 2700°C		(III) Light scattering; C, Na$_2$O Molecular absorption; NaCl, broad band maximum near 250 nm.	Spectral interferences

References pp. 430–438

B. Physical and chemical interferences

Physical interferences occur in the condensed phase due to variable surface tension and viscosity of the diluted sample solutions injected onto the graphite surface. Diluted whole blood, serum and urine all have different solution characteristics. Depending upon the diluent, variable diffusion of the sample along and into the graphite surface can lead to analyte loss by variable diffusion through the tube wall during atomization. Carbonaceous and inorganic residues from ashed blood and serum can cause problems with sample injection.

Chemical interferences occur both in the condensed and vapour phases. In the former, atomization rates are modified by bulk matrix constituents such as carbonaceous and inorganic residues from ashed blood. Interactions in the vapour phase between analyte atoms and gaseous molecular species alter the free atom population. Rates of atom formation, atom diffusion from the tube and the effects of interfering species dissociation equilibria are all temperature-dependent so that temporal and spatial changes in the temperature within the graphite tube will lead to variable absorption signals. Physical and chemical interferences in ETA-AAS can be controlled or overcome by a variety of procedures, which include dilution with acids or detergents to overcome sample viscosity, and addition of chemical modifiers such as ammonium dihydrogen phosphate to remove chloride as volatile NH_4Cl, and form thermally stable phosphates with volatile elements such as cadmium and lead. The use of oxygen as an ashing aid within the graphite tube prevents the formation and build up of a carbonaceous residue from clinical samples. Methods to reduce the temporal and spatial temperature variation within a graphite tube atomizer include the L'vov platform, probe atomizers and constant temperature atomizers. These reduce vapour phase interferences by delaying volatilisation of analyte atoms until the rate of change of tube temperature with time is minimal. Some of these techniques are discussed in detail below.

1. *Electrothermal control of interferences*

Electrothermal atomization and atomic absorption for clinical analysis during the late 70s was beset by severe interferences mainly because atomizers operated under non-isothermal conditions. The nature of these interferences has been extensively reviewed by Matousek [15], Slavin and Manning [16] and Frech et al. [17]. Since that time a number of attempts have been made to overcome the temporal and spatial temperature variation of the Massmann design atomizer. One approach has been to increase the heating rate of the graphite atomizer so that high and constant temperature conditions are achieved during the lifetime of the analyte species. Modern instruments using low voltage–high current flow can achieve heating rates up to $2 \times 10^{3}°C\ s^{-1}$. Chakrabarti et al. [18, 19] have used capacitive discharge heating with a bank of condensers to increase this rate up to $10^{5}°C\ s^{-1}$ and have shown a significant reduction in the interference effects from chloride salts.

L'vov [11] was the first to suggest placing a small platform of solid graphite within a tube furnace to delay atomization of the sample into a more stable thermal environment. The platform which has contact with the tube only at its edges is heated principally by radiation from the tube wall. Recent work [20] on measuring the temperature inside a graphite atomizer both with and without a platform using coherent anti-Stokes Raman scattering thermometry has shown that with rapid heating (1800°C s^{-1}) the temperature of the gas phase follows that of the tube wall very closely. There is a small (ca 100°C temperature) difference between the gas phase and the platform at the time the atomizer has reached its preset temperature. The analyte is vaporised at a later stage in the atomization cycle, when the tube wall and vapour phase temperatures are theoretically the same, but are substantially higher than that of the platform. The atoms are therefore released into a relatively more stable thermal environment, in which molecular dissociation is more complete and chemical interferences are significantly reduced [11, 21, 22]. Several authors have claimed interference-free determinations using L'vov platforms for various analytes, in samples as diverse as whole blood [23], seawater [24] and food digests [25]. The use of L'vov platforms is undoubtedly a simple means of reducing many of the interferences of ETA-AAS.

Frech et al. [13] developed a side-heated electrothermal atomizer which is spatially isothermal. They found for non-volatile elements that problems associated with signal tailing, condensation and memory effects were significantly reduced. Atomization temperatures were lower than with a Massmann atomizer, and use of a L'vov platform gave a further improvement in performance. The "two-step" atomizer of Frech and Jonsson [26] consists of a graphite tube which has a hole in the base into which is tightly inserted a graphite cup. The tube and cup are independently heated by means of two separate power supplies. The samples are (manually or automatically) injected into the cup. For atomization, the tube is heated to a preselected equilibrium temperature followed by heating of the cup. In principal, the system is similar to that proposed by L'vov, but has the advantage of a side-heated atomizer to overcome the spatial temperature distribution of L'vov's design. The separation of the volatilisation process from the atomization step results in a considerable reduction in interferences [26], although at the expense of two independent power supplies. This "two-step" atomizer has been used in comparison with a platform atomizer for the determination of lead in biological samples [27].

An alternative approach to the achievement of isothermal conditions is probe atomization. This was originally suggested by L'vov and Pelieva [28,

29], who used a tungsten wire mounted on the side arm of an autosampler. The sample was injected onto the wire and dried prior to introduction into the atomizer, at the optimum atomization temperature. Manning et al. [30] used a tungsten coil and found a reduction in interference effects for lead. However, these systems can only handle very small sample sizes (1–2 μl), and the tungsten wire or coil has a very short lifetime. The use of pyrolytic graphite probes has been extensively investigated by workers at the University of Strathclyde [12, 31–36]. With these, the sample is injected onto a graphite probe which is inserted through a slot cut into the side of the graphite tube. The sample is dried and ashed within the graphite tube. The probe is then removed and re-inserted into the tube after the selected atomization temperature has been reached. Interference effects are much reduced [31–33]. The determination of lead in whole blood has been achieved using aqueous standards for calibration [33]. Subsequent work has examined the use of automatic front- or side-entry probes [34], end-entry probes [35] and end-entry probes in the form of a tube [36]. While the use of probe atomization does lead to a reduction in interference effects, this system does not operate under true isothermal conditions because insertion of a cold graphite probe into a hot vapour will disturb to a small extent the thermal equilibrium established within the tube. The recent production of a commercial probe system [37] will enable this technique to be evaluated in routine trace element analysis.

C. Spectral interferences

The interference resulting from the occurrence of molecular absorption spectra at the resonance wavelength, and scatter of the resonance radiation by species other than the analyte atoms is one of the main potential sources of error in ETA-AAS. The main spectral interferences are:

(a) Spectral-line overlap due to the overlaying of an absorbing line from an atomic species produced from the sample matrix close to that of the analyte atomic line. This type of interference is quite rare.

(b) Molecular absorption due to broad band molecular spectra produced by gaseous molecules from the sample matrix, the spectra of which overlap the analyte atomic line.

(c) Radiation scatter due to solid particles produced in the gas phase from the sample matrix or the degradation of the graphite atomizer wall, which scatters the incident radiation leading to an apparent absorption. This interference is more serious at low wavelengths as from Rayleigh's law of light scattering; the scattered intensity is proportional to 2/4 [38].

Modern AA instruments employ one of three types of automatic background correction systems, whereby the background interference absorbance is electronically subtracted from the total absorbance of atomic absorption plus background absorbance. These systems are continuum source [39],

Zeeman-effect [40] and Smith-Hieftje [41] background correction, respectively. Background correction of transient signals obtained with electrothermal atomization gives rise to special problems due to differences in the spatial and temporal distributions of the atom population and of interfering species. These systems are discussed in some detail in Chapter 2, Section II. For further information excellent recent reviews by Slavin and Carnrick [42] and by de Galan and de Loos-Vollebregt [43] are recommended.

In practice the choice between background correction systems will be determined by the type of atomic absorption equipment used since at this time no manufacturers offer all three systems. The Zeeman-system is undoubtedly better in terms of wavelength proximity than continuum source correction and from the literature this appears also to be true for Smith-Hieftje systems [41]. However, it is our experience that continuum source background correction is perfectly adequate for all clinical analyses except whole-blood selenium and arsenic by ETA-AAS and in this case one can by-pass the problem by using hydride generation methods.

III. SAMPLE PREPARATION FOR ETA-AAS

A. Contamination of specimens

Inadequate control during specimen collection and subsequent preparative treatments generally produces one of two levels of elemental contamination. One is so gross that the elemental contamination would be incompatible with terrestrial life whereas the other increases to a pathological level what would have been a "normal" value. The former is so obvious that the result (after being repeated) would be rejected, but the latter, which is more commonly encountered, is potentially very serious. The only way to ensure freedom from trace element contamination is to check every stage of the analysis starting from specimen collection.

The following discussion indicates the range of elements likely to cause contamination problems in clinical analyses. For a more detailed consideration of these problems the reader is referred to the excellent book by Zieff and Mitchell [44] and to publications specifically concerned with clinical analysis [2, 7, 45, 46].

B. Contamination during collection of whole blood

Venesection using plastic disposable syringes fitted with stainless steel needles, after cleaning the skin with alcohol-impregnated swabs, is generally acceptable for most trace-element analyses. The notable exceptions are cobalt, chromium, manganese and nickel, which are alloying elements of stainless steel and which are present in low concentrations in human serum and whole blood [2]. The concentrations of these elements imparted to the blood/plasma specimens may exceed the physiological concentrations thus precluding detection of any deficiency states. Some authors [2, 4, 6] have eliminated this source of contamination by using for venesection a plastic cannula which is rinsed with the initial portion of blood. Volumes of up to 50 ml whole blood are taken. Such sample sizes are impractical for sick adults and impossible for sick children. However, even if this were not so, the sensitivity of ETA-AAS is not sufficient to detect abnormally low concentrations of these elements. Screening for excessive exposure can be done using blood collected with stainless steel needles. For example, the whole-blood concentrations of cobalt found in specimens from renal patients receiving cobalt sulphate as an erythropoietic stimulant are rarely less than $10 \mu l \ l^{-1}$ and can often exceed $100 \mu l \ l^{-1}$. These are distinctly higher than the "normal" levels which are invariably less than $2 \mu g \ l^{-1}$ of which at least $1.8 \mu g \ l^{-1}$ is contamination mainly from the syringe needles [47].

Collection of capillary blood using prick methods in finger, heel or even ear lobe, suffer more severely than venesection from trace element contamination, especially for ubiquitous elements such as lead and aluminium. Great care is needed when cleansing the skin prior to sampling in order to minimise contamination.

C. Contamination from anti-coagulants and collection tubes

Natural anti-coagulants such as heparin contain endogenous zinc and copper plus elements such as manganese, accumulated during chemical separations and purifications [48]. The chelating agent/anti-coagulant K_2H_2EDTA is difficult to obtain with acceptably low levels of manganese and cobalt and one must check each batch of this reagent. The literature abounds with references to trace element contamination in commercially available blood collection tubes, especially evacuated glass tubes with colour-coded rubber stoppers which can release cadmium, zinc and other elements. Aluminium is released from glass tubes including evacuated blood collection tubes [49].

D. Collection of urine and faeces

The only reliable procedures for collecting urine and faeces for trace element analyses involve the use of acid washed polythene or polypropylene containers. Urine samples are voided into acid-washed urinals, potties or jugs and transferred to acid-washed polythene bottles fitted with polythene screw-caps from which the "black-rubber" seals have been removed. The bottles should contain 100 ml of 10% v/v "AristaR" hydrochloric or nitric acid as a preservative. All containers should be stored inside polythene bags when not in use; this is especially important for collections over long periods, e.g. balance studies, when containers may be left within the ward.

E. Contamination during sample preparation

Trace element contamination can be reduced by washing all the laboratory ware including pipette tips in 10% v/v nitric acid, then deionised water and allowing to dry under dust free conditions. Reagents used in sample preparation should always be of the highest quality available, e.g. "SpectrosoL", "AristaR" or "AnalaR" grades to minimise blank values. In the main, most reagents such as salt solutions, phosphate buffers etc., can be purified by passing the solution repeatedly through a column of chelating ion-exchange resin such as "Chelex-100" until the blank levels are insignificant. Although this procedure takes longer than others, e.g. APDC/MIBK extraction, it generally produces a cleaner reagent and can easily be scaled-up to process large volumes.

The separation of plasma from red cells is best performed using an all-in-one polythene Pasteur pipette with bulb. Rubber teats fitted to glass Pasteur pipettes should never be used because talc and/or dust inside the teat contains zinc, either endogenous or abraded from the rubber, which can be blown into the pipette during separation thus contaminating the plasma specimen. The release of aluminium from the glass can invalidate assays of this element. Homogenization of diets, faeces etc., requires titanium-bladed blenders and homogenizers if contamination with chromium (from stainless steel) is to be avoided. Furthermore, it is necessary to replace copper bushes and some plastic components with components made from PTFE [50].

Surface contamination of lead on teeth can be removed without losses of endogenous lead by washing with hot non-ionic detergent [51] and should be generally applicable to similar solid samples. Alternatively ultrasonication in cold deionised water may be used. A variety of procedures for removal of surface trace element contaminants from scalp hair have been described [52]. However, it is our view that with few exceptions, e.g. monitoring exposure to methyl mercury, or arsenic, that trace element analyses of hair samples are of little or no clinical value. A detailed objective review of this subject has recently been reported by Taylor [53].

F. Sample preparation

The sample preparation procedure used for a given analysis will be determined mainly by the nature of the matrix, solid or liquid, and by the amount available for investigation. Generally, solids will require some form of digestion/dissolution stage, although direct analyses are feasible. While liquids are readily amenable to direct analysis, matrix interferences for some analytes require extensive pre-treatment. The various procedures currently used are discussed in two main sections: analyses requiring extensive pre-treatment, and analyses requiring minimal pre-treatment.

References pp. 430–438

G. Analyses requiring extensive chemical pre-treatment

1. *Decomposition and deproteinisation with acid*

Recent developments in methods for decomposing biological materials have been extensively and excellently reviewed by Knapp [54] and by Angerer and Schaller [55]. A wealth of practical detail is given in discussions of the relative merits (or otherwise) of: wet chemical decomposition both in open and in sealed containers; microwave heating and digestion in sealed vessels at low (up to 0.8 mPa) and high (up to 8.5 mPa) pressures; high-pressure (up to 12 mPa) digestion in sealed quartz or glassy carbon vessels; low-temperature ashing in radio-frequency excited oxygen atmosphere; combustion in oxygen at atmospheric pressure with reflux condensation of volatile material and subsequent acid dissolution, which for brevity is understandably termed "combustion in a dynamic system". The need for careful method development for pressurised digestion of matrices for which reagent systems have not yet been described, is emphasised by the observations of Angerer and Schaller [55] that although the peak pressure obtained on digesting 1 ml blood with HNO_3 was 1.5 mPa, a peak of 9 mPa (90 atm) was obtained on digestion of 1 g fish fillet. For decomposition of small ($\sim$5 mg) weights of tissues, Knapp recommends either microwave digestion in sealed vessels at low pressures or high-pressure acid decomposition in quartz vessels. The former is quicker but the latter is less prone to contamination. Our experience of analyses of tissue biopsies (<3 mg) is that HNO_3 digestion at atmospheric pressure and at 150°C in quartz conical flasks is rapid and efficient. The main problem is weighing the sample and using an endogenous element as internal standard as noted by Knapp [54] is the easiest solution. Stoeppler and Tonks [56] have also made an excellent review of sample preparation procedures. This paper together with that by Knapp [54], and the excellent book of Angerer and Schaller [55] contain all one needs to know of this subject.

Protein precipitation with HNO_3 as treatment for analysis of cadmium in whole blood has been improved by addition of Triton X-100 to allow easier separation by centrifugation of supernate [56]. This procedure has been applied to measuring manganese in whole blood [57] but there is the possibility that manganese bound to haem may not be completely released so that analyses will be biased low. This procedure cannot be recommended for analysis of manganese (or of iron) in serum because of the possible contamination from slight haemolysis of red blood cells.

Given the problems of contamination with acid digestion/precipitation, it would seem prudent to use where possible, the simplest procedure for sample preparation, i.e. dilution with or without a simple matrix modifier. For solid samples acid digestion should be used.

2. *Solvent extraction procedures*

It is likely that blank values from reagents and glassware will limit the use of solvent-extraction and concentration procedures for ETA-AAS. Indeed, there are few examples of single element analyses done this way which could not have been done as well by simpler procedures. These include measurements of physiological levels of Tl, V, and Co. Chandler and Scott [58] extracted Tl directly from 50 ml urine into 3 ml toluene using sodium diethyldithiocarbamate. Direct ETA-AAS analysis of the separated organic phase gave a detection limit of 0.5 nmol l^{-1} and an RSD of 3.5–4.4% at 26 nmol l^{-1}. This method allowed detection of slight increases in exposure of submariners to Tl. Ishizaki and Ueno [59] used 10-50 fold pre-concentrations of V from ashed blood or urine with N-cinnamoyl-N-2,3-xylylhydroxylamine in CCl_4 to determine down to 2 nmol l^{-1}. Very low levels of Co, down to 1 nmol l^{-1} were measured in serum by ETA-AAS [60], following a 12-fold pre-concentration involving $HNO_3/HClO_4$ digestion and APDC/MIBK extraction at pH 9.0. The unselective nature of this reagent suggests the possibility of multi-element pre-concentration.

H. Analyses requiring minimal chemical pre-treatment

1. *Direct injection*

Many of the earlier publications on the use of ETA-AAS to measure trace elements in body fluids advocated little or no sample pretreatment. For the large hospital or clinical laboratory, direct methods appear ideal because they are simple, rapid, involve little sample pretreatment, reduce the risks of contamination and do not require considerable operator skill. However, accurate direct injection of small 10 μl volumes of viscous blood or serum sample is difficult. Poor contact of the blood with the graphite surface can lead to poor precision, frothing can occur during the drying stage [8, 61] and carbonaceous residues can build within the furnace. The use of an ultrasonic probe [62] to ensure homogeneous mixing of the sample solution immediately prior to sample injection may well be an attractive proposition in the future for those elements whose physiological concentration is so low such that any form of dilution is impossible.

2. *Dilution*

There are time-dependent interferences associated with aqueous dilution of whole blood. Within 1–2 minutes a precipitate of the red cell membranes appears and reaches a maximum after about 15 minutes. This is of particular importance when large numbers of samples may be prepared hours in advance of the analysis. Dilution of serum and urine seldom presents many problems; however, increased acid concentrations will eventually cause the precipitation of the serum proteins. Dilution of blood samples with dilute acid followed by ultrasonic mixing is generally unsuccessful, because the combination of the protein/red cell precipitation and ultrasonic agitation generally results in a turbid mixture, which is more difficult to pipet than the original solution.

Triton X-100, a non-ionic surfactant, has found widespread use as a diluent because it produces cell lysis and a clear homogeneous solution from whole blood. Obviously it should be checked for possible contamination prior to use. However, this reagent can diffuse into the graphite, leading to variable rates of atomization. The use of pyrolytic graphite-coated tubes and/or L'vov platforms can reduce but not eliminate this problem. The concentration of Triton X-100 used is critical and the final concentration injected into the atomizer should be preferably less than 0.25% v/v. High >1% v/v solutions of Triton X-100 can completely degrade L'vov platforms after only approximately 100 firings.

An often overlooked reagent is dilute ammonia solution. A 1 + 19 dilution of whole blood, using 1% v/v ammonia solution provides complete cell lysis within minutes of mixing, and diluted samples are stable for up to 48 h when stored at 4°C.

3. Chemical modification—liquid phase

A few simple chemical reagents added either directly to biological samples or to digests minimise molecular absorption interferences by facilitating oxidation of the sample matrix within the atomizer. They can, in addition, minimise the effects of the matrix upon analyte sensitivity by modifying the volatility of the analyte, or of the matrix. Ediger [63] introduced the concept of chemical or "matrix" modification to ETA-AAS when he recommended NH_4NO_3 for removal of NaCl by conversion to NH_4Cl and $NaNO_3$, both of which decompose below 400°C. Since then considerable research has focussed on the role and mechanism of chemical modifiers in both liquid and gaseous phases. A chemical modifier generally enables formation of a more thermally stable, uniform analyte/modifier compound which allows the matrix to be separated from the analyte and removed from the atom cell prior to atomization. This reduces background absorption and gas phase interferences, and prevents the removal of analyte occluded with matrix particles during the initial gas expulsion. It can also delay atomization until a more stable thermal environment is achieved so that residual vapour phase reactions are better controlled. This may also improve the temporal separation of the analyte and background signals, thus reducing further background absorption. However, interactions of the chemical modifier with the graphite surface should not be ignored. By occluding the analyte in its crystal structure, the modifier can reduce the reactions of both analyte and matrix with the graphite surface.

Many reagents have been suggested as chemical modifiers for biological samples. The various ammonium phosphates, either alone or in combination with magnesium nitrate, have been used for the determination of lead and cadmium in blood [8, 64]. These enable pyrolysis temperatures of approximately 650 and 900°C to be used for cadmium and lead, respectively. The use of metals such as Ag, Ni, Cu, Pt and Pd [65–69] have been employed as chemical modifiers for the determination of volatile elements such as As, Se, Hg in whole-blood serum and urine. Some individual chemical modifiers are discussed in more detail under the relevant element sub-headings.

Attempts to find a chemical modifier which is applicable to as many elements as possible, have centred on the use of palladium, originally suggested by Shan and Ni [70], either alone or in combination with magnesium nitrate [71]. It is still unclear whether it is necessary to pre-reduce Pd(II) to Pd metal within the atomizer or whether this reduction can be done effectively using reducing agents such as ascorbic acid, hydroxylamine hydrochloride or even the graphite surface itself [72].

Modern, computer-controlled autosamplers often allowing separate addition of up to two chemical modifiers either before or after injection of the sample solution, and with the ability to thermally pretreat the chemical modifier, allow a considerable degree of flexibility and scope for experimentation.

It should be remembered that pyrolysis temperatures suggested by instrument manufacturers for "standard" chemical modifiers are often derived from studies on dilute (0.2% v/v) nitric acid solutions of the specified analyte. It is essential when developing and/or optimising any electrothermal atomization programme that the effects of chemical modifiers are examined in both aqueous and matrix solutions. For certain elements, i.e. Mn, Al and Cr, in diluted whole blood or urine the marginal advantages to be gained when using a chemical modifier, e.g. magnesium nitrate, are offset by the problems of reducing to an acceptable level the contamination in the chemical modifier solution.

Many authors claim that the use of chemical modification allied with an integrated approach to atomizer chemistry can allow calibration against aqueous standards. This may be so. However, our experience over many years of practical ETA-AAS on aqueous solutions, diluted whole blood, serum or urine has shown that accurate sample injection requires adjusting the distance between the tube wall/platform and the autosampler probe when changing from aqueous solutions to diluted biological samples. For this reason we strongly advocate using matrix-matched standards for calibration in routine analysis.

4. *Use of oxygen ashing; gaseous chemical modification*

The repeated injection of a diluted whole-blood or serum sample into an electrothermal atomizer, regardless of pyrolysis temperature, will eventually

394

result in the gradual deposition within the atomizer of a carbonaceous residue which is resistant to removal by use of a high-temperature "clean" stage in excess of 2700°C [1, 61, 64]. Accumulation of this residue leads to irreproducible sample deposition in the atomizer, variable rates of atomization from the residue and eventually partial occlusion of the light beam. The net result of these effects is a loss of accuracy and of precision. The use of air to prevent these problems was first suggested in 1978 for the determination of Ni in serum by Beaty and Cooksey [73]. The use in situ of oxygen ashing at 500–600°C is far more efficient. It prevents the formation of any carbonaceous residue and in addition, reduces the molecular signal on subsequent atomization [64, 73] (see also Fig. 1).

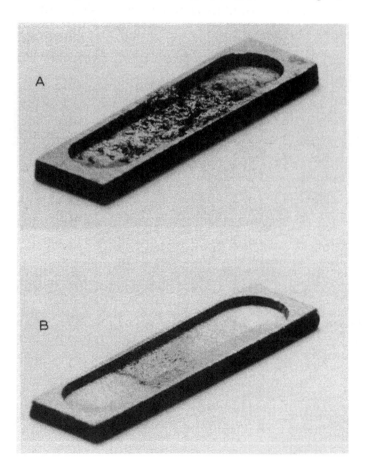

Fig. 1. Effect of pyrolysis in argon, or ashing in oxygen, on a diluted blood matrix on a L'vov platform: A. 50 firings of a diluted whole-blood sample without oxygen ashing; B. 300 firings of a diluted whole-blood sample with oxygen ashing. Samples diluted 1 + 19 with a solution containing 1% v/v ammonia solution, 0.33% m/v $NH_4H_2PO_4$ and 0.11% m/v $(NH_4)_2EDTA$; 10 μl injections for each firing.

The claim that up to 300–400 firings of a 1 + 5 diluted blood solution in the presence of $NH_4H_2PO_4$ is possible without any build-up of carbonaceous residue [74], is difficult to believe when others complain of a gradual build-up of ash [61, 75, 76]. Suggestions to overcome this problem include the use of a pipe cleaner [77], a miniature vacuum cleaner [76] and scraping the residue from a L'vov platform with a scalpel [75, 69, 78] before re-insertion into the tube. Apart from the problems of contamination and the effect on the precision of dispensing the sample into the atomizer, these techniques are hardly appropriate for reliable analyses.

The reaction between oxygen and graphite in an electrothermal atomizer has been well researched [61, 79–83]. The optimum temperature for chemisorption of oxygen on the active carbon surface sites, as the surface oxide, is 500–600°C and the optimum desorption temperature is 950°C [81]. Atomization without desorption of oxygen leads to severe surface attack, removal of carbon as carbon monoxide, and a short graphite tube lifetime. Desorbing oxygen at 950°C produces negligible attack on the graphite during subsequent atomization and can give tube lifetimes of up to 200–250 firings (Fig. 1). We have found [79] that the subsequent atomization temperature following desorption of oxygen at 950°C, has an effect on the tube lifetime. For the determination of lead with an atomization temperature of 1900°C tube lifetimes are 200–250 firings, but for manganese (atomization temperature 2300°C) tube lifetimes are only 100–120 firings. The removal of chemisorbed oxygen as CO_2 and CO at 950°C leaves a defective "active" graphite surface [61, 81]. The extent to which this defective carbon is vaporised during subsequent atomization, will depend on the atomization temperature. Although this is a disadvantage for some involatile elements, a reliable analytical tube lifetime of 100 firings is preferable to the inaccuracy associated with pyrolysis in argon.

Activation of the graphite surface which occurs when oxygen is used as an ashing aid, leads to increased thermal stability of the analyte enabling the matrix to be more efficiently removed at a higher pyrolysis temperature. It is clear that the role of oxygen within a graphite electrothermal atomizer is more than a "simple" ashing aid for biological samples. It is a gaseous chemical modifier, which mediates atom formation via activation of the graphite surface.

IV. ESSENTIAL TRACE ELEMENTS

The physiological concentrations of those elements generally considered as essential for man are listed in Table 2, together with some ETA-AAS performance data. These allow a perspective of the sample volumes required for assays of these elements. These measurements provide valuable data for studies of trace elements in nutrition and of inherited and acquired

TABLE 2

PHYSIOLOGICAL CONCENTRATIONS OF ESSENTIAL TRACE ELEMENTS IN BODY FLUIDS AND SAMPLE VOLUMES REQUIRED FOR ANALYSIS BY ETA-AAS

Element μmol l^{-1}	Matrix	Reference concentrations for healthy controls		ETA-AAS characteristic mass [a], m_0: pg/0.0044 A.s	Sample volume required for analysis [b] μl/0.0044 A.s
		μ/nmol l^{-1}	μg l^{-1}		
Zn	P/S	11–24	720–1570	0.4 [P, C] [c]	0.0006
	U	4.5–9.0	294–558		0.0013
Fe	S	11–36	614–2010	5.0 [P, C]	0.0013
	U	0.2–1.0	11–56		0.46
Cu	S	3.0–11 [d]	191–700	4.0 [P, C]	0.21
	S	12.0–26 [e]	763–1652		0.005
	S	27–40 [f]	1716–2542		0.002
	U	<0.8	<51		0.08
Mn	B	73–210	4–12	2.0 [P, C]	0.5
	S	9–24	0.5–1.3		4
	U	2–27	0.1–1.5		20
Se	B	1.2–2.2 [g]	95–174	28 [P, Zm]	0.30
	B	1.0–2.0 [h]	79–158		0.35
	S	1.1–1.9	87–150		0.32
	U	0.1–0.7	7.9–55		3.5
Cr	B/S	<20	<1.0	3.0 [P, C]	3.0
	U	<20	<1.0		3.0
Mo	S	2–12	0.2–1.2	9.0 [W, C]	45
Co	B	<17	<1.0	6.0 [P, C]	6
	U	<17	<1.0		6
V	B	<6	<0.3	30 [W, C]	100
	U	<6	<0.3		100

[a] Date from Perkin-Elmer data sheet: Part number 0993-8199, Rev B. [b] Calculated for the minimum value of reference range or less than value. [c] P=L'vov platform atomization; W = wall atomization using pyrolytic graphite coated tubes; C = continuum source background correction; Zm = Zeeman background correction. [d] Neonates. [e] Children over 6 months and adults. [f] Healthy pregnant women. [g] Adults. [h] Children: Se concentrations is age.
N.B. Urine values assuming 24 h collection.

metabolic disorders. Most elements may be determined directly by ETA-AAS using small sample volumes, but it is clear that pre-concentration is needed for measurements of physiological levels of V and Mo.

A. Zinc

ETA-AAS is only of value for measuring total zinc in serum when sample volumes are severely limited, e.g. studies of neonates. The very high analytical sensitivity (8.6×10^{-13} g/0.0044 A) combined with the relatively high zinc concentrations normally present in serum ($>10 \ \mu$mol 1^{-1}) would produce peak absorbances at 213.9 nm in excess of 1.0 A from only microlitre sample volumes. Analytical sensitivity at the alternate resonance line 307.6 nm is too low to be of any use in clinical studies so that all ETA-AAS methods involve some means of reducing sensitivity without compromising accuracy. Levi and Purdy [84] used manual injections of 800 nl undiluted serum for direct analysis by ETA-AAS, which has the disadvantage of measuring and injecting accurately nanolitre sample volumes. Viera and Hansen [85] used large (1 + 99) dilutions of serum and μl injection volumes which suffers from the problems of contamination associated with high dilutions. For example, at the lower limit of zinc concentration in serum for healthy controls, a 1 + 99 dilution yields a solution containing only 110 nmol 1^{-1}, which is within an order of magnitude of the detection limit. An alternative approach combined a smaller (1 + 24) aqueous dilution of serum with a gas phase dilution of zinc atoms by using an increased argon gas flow within the furnace during atomization [86]. Increasing the internal gas flow from 10 to 120 ml min^{-1} reduced the peak absorbance from 5 μl injections of a 1 + 24 dilution of serum from >1.0 to <0.25 A. This gas phase dilution combined with the 25-fold liquid phase dilution is equivalent to a 1 + 99 aqueous dilution but with a much reduced degree of contamination. The detection limit of 0.36 μmol 1^{-1} was 30 times lower than the lower limit of zinc in serum for healthy controls, 11 μmol 1^{-1}, and the method had a useful working range of 2.5–40 μmol 1^{-1} zinc in serum.

Matrix interferences in ETA-AAS prevent the use of simple aqueous solutions for calibration of serum zinc analyses even at very high dilutions. Viera and Hansen [85] were only able to compensate for matrix effects in 1 + 99 dilutions of sera by the addition to calibrating standards of albumin solution which had been dialysed against a complex mixture of electrolytes and by using a calibration algorithm to deal with the continuously curved calibration response. Similar interferences were observed by Foote and Delves [86] but these varied both in nature and in magnitude with the type of graphite used in the manufacture of the furnace. Using standard dectrographite tubes, but not pyrolytic graphite-coated tubes, it was possible to compensate for matrix interferences by adding to aqueous standards albumin at 4–40 g l^{-1} and sodium chloride at 0.14 M. Calibration graphs could be used for direct absorbance–concentration conversion without recourse to mathematical processing. This ETA-AAS method agreed excellently with FAAS for determinations of zinc in sera from 50 patients with a variety of clinical conditions giving rise to serum zinc levels ranging from 5.4 to 24.8 μmol l^{-1} (r = 0.94). This method has been applied successfully to studies of the distribution of zinc among serum proteins using affinity and gel-filtration chromatography [87, 88]. Measurements of low molecular weight zinc complexes separated from 400 μl serum using ultrafiltration were made using an ETA-AAS programme with an argon gas flow of 50 ml min^{-1} during atomization [89]. Using minimal dilution (1 + 3) of the serum ultrafiltrate a detection limit of 51 nmol l^{-1} and a working range of 50–300 nmol l^{-1} were obtained with a RSD at 90 nmol l^{-1} of 10.8%.

B. Copper

The most important clinical applications which require the use of ETA-AAS for measuring copper are in the analyses of specimens in which the concentration of copper is very low, e.g. urine, or solutions which are not suitable for nebulisation, such as seminal plasma tissue biopsies or where only small volumes may be available, e.g. sera from neonates.

Carrelli et al. [90] determined copper in urine by ETA-AAS by using standard additions to overcome the small but nonetheless significant variable suppression of analyte sensitivity by the urine matrix. Sample preparation was simply a 1 + 4 dilution with water. A detection limit of 3 nmol l^{-1} was obtained using 50 μl injection volumes and RSD's were less than 3% over the range 0.15–31 μmol l^{-1}. A similar methodology but using only 20 μl injection volumes has been used in the authors' laboratory for some years. The reliability and simplicity of the method more than compensates for the disadvantage of having to use standard additions. The sample solution (20 μl) is dried at 120°C, ashed at 900°C and atomized at 2000°C, followed by a clean-out step at 2700°C.

Pleban and Mei [91] measured copper at concentrations of 0.47–3.15 μmol l^{-1} in seminal plasma using ETA-AAS following a simple 1 + 4 dilution in water and calibration using matrix-based standards. Spermatozoa were also analysed but these required separation by centrifugation, lyophilization and digestion with nitric acid. The RSD's were 5% at 2.3 μmol l^{-1}, and 2% at 1.81 μg g^{-1}. Matrix-based standards were used for calibration at copper concentrations ranging from 0.15 to 0.63 μmol l^{-1}. The limited number of cells available for analysis, matrix viscosity and relatively low copper concentrations precluded the use of FAAS for these measurements.

A formidable challenge in trace element analysis has been the determination of very low concentrations of copper (and iron and zinc) in very small 150–200 μl volumes of aqueous humour [92, 93]. The latter workers used a 1 + 2 dilution (50 μl + 100 μl) with 1% v/v HNO_3 to overcome matrix interferences, and with 20 μl injection volumes a linear calibration from 0 to 79 nmol l^{-1} (0–5 μg l^{-1}) was obtained with a detection limit of <3 nmol l^{-1}. The electrothermal programme is very similar to that given above for urine copper analysis. The precision of better than 3% at 90 nmol l^{-1} was more than sufficient to demonstrate significant changes in copper concentrations under different storage conditions. The mean concentration of 142 nmol l^{-1} found in specimens was similar to, but lower than the 220 nmol l^{-1} reported by McGahan and Bito [92].

Determinations of total copper in serum by ETA-AAS represent some of the easiest of all graphite furnace methods. The concentration range 12–26 μmol l^{-1} is 150–330 times higher than the calibration range for the analysis of aqueous humour, and approximately 20 times higher than that for urine analysis. Accurate (100 $\pm$ 3%) and precise (1.4–2.1% RSD) data were reported in some of the better publications as long ago as 15 years [94, 95]. These analytical performances are similar to those currently obtained using simpler FAAS methods. It is probably that the most useful application of ETA-AAS to serum analysis will be in studies of the low molecular weight species present only at low concentrations. Kamel et al. [96] have already published an ETA method for this application in which ultrafiltratable (RMM < 1000) copper complexes at 63 nmol l^{-1}Cu were detected in plasma.

For most clinical purposes the determinations of copper and zinc in biological specimens are easily accomplished with good accuracy and precision using methods based on flame atomic absorption spectrometry. Electrothermal atomization and atomic absorption only becomes the method of choice when the sample sizes are severely limited and/or the elemental concentrations are below 1 μmol l^{-1}.

C. Iron

Serum iron and total iron binding capacity of serum may be measured by ETA-AAS, provided care is taken to eliminate potential interference from haemoglobin-bound iron by precipitation with tri-chloraceticacid [97]. However, these measurements are more easily done in routine clinical laboratories by using automated calorimetry. The main uses of ETA-AAS in clinical studies of iron are measurements of urinary iron and of iron in tissue biopsies. The former is easily accomplished by using standard additions in presence of 1% v/v HNO$_3$ as matrix modifier. Tissue samples require digestion with HNO$_3$ prior to analysis by standard additions. In our laboratory liver biopsy specimens of <1 mg wet weight, too small to be weighed accurately, have been analysed successfully using Mn, and Cu, as endogenous internal standards. The Fe/Mn and Fe/Cu ratios may be used as indices of excessive iron accumulation (haemachromatosis). The ETA programme is simply a ramp dry/ash to 250°C, cool to 20°C, then atomize at

2600°C. The RSD's are 1.0% to 2.4% for 20 to 50 $\mu g\ l^{-1}$ of Fe in solutions of ashed liver biopsies.

D. Selenium

The last few years have seen an enormous increase in Se determinations. "Atomic Spectrometry Updates for Clinical Materials, Foods and Beverages", published in 1986 [98] listed only two papers concerned with Se measurements by ETA-AAS, while in the 1989 review [99] there were 18 papers and 4 reviews.

There are four main problems with the determination of Se by ETA-AAS: (1) pre-atomization/volatilisation losses on pyrolysis; (2) spectral interferences, mainly due to Fe and P at the 196.0 nm Se resonance line; (3) chemical interferences; and (4) formation of carbonaceous deposits. In order to understand and to overcome these problems, fundamental studies have been made of reactions of Se with various chemical modifiers [100, 72], its interaction with the graphite [100, 101] and the nature of the spectral interferences [102] at the Se resonance line. Welz et al. [103] have shown that the valency state of Se affects the efficiency of the various chemical modifiers. Since Se(IV) is used exclusively for standards, severe errors could result when attempting to use an aqueous calibration. Using Ni alone or with $Mg(NO_3)_2$ it was not possible to stabilise Se(VI), whereas Cu and $Mg(NO_3)_2$ was able to stabilise organoselenium, Se(IV) and Se(VI) up to 1100°C.

Carnrick et al. [66] attempted to measure Se in urine and serum using a Ni chemical modifier with $Mg(NO_3)_2$ as an ashing aid, together with a L'vov platform and Zeeman-effect background correction. Their recoveries of added Se ranged from 89 to 104%, and they found variable losses of Se prior to atomization, mainly in matrices containing sodium and sulphate. The use of oxygen ashing, and other chemical modifiers, Ag, Mo, or Cu, did not improve the situation. Paschal and Kimberly [104] used only a Ni modifier with a 1 + 1 dilution of serum. Matrix-matched standards were employed along with a L'vov platform and Zeeman-effect background correction. With 10 μl injections a characteristic mass of 34 pg/0.0044 A was found, with between-run precisions of 3.8% at 1.5 $\mu mol\ l^{-1}$ and 4.8% at 4.6 $\mu mol\ l^{-1}$. A Ni modifier with Zeeman-effect background correction was also used by Lewis et al. [105] and by Edwards and Blackburn [106] for whole blood.

The combined Cu and $Mg(NO_3)_2$ chemical modifier has been used by several workers for the determination of Se in serum [67, 68] and whole blood [107]. Sampson [68] used a L'vov platform and deuterium background correction. Spectral interference from Fe and phosphate was minimised by using an 0.7 s read delay and peak height measurements. To reduce the overall background signals to within the capabilities of the D_2 system a 1 + 2 dilution with water was used. 20 μl of the diluted sample plus 10 μl of the modifier (0.19% m/v $Cu(NO_3)_2$, 1.09% m/v $Mg(NO_3)_2$ and 0.1% Triton X-100) were injected. An air ashing step was used to reduce carbon build-up on the platform. Even with this preparation calibration curves varied widely with different samples and the method of additions was employed. Recoveries were 91.7% to 111.8% over the concentration range 0.7 to 3.0 μmol l^{-1}. Between-batch precision was 9.7% at 1.3 μmol l^{-1}. This method has been validated against fluorometry and also hydride generation AAS [108]. All methods gave virtually identical results for certified reference materials. It is interesting to note that the fluorometric and hydride methods were able to assay Se in whole blood, washed and haemolyzed erythrocytes, plasma, urine, haemodialysis and water samples.

Saeed [69] determined Se in whole blood with deuterium background correction. This method used a chemical modifier of 0.1% m/v Ni and 0.4% m/v Pt (serum) or 0.8% m/v Pt (whole blood) in 3.5% v/v HNO$_3$. Bauslaugh et al. [109] had shown the addition of Pt to have two beneficial effects: it reduced the interference from phosphate presumably by converting the PO$^+$, or P$_2$ molecules into free phosphorus atoms, and it delayed the appearance time of iron through alloy formation, without influencing the Se signal. Serum was diluted 1 + 2, and whole blood 1 + 4 with 0.3% v/v Triton X-100. 10 μl injections were taken with 15 μl volumes of the chemical modifier. Wall atomization from pyrolytic graphite tubes with integrated absorbance measurements were chosen as this gave better sensitivity and precision. It was clear from this work that a problem was encountered with carbonaceous residues accumulating in the atomizer. L'vov platforms were not used due to the difficulty of scraping the residues from the platform and repeatedly re-inserting the platform into the tube. Less problems were encountered with wall atomization and it was easier to remove the residue. The method was validated against hydride generation AAS and neutron activation analysis, and through the use of certified reference materials. The between-batch precision was less than 10% for 1.2 μmol l^{-1}.

Eckerlin et al. [110] used a combined chemical modifier of 0.09% m/v PdCl$_2$ and 1.25% m/v Ni(NO$_3$)$_2$ in 0.25% HCl and 0.1% Triton X-100 with Zeeman-effect background correction to measure Se in bovine and equine blood. Samples were diluted 1 + 9 with the modifier and 20 μl aliquots taken. Calibration was against aqueous standards. Air ashing at 500°C reduced but did not completely remove the carbonaceous residue, which was in accordance with our experience. The use of oxygen can be far more effective and completely eliminates the build-up of carbon. Recoveries for 1.3 μmol l^{-1} additions of Se ranged from 95 to 117% and the detection limit was 0.13 μmol l^{-1}. The authors stress that this method has not been formally validated as it has only been used to monitor trends in blood Se levels resulting from supplementation studies. They found a consistent 5% negative bias against a fluorometric procedure, which could be due to inter-laboratory differences or loss of organoselenium which may not be stabilised by the Pd/Ni chemical modifier.

For laboratories involved in a wide range of routine trace element determinations in clinical materials, the methods discussed abqve all present a serious disadvantage. The chemical modifiers suggested all involve injecting μg amounts of routinely measured elements such as Cu, Ni and Pt. While there are suggested procedures to clean up the Ni contamination [111], the only reliable way is to change the atomizer contact cylinders/cones. This can become expensive when it has to be carried out on an almost daily basis. Not all laboratories can afford the luxury of an instrument dedicated to one analysis.

Since its first suggested use by Shan and Ni [70], many workers have investigated the chemical modification potential of Pd, either alone or in combination with $Mg(NO_3)_2$ or with various reducing agents. Voth-Beach and Shrader [72] have examined the effects of reducing agents such as 5% hydrogen in argon, hydroxylamine hydrochloride and ascorbic acid. It appears that the presence of a reducing agent ensures that the Pd is reduced to metal early in the atomizer programme, and provides a more consistent performance. Knowles and Brodie [112] applied a reduced Pd modifier and Zeeman-effect background correction to the determination of Se in whole blood. Whole blood was diluted $1 + 4$ with a solution containing 0.5% v/v Triton X-100, 0.125% v/v antifoam B and 0.25% m/v ascorbic acid in water. The diluted blood samples were vortex-mixed to produce lysis, centrifuged, and the clear red supernatant taken for analysis. A 20 μl aliquot of the modifier (500 μg ml^{-1} Pd as $PdCl_2$ was injected followed by 20 μl of the sample. Mixing the ascorbic acid and Pd solutions caused precipitation problems with the blood samples. A maximum ashing temperature of 1200°C was used. Calibration was by standard additions. Problems were encountered with the build-up of carbonaceous residues, although pre-injection of the modifier reduced this. It was claimed that introducing air for 5 s at 400°C after atomization assisted in removing this residue. From our experience the carbonaceous residue remaining after heating a blood matrix at 2700°C is extremely resistant to removal, even when oxygen is subsequently used. While this method appears promising, results are presented for only two samples from an inter-laboratory study. The within-batch precision was between 2.5 and 8.7% for Se levels in whole blood of 1.0–2.0 μmol l^{-1}. These same workers have applied this method to the determination of Se in serum [113]. Ten μl aliquots of $1 + 1$ diluted samples were injected onto a 10 μl aliquot of pre-injected Pd modifier. It would appear that direct calibration can be used. Analysis of a Seronorm reference material gave 1.09 ± 0.02 μmol l^{-1} ($n=5$), compared with a certified value of 1.14 μmol l^{-1}.

Neve et al. [114] measured Se in human seminal plasma. Samples were diluted 300 μl + 700 μl with a chemical modifier containing 0.05% m/v Cu, 0.1% m/v Mg (as the nitrates) and 0.015% Triton X-100 in 0.25% m/v HNO_3. 15 μl aliquots were taken and ashed in situ, with oxygen at 450°C. Zeeman-effect background correction was used. A matrix-matched calibration curve gave data that agreed well with results obtained by standard additions and an established acid digestion and liquid–liquid extraction method [115]. Within- and between-batch precisions were 8% and 9.7% for 0.2 μmol l^{-1} Se and 4.7% and 5.6% for 0.60 μmol l^{-1} of Se, respectively. For 46 normal subjects a mean Se concentration of 0.54 ± 0.2 μmol l^{-1} was found with a range of 0.14–0.94 μmol l^{-1}.

Subramanian [9] has indicated the difficulties of measuring Se in urine by ETA-AAS. The work of Carnrick et al. [66] has been discussed earlier. Saeed [69] applied the serum and whole-blood method using a Ni/Pt/ HNO$_3$ modifier. Urine was diluted 5-fold with distilled water and 10 μl injections were taken plus 10 μl of modifier. For some urine samples the Pt concentration had to be increased, which as Bauslaugh et al. [109] have shown, suppressed the Se signal.

Lindberg et al. [116] have measured Se in pancreatic ß-cell pellets. Solid sampling was used with the Perkin-Elmer "cup-in-tube" solid sampling accessory along with Zeeman-effect background correction. Using radio-active tracer studies with ^{75}Se, Pd was found to be a more efficient modifier than Ni. While for a standard reference material (NIST 1577a Bovine Liver), calibration was only possible by aqueous additions to the solid sample, the cell pellets presented a more "favorable" matrix and calibration was possible against aqueous standards. Reasonable agreement was found for 4 certified reference materials. The detection limit with a typical sample mass of 0.5 mg was 150 μg g^{-1}. The overall precision was of the order of ±10% at 1000 μg g^{-1} and ±30% at 280 μg g^{-1}.

The use of graphite furnace ETA-AAS for the determination of Se in biological body tissues and fluids presents a considerable challenge to the analytical chemist. Hydride generation AAS procedure following acid digestion can measure accurately Se in whole blood, serum, cell fractions, urine, faecal material, muscle and skin tissues and mixed diet samples [117, 118] and in our view remains, for the time being, the technique of choice.

E. Manganese

Measurements of manganese are important in monitoring occupational exposure. Chronic toxicity from inhalation of manganese dusts is char-acterised by Parkinson-like neurological and physiological manifestations. Occupational exposure to manganese is accompanied by increased blood and urinary concentrations of this element, but at present there are no recommended action levels for these biological indicators [119]. Whole-blood manganese levels of 26–32 nmol l^{-1} in welders of stainless steel were significantly greater ($P < 0.05$) than the 14 ± 2 nmol l^{-1} found in control subjects [120]. These concentrations are much lower than data from the United Kingdom: for whole blood 90–270 nmol l^{-1} [121], 18–180 nmol l^{-1} [47]; and 13–23 nmol l^{-1} in plasma [1].

Measurements of manganese are also becoming increasingly important in clinical conditions not associated with occupational exposure. Elevated plasma manganese levels were seen in patients with liver metastases: mean 34 nmol l^{-1} compared with less than 18 nmol l^{-1} in controls [122]. Very high concentrations, up to 300 nmol l^{-1} in whole blood and up to 100 nmol l^{-1} in plasma, have been observed in patients receiving total parenteral nutrition [47]. Children with Perthes disease—a congenital hip malformation—had concentrations of manganese in whole blood which, although within the reference range (between 73 and 210 nmol l^{-1}), were significantly lower than an age- and sex-matched control group [123]. Clearly, the roles of manganese in both whole blood and in serum need to be evaluated in health and disease.

The literature concerning the determination of manganese in body tissues and fluids by ETA-AAS has been extensively reviewed [8, 9, 10, 124, 125], with particular attention given to the problems of contamination during collection and preparation of samples. Sample pre-treatments are the following: direct injection [126, 127]; dilution with water [128], Triton X-100 [4, 129, 130, 131, 132], ethylene glycol [133]; chemical modification with ammonium oxalate [134], ammonium acetate [132] or magnesium nitrate [135]; and de-proteinisation with nitric acid [5]. Direct injection of serum or whole blood usually leads to frothing and spluttering of the sample during drying and offers no practical advantages over dilution. Apart from de-proteinisation with nitric acid [5] or trichloracetic acid [136] most of the other treatments will lead to the deposition of a carbonaceous residue in the graphite tube. However, neither Halls and Fell [128] nor Uschida and Vallee [130] appear to have experienced this problem when injecting 1 + 1 diluted serum.

Char/ashing temperatures which have been suggested for the determination of manganese in serum and whole blood range from 450°C to 1400°C in the presence of Mg(NO$_3$)$_2$ [4, 135]. We have found that without chemical modification, Mn in whole blood and serum is thermally stable to 1100°C [79]. Slavin et al. [135] have recommended the use of Zeeman-effect background correction since deuterium-source background correction may not be totally effective at the 279.5 nm line. This may be true for older instruments [5]. However, we have found no difference between analyses made using a Perkin-Elmer 4000 instrument with D$_2$ background correction and a Zeeman/5000 instrument for blood manganese measurements following oxygen ashing at 550°C and desorption at 950°C.

Most published work uses calibration by standard additions [129, 130, 131], to overcome the residual matrix interferences, with wall atomization from pyrolytically coated graphite tubes. Halls and Fell [128] prefer to use electrographite tubes and calibration against aqueous standards using a 1 + 1 aqueous dilution of serum or urine and 20 μl injections. The within-batch precision for serum analysis was 7.5% at 11 nmol l^{-1} and 2.4% for urine at 211 nmol l^{-1}. The lack of interferences could be due partly to the use of an argon gas flow rate of 10 to 30 ml min^{-1} during atomization.

Subramanian and Meranger [5] used a nitric acid de-proteinisation of serum to overcome the problems of carbonaceous residues forming in the tube and still found it necessary to use matrix-matched standards prepared in bovine serum to obtain reliable data. Their detection limit was 2 nmol l^{-1} with a between-batch precision of 17.5% to 3.8% over the range 11–37 nmol l^{-1} serum manganese. Allain et al. [57] warn that the use of acid precipitation for determining whole blood Mn may give low results because endogenous Mn remained bound to haem, which was precipitated by the acid. They recommend the dilution of whole blood with water and the use of oxygen ashing.

Subramanian and Meranger [5] found no improvement in sensitivity with use of a L'vov platform but gave no indication of whether peak or integrated absorbance measurements were compared. Bayer [131] used a L'vov platform to determine manganese in whole blood diluted 1 + 4 with 0.5% m/v Triton X-100, and still needed to use standard additions for calibrations even when magnesium nitrate was used as a chemical modifier. Others [4], however, used platform atomization and reported identical slopes for serum and aqueous calibration curves. They used ramp heating (over 1 second) rather than the maximum heating rate to reach the atomization temperature. This is not the usual mode of heating, and from our experience can give poor atomic signals. The within-batch precisions of 5–15% at high serum Mn levels of 45–89 nmol l^{-1} [4] are perhaps indicative of this.

The prevention of accumulation of carbonaceous residues for whole-blood analysis has been achieved using in-situ ashing with air [137] or oxygen [57, 79, 138]. In our procedure [79] no chemical modification is required other than in-situ oxygen ashing at 550°C to remove the organic sample matrix, followed by desorption at 950°C; atomization is from a L'vov platform. Sample preparation is a simple dilution: 1 + 3 for whole blood and 1 + 1 for serum, with 0.5% v/v Triton X-100 solution. Tube lifetimes are typically 100 firings. The calibration graph is established with matrix-matched standards and using integrated absorbance measurements. The detection limit is 4 nmol l^{-1} in serum. The within-run precisions are 7% for serum manganese at 14 nmol l^{-1} and 5% for 150 nmol l^{-1} of whole-blood manganese. The accuracy of the method was demonstrated by satisfactory recovery studies: 104% for additions to serum of 9–36 nmol l^{-1} of manganese, and 100% for additions to whole blood of 36–182 nmol l^{-1} manganese. The use of conventional needle and syringe sampling techniques was shown to give whole-blood and serum values within the accepted normal ranges (for serum manganese 7–18 nmol l^{-1})

F. Chromium

Although chromium is an essential trace element, it is toxic at high levels of exposure. The toxic nature of Cr(VI) is now well established and the measurement of Cr at industrial levels presents relatively few problems. There is now general consensus that normal unexposed levels of Cr in urine and serum are lower than 20 nmol l^{-1} [139]. In the past low results were claimed to be due to volatilisation losses of organically bound Cr. However, recent studies have not substantiated this and from careful radio-tracer studies significant losses of Cr only occur above 1200°C [140, 141].

The determination of Cr in whole blood has been reviewed by Subramanian [8, 10]. Ottaway and Fell [139] in an excellent review provide detailed sampling precautions and critically evaluate many recent publications. McAughey and Smith [142] developed a method for the determination of whole-blood Cr in industrial workers as it is known that Cr(VI) is taken up by erythrocytes where it binds to haemoglobin; consequently blood Cr may reflect a better measure of long term exposure, while urinary Cr reflects recent uptake. The data from Morris et al. [143] on plasma, urine and erythrocyte Cr levels of welders supports this hypothesis. McAughey and Smith [142] diluted blood 1 + 2 with 0.1% v/v Triton X-100 and measured integrated absorbance signals following atomization from the wall of a pyrolytically coated tube. A gas flow of 20 ml min^{-1} was found to reduce readsorption of Cr at the cooler tube ends. Background correction was achieved using a tungsten-iodide lamp as the energy output of the D$_2$ lamp at 357.9 nm was too low. Matrix-matched standards prepared in human whole blood were used for calibration. With a 20 μl injection the detection limit was 11.5 nmol l^{-1}, with recoveries of 102 and 101% at 38.5 and 96.0 nmol l^{-1} of added Cr, respectively. Within- and between-batch precisions were respectively 3.7% and 3.6% at 143 nmol l^{-1} Cr.

Schermaier et al. [144] diluted whole blood 1 + 3 and serum 1 + 1 with 0.2% Triton X-100 and used air ashing at 700°C. With 40 μl injections the detection limits were 2 nmol l^{-1} and 0.7 nmol l^{-1} for whole blood and serum, respectively. The within- and between-batch precisions were 4.65% for 3.3 nmol l^{-1} in whole blood and 5.26% for 3 nmol l^{-1} in serum.

Normal serum Cr concentrations are in the range of 2 to 4 nmol l^{-1} (0.1–0.2 μg l^{-1}) [139] which is close to the detection limit of ETA-AAS of 1–3 nmol l^{-1} (0.05–0.16 μg l^{-1}). Many workers use standard additions to overcome the matrix interferences [143, 145, 146, 147]. Taylor and Green [148] found improved sensitivity for Cr in serum by using totally pyrolytic graphite tubes. Samples were diluted 1 + 1 in 0.1% m/v Triton X-100 and 0.1% v/v HNO$_3$. No advantages were found with the use of either Mg(NO$_3$)$_2$ or (NH$_4$)$_2$HPO$_4$ as chemical modifiers. Calibration was achieved using matrix-matched standards in the range 0–96 nmol l^{-1} and standard additions were used for higher levels. The detection limit (2σ) using 10 μl injection volumes was 2 nmol l^{-1}. The within- and between-batch precisions were 4.8% at 21 nmol l^{-1} and 9.7% at 8.9 nmol l^{-1}, respectively.

Veillon et al. [149] lyophilised and dry-ashed sera in silanised quartz tubes using $Mg(NO_3)_2$ as an ashing aid. The ash was re-dissolved in 0.1 M HCl and Cr determined with the included $Mg(NO_3)_2$ as a modifier. A mean normal serum Cr level of 0.21 ± 0.013 nmol l^{-1} was found for 15 adults. Lewis et al. [150] also used this dry-ashing procedure for determination of Cr by ETA and a simultaneous multi-element atomic absorption spectrometer. The marginal increase in pyrolysis temperature afforded by the use of chemical modifiers such as $Mg(NO_3)_2$ has to be balanced against the problems of reducing the Cr contamination in the modifier. While several reviewers [139, 151] argue that the use of a L'vov platform rarely gives any increase in detection limit or sensitivity, it should not be overlooked that one advantage of the L'vov platform is that atomization is taking place under more stable conditions. This often results in more reliable methods.

Urine Cr is by far the more frequently used index of exposure and the suggested procedures range in their degree of complexity. The literature up to 1987 has been reviewed by Subramanian [9]. The determination of urine Cr has been extensively investigated by Halls and Fell, who used both D_2 and Zeeman-effect background correction [152] to show that the interference experienced by earlier workers using D_2 background correction systems was due to emission from the chromium, potassium and sodium in the urine matrix. They found that the interference could be easily eliminated by using a lower atomization temperature of 2400°C with only a small reduction in sensitivity [152, 153]. By diluting urine $1 + 1$ with 1% v/v HNO_3 and 0.25% v/v Triton X-100, and taking 20 μl injections, a detection limit (2σ) of 0.19 μg l^{-1} was obtained. The within-batch precision was 4.5% at 1.3 μg l^{-1}. Calibration was against aqueous standards. Comparison of this method with that using Zeeman-effect background correction and ETA-AES with wavelength modulation has given good agreement [154].

Using a tungsten-iodide lamp for background correction Marks et al. [155] analysed urine diluted $1 + 1$ with 0.1% v/v HNO_3 and 0.05% v/v Triton X-100 and calibrated analyses using matrix-matched standards. With 20 μl injections a detection limit of 10 nmol l^{-1} was obtained. Excellent agreement was found between this method and that used by Halls and Fell [153]. The former workers were able to double the number of samples analysed per day compared to their older standard additions procedure. McAughey and Smith [142] used a direct method without dilution: a withan instrument with a tungsten-iodide background correction lamp, an atomization temperature of 2500°C and an internal gas flow during atomization of 50 ml min^{-1}. Calibration was achieved using matrix-matched standards and integrated absorbance measurements. With 20 μl injections the detection limit (3σ) was 3.8 nmol l^{-1}. Within- and between-batch precisions were 3.4% at 150 nmol l^{-1} and 5.4% at 153 nmol l^{-1}, respectively.

Dube [156] using Zeeman-effect background correction found a detection limit (3σ) with 40 μl injections of 1.7 nmol l^{-1}. Matrix-matched standards were used for calibration. The within- and between-batch precisions were in the range 7–12%, respectively, but no indication is given of the relative concentrations. Good agreement was found in a comparison with four other laboratories and an external quality assurance (EQA) scheme. Minor modifications to the atomizer programme, including air ashing at 500°C, enabled the method to be used for the determination of whole-blood and serum Cr.

G. Cobalt

Cobalt is an essential constituent of the cyanocobalamine complex B$_{12}$. At pharmacological doses, oral cobalt sulphate stimulates erythropoiesis and has been used to correct anaemia in uraemic patients. Although it is now rarely prescribed, Co levels in blood need to be monitored because of possible toxic side effects, including fatal cardiomyopathy. There is also a need to monitor persons who are occupationally exposed to this element either from welding fumes or from dusts/powders.

The normal range for whole-blood and urine cobalt is less than 17 nmol l^{-1}, most of which is cobalt contamination from the stainless steel needles used during sampling. Versieck and Cornelius [7] have shown the true normal range for serum cobalt to be 3–5 nmol l^{-1}, which is near the detection limits of ETA-AAS. The reviews by Subramanian [8, 9] give a detailed overview of recent analytical work.

Anderson and Hoegetveit [60] determined low levels of Co in serum using a liquid–liquid extraction procedure with APDC/MIBK at pH 9 ± 0.1, after digesting the samples in a mixture of 14 M nitric acid and 12 M perchloric acid. Six ml of serum was taken and the final MIBK extraction volume was 0.5 ml. Total volumes of 100 μl were injected into the furnace in 20 μl aliquots. These injections had to be performed manually but modern ETA autosamplers would allow automatic repeated injections. The recovery of the procedure, using ^{60}Co was 97.1 ± 1.24%, the detection limit was 1 nmol l^{-1} and a precision of 9.5% was obtained at 3.4 nmol l^{-1}. Serum Co concentrations were 2.5 ± 1.2 nmol l^{-1} for 32 unexposed controls but refinery workers exposed to CoCl$_2$ aerosols had levels of 164 ± 120 nmol l^{-1}. Although such large volumes of serum are rarely available for routine measurements, this procedure shows the improved detection limit achieved by liquid–liquid extraction/concentration for ETA-AAS. Furthermore, other elements could easily have been determined in the same MIBK extract.

Angerer and Heinrich [157] developed a direct ETA-AAS method for the occupational monitoring of Co in whole blood using a 1 + 8 dilution of whole blood with 0.001% Triton X-100. This large dilution was necessary to overcome matrix interferences. Good agreement was found with a polarographic method but the detection limit of 34 nmol l^{-1} was too high to allow normal levels to be assessed. These authors measured urine cobalt [158] by extraction of 4 ml urine with 1 ml of hexamethylene–ammonium hexamethylenedithiocarbamidate in a (3 + 7) mixture of xylene and di-isoproyl ketone prior to ETA-AAS. Using 50 μl aliquots of the organic extract the detection limit was 1.7 nmol l^{-1}. Subsequently these authors [159] compared the determination of Co in whole blood by two acid deproteinisation procedures, one using HNO_3 and the other a HNO_3–$Cr(NO_3)_3$ mixture. Both gave low results, 64% and 93–100% losses of cobalt, respectively, and only a direct method gave satisfactory recoveries. It would appear from this work that protein precipitation with HNO_3 does not appear to be a viable method for determining Co in whole blood.

For the determination of cobalt in urine, Bouman et al. [160] compared a direct method using a Zeeman-effect background correction instrument and a liquid–liquid extraction procedure. Both gave good detection limits of respectively 8.5 nmol l^{-1} and 3.4 nmol l^{-1}. Matrix-matched standards were used with the direct procedure but the extraction procedure required the use of standard additions.

Although pre-concentration procedures give reliable results under carefully controlled conditions, direct analysis is to be preferred, for increased speed of analysis and reduced contamination. Delves et al. [161] diluted whole blood 1 + 1 + 1 with 1% v/v $NH_4H_2PO_4$ and 0.5% v/v HCl to measure occupational exposure and assess the clinical management of patients on $CoSO_4$ therapy. Substituting 0.5% Triton X-100 for the HCl gave an improvement in drying and pipetting precision. The organic matrix was removed by the use of oxygen ashing at 600°C followed by a subsequent pyrolysis at 1000°C. During atomization at 2300°C the small residual molecular signals (<0.1 A) were easily compensated for by continuum source background correction. Calibration was achieved by matrix-matched standards. The detection limit was 3.4 nmol l^{-1} and the RSD at 339 nmol l^{-1}, 4.8%. Using this method with a Zeeman-effect instrument, the authors have found no difference in analytical results obtained using wall or platform atomization; however, the use of a platform gives improved graphite tube lifetimes. Although Slavin [162] recommends using the slightly less sensitive line at 242.5 nm for Zeeman-effect instruments, we have worked successfully at 240.7 nm. Though this line is subject to "roll-over" above 0.6 A, this magnitude of signal is unlikely to be encountered even when measuring Co in exposed populations.

Kimberley et al. [163] used ETA-AAS with Zeeman-effect background correction to measure urine cobalt levels in workers in the tungsten carbide industry. Samples were diluted 1 + 1 with a chemical modifier of 1% m/v $Mg(NO_3)_2$ in 4% v/v HNO_3. A pyrolysis temperature of 1450°C was followed by wall atomization at 2400°C. The calibration curve was established using integrated absorbance measurements and matrix-matched standards in a urine containing no detectable Co. The variation in calibration slopes in different urines was less than 3%. A detection limit (3σ) of 44 nmol l^{-1} was reported.

Sampson [164] examined the use of Pd as a chemical modifier for Co in plasma and urine. The use of Pd allowed an increase in pyrolysis temperature up to 1250–1300°C, compared with 1000°C with $NH_4H_2PO_4$. Air ashing at 600°C was used to destroy the organic matrix and prevent the build-up of carbonaceous residues. However, the Pd chemical modifier gave a background signal of approximately 0.3 A. To 400 μl of sample were added 25 μl of the chemical modifier consisting of 9 g l^{-1} Pd, 1% v/v Triton X-100 and antifoam; 40 μl volumes of this mixture were taken for analysis. Matrix-matched standards were used to establish a calibration with wall atomization at 2500°C. A detection limit $(3\ \sigma$ of 2.5 nmol l^{-1} was found. The within- and between-batch precisions at 20 nmol l^{-1} were 3.5% and 17.7%, respectively. For the determination of Co in urine the simple procedure developed by Smith [165] can be recommended. 20 μl aliquots of urine are injected directly into the atomizer and analysed against a calibration curve prepared in urine from a non-occupationally exposed subject. A pyrolysis temperature of 1200°C and a wall atomization temperature of 2400°C are used with integrated absorbance measurements.

H. Molybdenum

The concentrations of molybdenum in blood/serum/urine of healthy unexposed subjects (26 nmol l^{-1}) are too low to be measured accurately by ETA-AAS. Typical detection limits range from 3 to 20 nmol l^{-1} [166]. Himmel et al. [167] reported a detection limit of 10 nmol l^{-1} for direct injections of 10–20 μl volumes of plasma. The samples were dried at 120°C, pyrolysed at 1800°C and atomized at 2650°C. Most (69%) healthy blood donors examined were below this level and nearly all (95%) were below 20 nmol l^{-1}. However, dialysis patients had much higher concentrations: 2100 nmol l^{-1} pre-dialysis and approximately 50 nmol l^{-1} post-dialysis. One possible way of increasing analytical sensitivity which would be suitable for urine analysis is preconcentration by anion exchange separation. Valazquez-Gonzalez et al. [168] used this technique to obtain a 10-fold concentration step for measuring 3 to 7 nmol l^{-1} Mo in water with RSD's of 9.6% and 4.0%, respectively. The electrothermal programme was similar to that used by Himmel et al. [167], and neither group appeared to have experienced losses of MoO_3 during ashing at >1100°C [169]. Another way to increase ETA-AAS sensitivity for Mo is to use chemical modification with BaF_2. This has reduced the characteristic mass from 12.9 to 5.6 pg, and was applied successfully to analysis of digested NBS bovine liver SRM [169].

I. Vanadium

With few exceptions, measurements of V in body tissues and fluids by ETA-AAS are restricted to those involving increased exposure. The normal concentrations in blood/urine of healthy unexposed people are about 5–9 nmol l^{-1} and with a characteristic mass of 30 pg, at least 100 μl is needed to produce only 0.0044 A. Missenard et al. [170] used a direct ETA-AAS method in which sera were diluted $1 + 1$ with 0.1% m/v Triton X-100, and three consecutive 30 μl volumes were deposited into the graphite tube. The samples were dried and ashed at 1400°C after each addition, with a final atomization at 2700°C. Even with this large sample size ($\equiv$45 μl serum) all unexposed subjects had serum concentrations below the detection limit of 40 nmol l^{-1}. However, occupationally exposed subjects had much higher levels, 60–740 nmol l^{-1}. The within-batch RSD at 300 nmol l^{-1} was better than 5%. An important observation was the strong positive correlation between serum V concentrations and zinc protoporphyrin levels, indicating inhibition of haem biosynthesis by vanadium. Accurate analysis of the low levels in unexposed subjects would require a much greater pre-concentration, e.g. 10- to 50-fold as reported by Ishizaki and Ueno [59]. These workers used liquid–liquid extraction/preconcentration from large volumes of blood (10 ml) or urine (50 ml) prior to analysis by ETA-AAS. They achieved RSD's of 4.3% at only 9 nmol l^{-1} in blood and 11.7% at only 5 nmol l^{-1} in urine.

V. THERAPEUTIC TRACE ELEMENTS

Non-essential elements whose compounds have therapeutic uses are listed in Table 3. No discussions are presented here of ETA-AAS analyses of Au, Bi, Ga, either because there is little which is new, or because alternative techniques are preferable, e.g. FAAS for Au and hydride generation (AAS for Bi). The emphasis given to Al and Pt is in our view deserved because of their increasingly important clinical interest and because of the analytical challenge in measuring them.

A. Aluminium

Routine analytical services for monitoring potential aluminium toxicity are provided by many hospital pathology laboratories. The patients most at risk are those who have chronic renal disease. As part of their treatment they receive oral doses of aluminium hydroxide gels to inhibit intestinal absorption of phosphate and they may also be exposed to aluminium from their dialysis fluids [171]. Aluminium toxicity has also occurred from the use of some infant formula feeds and intravenous fluids [172] and there are reports of very high concentrations of aluminium in some protein infusion fluids [173, 174]. The possible link between aluminium and Alzheimer

TABLE 3

REFERENCE AND THERAPEUTIC LEVELS OF TRACE ELEMENTS IN BODY FLUIDS AND TISSUES

Element	Matrix	Concentrations in tissues/fluids				ETA-AAS characteristic mass [a]	Sample volume required for analysis [b]
		Reference levels		Therapeutic levels			
		μmol l^{-1}	μg l^{-1}	μmol l^{-1}	μg l^{-1}	m_0: pg/0.0044 A.s	μl/0.0044 A.s
Al	S	<0.4	<11	1.8–22	48–593	10 [P, C]	0.9
	U	<2.0	<54	>15	405		0.2
	CSF	0.15–0.19	4–5	0.22–1.26	6–34		2.5
Au	S	<0.01	1.97	5–40	985–7879	8.5 [P, C]	4.3
Bi	S	<0.05	<10	0.2–14	42–2926	19.0 [P, C]	1.9
	U	<0.05	<10	0.2–14	42–2926		
Ga	S	<1	70	up to 14	976	12.0 [P, C]	0.17
	U	<1	70				
Pt	B	<0.005	<1	0.1–50	20–9755	95 [W, C]	95
	S	<0.005	<1	0.1–50	20–9755		
Ag	B	<0.01	<1	–	–	1.5 [P, C]	1.5

[a] Data from Perkin-Elmer data sheet: Part Number 0993-8199 Rev B.
[b] Calculated from the lower level of the reference range.

disease has stimulated environmental studies at lower exposure levels and assessments of the general levels of dietary intakes of aluminium from foods and beverages.

1. *Measurement of aluminium*

The high analytical sensitivity of ETA-AAS (10 pg Al/0.0044 A) enables the very low concentrations of aluminium normally present in human tissues (<1 μmol kg^{-1}) to be measured accurately with very small sample sizes, e.g. 1–5 μl serum. The main difficulties with clinical aluminium analyses by ETA-AAS are the effect of matrix composition on analyte sensitivity and contamination.

It is essential to maintain a high standard of cleanliness and to pay strict attention to practical details, no matter how apparently insignificant, in order to eliminate adventitious contamination from aluminium. For example, contamination caused by tearing a paper towel in the vicinity of uncapped solutions has ruined a whole days work at low concentrations $<$ 0.4 μmol l^{-1} (<10 μg l^{-1}). Laboratory dust can contain 0.2 to 2.0% w/w of aluminium so that a few dust particles of 50 μm diameter falling into a 1 ml volume of solution would increase the concentration by 1 μg l^{-1}. Furthermore, single particles of 20–100 μm diameter entering the furnace atomizer itself would produce absorbance signals at 309.3 nm of 0.010–1.100 A. For these reasons all analyses in the authors laboratory are done in triplicate and a good within-sample RSD (usually $<$3%) can be taken as freedom from random contamination. Skelly and Di Stefano [175] minimised background contamination for analysis of Al in brain tissue and in bone by using a "clean room" and sealed vessel microwave digestion with HNO$_3$. A 10-fold reduction in blank value lowered the detection limit from 6.4 to 0.6 μg kg^{-1} and reduced the repeat rate needed because of spurious contamination from 12.5% to 1.9% of samples analysed.

Matrix interferences are more easily controlled. The main interferences from biological samples are volatilisation losses of Al$_2$Cl$_6$ from chloride-rich media, and variable enhancement of analyte atomic signals from PO$_4^{3-}$, from Ca^{2+}, and from carbonaceous residues formed during electrothermal decomposition. A variety of procedures, such as liquid–liquid extraction, acid digestion, and matrix modification, have been used to reduce, as far as possible, levels of chloride and of carbonaceous residue before atomization of aluminium. Buratti et al. [176] extracted aluminium directly from 0.5 ml whole blood diluted 1 + 4 with H$_2$O, or 0.5 ml undiluted serum, or 0.5 ml undiluted urine using 2,4-pentandione (0.1 ml) and 4-methyl-2-pentanone (0.5 ml). The extraction was quantitative (97%) from pH 3.0 to pH 9.0 so that no buffering of specimens was required. The organic phase was separated by centrifugation and 40 μl volumes were injected into a graphite tube furnace for analysis by ETA-AAS. Calibration curves established by extracting aqueous standards (0–3 μmol l^{-1}) were parallel to those constructed by standard additions to blood or plasma or urine. Results for Al in plasma obtained by solvent extraction agreed excellently with those obtained by aqueous standard additions: $r = 0.992$, $n = 20$; concentration range studied 0.4–3.4 μmol l^{-1}. The detection limit was 0.02 μmol l^{-1} and the within- and between-run RSD's were 5.0% and 5.4%, respectively, for Al in plasma in the range 0.2–2.0 μmol l^{-1}. Frech et al. [177] evaluated a range of sample preparation procedures and ETA conditions for measuring Al in whole blood, plasma/serum. They recommended HNO$_3$ digestion for removal of the organic matrix and minimised blank values to 0.04 ± 0.02 (mean $\pm$ SD) μmol l^{-1} by using relatively large sample volumes (1 ml). Their ETA conditions included using a L'vov platform and H$_2$/Ar gas mixtures during ashing and atomization stages. The importance of using background correction at low analyte concentrations was stressed. Calibration was by standard additions. The concentrations of Al in whole blood of healthy controls ($N = 22$) of 0.06 ± 0.07 μmol l^{-1} (mean $\pm$ SD) were generally lower than other published values. The recovery of Al added to blood at 0.34 to 1.5 μmol l^{-1} was $98 \pm 9.6\%$ (mean $\pm$ SD). Brown et al. [178] precipitated serum proteins using HNO$_3$, yet still found that standard additions to different sera, especially uraemic sera, gave different calibration slopes. Others have used a variety of procedures to enable direct calibration with similarly treated aqueous standard solutions. Alderman and Gitelman [179] diluted sera, from 1 + 1 to 1 + 4, with a diluent containing NH$_4$OH/Na$_2$H$_2$EDTA/H$_2$SO$_4$/Triton X-100, ashed the sample in situ in argon at 1580°C and atomized at 2600°C using a molybdenum-coated graphite tube. Others have used atomization from a L'vov platform following matrix modification with Mg(NO$_3$)$_2$/Triton X-100 [180] or Mg(NO$_3$)$_2$ with in-situ oxygen ashing [181]. It is our experience that in-situ oxygen ashing is essential to achieve parallel slopes for different sera. This removes completely the carbonaceous residues from ETA-AAS analyses of sera, which would otherwise enhance in an uncontrolled way the rates of formation of aluminium atoms via reduction of the oxide. It does not, however, compensate for the effect of matrix on analyte sensitivity, and matrix-matched standards are needed. Although matrix modification using either (NH$_4$)$_2$HPO$_4$ or Mg(NO$_3$)$_2$ can extend the ashing temperatures to 1300–1400°C before losses of Al are observed, they offer no overall practical advantage. A simple 1 + 3 dilution with water and ashing (in Ar) at 1200°C following oxygen ashing at 550°C and desorption at 650°C, minimises contamination from reagents and provides accuracy of routine analyses [182]. Calibration is achieved using matrix-matched standards prepared in diluted bovine serum. The detection limit of 0.02 μmol l^{-1} and between-run RSD's of 1.1% to 3.3% at 0.5 to 2.9 μmol l^{-1} compare favourably with the more complicated procedures outlined above. As with most trace element determinations the simpler the procedure the more reliable it is in routine analyses.

Aluminium assays currently constitute the most commonly requested clinical trace element assays in the United Kingdom. These provide useful but limited information on monitoring excessive exposure to this element and the biochemical and clinical sequelae. The high levels of serum aluminium in sera of some pre-term infants fed parenterally [183, 184] give cause for concern over the aluminium content of these infusion fluids, as do the high levels of aluminium in conventional feeds and beverages [185] which could be given to young children. In spite of such studies and earlier evidence of aluminium toxicity as an iatrogenic complication of chronic renal disease, there is still no good biological index of potential aluminium toxicity. Total serum concentrations of 10 to 20 times the upper limit for healthy controls (3.7–7.4 μmol l^{-1}) are often encountered in patients with chronic renal failure on dialysis and receiving aluminium hydroxide gels, yet these values are of little predictive value for bone disorders or neurotoxicity [186]. Although aluminium concentrations in cerebrospinal fluid are reported to give a better index of neurological impairment [187] there remains a need for a good diagnostic indicator of early risk of aluminium toxicity.

Measurements of total serum aluminium concentrations provide limited information because there are at least three important carrier species. About 20% of serum aluminium is bound to low molecular weight species, RMM < 5000–10,000 Da, whereas the remainder is bound to proteins of molecular weights between 60 and 60 kDa, mainly transferrin but also albumin [188]. It is probable that the low molecular weight species will play an important role in the transport of aluminium to the brain and other tissues and that measurements of these species will be of greater clinical value as an index of potential aluminium toxicity than a total serum concentration. Evidence that dietary citrate can enhance the intestinal absorption of aluminium from diets and from aluminium hydroxide antacid tablets [189] is one example of the importance of speciation studies. Another, which indicates a need to study tannin complexes and bioavailability is the increased urinary excretion of aluminium in healthy persons drinking tea, known to accumulate naturally high levels of aluminium, rather than coffee or water [190]. Additional information on aluminium exposure has been provided by analyses of bone [191]. Bone samples are usually extracted with petroleum ether to remove fatty material before digestion with HNO_3. The concentrations of aluminium in bone biopsies (usually iliac crest) in the absence of increased exposure are usually in the range 2–10 mg kg^{-1} (dry weight). A higher concentration of 24 mg kg^{-1} was seen, however, in a patient taking aluminium preparations for treatment of peptic ulcer [192] and grossly elevated levels up to 500 mg kg^{-1} can be seen in renal patients on dialysis [193].

Monitoring dialysis fluids to prevent excessive exposure to aluminium is becoming increasingly important. Halls and Fell [194] successfully analysed a range of dialysates by ETA-AAS using addition of 2% v/v HNO_3 as modifier to prevent chloride mediated losses of aluminium during electrothermal decomposition. Although most dialysates contained <0.04 μmol l^{-1} (1 μg l^{-1}) some contained up to 1.2 μmol l^{-1}. Analysis of dialysis concentrates which contain approximately 40% m/v dissolved solids cannot be done by direct dilution without losing analytical sensitivity. However, a simple chelating resin ion-exchange separation [195] allows analyses down to 0.08 μmol l^{-1} (2 μg l^{-1}).

Concern over the possible relation between environmental aluminium exposure and Alzheimer disease [196] has prompted studies of all forms of input of this element including that from foods. Some foods, such as tea, accumulate naturally high levels of aluminium. Others contain aluminium as a result of adventitious contamination during processing and packaging and from the use of permitted additives such as aluminium phosphates, aluminosilicates and aluminium colorants [197, 198]. The concentration of aluminium in food as purchased ranges from <0.7 nmol kg^{-1} (<0.02 μg g^{-1}) up to 60 mmol kg^{-1} (0.16% m/v), the latter figure being the addition of aluminosilicates to some icing sugars [185]. The current situation regarding aluminium in food and the environment is presented in an excellent Royal Society of Chemistry publication [199].

2. *Quality of aluminium analysis*

At a recent consultative meeting on the quality of clinical aluminium analysis it was reported that only about one-third of laboratories participating in a UK-EQA Scheme were competent in doing serum aluminium analyses [200]. This situation is worse for water and dialysis fluids. The situation, which is equally bad elsewhere in Europe [201], is nothing short of disgraceful. Twenty years after L'vov's pioneering work on ETA-AAS the avant garde of atomic spectroscopy debates the possibility of standardless analysis; it would appear that some aluminium measurements are already performed this way.

B. Platinum

Platinum co-ordination complexes have been investigated as potential anti-cancer agents since 1972. The most successful of these is Cisplatin: cis-diamminedichloroplatinum(II). This provides effective treatment of testicular and ovarian cancers, and head and neck malignancies. The high intravenous doses required for effective therapy, typically of the order of 4 μmol (825 μg kg^{-1}), produce harmful side-effects, particularly nephrotoxicity and nausea. Consequently, the determination of Pt in whole blood, serum and urine is important in studying its mode of action and controlling the toxic side effects.

Direct measurements of Pt by ETA-AAS are restricted to concentrations greater than 0.25–0.51 μmol l^{-1} due to the relatively poor sensitivity of the technique. Therapeutic levels, up to 0.1–50 μmol l^{-1}, can be easily measured; however, characterisation of the various Pt binding species that may be present, is difficult if not impossible with currently available instrumentation.

Delves [1] and Subramanian [8, 9] have reviewed the earlier sample digestion procedures. Matsumoto et al. [202] showed that the presence of nitric acid depressed the Pt atomic signal, but that the addition of ammonia removed the nitrate interference. Others [203, 204] have shown that both HCl and HNO$_3$ have some depressive effect, up to 30% on the Pt signal from ashed blood samples. The use of a chemical modifier consisting of 0.42 M ammonia, 0.33% m/v NH$_4$H$_2$PO$_4$ and 0.11% m/v (NH$_4$)$_2$H$_2$EDTA overcame this interference and gave parallel calibration curves for ashed blood samples and aqueous standards [203].

McGahan and Tyczowska [204] injected 10 μl aliquots of undiluted plasma, urine or tissue digests directly into the atomizer pyrolysis at 1500°C and atomization from a L'vov platform at 2650°C. No problems appear to have been experienced with the build-up of carbonaceous residues within the atomizer. The detection limit was 0.029 μmol l^{-1} with a within-batch RSD of 2.6% at 10.3 μmol l^{-1}. These authors found low results, of the order of 50%, on dry-ashing tissue samples, but could offer no explanation other than that the same loss was experienced in the ashing tubes, even when no tissue samples were present. Delves and Shuttler [203] found poor sensitivity using a L'vov platform and preferred wall atomization with maximum power heating to 2700°C. Blood and serum samples were diluted 1 + 2 with water and 10 μl injections taken. After five injections, the accumulated carbonaceous residue severely degraded the analytical signal. While oxygen ashing overcame this problem, very short tube lives were experienced (20–25 firings) due to increased rate of removal of the activated graphite surface at the high atomization temperature. For this reason dry ashing of the blood samples prior to analysis was preferred. The detection limit was of the order of 0.05–0.08 μmol l^{-1}. Queralto and Rodriguez [205] diluted plasma 1 + 3 with 25% Triton X-100 and used Zeeman-effect background correction. Surprisingly, these authors did not report any problems with graphite tube lifetimes at this high concentration of Triton X-100. The high pyrolysis temperatures that can be used with Pt, produce low background signals which are easily compensated with continuum source correction. Erythrocyte [206] and salivary [207] Pt levels have been measured after patients were treated with Carboplatin.

Reed et al. [208] used the repeat injection facility of a modern ETA autosampler to measure Pt in DNA extracts. It is important to know how much Pt is bound to DNA since this is directly related to tumour cell kill. Calibration was against aqueous standards, using wall atomization. Recoveries greater than 90% were obtained from spiked DNA extracts using a 0.05 μmol l^{-1} Pt solution. The detection limit depended upon the mass of

DNA taken into the atomizer, but these authors were able to measure one Pt atom per million nucleotides using less than 200 μg of DNA.

C. Silver

The marked antibacterial properties of silver, its compounds and alloys have been exploited widely in dental and pharmaceutical preparations and in implanted prostheses. Measurement of the normal unexposed concentration of silver in whole blood, less than 10 nmol l^{-1}, requires sensitive analytical methods and removal of chloride interference without losing silver from the graphite atomizer at temperatures greater than 400°C. Most authors have used chemical modification to overcome these problems. Starkey et al. [209] diluted whole blood 1 + 2 with a diluent containing 0.1% Triton X-100 v/v and 0.4% m/v NH$_4$H$_2$PO$_4$ and used 20 μl volumes for peak absorbance measurements using wall atomization. They found little variation (<7%) in standard addition slopes prepared in 15 different blood samples, hence they used matrix-matched standards for calibration. The detection limit was 1.7 nmol l^{-1}. 51 non-exposed subjects had a mean Ag concentration of 3.2 nmol l^{-1} with a range of 0–11.1 nmol l^{-1}. Anderson et al. [210] and Julshamn et al. [211] measured Ag in serum, hip-joint fluid and tissue digests which were diluted 1 + 4 with NH$_4$H$_2$PO$_4$ chemical modifier. Platform atomization, integrated absorbance measurements and Zeeman-effect background correction were used, but it was still necessary to use the method of standard additions for accurate results.

Nitric acid deproteinisation has been used [212] to study whole-blood and urine Ag levels in rats which had been implanted with different silver materials. The detection limits were 22 nmol l^{-1} and 13 nmol l^{-1} in whole blood and urine, respectively. These authors found from "in vitro" studies that acid deproteinisation did not release all the silver, which is tightly protein-bound, and used matrix-matched standards in equine blood for calibration. This probably explains the higher detection limit obtained in whole blood, and indicates that one of the direct procedures described above may be more reliable. Manning and Slavin [213] have suggested that Pd may be a better chemical modifier than ammonium dihydrogen phosphate, but to date no applications have appeared with respect to clinical analysis.

VI. TOXIC TRACE ELEMENTS

Non-essential trace elements and toxic trace elements are listed in Table 4. None have any known biological functions which are essential to man, and all can induce toxicity states when present in body tissues and fluids at concentrations in excess of the stated reference levels. No discussions are given here of those elements which are more easily determined by cold-vapour AAS (Hg) or hydride generation—AAS (As, Sb, Te).

TABLE 4
CONCENTRATIONS OF NON-ESSENTIAL, TOXIC TRACE ELEMENTS IN BODY TISSUES AND FLUIDS

Element (unit)	Matrix	Reference concentrations for exposed healthy controls		ETA-AAS characteristic mass [a]	Sample volume required for analysis [b]
		μmol/nmol l^{-1}	μg l^{-1}	m_0: pg/0.0044 A.s	μl/0.0044 A.s
Pb (μmol l^{-1})	B	0.2–1.0[c]	41–207	10.0 [P, C]	0.24
	B	0.2–1.2[d]	41–249		0.24
	U	0.05–0.4[e]	10.83		1.0
Cd (nmol l^{-1})	B	1.8–27[f]	0.2–3		1.8
	B	1.8–54[e]	0.2–6	0.35 [P, C]	1.8
	U	<27	<3		0.12
Be (nmol l^{-1})	U	44–100	0.4–0.9	0.5 [P,C]	1.25
Ni (nmol l^{-1})	S	19–14	1.1–0.8	13.0 [P, C]	11.8
	U	7.68	0.4–4.0		
As* (nmol l^{-1})	B	7–125	0.5–9.4	15.0 [P, Zm]	30
	U	40–700[g]	3–53		5
Hg* (nmol l^{-1})	B	<30	<6	72.0 [P, C]	12
	U	<50	<10		7.2
Sb* (nmol l^{-1})	B	8	1.0	22.0 [P, Zm]	22
Te* (nmol l^{-1})	B	2	0.26	17.0 [P, Zm]	65
Tl* (nmol l^{-1})	U	<1.0	<0.2	19.0 [P, Zm]	95

[a] Data from Perkin Elmer data sheet: Part number 0993-8199, Rev B. [b] Calculated from the lowest level of the reference concentration or less than value. [c] Children and men. [d] For women. [e] Men. [f] Non-smokers. [g] Smokers. N.B. Urine values assuming 24 h collection.

* Elements measured more easily by alternative techniques, i.e. cold vapour (Hg) and hydride generation (As, Te, Sb).

A. Lead

Lead is a non-essential toxic trace element with no known beneficial biological role. Much debate [214, 215] has centred round the most suitable tissue samples to assess the body burden of lead. The most useful index of environmental and other excessive exposure to inorganic lead is whole-blood lead analysis [216, 217], and the majority of methods discussed here are concerned with this measurement.

The number of papers published on the determination of lead in body tissues and fluids is extensive (see reviews: [8, 9, 10]). The proliferation of methods for the determination of lead in whole blood shows no signs of abating, even though many recent papers offer little that is fundamentally new. Few show comparisons of results with other methods, rarely they have more than ten "real" samples analysed, and seldom are there results for the many certified reference materials now available for lead.

The measurement of lead in whole blood and urine requires the elimination, or strict control, of the chemical effects of the matrix upon the analyte sensitivity and of the considerable molecular absorption interferences that occur during atomization at the most suitable wavelengths for lead. The last six years have seen the introduction of a number of techniques, such as Zeeman-effect background correction, the L'vov platform, sophisticated chemical modification schemes designed to overcome or reduce the problems inherent in the determination of lead in body tissues and fluids, and application of the "Stabilised Temperature Platform Furnace" (STPF) concept.

One of the earliest, reliable methods for determining lead in blood [218], involved deproteinising blood with nitric acid to overcome the problems associated with sample dilution, residue build-up and background absorption. Subsequent modifications have included the addition of Triton X-100 to achieve complete separation of the proteins from the supernatant fluid [56]. This method is in wide use in a number of reputable laboratories and produces good results [219, 220]. Halls [221] and Subramanian [220] have shown for these deproteinised solutions that the ashing stage in the atomizer programme can be omitted to give much shorter atomizer programmes, without loss of accuracy.

Direct analysis of diluted blood requires reduction of the molecular absorption interferences by a pyrolysis or ashing stage prior to atomization. Losses of lead occur above 350–400°C with untreated blood samples so that some form of chemical modification is essential. The more common reagents suggested for the determination of lead in blood are: $(NH_4)_2HPO_4$ [61, 222, 223, 226], $NH_4H_2PO_4$ [74, 224, 225]; NH_4NO_3 [226]; and $NH_4H_2PO_2$ combined with $Mg(NO_3)_2$ [225, 227]. From combined atomic absorption and mass spectrometry results Bass and Holcombe [228] have suggested that $Pb_2P_2O_7$ is the most probable species formed during pyrolysis. Ammonium dihydrogen phosphate with or without magnesium nitrate has rapidly become the most popular chemical modifier. However, the maximum ash/char temperatures which can be attained within the atomizer before loss of lead, vary from paper to paper and cover the range 550–950°C.

For assessing environmental blood lead levels as part of a UK national survey to ascertain the effect of reducing petrol lead levels, Shuttler and Delves [229] developed an ETA-AAS procedure using all the STPF parameters. Sample preparation was a simple 1 + 19 dilution with a chemical modifier consisting of 0.14 M ammonia, 0.003 M $(NH_4)_2H_2EDTA$ and 0.029 M $NH_4H_2PO_4$. The dilute ammonia solution was used to produce cell lysis, the addition of $(NH_4)_2H_2EDTA$ stabilised the lead solutions at the elevated pH (8.5–9.0). Further chemical modification was achieved with $NH_4H_2PO_4$ and using in-situ oxygen ashing during electrothermal sample decomposition at 550°C. This reduced background levels to less than 0.2 A. Desorption of oxygen at 950°C without loss of lead gave graphite tube lifetimes of 200–250 firings. The calibration curve was established with matrix-matched standards and integrated absorbance measurements, and was found to be linear up to 5 μmol l^{-1}. The detection limit was 0.08 μmol l^{-1} and within- and between-batch precisions were 6.9% and 7.3%, respectively, at 0.5 μmol l^{-1}. The performance of this method was monitored continuously by comparison with MSFAAS [230], by use of internal quality control samples, by participation in EQA schemes and by exchange of samples with another laboratory.

It is interesting to compare this method with that developed by Miller et al. [231] for the third American National Health and Nutrition Examination Survey (NHANES III). The objective of these authors was to establish a simple, rugged method, that could be transferred to other laboratories and enable small changes in blood lead concentrations to be measured. Sample preparation consists of a simple dilution (1 + 9) with a chemical modifier containing 0.5% v/v Triton X-100, 0.2% v/v 16 M HNO_3 and 0.2% m/v $(NH_4)_2HPO_4$. The method has a detection limit of 0.07 μmol l^{-1} and a precision of $\pm$2–5% in the 0.24–2.4 μmol l^{-1} range and was linear up to 4 μmol l^{-1}. Atomization was from a L'vov platform, using a 1 second ramp to 2400°C, which was claimed to give better precision, and with a 20 ml min^{-1} argon gas flow during atomization to reduce the background signal. The higher than optimum atomization temperature (2400° vs 1800°C) was used because at 1800°C the build-up of carbonaceous residues was prohibitively high. These authors also established the accuracy of their method by analysing reference materials, participation in EQA schemes and by comparison of results obtained within the same laboratory using MSFAAS.

Novel AAS methods suggested for the determination of lead in whole blood have included the use of a constant temperature furnace [27], probe atomization [33]. The blood lead determinations on six reference samples by constant temperature furnace atomization [27] were obtained using

References pp. 430–438

an aqueous calibration curve and showed good agreement with the mean values. The use of probe atomization with aqueous calibration solutions gave good agreement for eleven quality control samples [33]. However, the advantages claimed for all of these techniques have to be weighed against the disadvantages of their greater complexity, compared to the L'vov platform approach. As a commercial probe atomizer has now been introduced [37], it will be interesting to see what advantages this has over the use of a L'vov platform.

The literature up to 1987 for the determination of Pb in urine has been reviewed by Subramanian [9]. Slavin [162] claims that urine lead down to 0.005 μmol l^{-1} can be successfully measured in 20 μl of a 1 + 1 dilution of urine in a chemical modifier of $NH_4H_2PO_4/Mg(NO_3)_2$ using STPF conditions. This has not been our experience. Although in-situ oxygen ashing, chemical modification and a L'vov platform do allow matrix-matched standards to be used on occasions, urine matrices are so variable that some samples give anomalous results due to non-parallel calibration slopes. Calibration by standard additions to each sample has been found to be necessary.

Xaio-Quan and Zhe-Ming [232] used a 2.5-fold dilution with 0.05 M HNO_3 and chemical modification using Pd and La. The Pd stabilised the lead to 1200°C and the La removed interferences due to sulphate. Even so recoveries were only 80–87% and calibration by standard additions was necessary. No one has yet appeared to try using a reduced Pd modifier for the determination of lead in urine. Paschal and Kimberley [233] used chemical modification with 1% HNO_3 and a 1 second ramp, with a gas flow, during atomization. However, matrix-matched calibration was necessary as only 85% recoveries were obtained. Lian [234] applied a similar method but used La/HNO_3 chemical modification.

For the determination of lead in teeth and bones the method of Keating et al. [235] is simple, straightforward and of proven performance. Samples are dissolved in HNO_3, evaporated to dryness and redissolved in 5% v/v HNO_3 for direct analysis against standards prepared in the same concentration of acid. Accuracy was demonstrated by recovery studies, analysis of IAEA RM-H5 animal bone and by inter-laboratory comparison.

A number of workers have used solid sampling to determine lead in tissues. The paper by Chakrabarti et al. [236] is worth examination by those interested in this technique. Often these methods assay reference materials, which are already in a dry, homogeneous state and do not address the problems of obtaining representative small, approximately 1 mg samples from tissues.

B. Cadmium

Cadmium is a non-essential toxic trace element, that has no known biological function. Absorbed cadmium is excreted very slowly (approximate half life 37 years), and continued exposure is associated with pulmonary, renal and hepatic damage. Measurements of whole-blood and urine cadmium are important for monitoring human exposure to cadmium. Whole-blood levels provide an indication of recent exposure. Urine levels may give an indication of the body burden when exposure is low, e.g. environmental exposure; however, exposure to higher levels with excessive accumulation in renal cortex is accompanied by greatly increased urine cadmium concentrations.

The determination of whole-blood and urine cadmium has received much attention in the literature and has been comprehensively reviewed [1, 8, 9, 10, 56]. The levels of cadmium in whole blood and urine for non-smoking, non-occupationally exposed subjects are less than 27 nmol l^{-1} and often in the range 1.8–5 nmol l^{-1}. Numerous procedures have been suggested for the analysis of cadmium, ranging from acid digestion with liquid–liquid extraction to direct methods.

The acid precipitation method for determining whole-blood cadmium, first proposed by Stoeppler and Brandt [237], and later modified by Stoeppler and Tonks [56] with the addition of Triton X-100 to aid in the separation of the precipitated protein and supernatant, is still in wide use. One part whole blood is mixed with 3 parts 1 M HNO_3 and 0.1% v/v Triton X-100, mixed well, centrifuged and 25 μl of the supernatant taken for ETA analysis. Matrix-matched standards are used for calibration. The within- and between-batch precisions are 16 and 18%, respectively, for a whole-blood cadmium concentration of 6–7 nmol l^{-1}. The detection limit (3σ) is $\leq$1.8 nmol l^{-1}. Halls [221] has shown that the furnace programme for this method can be reduced to 62 s (of which 26 seconds was the autosampler cycling time), and give within-batch precisions of 1.2% for 71 nmol l^{-1} of cadmium.

Black et al. [238] used a modification of this procedure to measure cadmium in the plasma obtained from rats during an investigation of the effects of chronic cadmium toxicity. After acid precipitation of the proteins, using an equal volume of 10% v/v HNO_3, Cd was measured in the supernatant by ETA-AAS, using wall atomization at 1400°C. Calibration was against aqueous standards in 5% v/v HNO_3 using integrated absorbance measurements. The detection limit (2σ) was 0.6 nmol l^{-1} and recoveries of the order of 99.7 to 102.7% for added cadmium concentrations of 8.22 nmol l^{-1} were found. Good agreement was obtained against an atomic fluorescence procedure and graphite probe AAS.

A variety of digestion and extraction procedures prior to ETA-AAS determinations have been proposed for the determination of cadmium in body tissues and fluids. These range from digestion with HNO_3/H_2O_2 [239], extraction with dithizone into benzene followed by back-extraction into 0.5 M HNO_3 [240], extraction into toluene [241] and extraction with trioctylamine into IBMK [242].

Various chemical modifiers have been suggested for the direct analysis of whole-blood, plasma and urine cadmium, ranging from none [243, 244], dilute HNO_3 [245, 246], $(NH_4)_2HPO_4$ [245, 247], $(NH_4)_2HPO_4$ + HNO_3 [248], $NH_4H_2PO_4$ [244, 249], NH_4F [245], to Pd combined with NH_4NO_3 [250]. Diammoniumhydrogen phosphate appears to be the most popular. Delves and Woodward [249] developed a simple procedure. Whole blood was diluted $1 + 1 + 1$ with 0.05% v/v HNO_3 and 1% m/v $NH_4H_2PO_4$ in acid-washed autosampler cups. After mixing, 20 μl aliquots were taken into the atomizer and calibration was against matrix-matched standards prepared in bovine blood. The organic matrix was destroyed using oxygen ashing at 600°C with atomization at 2100°C from the tube wall. This procedure reduced the molecular absorption signal to less than 0.2 A. The recoveries of 36 and 89 $nmol^{-1}$ of cadmium were 98.8 to 100.2%, respectively. The within-batch precisions were 3.4 and 4.6% at whole-blood cadmium concentrations of 14 and 103 nmol l^{-1}. The detection limit (3σ) is 1.3 nmol l^{-1}. This method has been used successfully for environmental surveys in our laboratory for over eight years.

Yin et al. [250] used a $1 + 3$ dilution of whole blood with 0.05% Triton X-100, chemical modification with Pd and $Mg(NO_3)_2$ plus oxygen ashing at 400°C, followed by a further pyrolysis step at 800°C. Atomization from a L'vov platform at 1700°C allowed calibration against aqueous standards to be achieved and also reduced background signals to less than 0.2 A. Problems of protein precipitation were encountered with pre-mixing of the modifier and sample. Mixing on the platform was successful, provided that the modifier was pipetted first, followed by the sample. For a reference whole blood containing 44.5 nmol l^{-1} cadmium a value of 45.4 $\pm$ 1.8 nmol l^{-1} ($n = 10$) was found. The detection limit (2σ) was 3.6 nmol l^{-1}; no doubt this could be improved by reducing the sample dilution ratio.

There are now a number of reference and quality control materials available for the determination of cadmium in whole blood, yet many recent papers fail to include analyses of such samples. There is, however, little room for complacency in those laboratories involved with this determination. Starkey et al. [251] presented data from an external quality assessment scheme which showed the inter-laboratory variation to be greater than 20% at cadmium concentrations below 100 nmol l^{-1}.

The determination of cadmium in urine appears to still present a considerable challenge to the analyst because the major organic and inorganic species present at variable concentrations in urine, produce chemical and spectral interferences. The problems of large and variable background can be illustrated by an example from our laboratory. A series of sequential 24-h urines were collected from the same subject. One mid-week sample produced a background signal of less than 0.1 A, whereas the following days samples gave a background signal greater than 1.5 A. Subramanian [9] has reviewed the literature on the determination of cadmium in urine. In general there are two approaches to overcoming these interferences: (1) selective volatilisation of the cadmium from the molecular background signal by using a low atomization temperature; and (2) chemical modification, to enable destruction of the matrix at elevated temperatures and retention of cadmium prior to atomization.

Halls et al. [246] used both of the above approaches. Nitric acid was found to be the most useful, and contamination-free, of a range of chemical modifiers. They were able to separate the cadmium and background peaks by diluting urine 1 + 4 with 6% v/v HNO_3, using a pyrolysis temperature of 500°C and atomizing from a L'vov platform at 1200°C with a 2-second ramp. Using integrated absorbance measurements, calibration against aqueous standards was possible. The detection limit (2σ) was 0.6 nmol l^{-1} and within-batch precisions were 5% RSD in the range 18–89 nmol l^{-1} and better than 2% RSD between 27 and 71 nmol l^{-1}. Between-batch precisions were 15–20% RSD. A second approach was to use tube wall atomization from a standard electrographite tube without a pyrolysis step. Urine samples already acidified to 1% v/v HNO_3 were diluted 1 + 1 with 6% v/v HNO_3. This reagent and a gas flow of 30 ml min^{-1} during atomization at 2400°C with a 1-second ramp reduced background signals. Calibration was against aqueous standards using peak absorbance measurements. The detection limit (2σ) was 1.2 nmol l^{-1}, the within-batch precision was approximately 1% RSD over the range 44.5–89 nmol l^{-1} and better than 2% RSD over the range 18–89 nmol l^{-1}. Between-batch precision data obtained using a Lanonorm quality control material with an assigned value of 25.8 nmol l^{-1} was 9.9% with a measured value of 26.7 ± 2.7 nmol l^{-1}. Even though a very short atomizer programme was used (21 s), these results indicate that accuracy and precision were not compromised. Both of these methods gave good agreement against a direct APDC/MIBK extraction method.

McAughey and Smith [244] diluted urine 1 + 1 with water and used selective atomization from a L'vov platform at 800°C. Using zero gas flow during atomization they achieved a detection limit (2σ) of 0.5 nmol l^{-1}. However, calibration by standard additions was necessary to overcome the large matrix variations in the urine samples. This method gave reasonable agreement against an established anodic stripping voltammetric procedure. Within- and between-batch precisions were 4.9% RSD and 11.3% RSD, respectively, for a cadmium concentration of 7.5 nmol l^{-1}. Urine cadmium levels in 29 unexposed subjects averaged 3.6 nmol l^{-1}, with a range of 0.7–8.4 nmol l^{-1}, which is in good agreement with other studies. In our experience the success of selective atomization methods is highly dependent upon the careful optimisation of the ashing and atomization temperatures. Small changes in operating conditions caused by changes in contact resistance between graphite tubes and electrical supply can have a disproportionate effect on the results. The use of quality control or reference materials for regular assessment of the accuracy of the procedure is essential.

Using a L'vov platform and chemical modification with $(NH_4)_2HPO_4$, Pruszkowska et al. [247] were able to use a pyrolysis temperature of 700°C and an atomization temperature of 1600°C. Urine samples were diluted 1 + 4 with deionised water and 20 μl aliquots plus 5 μl aliquots of the chemical modifier (which contained 200 μg $(NH_4)_2HPO_4$ in 3% v/v HNO_3 in 5 μl) dispensed into the atomizer. It was necessary to add 40 μg NaCl on the platform for analysis of the aqueous standards. These authors used Zeeman-effect background correction because background signals were of the order of 0.7 to 2 A. A pyrolysis temperature of 900°C could have been used with this chemical modifier to further reduce the background signals [252]. In view of the five-fold urine dilution the claimed detection limit (2σ) of 0.36 $nmol^{-1}$ (0.04 $\mu g\ l^{-1}$) is impressive. However, this was obtained from 10 replicate measurements of a normal urine containing approximately 2.7 $nmol\ l^{-1}$ Cd, and not from measurements at or near the blank value.

Yin et al. [250] diluted urine 1 + 1 with 0.05% Triton X-100 and used a chemical modifier on the platform of 50 μg Pd plus 500 μg NH_4NO_3. With a pyrolysis temperature of 850°C and platform atomization at 1600°C, it was possible to calibrate with aqueous standards, and the background signals were less than 0.5 A. A detection limit (2σ) of 0.9 $nmol\ l^{-1}$ was found and for a human quality control urine with a recommended value of $58.7 \pm 8.9\ nmol\ l^{-1}$, a value of $57.8 \pm 9.8\ nmol\ l^{-1}$ was found. The results for a group of randomly selected individuals, both smokers and non-smokers, are in agreement with other studies. The ability to use the same atomizer programme, diluent and chemical modifier for the analysis of whole blood, serum and urine makes this method very attractive for clinical studies.

Both digestion and solid sampling approaches have been examined for the determination of cadmium in tissue samples. Herber et al. [253] compare both types of approach for assessing cadmium in freeze-dried placental tissue. Both techniques gave similar results. The digestion technique had better precision, but was slower and suffered occasionally from contamination problems. The problems of homogeneity when considering solid sampling were used to advantage by Lucker et al. [254], who used direct solid sampling to measure the distribution of Cd in bovine renal tissue. The cadmium distribution in the renal cortex was found to be homogeneous. Samples from this area gave results which correlated well with results from the whole kidney obtained by acid digestion ETA-AAS.

C. Nickel

Nickel is a constituent of stainless steel and consequently humans are in almost daily contact with this element. Nickel is known to be allergenic and some of its compounds are carcinogenic [255]. There is a need to assess both environmental and occupational levels of exposure. The concentration of nickel in serum or plasma is a measure of the body burden of the element, whereas the level in urine is used for assessing occupational exposure. The normal serum nickel concentration for non-exposed subjects is considered to be less than 17 nmol l^{-1}, and no doubt most of this is due to contamination from stainless steel needles during sample collection. It is disappointing to see some recent publications quoting higher mean values for groups of unexposed subjects.

Most of the early procedures required acid digestion of the samples prior to liquid–liquid extraction and ETA-AAS or absorption voltammetry. All of these methods are time-consuming and prone to contamination and have been discussed by Delves [1] and Subramanian [8, 9]. Several workers have attempted direct methods of analysis, but as Table 4 shows, the current sensitivity of modern atomizers is such that large volumes of undiluted serum or urine are required to produce a signal of 0.0044 absorbance units.

Sunderman et al. [256] measured nickel in whole blood by using nitric acid deproteinisation, followed by ETA-AAS on the supernatant. A detection limit of 1.7 nmol l^{-1} was found and a normal serum range of 7.8 ± 4.4 nmol l^{-1} was established from 39 subjects. Other workers dilute serum $1 + 1$ with Triton X-100 [257, 258]. Andersen et al. [258] used 0.1% Triton X-100 with 50 μl injections, a pyrolysis temperature of 1250°C and wall atomization from a pyrolytically coated tube at 2700°C. The detection limit (3σ) was 1.5 nmol l^{-1} and a mean serum nickel value of 11 ± 6.0 nmol l^{-1} was found for six healthy volunteers, which is in good agreement with other studies. Calibration against matrix-matched or aqueous standards was possible. Even though a low value of 8.2 ± 0.7 nmol l^{-1} was produced for the NIST SRM 8419 Bovine Serum, recommended value 30.7 ± 10.2 nmol l^{-1}, subsequent analysis of a Seronorm freeze-dried serum with a recommended value of 52.8 nmol l^{-1} gave a value of 49.9 ± 5.8 nmol l^{-1}. This evidence suggests that the nickel value for the NIST material is too high.

Sunderman et al. [259] measured urine nickel using a $1 + 1$ dilution with 0.64% nitric acid followed by pyrolysis at 1200°C and wall atomization at 2600°C. The method gave a detection limit of 8.5 nmol l^{-1} and good agreement with the recommended IUPAC liquid–liquid extraction ETA-AAS method [260]. Recovery at the 85.2 nmol l^{-1} level was $98 \pm 3.4\%$ and the within-batch precision was 7.8% RSD for 71.6 nmol l^{-1} nickel. For 34 non-exposed controls a mean urine nickel concentration of 34.1 nmol l^{-1} was found. A method for tissue analysis was also developed by these authors [261]. Tissue samples were homogenised in plastic bags and digested with a mixture of HNO_3, H_2SO_4 and $HClO_4$, prior to ETA-AAS.

Gawkrodger et al. [262] monitored whole-blood and urine nickel levels from a nickel-allergic patient commencing a low nickel diet. Urine or whole blood was diluted $1 + 1$ with 0.1% Triton X-100 and 20 μl injections taken. Oxygen ashing at 550°C was used to destroy the organic matrix. The method gave a detection limit (3σ) of 5 nmol l^{-1}; within-batch precision was 11% RSD and 12% RSD for 33 nmol l^{-1} and 20 nmol l^{-1} of urine and whole blood, respectively. The pre-diet whole-blood and urine nickel levels were not excessive, but fell to approximately half and one-third to one quarter of their original value within ten days of commencing a low Ni diet, during which time the patients eczema cleared.

Alder et al. [263] describe the results of a novel study to obtain Ni (and Al) depth profiles through human skin biopsies. Skin layers of 20–30 μg were taken and introduced as solid samples into the atomizer. Calibration was against aqueous standards and a bulk sample of skin used as a control material.

D. Beryllium

Beryllium is a highly toxic metal which can cause extensive damage on contact with the skin and eyes. If ingested or inhaled it can cause dyspnea on exertion, coughing and weakness. The determination of beryllium in urine is the normal method of assessing exposure to this element. Normal unexposed levels are less than 56 nmol l^{-1}. Earlier work has been reviewed by Delves [1] and Subramanian [9].

For occupational monitoring, Paschal and Bailey [264] diluted urine 1 + 3 with a chemical modifier containing 0.25% m/v $Mg(NO_3)_2 \cdot 6H_2O$, 0.1% v/v Triton X-100 and 1% v/v HNO_3. Aqueous standards were also prepared in the chemical modifier. A pyrolysis temperature of 1400°C enabled most of the matrix to be removed and typical background signals during atomization from the L'vov platform at 2400°C were generally less than 0.1 A. The detection limit (3σ) of the method was 5.5 nmol l^{-1}, within- and between-batch precisions were 4.2% RSD and 9.8% RSD, respectively, for a urine containing 533 nmol l^{-1} Be. The accuracy of the method was assessed using NIST SRM 2670 Urines. While these SRM's do not have certified values, good agreement was achieved with the informational values. The SRM 1643a "Trace elements in water", which does have a certified value of 2.11 ± 0.11 μmol l^{-1}, was found to give a value of 2.26 μmol l^{-1}. No difference in results was found using either continuum source or Zeeman-effect background correction.

Zorn et al. [265] used a direct ETA-AAS procedure to measure Be in serum to monitor the progress of 25 people accidentally exposed. The detection limit of this method was 66.6 nmol l^{-1} and from their results Be clearance appears to have a half-life of 2–8 weeks.

E. Thallium

Thallium and its salts are known to be toxic. While in earlier times thallium salts were used for therapeutic purposes, they are now banned. Thallium salts are well absorbed via the skin and gut and the main excretory routes are urine and faeces. Given the substantial toxicity of thallium it is important to monitor those who are occupationally exposed. The concentration of thallium in normal unexposed subjects is considered to be less than 1 nmol l^{-1}. Many of the earlier procedures were developed to deal with cases of severe poisoning or for animal toxicological studies and therefore do not have sufficiently low detection limits for occupational monitoring. the literature has been reviewed by Leloux et al. [266] and Subramanian [9].

Thallium is subject to severe interferences in a graphite atomizer particularly in matrices with high chloride concentrations. Many early workers used chelation and liquid–liquid extraction procedures to overcome these interferences [9]. With the use of L'vov platforms and chemical modification more direct analyses have been attempted. Grobenski et al. [267] diluted urine 1 + 1 with 2% HNO$_3$ and took 10 μl injection volumes plus 10 μl of 2% m/v NH$_4$H$_2$PO$_4$ in 0.03% v/v Triton X-100 as a chemical modifier. This procedure enabled a pyrolysis temperature of 700°C to be used with atomization from the L'vov platform at 1500°C. With 50 μl sample volumes a detection limit (2σ) of 0.64 nmol l^{-1} was found. The high background signals of 1.5 A were corrected using a Zeeman-effect instrument. Leloux et al. [268] determined thallium in a variety of biological materials obtained from rats. A L'vov platform was used with 10% m/v ascorbic acid and 4% v/v H$_2$SO$_4$ as a chemical modifier which enabled a pyrolysis temperature of 800°C. Detection limits ranged from 24.5 to 58.7 nmol l^{-1}. Though this method may be of some use for a toxicological study, it is not appropriate for occupational monitoring. We would also question the validity of analytical results produced by the method of standard additions using only one addition. Problems were also encountered with the build-up of a carbonaceous residue upon the platform, which is not helped by the use of 10% m/v ascorbic acid as a chemical modifier.

Paschal and Bailey [269] used a similar approach, but with a chemical modifier of 4% v/v HNO$_3$, Mg(NO$_3$)$_2$ and 0.1% v/v Triton X-100. Urine samples were diluted 1 + 1 with the modifier. Zeeman-effect background correction was used with platform atomization and a matrix-matched calibration. Using different urine samples only a 4–5% variation in the standard addition slopes was found. A detection limit of 2.5 nmol^{-1} was obtained using 40 μl injections.

Welz et al. [270] showed the chloride interferences on Tl to be caused by both a condensed phase chemical interference of volatilisation of thallium chloride during pyrolysis, and a vapour phase interference due to formation of TlCl during atomization. The use of a Pd chemical modifier pre-treated at 1000°C prior to introduction of the sample, allied with the use of 5% hydrogen in argon as the internal gas, gave an interference-free determination. Using 10 μl injections of an undiluted urine, a detection limit (2σ) of 9.8 nmol l^{-1} was found. While this procedure appears promising, it has yet to be applied to routine clinical analysis.

It is clear from the above that few methods come near to achieving the necessary detection limit for routine occupational monitoring. There may be no option but to use liquid–liquid extraction procedures prior to ETA-AAS to achieve this. Chandler and Scott [58] developed a method involving the direct extraction of thallium from urine with diethyldithiocarbamate at pH 7.0 into toluene. 50 ml of urine was extracted with 3 ml of toluene and matrix-matched standards were prepared from a urine with a low thallium content for calibration. The emulsion was broken down by centrifugation and the organic extract taken for ETA analysis using a pyrolysis temperature of 550°C and an atomization temperature of 1300° with standard electrographite tubes. Recoveries using ^{204}Tl were found to be $95.6 \pm 2.5\%$. The detection limit (3σ) was 0.5 nmol l^{-1}, within- and between-batch precisions were 3.5% and 4.4% RSD, respectively, for a urine thallium concentrations of 5.7–5.9 nmol l^{-1}. Background signals were less than 0.1 A and pyrolytic graphite-coated tubes were found to give worse sensitivities.

VII. CONCLUSION

The past decade has witnessed substantial improvements in instrumentation for ETA-AAS. These have been at least partly responsible for the reported decreases, over that period, in reported data for reference concentrations of trace elements in body tissues and tissue fluids; chromium is a particularly good example. The maturity of ETA-AAS is now well established [271] and substantial improvements in inter-laboratory agreements of clinical assays, e.g. Pb and Cd in blood, are evidence of this. It is likely that new developments in chemical modifiers, both liquid and gaseous, will enhance further the quality of clinical data obtained by ETA-AAS. A potential problem to future developments in ETA-AAS is the variation in graphite components. This has long been suspected but is only recently confirmed. Until the manufacturing quality of these components is improved or maintained, there will be little progress in development of reference methods and in "standardless" analyses.

Notwithstanding this, clinical analyses by ETA-AAS have revealed much about the roles of trace elements in human health and disease. This progress will undoubtedly continue over the next decade, particularly in studies of trace element carrier species.

REFERENCES

1　H.T. Delves, in J. Cantle (Ed.), Atomic Absorption Spectrometry: Techniques and Instrumentation in Analytical Chemistry, No. 5, Elsevier, Amsterdam, 1982, pp. 341–380.
2　J. Versieck, C.R.C. Crit. Rev. Clin. Lab. Sci., 22 (1985) 97–184.
3　D.J. Halls and G.S. Fell, Anal. Chim. Acta, 129 (1981) 205–211.

431

4   D.C. Paschal and G.C. Bailey, At. Spectrosc., 8 (1987) 150–152.
5   K.S. Subramanian and J.C. Meranger, Anal. Chem., 57 (1985) 2478–2481.
6   J. Versieck, F. Bartbier, R. Cornelis and J. Hoste, Talanta, 29 (1982) 973–984.
7   J. Versieck and R. Cornelis, Anal. Chim. Acta, 116 (1980) 217–254.
8   K.S. Subramanian, Prog. Anal. At. Spectrosc., 9 (1986) 237–334.
9   K.S. Subramanian, Prog. Anal. At. Spectrosc., 11 (1988) 511–608.
10   K.S. Subramanian, At. Spectrosc., 9 (1988) 169–178.
11   B.V. L'vov, Spectrochim. Acta, Part B, 33 (1978) 153–193.
12   J.M. Ottaway, J. Carroll, S. Cook, S.P. Corr, D. Littlejohn and J. Marshall, Fresenius
     Z. Anal. Chem., 323 (1986) 742–747.
13   W. Frech, D.C. Baxter and B. Hutsch, Anal. Chem., 58 (1986) 1973–1977.
14   H.T. Delves, Ann. Clin. Biochem., 24 (1987) 529–551.
15   J.P. Matousek, Prog. Anal. At. Spectrosc., 4 (1981) 247–310.
16   W. Slavin and D.C. Manning, Prog. Anal. At. Spectrosc., 5 (1982) 243–340.
17   W. Frech, E. Lundberg and A. Cedergren, Prog. Anal. At. Spectrosc., 8 (1985)
     257–370.
18   C.L. Chakrabarti, H.A. Hamed, C.C. Wan, W.C. Li, P.C. Bertels, D.C. Gregoire and
     S. Lee, Anal. Chem., 52 (1980) 167–176.
19   C.L. Chakrabarti, C.C. Wan, H.A. Hamed and P.C. Bertels, Anal. Chem., 53 (1981)
     444–450.
20   B. Welz, M. Sperling, G. Schlemmer, N. Wenzel and G. Marowsky, Spectrochim.
     Acta, Part B, 43 (1988) 1187–1207.
21   W. Slavin and D.C. Manning, Anal. Chem., 51 (1979) 261–265.
22   W. Slavin and D.C. Manning, Spectrochim. Acta, Part B, 35 (1980) 701–714.
23   E. Pruszkawska, G.R. Carnrick and W. Slavin, At. Spectrosc., 4 (1983) 59–61.
24   W. Slavin, D.C. Manning and G.R. Carnrick, At. Spectrosc., 2 (1981) 137–145.
25   F.J. Fernandez, M.M. Beaty and W.B. Barnett, At. Spectrosc., 2 (1981) 16–21.
26   W. Frech and S. Jonsson, Spectrochim. Acta, Part B, 37 (1982) 1021–1028.
27   E. Lundberg, W. Frech and I. Lindberg, Anal. Chim. Acta, 160 (1984) 205–215.
28   B.V. L'vov and L.A. Pelieva, Zh. Anal. Khim., 33 (1978) 1572–1575.
29   B.V. L'vov and L.A. Pelieva, Zh. Anal. Khim., 34 (1979) 1744–1755.
30   D.C. Manning, W. Slavin and S. Myers, Anal. Chem., 51 (1979) 2375–2378.
31   S.K. Giri, D. Littlejohn and J.M. Ottaway, Analyst, 107 (1982) 1095–1098.
32   J. Marshall, S.K. Giri, D. Littlejohn and J.M. Ottaway, Anal. Chim. Acta, 147
     (1983) 173–182.
33   S.K. Giri, C.K. Shields, D. Littlejohn and J.M. Ottaway, Analyst, 108 (1983) 244–
     253.
34   D. Littlejohn, J. Marshall, J. Carroll, W. Cormack and J.M. Ottaway, Analyst, 108
     (1983) 893–896.
35   J. Carroll, J. Marshall, D. Littlejohn and J.M. Ottaway, Fresenius Z. Anal. Chem.,
     322 (1985) 145–150.
36   J. Carroll, J. Marshall, D. Durie, D. Littlejohn and J.M. Ottaway, Spectrochim.
     Acta, Part B, 41 (1986) 751–759.
37   D. Littlejohn, Lab. Pract., 36 (1987) 121–126.
38   R.A. Newstead, W.J. Price and P.J. Whiteside, Prog. Anal. At. Spectrosc., 1 (1978)
     267–298.
39   H.L. Kahn, At. Absorpt. Newsl., 7 (1968) 40–43.
40   K. Yasuda, H. Koizumi, K. Ohishi and T. Noda, Prog. Anal. At. Spectrosc., 3 (1980)
     299–368.

432

41 S.B. Smith, Jr. and G.M. Hieftje, Appl. Spectrosc., 5 (1983) 419–424.
42 W. Slavin and G.R. Carnrick, C.R.C Crit. Rev. Anal. Chem., 19 (1988) 95–134.
43 L. de Galan and M.T.C. de Loos-Vollebregt, Spectrochim. Acta, Part B, 39 (1984) 1011–1019.
44 M. Zief and J.W. Mitchell, Contamination Control in Trace Element Analysis, Wiley, New York, N.Y., 1976.
45 M. Stoeppler, P. Valenta and H.W. Nurnberg, Fresenius Z. Anal. Chem., 297 (1979) 22–34.
46 A. Taylor and V. Marks, Ann. Clin. Biochem., 10 (1973) 42–46.
47 H.T. Delves, in A. Tayler (Ed.), Clinics in Endocrinology and Metabolism, Eastbourne, Saunders, 14 (1985) 725–760.
48 H.T. Delves, Determination of Trace Elements in Tissues and Excretia of Children, PhD Thesis, University of London, London, 1970.
49 O. Guillard, A. Piriou, P. Mura and D. Reiss, Clin. Chem., 28 (1982) 1714–1715.
50 V.W. Bunker, H.T. Delves and R.F. Fautley, Ann. Clin. Biochem., 19 (1982) 444–445.
51 H.T. Delves, B.E. Clayton, Carmichael et al., Ann. Clin. Biochem., 19 (1982) 329–337.
52 D.C. Hildebrand and D.H. White, Clin. Chem., 20 (1986) 148–151.
53 A. Taylor, Ann. Clin. Biochem., 23 (1986) 364–378.
54 G.J. Knapp, Environ. Anal. Chem., 22 (1985) 71–83.
55 J. Angerer and K.H. Schaller, Analysis of Hazardous Substances in Biological Materials, Vol. 2, Methods for Biological Monitoring, Weinheim, FDR, VCH, 1988.
56 M. Stoeppler and D.B. Tonks, IUPAC-activities: co-operative laboratory surveys, proposed working method for cadmium in whole blood, paper presented at International Workshop on Biological Indicators of Cadmium, Exposure, Diagnostic and Analytical Reliability, 1982.
57 P. Allain, Y. Mauras and C. Grangeray, Ann. Clin. Biochem., 24 (1987) 518–519.
58 H.A. Chandler and M. Scott, At. Spectrosc., 5 (1984) 230–233.
59 M. Ishizaki and S. Ueno, Talanta, 26 (1979) 523.
60 I. Andersen and A.C. Hoegetveit, Fresenius Z. Anal. Chem., 318 (1984) 41–44.
61 D.K. Eaton and J.A. Holcombe, Anal. Chem., 55 (1983) 946–950.
62 N.J. Miller-Ihli, J. Anal. At. Spectrom., 4 (1989) 295–297.
63 R.D. Ediger, At. Absorpt. Newsl., 14 (1975) 127–130.
64 H.T. Delves and J. Woodward, At. Spectrosc., 2 (1981) 65–67.
65 G. Moris, M. Patriarca and A. Menotti, Clin. Chem., 34 (1988) 127–130.
66 G.R. Carnrick, D.C. Manning and W. Slavin, The Analyst, 108 (1983) 1297–1312.
67 J. Neve and L. Molle, Acta Pharmacol. Toxicol. (Suppl. VII), 59 (1986) 606.
68 B. Sampson, J. Anal. At. Spectrom., 2 (1987) 447–450.
69 K. Saeed, J. Anal. At. Spectrom., 2 (1987) 151–155.
70 X.Q. Shan and Z. Ni, Acta Chim. Sinica, 37 (1979) 261.
71 G. Schlemmer and B. Welz, Spectrochim. Acta, Part B, 41 (1986) 1157–1165.
72 L.M. Voth-Beach and D.E. Shrader, J. Anal. At. Spectrom., 2 (1987) 45–50.
73 R.D. Beaty and M.M. Cooksey, At. Absorpt. Newsl., 17 (1978) 53–58.
74 F.J. Fernandez and D. Hilligoss, At. Spectrosc., 3 (1981) 130–131.
75 C.G. Bruhn, J.M. Piwonka, M.O. Jerardino, G.M. Navarrete and P.C. Maturana, Anal. Chim. Acta, 198 (1987) 113–123.
76 N.W. Alcock, At. Absorpt. Newsl., 18 (1979) 37.
77 D.F. Sinclair and B.R. Dohnt, Clin. Chem., 30 (1984) 1616–1619.
78 L. Novak, P. Ostapczuk and M. Stoeppler, in B. Welz (Ed.) 5th Colloquium Atom-

spektrometrische Spurenanalytik, Bodenseewerk Perkin-Elmer GmbH, 1989.

79   I.L. Shuttler, PhD Thesis, University of Southampton, 1988.
80   C.W. Fuller, Electrothermal Atomization for Atomic Absorption Spectrometry, Analytical Sciences Monograph No. 4, The Chemical Society, London, 1977.
81   S.G. Salmon, R.H. Davies, Jr. and J.A. Holcombe, Anal. Chem., 53 (1981) 324–330.
82   R.E. Sturgeon and C.L. Chakrabarti, Prog. Anal. At. Spectrosc., 1 (1978) 5–99
83   A. Cedergren, W. Frech and E. Lundberg, Anal. Chem., 56 (1984) 1382–1387.
84   S. Levi and W.C. Purdy, Clin. Biochem., 13 (1980) 253–256.
85   N.E. Viera and J.W. Hansen, Clin. Chem., 27 (1981) 73–77.
86   J.W. Foote and H.T. Delves, Analyst, 107 (1982) 121–124.
87   J.W. Foote and H.T. Delves, Analyst, 108 (1983) 492–504.
88   J.W. Foote and H.T. Delves, Analyst, 109 (1984) 709–711.
89   J.W. Foote and H.T. Delves, Analyst, 113 (1988) 911–915
90   G. Carelli, M.C. Atavista and F. Aldrighetti, At. Spectrosc., 3 (1982) 200–202.
91   P.A. Pleban and D.S. Mei, Clin. Chim. Acta, 133 (1983) 433–50.
92   M. McGahan and L.Z. Bito, Anal. Biochem., 135 (1983) 18–192.
93   I.L. Shuttler and H.T. Delves, J. Anal. At. Spectrom., 4 (1989) 137–142.
94   J.P. Matousek and B.J. Stevens, Clin. Chem., 17 (1971) 363–365.
95   O. Wawschinek and H. Hofler, At. Absorpt. Newsl., 18 (1979) 97.
96   H. Kamel, J. Teape, D.H. Brown, J.M. Ottaway and W.E. Smith, Analyst (London), 103 (1978) 921–927.
97   E. Berman, Toxic Metals and their Analysis, Heyden, London, 1980.
98   A.A. Brown, D.J. Halls and A. Taylor, J. Anal. At. Spectrom., 1 (1986) 29R.
99   A.A. Brown, D.J. Halls and A. Taylor, J. Anal. At. Spectrom., 4 (1989) 47R–87R.
100 J. Dedina, W. Frech, A. Cedergren and I. Lindberg, J. Anal. At. Spectrom., 3 (1987) 465–469.
101 J. Dedina, W. Frech, I. Lindberg, E. Lundberg and A. Cedergren, J. Anal. At. Spectrom., 2 (1987) 287–291.
102 I. Martinsen, B. Radziuk and Y. Thomassen, J. Anal. At. Spectrom., 3 (1988) 1013–1022.
103 B. Welz, C. Schlemmer and U. Voellkoff, Spectrochim. Acta, Part B, 39 (1984) 501–510.
104 D.C. Paschal and M.M. Kimberley, At. Spectrosc., 7 (1986) 75–78.
105 S.A. Lewis, N.W. Hardison and C. Veillon, Anal. Chem., 58 (1986) 1272–1273.
106 W.C. Edwards and B.S. Blackburn, Vet. Hum. Toxicol., 28 (1986) 12–13.
107 P.L.H. Jowett and M.I. Banton, Anal. Lett., 19 (1986) 1243–1258.
108 A.K. MacPherson, B. Sampson and A.T. Diplock, Analyst, 113 (1988) 281–283.
109 J. Bauslaugh, B. Radziuk and K. Saeed, Thomassen. Anal. Chim. Acta, 165 (1984) 149–157.
110 R.H. Eckerlin, D.W. Hoult and G.R. Carnick, At. Spectrosc., 8 (1987) 64–66.
111 B.E. Manning, At. Spectrosc., 4 (1983) 159.
112 M.B. Knowles and K.G. Brodie, J. Anal. At. Spectrom., 3 (1988) 511–516.
113 M.B. Knowles and K.G. Brodie, J. Anal. At. Spectrom., 4 (1989) 305.
114 J. Neve, S. Chamart, P. Trigaux and F. Vertongen, At. Spectrosc., 8 (1987) 167–169.
115 J. Neve, M. Hanocq, L. Molle and G. Lefebvre, Analyst, 107 (1982) 934–941.
116 I. Lindberg, E. Lundberg, P. Arkhammar and P.O. Berggren, J. Anal. At. Spectrom., 3 (1988) 497–501.
117 B. Lloyd, P. Holt, P. and H.T. Delves, Analyst, 107 (1982) 927–933.
118 V.W. Bunker and H.T. Delves, Anal. Chim. Acta, 201 (1987) 331–334.

434

119   WHO Study Group, Recommended Health-Based Limits in Occupational Exposure
      to Heavy Metals, WHO, Geneva, 1980.
120   J. Jarvisalo, M. Olkiuora, A. Tossavainen, M. Tirtamo, P. Ristola and A. Aito, in
      S.S. Brown and J. Savoy (Eds.), Chemical Toxicology and Clinical Chemistry of
      Metals, Academic Press, London, 1983. pp. 123–126.
121   D. Gompertz, Laboratory Methods for Biological Monitoring, Occupational Medicine
      and Hygiene Laboratories, Health and Safety Executive, London, 1985, 3rd ed.
122   J. Versieck, J. Hoste and F. Barbier, Acta Gastro-Enterol., Belgium, 39 (1976)
      340–349.
123   A.J. Hall, B.M. Margets, D.J.P. Barker, H.P.J. Walsh, T.R. Redfern, J.R. Taylor, P.
      Dangerfield, I.L. Shuttler and H.T. Delves, Paediatr. Perin. Epidemiol., 3 (1989)
      131–136.
124   F. Baruthio, Q. Guillard, J. Arnaud, F. Pierre and R. Zawislak, Clin. Chem., 34
      (1988) 227.
125   J.M. Ottaway and D.J. Halls, Pure Appl. Chem., 58 (9) (1986) 1307–1316.
126   D.J. D'Amico and H. Klawans, Anal. Chem., 48 (1976) 1469–1472.
127   V. Hudnick, M. Marldt-Gomiscek and S. Gomiscek, Anal. Chim. Acta, 157 (1984)
      143–150.
128   D.L. Halls and G.S. Fell, Anal. Chim. Acta, 129 (1981) 205–211.
129   K.G. Brodie and M.W. Routh, Clin. Biochem., 17 (1984) 19–26.
130   T. Uchida and B.L. Vallee, Anal. Sci., 2 (1986) 71.
131   W. Bayer, Fortschr. Atomspektr. Spurenanal., 2 (1986) 197–206.
132   M.S. Clegg, B. Lonnerdal, L.S. Hurley and C.L. Keen, Anal. Biochem., 157 (1986)
      12.
133   A. Favier, D. Ruffieux, A. Alcaraz and B. Maljournal, Clin. Chim. Acta, 124 (1982)
      239–244.
134   F.S. Wei, W.O. Ou and F. Yin, Anal. Lett., 15B (1982) 721.
135   W. Slavin, G.R. Carnrick and D.C. Manning, Anal. Chem., 54 (1982) 621.
136   E. Preu, H. Schroter, K. Winnefeld and U. Schmidt, Bestimmung von Mangan in
      Serum mittels ETA-AAS, IX CANAS, GDR, 1986.
137   S.P. Ericson, M.L. McHalsky, B.E. Rabinow, K.G. Kronholm, C.S. Arceo, J.A.
      Weltzer and S.W. Ayd, Clin. Chem., 32 (1986) 1350–1356.
138   G.S. Hams and J.K. Fabri, Clin. Chem., 34 (1988) 1121–1123.
139   J.M. Ottaway and G.S. Fell, Pure Appl. Chem., 58, 12 (1986) 1707–1720.
140   C. Veillon, B.E. Guthrie and N.R. Wolf, Anal. Chem., 52 (1980) 457–459.
141   S. Arpadjan and V. Krivan, Fresenius Z. Anal. Chem., 329 (1988) 745.
142   J.J. McAughey and N.J. Smith, Anal. Chim. Acta, 10 (1989) 1–3.
143   B.W. Morris, H. Griffiths, C.A. Hardisty and G.J. Kemp, At. Spectrosc., 10 (1989)
      1–3.
144   A.J. Schermaier, L.H. O'Connor and K.H. Pearson, Clin. Chim. Acta, 152 (1985)
      123–124.
145   J. Christensen, C.J. Molin and M. Kirchoff, Int. Congr. Ser. Excerpta Med., 676
      (1986) 181.
146   C.J. Molin and P.L. Milling, Acta Pharmacol. Toxicol. Suppl., 59 (1986) 399-.
147   E.G. Offenbacher, H.J. Dowling, C.J. Rinko and F.X. Pi-Sunyer, Clin. Chem., 32
      (1986) 1383–1386.
148   A. Taylor and P. Green, J. Anal. At. Spectrom., 3 (1988) 155–1128.
149   C. Veillon, K.Y. Patterson and N.A. Bryden, Anal. Chim. Acta, 164 (1984) 67–76.
150   S.A. Lewis, T.C. O'Haver and J.M. Harnly, Anal. Chem., 57 (1985) 2–5.

151 A.A. Brown, D.J. Halls and A. Taylor, J. Anal. At. Spectrom., 3 (1988) 45R–78R.
152 D.J. Halls and G.S. Fell, J. Anal. At. Spectrom., 1 (1986) 135–139.
153 D.J. Halls and G.S. Fell, J. Anal. At. Spectrom., 3 (1988) 105–109.
154 J.N. Egila, D. Littlejohn, J.M. Ottaway and S. Xiao-Quan, Anal. Proc., 23 (1986) 426.
155 J.N. Marks, M.A. White and A.R. Boran, At. Spectrosc., 9 (1988) 73–75.
156 P. Dube, Analyst, 113 (1988) 917–921.
157 J. Angerer and R. Heinrich, Fresenius Z. Anal. Chem., 318 (1984) 37–40.
158 R. Heinrich and J. Angerer, Int. J. Environ. Anal. Chem., 16 (1984) 305–314.
159 R. Heinrich and J. Angerer, Fresenius Z. Anal. Chem., 322 (1985) 772–774.
160 A.A. Bouman, A.J. Platenkamp and F.D. Posma, Ann. Clin. Biochem., 23 (1986) 346.
161 H.T. Delves, R. Mensikov and L. Hinks, in P. Bretter and P. Schramel (Eds.), Trace Element—Analytical Chemistry in Medicine and Biology, Vol. 2, Walter de Gruyter, Berlin, 1983, pp. 1123–1127.
162 W. Slavin, Sci. Total Environ., 71 (1988) 17–35.
163 M.M. Kimberly, G.G. Bailey and D.C. Paschal, Analyst, 112 (1987) 287–290.
164 B. Sampson, J. Anal. At. Spectrom., 3 (1988) 465–469.
165 N.J. Smith, HSE Occupational Medicine and Hygiene Laboratory, Analytical Method No. 6, November 1987.
166 S.A. Lewis, T.C. O'Haver and J.M. Harnly, Anal. Chem., 57 (1987) 2–5.
167 A. Himmel, Y. Schmitt and J.D. Krusc-Jones, in P. Bratter and P. Schramel (Eds.) Trace Element Analytical Chemistry in Medicine and Biology, Vol. 5, de Gruyter, Berlin, 1988, 440.
168 J.F. Vazquez-Gonzalez, P. Bermejo-Barrere and F. Bermejo-Martinez, At. Spectrosc., 8 (1987) 159–160.
169 S.P. Ericson, M.L. McHalsky and B. Jaselskis, At. Spectrosc., 8 (1987) 101–104.
170 C. Missenard, G. Hansen, D. Kutter and A. Kremer, Br. J. Ind. Med., 46 (1989) 744–747.
171 D.N.S. Kerr and M.K. Ward, in A. Taylor (Ed.), Aluminium and Other Trace Elements in Renal Disease, Baillier Tindall, London, 1986, 1.
172 M. Freundlich, G. Zillereulo, C. Abitnol and J. Strauss, Lancet, (1985) 527.
173 E.R. Maher, E.A. Brown, J.R. Curtis et al., Br. Med. J., 292 (1986) 306.
174 D. Maharaj, G.S. Fell, D.F.Boyce et al., Br. Med. J., 205 (1987) 693.
175 E.M. Skelly and F.T. Di Stefano, Appl. Spectrosc., 42 (1988) 1302–1306.
176 M. Buratti, G. Caravelli, G. Calzaferri and A. Colombi, Clin. Chim. Acta, 141 (1984) 253–259.
177 W. Frech, A. Cedergren, C. Cedeberg and J. Vessman, Clin. Chem., 28 (1982) 2259–2263.
178 S. Brown, R.L. Bertholf, M.R. Wills and J. Savory, Clin. Chem., 20 (1984) 1216.
179 F.R. Alderman and H.J. Gitleman, Clin. Chem., 26 (1980) 258–260.
180 F.Y. Leung, C. Bradley, W. Slavin and A.R. Henderson, in A. Taylor (Ed.), Aluminium and Other Trace Elements in Renal Disease, Bailliere Tindall, London, 1986, p. 296.
181 M. Bettinelli, U. Baroni, F. Fontant and P. Poisetti, Analyst, 110 (1985) 19–22.
182 C.S. Fellows, M. Phil. Transfer Thesis, University of Southampton, 1989.
183 M.J. Robinson, S.W. Ryan, C.J. Newton, J.P. Day et al., Lancet, i (1987) 1206.
184 A.B. Sedman, G.L. Klein, R.J. Merritt, N.L. Miller et al., N. Engl. J. Med., 312 (1985) 1337–1343.

436

185 H.T. Delves and B. Suchak, in R. Massey and D. Taylor (Eds.), Aluminium in Food and Environment, Royal Soc. Chem., Cambridge, 1989, pp. 52–67.

186 A.W. Walker (Ed.), Specialised Assay Services for Hospital Laboratories, Handbook, SAS Group of National Health Service, 1986.

187 D. Brancaccio, O. Bugrami, L. Pacini et al., in A. Taylor (Ed.), Aluminium and Other Trace Elements in Renal Disease, Bailliere Tindall, London, 1986, pp. 19–23.

188 M.R. Wills and J. Savory, Aluminium homeostasis, 1986, ibid., pp. 24–37.

189 P. Salina, W. Frech, L-G. Ekstron, S. Slorach and A. Cedergren, Clin. Chem., 32 (1986) 539–541.

190 K.R. Koch, M.A.B. Pougnet and S. De-Villiers, Nature, 332 (1988) 122.

191 A.C. Alfrey, G.R. Le Grendre and W.D. Kaehny, N. Engl. J. Med., 294 (1976) 184–188.

192 R.R. Recker, A.J. Blotcky, J.A. Leiffler and E.P. Rack, J. Lab. Clin. Med., 90 (1977) 810–815.

193 P.J. Day, P. Acrill, F.M. Garstang, K.C. Hodge, P.J. Metcalfe et al., in S.S. Brown and J. Savory (Eds.), Chemical Toxicology and Clinical Chemistry of Metals, Academic Press, London, 1983, pp. 353–356.

194 D.J. Halls and G.S. Fell, Analyst, 100 (1985) 243.

195 B. Suchak, C.S. Fellows and H.T. Delves, in preparation.

196 C.N. Martyn, D.J.P. Barker, C. Osmond, E.C. Harris, J.A. Edwardson and R.F. Lacey, Lancet, i (1989) 59–62.

197 S. Fairweather-Tait, R.M. Faulks, S.J.A. Fatemi and G.R. Moore, Hum. Nutr. Food Sci. Nutr., 41F (1987) 183.

198 J.E.A.T. Pennington, Food Addit. Contam., 5 (1988) 161.

199 R.C. Massey and D. Taylor, Aluminium in Food and Environment, Royal Society of Chemistry, Cambridge, 1989.

200 A. Taylor, paper presented at "Consultative Mereting on Clinical Aluminium Analyses", organised by D.H. Steeting Committee for Trace Elements for Advisory Committee on Laboratory Standards, Charing Cross, London, 1989.

201 O. Guillard, A. Pineau, F. Baruthio and J. Arnaud, Clin. Chem., 34 (1988) 1603–1604.

202 K. Matsumoto, T. Solin and K. Fuwa, Spectrochim. Acta, B, 39 (1984) 481–483.

203 H.T. Delves and I.L. Shuttler, in D.C.H. McBrien and T.F. Slater (Eds.), Biochemical Mechanisms of Platinum Antitumour Drugs, IRL Press, Oxford, 1985, No. 4, pp. 329–346.

204 M.C. McGahan and K. Tyczkowska, Spectrochim. Acta, B, 42 (1987) 665–668.

205 J.M. Queralto and J.M. Rodriguez, Ann. Clin. Biochem., 24 (1987) 71 (Suppl. 2).

206 M.D. Shelley, R.G. Fish, A. Brewster and M. Adams, Ann. Clin. Biochem., 27 (1987) 83 (Suppl. 1).

207 M.D. Shelley, R.G. Fish and M. Adams, Ann. Clin. Biochem., 24 (1987) 84 (Suppl. 1).

208 E. Reed, S. Sauerhoff and M.C. Poirier, At. Spectrosc., 9(3) (1988) 93–95.

209 B.J. Starkey, A.P. Taylor and A.W. Walker, Ann. Clin. Biochem., 24 (1987) 91 (Suppl. 1).

210 K.J. Anderson, A. Wikshaaland, A. Utheim, K. Julshamn and H. Vik, Clin. Biochem. (Ottawa), 19 (1986) 166.

211 K. Julshamn, K.J. Andersen and H. Vik, Acta Pharmacol. Toxicol. Suppl., 59 (1986) 613.

212 D.G. Vince and D.F. Williams, Analyst, 112 (1987) 1627–1629.
213 D.C. Manning and W. Slavin, Spectrochim. Acta, B, 42 (1987) 755–763.
214 Medical Research Council, The Neuropsychological Effects of Lead in Children: A Review of Recent Research 1979–1983, MRC, S1102/20.
215 M. Smith, Clin. Endocrinol. Metabol., 14, 3 (1985) 657–680.
216 H.T. Delves, J.C. Sherlock and M.J. Quinn, Hum. Toxicol., 3 (1984) 279–288.
217 H.T. Delves, P. Harvey and M.J. Quinn, in preparation.
218 M. Stoeppler, K. Brandt and T.C. Rains, Analyst, 103 (1978) 714–722.
219 D.J. Halls and G.S. Fells, Royal Infirmary, Glasgow, pers. commun., 1988.
220 K.S. Subramanian, At. Spectrosc., 8 (1987) 7–11.
221 D.J. Halls, Analyst, 109 (1984) 1081–1084.
222 K.S. Subramanian and J.C. Meranger, Clin. Chem., 27 (1981) 1866–1871.
223 S.T. Wang, G. Strunc and F. Peter, in S.S. Brown and J. Savory (Eds.), Chemical Toxicology and Clinical Chemistry of Metals, Academic Press, London, 1983, pp. 57–60.
224 E.J. Hinderberger, M.L. Kaiser and S.R. Koityohann, At. Spectrosc., 2 (1981) 1–7.
225 Perkin-Elmer, Techniques in Graphite Furnace Atomic Absorption Spectrometry, April 1985, PE-Part No. 0993-8150.
226 F. Alt, Z. Anal. Chem., 290 (1978) 108–109.
227 E. Pruozkawska, G.R. Carnick and W. Slavin, At. Spectrosc., 4 (1983) 59–61.
228 D.A. Bass and J.A. Holcombe, Anal. Chem., 59 (1987) 974–980.
229 I.L. Shuttler and H.T. Delves, Analyst, 111 (1986) 651–656.
230 H.T. Delves, Analyst, 95 (1970) 431–438.
231 D.T. Miller, D.C. Paschal, E.W. Gunter, P.E. Stroud and J. D'Angelo, Analyst, 112 (1987) 1701–1704.
232 S. Xaio-Quan and N.I. Zhe-Ming, Can. J. Spectrosc., 27 (1982) 75.
233 D.C. Paschal and M.M. Kimberley, At. Spectrosc., 6 (1985) 134–136.
234 L. Lian, Spectrochim. Acta, 41B (1986) 1131–1135.
235 A.D. Keating, J.L. Keating, D.J. Halls and G.S. Fell, Analyst, 112 (1987) 1381–1385.
236 C.L. Charrabarti, R. Karwowska, B.R. Hollebone and P.M. Johnson, Spectrochim. Acta, B, 42 (1987) 1217.
237 M. Stoeppler and K. Brandt, Fresenius Z. Anal. Chem., 300 (1980) 372–380.
238 M.M. Black, G.S. Fell and J.M. Ottaway, J. Anal. At. Spectrom, 1 (1986) 369–372.
239 C.A. Roberts and J.M. Clark, Bull. Environ. Contam. Toxicol., 36 (1986) 496.
240 S. Matsushita, Nippon Koshu. Eisei. Zasshi, 32 (1985) 247.
241 S. Suna and T. Nakajima, Sagyo. Igaku., 29 (1987) 292.
242 E.A. Piperaki, Chem. Chron., 14 (1985) 57.
243 K.R. Lum and D.G. Edgar, Inst. J. Environ. Anal. Chem., 33 (1988) 13.
244 J.J. McAughey and N.J. Smith, Anal. Chim. Acta, 156 (1984) 129–137.
245 K.G. Feitsma, J.P. Franke and R.A. de Zeeuw, Analyst, 156 (1984) 129–137.
246 D.J. Halls, M.M. Black, G.S. Fell and J.M. Ottaway, J. Anal. At. Spectrom. 2 (1987) 305–309.
247 E. Pruszkowska, G.R. Carnick and W. Slavin, Clin. Chem., 29/3 (1983) 477–480.
248 K.S. Subramanian, J.C. Meranger and J.E. Mackeen, Anal. Chem., 55 (1983) 1064–1067.
249 H.T. Delves and J. Woodward, At. Spectrosc., 2 (1981) 65–67.
250 X. Yin, G. Schlemmer and B. Welz, Anal. Chem., 59 (1987) 1462–1466.
251 B.J. Starkey, A. Taylor and A.W. Walker, J. Anal. At. Spectrom., 1 (1986) 397–400.

438

252 U. Voellkopf and Z. Grobenski, At. Spectrosc., 4 (1984) 115–122.
253 R.F.M. Herber, A.M. Roelofsen, W. Hazelhoff Roelfzema and J.H.J.C. Peereboom-Stegeman, Fresenius Z. Anal. Chem., 322 (1985) 743–746.
254 E. Lucker, A. Rosopulo and W. Kreuzer, Fresenius Z. Anal. Chem., 328 (1987) 370.
255 F.W. Sunderman, Jr., in S.S. Brown (Ed.), Clinical Chemistry and Chemical Toxicology of Metals, Elsevier, Amsterdam, 1977, pp. 231–260.
256 F.W. Sunderman, Jr., M.C. Chrisostomo, M.C. Reid, S.M. Hopfer and S. Nomoto, Ann. Clin. Lab. Sci., 14 (1984) 232.
257 M. Drazniosky, I.S. Parkinson, M.K. Ward, S.M. Channon and D.N.S. Ferr, Clin. Chim. Acta, 145 (1985) 219.
258 J.R. Andersen, B. Gammelgaard and S. Reimert, Analyst, 111 (1986) 721–722.
259 F.W. Sunderman, Jr., S.M. Hopfer, M.C. Chrisostomo and M. Stoeppler, Ann. Clin. Lab. Sci., 16 (1986) 219.
260 S.S. Brown, S. Nomoto, M. Stoeppler and F.W. Sunderman, Jr., Pure Appl. Chem., 53 (1981) 773.
261 F.W. Sunderman, Jr., A. Marzouk, M.C. Chrisostomo and D.R. Weatherby, Ann. Clin. Lab. Sci., 15 (1985) 299.
262 D.J. Gawkrodger, I.L. Shuttler and H.T. Delves, Acta Derm. Venereol., 68 (1988) 453–455.
263 J.F. Alder, M.C.C. Batoreu, A.D. Pearse and R. Marks, J. Anal. At. Spectrom., 1 (1986) 365–367.
264 D.C. Paschal and G.G. Bailey, At. Spectrosc., 7(1) (1986) 1–3.
265 H. Zorn, Th. Stiefel and H. Porcher, Toxicol. Environ. Chem., 12 (1986) 163.
266 M.S. Leloux, N.P. Lich and J-R. Claude, At. Spectrosc., 8(2) (1987) 71–75.
267 Z. Grobenski, R. Lehmann, B. Radziuk and U. Voellkopf, At. Spectrosc., 7 (1986) 61–63.
268 M.S. Leloux, N.P. Lich and J-R. Claude, At. Spectrosc., 8(2) (1987) 7555–7577.
269 D.C. Paschal and G.G. Bailey, J. Anal. Toxicol., 10 (1986) 252.
270 B. Welz, G. Sclemmer and J.R. Mudakavi, Anal. Chem., 60 (1988) 256.
271 W. Slavin, Graphite Furnace AAS —A Source Book, Perkin-Elmer, Norwalk, USA, 1984, Part No. 0993-8139.

Chapter 4l

Forensic Science

I.M. DALE

Greater Glasgow Health Board, Occupational Health Service,
Glasgow G1 1JA (Great Britain)

I. INTRODUCTION

In 1752 Mary Blandy was tried for the murder of her father by poisoning him with arsenic. During the course of the trial one of the chief medical witnesses for the prosecution, Dr. Anthony Addington, gave evidence regarding his examination of a white powder found in the possession of the accused. He stated that in his opinion the poison administered was white arsenic for the following reasons: (a) this powder has a milky whiteness, so has white arsenic; (b) this is gritty and almost insipid, so is white arsenic; (c) part of it swims on the surface of cold water like a pale, sulphurous film, but the greatest part sinks to the bottom and remains there undissolved, the same is true of white arsenic, etc. The organs of the deceased were not analysed for the presence of arsenic and in fact appropriate chemical tests would not be available for some one hundred years when the Marsh (1836) and Reinsch (1841) tests were developed. In spite of the less than perfect scientific evidence, Mary Blandy was found guilty and executed [1].

It is fortunate, both for the accused and for the integrity of the legal system, that modern forensic science is more exact. However, this account does demonstrate one aspect of forensic science, that of comparing certain, hopefully characteristic, attributes of evidence found at or near the scene of the crime with similar items associated with the accused. The two most obvious examples of this type of investigation are in the use of fingerprints (physical characteristics of the skin, first employed in Britain in 1902) and blood grouping (biochemical characteristics). In recent years chemical techniques, particularly involving trace element analysis, have been applied successfully in the elucidation of forensic science problems.

The forensic scientist employed in the analysis of specimens for elemental concentrations is involved generally in two main areas of investigation. The first is the determination of toxic elements in biological tissues in order to ascertain the cause of death or injury (homicidal or suicidal) in suspected poisoning cases. The second is to compare certain characteristic trace element concentrations in materials found at the scene of the crime with the same type of material found in the possession of the accused. A special case of this second approach is in the analysis of the elements barium,

References pp. 460–462

antimony and lead, deposited on people's hands after they have discharged a firearm. With the introduction in recent years of comprehensive legislation relating to health and safety in industry, the forensic scientist can be called upon to examine possibly hazardous concentrations of dusts and fumes in the factory environment and to give evidence on the results.

II. BIOLOGICAL MATERIAL AND POISONING

A. Sampling

It is not normal for the analyst to be involved in the collection of specimens during the post mortem dissection of the victim. Accordingly it is imperative that the pathologist carrying out the examination is given clear instructions of the type of specimens required. A supply of clean plastic or glass jars is essential for the transport of portions of the organs to the laboratory. Blood and tissues can be obtained using stainless steel instruments except when cobalt, chromium, manganese and nickel are to be determined. These elements are in very low concentrations in normal human blood and the use of stainless steel needles only gives an assessment of the quantities of these metals which can be extracted from steel by body fluids. There is considerable evidence that certain commercially available blood collection tubes with colour coded stoppers can release significant quantities of metals (e.g. zinc from rubber stoppers, cadmium from orange plastic) [2]. Schmitt [3] considered the factors influencing precision and accuracy in the determination of a number of trace elements by atomic absorption spectrometry, including increased concentration from syringes and decreased values associated with storage of the sample.

Most toxic heavy metals, such as mercury, arsenic, antimony and thallium, are known to accumulate in the kidney, liver and, to a lesser extent, the brain. If the poison was administered shortly before death substantial quantities will also be found in the stomach. Specimens of kidney, liver, brain and stomach contents should certainly be obtained in any case of suspected exposure to toxic metals, together with samples of blood and urine if these can be collected.

If it is considered that the deceased had been poisoned over an extended period prior to death, it may be of value to collect hair (plucked from head with the ends tied with thread at the scalp end) and nail specimens. The growing hair and nail can absorb many trace elements from the blood stream and these elements are permanently bound to the keratin. Knowing that the average rate of growth of human head hair is approximately 1 cm per month, sectional analysis of the hair from the root to the tip will produce a "calendar" of exposure to the element.

B. Integrity of samples

It is absolutely essential that all samples can be directly related back to the scene of the crime or the accused person. The identifying labels must be signed by all staff handling the specimens as these will require to be produced when giving evidence in court.

C. Infection risks

In recent years there has been increased concern regarding the possible risks to pathologists and laboratory staff from handling infected human and animal tissues. The accidental puncturing of the skin by hypodermic needles, other instruments or broken glass which have been in contact with micro-organisms can transfer a wide range of diseases. A review giving details of 22 occupationally acquired infections, ranging from AIDS to tuberculosis, has been published [4]. It is essential that all staff are aware of the real risks which can be present during the handling of specimens from subjects with often unknown medical histories and take appropriate precautions to ensure the safe manipulations of all materials both for themselves and other persons working in the laboratory. In many instances it may be considered of value to check the immunological status of all staff and offer vaccinations (e.g. hepatitis B and BCG for tuberculosis).

D. Pretreatment

As in all fields of trace element analysis it is imperative that no contamination of the specimen is allowed to take place prior to examination. There is always the danger that during the dissection of the organs some contamination of the tissue could occur from metal knives and scalpels; added to this are problems of dust and flaking paint and plaster from the mortuary walls and ceilings. Once the material has arrived in the laboratory it is advisable to remove the cut and exposed portions of the organs with a non-metallic knife. A suitable cutting edge can be prepared from broken pieces of clean laboratory glassware and such a glass knife will cut most types of biological tissue with minimal risk of contamination.

It has to be admitted that decaying human organs have a rather unpleasant odour and for this reason, if for no other, some form of preservation is required. Placing the tissue in preserving liquids such as formalin is not recommended due to the contamination of the sample from trace elements found in commercial preserving fluids, and to the possible leaching of elements from the sample into the liquid. An equivalent problem arises in the examination of bodies which have been embalmed with chemical solutions.

References pp. 460–462

TABLE 1

CONCENTRATION FACTORS OF TRACE ELEMENTS IN DRIED AND ASHED ORGANS AND BODY FLUIDS (I.C.R.P.–23 REPORT [6])

Sample	Concentration factor		Sample	Concentration factor	
	ashed	dried		ashed	dried
Blood, whole	100	5.0	Kidney	91	4.4
Brain	67	4.7	Liver	78	3.6
Stomach	125	3.7	Lung	91	4.5
Hair	200	1.1	Spleen	72	4.5
Heart	92	3.7	Urine	93	14.6

Dry-ashing of the sample is not suitable because elements of interest such as mercury, arsenic, lead and antimony can be lost from the sample at the temperatures employed in the furnace; however, the addition of magnesium and nickel nitrates has been shown to prevent the loss of arsenic and antimony by the formation of arsenides, arsenates and antimonides [5]. There is also the possibility of contamination of the sample from dust from the internal surfaces of the furnace. The two most suitable procedures for the pretreatment of biological material are freeze-drying (lyophilisation) and low-temperature ashing. The added advantage of such treatment is the concentration of the toxic element in the final sample as shown in Table 1 [6]. A further advantage is in the ease of comparing results with previously reported data. Low-temperature ashing of biological material gives a white or grey powder soluble in hydrochloric acid, while the freeze-dried sample requires to be dissolved in concentrated nitric or sulphuric acids.

E. Freeze drying

De Goeij et al. [7] reported on an extensive investigation into the possible loss of trace elements from biological tissue during freeze-drying at 10^{-5} atm for 48 h. They found no significant loss of antimony, arsenic, bromine, cadmium, chromium, cobalt, copper, iron, mercury, molybdenum, selenium or zinc. Thus freeze-drying of human tissue is a suitable treatment prior to analysis, together with the fact that most laboratories have access to appropriate facilities.

Thin slices of tissue are placed in clean plastic petri dishes, the loose fitting lids replaced and the dishes inserted into the chamber of the freeze drier. Blood specimens can be poured directly into the dish. It is vital that the tissue be thoroughly frozen prior to evacuating the chamber. The time required for complete removal of water from the sample depends on the nature and weight of the material and the type of equipment employed.

Drying is usually completed in 12–48 h. The dried tissue is not susceptible to decay and can be stored at room temperature in sealed plastic bags.

F. Low-temperature ashing

A radio-frequency coil is used to dissociate oxygen molecules in a suitable chamber for the ashing of the sample. A number of systems are commercially available. The sample is placed in a pure quartz boat and introduced into the oxidation chamber. The chamber is kept under vacuum during the ashing and oxidation products are vented to the outside atmosphere. The treatment of the tissue takes 24–48 h, depending on the weight of the sample and the power and design of the low-temperature asher. A clean, dry powder easily dissolved in dilute nitric or hydrochloric acid is obtained and is entirely suitable for subsequent atomic absorption analysis by either flame or non-flame systems.

Locke et al. [8] have investigated the possible loss of trace elements during the low-temperature ashing of human liver samples. No losses of magnesium, calcium, manganese, iron, copper, zinc, rubidium or cadmium were noted. Greater than 90% recovery of 50 μg g^{-1} added elements were observed for antimony, barium, thallium, bismuth and lead; however the recovery of arsenic was 64%, selenium 18% and tellurium 15%. In contrast, Gleit and Holland [9] reported good recovery of selenium in alfalfa grass after low-temperature ashing.

G. Sample digestion

Most freeze-dried biological samples can be digested by treatment with high-purity concentrated nitric and sulphuric acids, although tissues with high fat content can prove to be difficult. The addition of concentrated perchloric acid to the digestion mixture can be of advantage, but the design of the fume cupboard and the exhaust ducting must be carefully examined because of the very real dangers caused by the production of explosive perchlorates. The use of PTFE-lined stainless steel pressure digestion vessels have much to recommend them for the rapid dissolution of samples [10]. Welz and Melcher [11] have examined the decomposition of marine biological tissues prior to the determination of arsenic, selenium and mercury using hydride generation and cold vapour atomic absorption spectrometry. It was reported that nitric acid in a PTFE vessel was satisfactory for mercury but low values were found for arsenic and selenium. They recommend pressure digestion for mercury followed by sulphuric/perchloric acid mixture for arsenic and selenium.

The laboratory application of microwave ovens for the heating of digestion mixtures in suitable sealed plastic vessels has been investigated by a number of workers [12]. This technique has much to recommend it since it can produce a solution suitable for atomic absorption spectrometry very quickly. For example Schnitzer et al. [13] found that 1 g fish muscle treated with 2 ml 65% nitric acid and 1 ml conc. sulphuric acid irradiated for 5 min, followed by the addition of 1 ml 30% hydrogen peroxide and a further irradiation for 5 min gave a digestion solution for subsequent analysis for mercury content. Blust et al. [14] also used a microwave technique for the analysis by graphite furnace atomic absorption spectrometry of fish samples for iron, copper and cadmium.

H. Quality control

It is inevitable that the analyst may be required to examine tissue specimens for elements that are not normally determined and in these circumstances it is essential that precautions are taken to ensure that the results are correct, particularly as the evidence will be presented and defended in court. The use of previously analysed international standard samples can be of considerable value in ensuring the accuracy of the results obtained. Unfortunately there are no reference samples having toxic concentrations of many of the possible poisonous trace elements. Accordingly the use of standard addition and other techniques will be required in such circumstances. Care should also be taken to ensure that the results are quoted in the same units as the values obtained by other workers and which may be referred to in the assessment of possible toxic dose, e.g. fresh weight, dry weight, ash weight, parts per million, μg l^{-1}, meq l^{-1}, etc. Reference values for the normal concentrations of a range of elements have been published by a number of researchers [15, 16, 17, 18]. Table 2 presents trace element values found in five normal liver specimens [18].

TABLE 2

MEAN VALUES OF TRACE ELEMENTS IN EIGHT SEGMENTS OF FIVE LIVERS (μg g^{-1} fresh weight)

Element	Mean	Range	Element	Mean	Range
As	0.0065	0.002–0.012	Hg	0.077	0.055–0.108
Br	2.06	0.7–9.1	K	3267	3013–3543
Cd	2.61	0.8–7.1	La	0.02	0.003–0.043
Cl	838	740–936	Mg	171	148–183
Co	0.034	0.023–0.039	Sb	0.011	0.003–0.020
Cr	0.0054	0.002–0.010	Se	0.26	0.22–0.33
Cs	0.012	0.006–0.025	Sn	–	<0.097–0.32
Cu	5.98	3.9–7.7	Zn	59.0	53–66
Fe	205	21–450			

Lievens et al. [18].

III. TOXIC ELEMENTS

A. Arsenic

1. *Introduction*

This element can be considered to have the longest history of pharmaceutical and homicidal use of any substance; in fact, for many people the word "poison" is synonymous with arsenic. Its properties were described in detail by Greek physicians, the Roman physician Paulus Aegineta and in Chinese, Indian and Tibetan manuscripts. Arsenious oxide was employed on a grand scale during the middle ages as a murder weapon, and since the symptoms of acute poisoning (violent stomach pains and vomiting) are similar to the common disease of cholera, the homicide often went undetected [19]. A further advantage was that in amounts necessary for death, arsenic is practically tasteless, particularly when administered in food or drink. In the nineteenth and early twentieth century, arsenic was readily available to the poisoner either as a rat poison or as a constituent of fly paper, in which case the arsenic was extracted into hot water and then added to the victim's food. Famous British trials of people accused of using arsenic include those of Mary Bateman (1807, home produced "medicines"), Madeleine Smith, 1857, cosmetics), Elizabeth Frances Maybrick (1889, fly paper), Frederick Harry Seddon (1911, fly paper) and Herbert Rouse Armstrong (1921, weed killer). In recent years legislation restricting the use of arsenic in household products has reduced the incidence of arsenic poisoning. However, the forensic scientist is often asked to examine specimens to eliminate the possibility of poisoning and arsenic is one such element of interest to the pathologist. Various compounds of arsenic are used in industry, particularly electronics, wood processing and glass making. Arsenious oxide has been employed in taxidermy for the prevention of fungal attack of bird feathers and museum workers can be exposed to high concentrations of arsenic in dust [20]. Occasionally, chronic arsenic poisoning has been associated with the excessive ingestion of arsenic containing tonics [21]. Fatalities have occurred when workers have been exposed to arsine gas liberated during metal processing [22].

2. *Symptoms*

The symptoms of acute and chronic poisoning by arsenic have been described by Rentoul and Smith [23], Davidson and Henry [24] and Fowler [25]. Acute symptoms included gastrointestinal damage, convulsions and haemorrhage; at autopsy, fatty degeneration of the liver and kidneys is noted frequently. Acute inhalation of arsine is followed by extensive haemolysis, haemoglobinuria and death from renal failure.

In chronic arsenic poisoning anorexia, anaemia, disturbances in renal and hepatic function, dermatitis, skin hardening (hyperkeratosis) and pigmentation can occur. Arsine gas is a powerful haemolytic agent and symptoms of poisoning are due to its action on the blood.

3. *Fatal dose*

The smallest recorded fatal dose is 125 mg [23]. The action of vomiting has resulted in the recovery of a woman following the deliberate swallowing of 14 g of arsenious acid. However, in another case 13 g of arsenious acid caused death after seven days. Arsine gas is far more toxic and exposure to concentrations in excess of 30 ppm for an hour or more is dangerous.

4. *Sampling*

Arsenic is concentrated in the liver, spleen and kidney and is bound preferentially to sulphydryl groups in skin, hair and nails. Urine and gastric contents should also be obtained. The sequential analysis of hair sections has been shown to be of value in the examination of prolonged exposure to arsenic [21].

5. *Analytical methods*

Direct flame atomization of the dissolved biological material has not been found suitable for the estimation of arsenic. Instead, arsenic has been determined following the generation of arsine gas by the action of a reducing solution (e.g. sodium borohydride), and subsequent analysis of the arsenic in an argon–hydrogen entrained air flame or a heated quartz cell. Electrothermal atomization–atomic absorption determination of arsenic in biological samples requires the addition of nickel or copper nitrate as a matrix modifier. The use of an ion exchange column prior to hydride generation has permitted the determination of As(V), As(III), monomethylarsonic acid and dimethylarsinic acid species in human urine following suicide attempts [26].

6. *Hydrogen-entrained air flame*

Fiorino et al. [27] have described the use of a shielded hydrogen (nitrogen-diluted) entrained air flame for the analysis of food and animal tissue for arsenic.

Dry tissue (1–3 g) was digested with 30 ml of nitric, sulphuric and perchloric acids (4:4:1). After heating for approximately 1.5 h, the cooled solution was transferred to a 100 ml flask, 30 ml hydrochloric acid added and diluted to 100 ml with water. Suitable aliquots, up to 20 ml, of this solution were used for the analysis. 0.5 ml sodium iodide (10% w/v) was added as a pre-reductant, and the arsine generated by the addition of 6–8 ml alkaline sodium borohydride solution (4 g in 100 ml sodium hydroxide, 10% w/v). The arsine gas was led to the burner and the absorption signal measured at 197.2 nm. For a 5 g sample the detection limit was quoted as 7 ppb.

7. Heated quartz cell

Standard techniques are described by all the major manufacturers of atomic absorption spectrometers for the analysis of arsenic in a heated quartz cell.

Peats [28] described a method for the analysis of arsenic in dried algae and similar biological material. 100 mg dried sample was digested with 1 ml of nitric, perchloric and sulphuric acids (23:23:1) by heating first to 150°C for one hour then to near dryness, and the material allowed to cool. A pre-reductant solution (0.5 ml) of 10% potassium iodide in 5% hydrochloric acid, stabilised with 1% ascorbic acid was then added and the solution made up to 5 ml with 5% hydrochloric acid. The solution was left for at least four hours to assure quantitative reduction of As(V) to As(III). 200 μl of this solution was added to 10 ml of 5% hydrochloric acid in the reaction vessel and 3% sodium borohydride in 1% sodium hydroxide was used as the reducing solution. The generated arsine was carried to the quartz cell by high-purity nitrogen. The quartz cell was heated by the air/acetylene flame, an arsenic electrodeless discharge lamp used as the source and the 193.7 nm line measured for analysis. The absolute lower limit of detection was quoted as 0.45 ng arsenic, with the relative lower limit as 0.11 μg arsenic g^{-1}.

Lovell and Farmer [26] examined the speciation of arsenic in urine samples after unsuccessful suicide attempts by the ingestion of arsenic trioxide and sodium orthoarsenate. In this example 1 ml urine was diluted with 1 ml water and the separation of As(III), monomethylarsonic acid (MMAA), As(V), and dimethylarsinic acid (DMAA) was achieved on a combined cation/anion exchange resin column (Dowex AG50W-X8 over Dowex AG1-X8). The eluants were successively 55 ml of 0.006 M trichloroacetic acid (2 ml min^{-1}), 8 ml of 0.2 M trichloroacetic acid (2 ml min^{-1}), 55 ml of 1 M ammonium hydroxide (6 ml min^{-1}), and 50 ml of 0.2 M trichloroacetic acid (6 ml min^{-1}). The eluant was collected in 3 ml fractions for analysis by arsine generation. A 1 ml portion of each fraction was pre-reduced with 1 ml of a solution 10% in potassium iodide, 10% in ascorbic acid and 20% (w/v) in hydrochloric acid made up to 10 ml with distilled deionised water before hydride generation using 3% (w/v) sodium borohydride in 1% sodium hydroxide solution in an argon-purged Perkin-Elmer MHS-10 system connected to a quartz tube heated in the air/acetylene flame of a Perkin-Elmer PE-306 atomic absorption spectrometer. Deuterium arc background correction was employed and the electrodeless discharge arsenic lamp operated at 8 W and measured at 193.7 nm. The absolute detection limit was 1 ng. This technique has been shown to be of value in the examination of the biochemical transformation of arsenic compounds.

8. Electrothermal atomization

Again standard techniques have been described for this form of analysis using either nickel nitrate or copper nitrate as the matrix modifier to prevent the volatilisation of the arsenic during the heating steps prior to the atomization stage.

Haynes [5] described a method involving the dry-ashing of a 2 g sample with magnesium nitrate and nickel nitrate followed by heating with nitric acid (20–25 ml) and hydrofluoric acid (1–3 ml), taken to dryness and further heated with nitric acid (10 ml) and 30% hydrogen peroxide (10 ml). The resulting solution was filtered and made up to 100 ml with 0.5% (v/v) nitric acid. 20 μl was taken for the arsenic analysis, employing the following instrument parameters: drying 30 s at 110°C, ashing 30 s at 1200°C, atomization 8 s at 2700°C. The sensitivity was given as 100 ng arsenic 1 g sample.

Solomons and Walls [29] described a rapid technique for the analysis of arsenic in forensic cases using digestion of the sample, to which standard amounts of arsenic had been added, with nitric acid and nickel as the matrix modifier. The instrument parameters were: drying 30 s at 110°C, ashing 40 s at 1150°C, atomization 7 s at 2600°C.

A standard method [30] for the determination of total arsenic content of urine has been described, in which a 20 ml urine sample was wet-ashed with a nitric–perchloric–sulphuric acid mixture. The resulting solution was then cooled and cyclohexane and potassium iodide or sodium iodide solution added. An aliquot of the organic phase was mixed with ethanolic copper nitrate as matrix modifier and the arsenic determined by furnace atomic absorption spectrometry (dry at 120°C, ash 1300°C, atomization 2600°C).

Subramanium [31] repeated the determination of the arsenic content of blood by digestion of 0.2 ml blood, diluted with 0.68 ml water, with 0.1 ml of 50% (v/v) nitric acid and 20 μl 5% nickel nitrate being added as modifiers. Further portions were treated with arsenic as standard additions. The detection limit was given as 0.8 ng ml^{-1}.

9. Arsenic in human tissue—normal subjects

Cross et al. [20] reported on the arsenic concentrations in tissue samples from people who died as a result of violence and had no known exposure to arsenic other than that in the general environment. Their results are given in Table 3 and are in general agreement with values reported by McKenzie and Neallie [32], Larsen et al. [33], and Gordus [34].

TABLE 3

ARSENIC CONCENTRATIONS IN UNEXPOSED SUBJECTS (ppm, dry weight)

Sample	n	Range	Arithmetic		Geometric	
			Mean	SD	Mean	SD
Brain	19	0.001–0.036	0.016	0.010	0.012	2.30
Blood, whole	12	0.001–0.920	0.147	0.270	0.036	7.04
Hair	52	0.01 –0.40	0.125	0.102	0.085	2.66
Heart	23	0.002–0.078	0.027	0.023	0.021	2.69
Kidney	25	0.002–0.363	0.050	0.075	0.026	3.45
Liver	27	0.005–0.246	0.057	0.059	0.034	2.84
Nail	124	0.02 –2.90	0.362	0.313	0.283	2.04
Spleen	23	0.001–0.132	0.032	0.035	0.017	3.62
Stomach	21	0.003–0.104	0.037	0.034	0.022	3.28
Urine ($\mu g\ ml^{-1}$)	25	0.006–0.77	0.114	0.164	0.053	3.53

10. *Arsenic in human tissue—exposed subjects*

The most popular era for homicidal arsenic poisoning occurred prior to the development of modern, accurate chemical analysis techniques. However, up-to-date information on the relative concentrations of arsenic in tissue samples is available in a number of cases of industrial, suicidal and homicidal exposure to arsenic.

Cross et al. [20] described cases of arsenic poisoning due to exposure to a wood preservative containing arsenic, to arsine gas liberated during metal processing and to arsine gas leaking from gas cylinders. Representative values are given in Table 4, for workers exposed to arsine in two metal-processing factories and to arsenic in wood-preserving solutions. Subject G died and the following arsenic concentrations were found in tissue specimens: brachial plexus 0.99 ppm (dry weight), heart 2.2 ppm, kidney 6.6 ppm and popliteal nerve 0.61 ppm.

A case of suicide due to the ingestion of a wood-preserving solution was investigated by Cross et al. [35]. The subject swallowed an unknown quantity of a corrosive solution containing copper sulphate, sodium dichromate and an arsenic compound. Death occurred 36 h after ingestion and the cause of death was reported as poisoning by a corrosive substance. The results of the arsenic analysis are given in Table 5.

The attempted suicide cases described by Lovell and Farmer [26] are summarised as follows:

Patient 1. A 41-year old male, who, in the course of consuming a considerable quantity of alcohol, had ingested an unknown quantity of rat poison containing arsenic trioxide, was admitted to hospital with typical symptoms of severe gastrointestinal irritation including vomiting. Dimercaprol was administered intramuscularly at regular intervals and

TABLE 4

ARSENIC LEVELS IN EXPOSED SUBJECTS (ppm, dry weight)

Source	Subject	Hair		Nail		Urine ($\mu g\ ml^{-1}$)
		Head	Pubic	Finger	Toe	
Steel bronze factory	A	71.5	31.5	15	51.9	–
	B	76.1	63.1	188	60.9	–
	C	10.8	8.6	18.6	10.8	–
	D	64.8	27.1	123	37.1	–
Zinc dross factory	E	2.1	6.3	2.2	–	–
	F	0.5	0.2	3.5	–	–
	G	14.0	4.4	3.2	–	–
Wood preservers	H	12.1	10.6	44.7	26.7	0.88
	I	37.4	16.7	–	32.2	0.39
	J	4.0	–	–	1.5	0.42
	K	25.2	17.9	22.5	0.8	–
	L	21.6	38.8	13.7	24.2	0.35

TABLE 5

ARSENIC LEVELS FOLLOWING INGESTION OF WOOD PRESERVATIVE SOLU-
TION (ppm, dry weight)

Tissue	Concentration
Blood, whole	2.0
Brain	2.7
Heart	7.7
Kidney	17.5
Liver	13.9
Spleen	7.4
Stomach	21.1

urine specimens collected over a period of five days. The patient discharged himself thereafter. Total inorganic arsenic decreased from 88% of total arsenic for the 0–12 h post-ingestion period to 28% for the 85–109 h period, while of the methylated forms, DMAA increased from 3% to 51% after four days. Over the period for which urine samples were continuously available As(V) contributed 3.5%, As(III) 39.3%, MMAA 25.6% and DMAA 31.6% of the 11.5 mg excreted.

Patient 2. A 25-year old female, who had ingested "two teaspoonfuls" of sodium orthoarsenate (corresponding to approximately 1.7 g of As(V)), was admitted to hospital 38 h later. She was treated with dimercaprol and discharged herself after two days. The total arsenic excreted in urine during the period 45.5–69.5 h after ingestion was 10.1 mg. The methylated forms,

MMAA and DMAA, increased from 37% to 48% of total arsenic, the DMAA increasing from 8 to 22%. Of the inorganic forms, As(III) was much more significant, constituting 49–56% of total arsenic over the 24 h period, As(V) declining from 10 to 3%.

In one of the suicide cases described by Solomons and Walls [29], the victim had ingested unknown quantities of arsenic over a period of time. However, the analysis for arsenic in the body was not carried out until after the remains had been interred and the internal organs of the thoracic and abdominal cavities had been discarded. An interesting finding was the high concentration of arsenic found in the maggots located in the hair. Results are given in Table 6.

A case of murder by the administration of arsenic was reported by Barrowcliff [36]. On 8th September 1969, Mrs. Waite died of acute gastro-enteritis with allergic polyneuropathy. Previous symptoms included diarrhoea, vomiting, peripheral neuritis, scaly skin, loss of hair and appetite, abdominal pain and shingles. Substantial concentrations of arsenic were found in the liver (40 ppm), duodenum (15 ppm) and small intestine (332 ppm). Sectional analysis of the hair showed variations in arsenic level corresponding with the history of her illness. Values ranged from 2 ppm (during her stay in hospital) to 209 ppm (hair roots at post mortem). Her husband was arrested, charged with murder and found guilty.

Gerhardsson et al. [37] described a fatal arsenic poisoning when a worker was buried in a mass of crude arsenic, containing more than 90% arsenic trioxide. A nearby worker saw the accident and managed to free his head within 1–2 min. However, the protective mask was clogged and the half-suffocated worker tore it off, with the result that there was a massive

TABLE 6

ARSENIC CONCENTRATIONS IN CASE OF SUICIDE

Specimen	Concentration (mg/100 g)
Fingernail: proximal (one half)	28.0
distal (one half)	3.9
Hair segments: 1st, 1 cm (including roots)	3.7
2nd, 1 cm	1.3
3rd, 1 cm	1.0
4th, 1 cm	1.0
5th, 1 cm	0.92
6th, 1 cm	0.74
Gastric contents	109 μg arsenic total
Maggots from scalp	3.8

Solomons and Walls [29].

References pp. 460–462

TABLE 7

CONCENTRATIONS OF TOTAL ARSENIC IN BODY FLUIDS SHORTLY AFTER ADMITTANCE TO THE HOSPITAL AND TISSUE CONCENTRATIONS OF ARSENIC AT AUTOPSY

Fluid	Concentration at admittance ($mg\,l^{-1}$)	Tissue	Concentration at autopsy ($\mu g\,g^{-1}$ wet weight)
Urine	1.9	Brain (front cortex)	0.3
Blood	3.4	Myocardium (left ventricle)	1.2
Gastric fluid	550.0	Kidney (cortex)	1.4
		Lung (peripheral)	2.9
		Liver	3.8
		Blood	2.3

inhalation of arsenic dust. He was admitted to hospital within 40 min and treated by gastric lavage and with activated charcoal, British Anti-Lewisite (2,3-dimercapto-1-propanol) and other drugs. In spite of all the medical care he died 6 h after the accident. The results of the chemical analysis (mostly performed by atomic absorption spectroscopy) are given in Table 7.

B. Thallium

1. *Introduction*

Salts of this element are widely used for the destruction of vermin, since it is highly toxic but has no taste or odour. In the past it was employed in medicine to treat tuberculosis, gonorrhoea, syphilis and ringworm of the scalp, but many cases of thallium poisoning were reported and thallium therapy has been banned in many countries. However, its availability as a rodenticide has resulted in a number of cases of criminal or suicidal ingestion. Various compounds of thallium are used in the manufacture of special glasses and infrared detectors.

2. *Symptoms*

The symptoms of acute and chronic poisoning have been recorded in the standard textbooks on toxicology. When a large dose is ingested, irritable gastrointestinal symptoms tend to occur before neurological findings [38]. The most distinctive feature of thallium poisoning is hair loss.

3. *Fatal dose*

The estimated lethal dose in humans is about 8–12 mg kg^{-1} for an adult and death may not occur for eight to twelve days [39].

4. *Sampling*

Thallium is concentrated in the heart, kidney, muscle, testis, thyroid gland and brain [40, 41].

5. *Analytical methods*

Aoyama et al. [40] used the method of Wall [42] and digested 1–2 g tissue samples in a PTFE pressure vessel with 5 ml concentrated nitric acid at 130–140°C for 90 min. The solution was adjusted to pH 8.5–9.0 by the addition of 25% (v/v) ammonia/water. 2 ml of a 5% (w/v) aqueous solution of sodium diethyldithiocarbamate and 10 ml of water-saturated methyl isobutyl ketone. After shaking for 15 min the solution was centrifuged for 1 min at 2000 g. Thallium concentrations in the organic phase were measured by flame atomic absorption spectrometry. Paschal and Bailey [43] determined thallium in urine samples following the addition of magnesium nitrate as a matrix modifier by graphite furnace atomic absorption with Zeeman background correction. The instrument parameters were dry 180°C (5 s ramp, 55 s hold), char 650°C (5 s ramp, 20 s hold) and atomize at 2000°C (0 s ramp, 5 s hold). The detection limit was calculated to be 0.5 μg l^{-1} and the sensitivity given as 35 pg.

6. *Thallium in human tissue—exposed subjects*

Aoyama et al. [40] investigated the suicide of a 26-year old man who had ingested 10 g of thallous malonate (corresponding to 8 g of thallium). Acute renal failure developed and he was treated by combined haemoperfusion and haemodialysis. In spite of the medical care he died 40 h after the ingestion of the thallium. The thallium concentrations in the tissue samples are given in Table 8 (case 1).

An astonishing case of homicidal use of thallium occurred in a photographic equipment firm in England in 1971. Two members of staff died and two others showed signs of thallium poisoning. Police attention was focused on the young storeman who seemed unaffected by the sickness involving his fellow workers. This man, Graham Young, proved to have an incredible knowledge of poisons and their effects and when the police examined his bedroom they discovered numerous textbooks on forensic medicine (including Glaister's *Medical Jurisprudence and Toxicology*), a collection of some 70 chemicals (including one marked as his "exit dose") and most incriminating a diary listing the quantities given to the men and the symptoms as they developed. It was subsequently discovered that Young had a long experience with poisons, having been convicted at the age of 14 years of poisoning his step mother (fatally) and two other people. He had been committed to Broadmoor Hospital for the criminally insane and was released after nine

TABLE 8

THALLIUM CONCENTRATIONS (μg g^{-1} wet weight) IN EXPOSED HUMAN TISSUE SAMPLES

Organ	Case 1	Case 2
Adrenal gland	72.2	
Cerebellum: grey matter	60.2	
white matter	55.0	
Cerebrum: grey matter	83.9	
white matter	69.5	
Brain: grey matter		10.0
white matter		3.0
Colon	74.4	
Large intestine		120.0
Heart	142.0	13.3
Kidney		20.0
Cortex	119.0	
Mendulla	115.0	
Liver	60.5	5.0
Lung	37.1	1.8
Muscle	106.0	5.0
Testis	104.0	
Whole blood	32.7	3.4
Urine		6.0

years [41, 44]. The thallium levels found in one of the victims is given in Table 8 (case 2) [41].

C. Potassium

1. Introduction

Potassium salts have a high potential for causing death if administered intravenously due to the depolarisation of muscle potential and sudden cardiac contraction which will stop the heart. Two cases recently reported [45, 46] have both involved health care workers who have committed suicide. The relatively high concentration of potassium normally present in human tissue samples could prove difficult in deciding on the cause of death. However, in both these cases death occurred so rapidly that the evidence of intravenous administration was still present.

2. Fatal dose

The estimated fatal dose for humans is 30–35 mg kg^{-1}, but in the case described by Chaturvedi et al. [46] the amount of injected K^{+} was calculated to be 13.4 mg kg^{-1}, approximately 2.2 to 2.6 times lower than previously considered necessary. However, the rate of injection is of paramount importance in deciding the lethal dose since the bolus of potassium containing solution causes death when it reaches the heart.

3. Sampling

As potassium is a normal constituent of the body, it is difficult to obtain meaningful results from the analysis of tissue samples. The determination of potassium in the vitreous humour by atomic absorption spectrometry has been attempted in such cases but the evidential circumstances leading to death are of vital importance in confirming potassium poisoning as the cause of death.

D. Lithium

1. Introduction

Lithium carbonate has been used as an effective treatment for manic depressives since its introduction by Cade in 1949 [47]. During therapy the concentration of lithium in blood can be monitored by atomic absorption spectrometry. Scott [48] has described a method for the analysis of blood samples from forensic cases.

2. Therapeutic and toxic levels

Thompson and Reynolds [49] quote a therapeutic level for lithium in serum of 3.5 μg ml^{-1} to 12 μg ml^{-1} with a toxic level above 14 μg ml^{-1}.

Dawson [50] gives background levels of lithium in blood serum of 0.002 μg ml^{-1} and 0.005 μg ml^{-1} in urine.

3. Sampling

Lithium is normally determined in blood serum and urine as a measure of the course of the therapy.

4. Analytical methods

The standard method for lithium analysis by atomic absorption requires a simple dilution of the sample with deionised water [49]. Scott [48] used an ionisation buffer (potassium salt) for the determination of lithium by atomic absorption. The method involved the pipetting of 0.5 ml of 20000 ppm potassium solution into a disposable centrifuge tube, adding 200 μl saponin solution in Triton X-100 (5 g saponin in 25 ml 4% aqueous Triton

X-100) and the volume made up to 5 ml with whole blood. The liquids are vortex mixed for 5 min, centrifuged for 15 min at 3000 rpm. The analysis is carried out in an air/acetylene flame at a wavelength of 670.8 nm, with the burner rotated 15° to the normal axis. The standards range from 1 to 20 μg ml^{-1} in whole blood.

IV. GLASS ANALYSIS

A. Introduction

Glass is a fusion of the common materials of sand, soda and lime. It has been shown by high-speed photography that during the breaking of glass, whether by a bullet or a blunt instrument, that a considerable cloud of glass fragments fly backwards from all parts of the window. This breakage results in the person being showered with minute glass particles. The main task for the forensic scientist is in attempting to match the glass found on the accused with the glass found at the scene of the crime. The primary discrimination test is based on an examination of the refractive index when differences as small as 0.0001 R.I. units can be measured. However, the overlap for different types of glass (mainly flat and container) is considerable with the result that this single measurement is of limited evidential value. The chemical analysis of the glass fragments can be of value if suitable discriminating elements can be determined. Hickman [51] has discussed a number of interesting forensic cases in which the analysis of glass fragments for elemental composition has proved to be suitable for presentation in court.

B. Discriminating elements

Hickman et al. [52] have shown that magnesium, lithium, cobalt, strontium, iron and arsenic are good classifying elements for colourless, flat, container and tableware glass having a refractive index range of 1.5177 to 1.5183 and the authors considered barium, rubidium, strontium, iron, potassium, manganese and lithium to be good discriminating elements [53].

C. Analysis

Catterick and Wall [54] used an atomic absorption analysis technique for the analysis of small samples of glass fragments. The glass pieces (250–500 μg) were cleaned by soaking in concentrated nitric acid for 30 min, rinsed three times with distilled water, then ethanol and dried at 60–80°C, 0.5 ml hydrofluoric–hydrochloric acid (1:2) added and the samples agitated in an ultrasonic bath for 30 min. The resulting solution was diluted and analysed for magnesium, iron and manganese. Typical concentrations for sheet glass are Mg 2.03%, Fe 0.057% and Mn 84 ppm.

V. COUNTERFEIT WHISKY

A. Introduction

In a number of countries the production of local whisky and its distribution as the genuine Scotch whisky is not uncommon. Illicit distilleries are often in places where soil and plant debris can fall into the fermenting bath and elemental analysis of the liquor can be of value in discriminating between genuine and homemade spirit. Also the quality control of the final product may not be quite as good as that produced by reputable manufacturers, since old automobile radiators are used for the condenser! Hoffman [55] found concentrations as high as 143 ppm for zinc, 28 ppm for lead, 37.6 ppm for cadmium and 15.8 ppm for mercury—an exciting mixture.

B. Case report

In one case report Lima et al. [56] investigated the substitution of genuine Scotch whisky by liquor of local manufacture (Brazilian). The analysis of the bottle caps for tin and antimony content indicated that the locally produced caps were significantly different from those used by the genuine distiller. The local caps had a tin content of 3500 to 4350 ppm while Scottish caps had 12700 ppm; the antimony range was 2300 to 2400 ppm—genuine caps 560 ppm.

VI. GUNSHOT RESIDUE ANALYSIS

A. Introduction

This subject is one of the most controversial topics in forensic science, with considerable debate as to the applicability and interpretation of the results. No matter how perfectly a firearm is constructed, there will always be some release of the products of the explosion onto the hands or face of the person firing the weapon. The diphenylamine–sulphuric acid dermal nitrate test, first introduced in the 1930's, was supposed to detect the presence of nitrates and nitrites from gunpowder discharge residues [57]. In this test, successive layers of molten paraffin wax were applied to the hands of the suspect, the wax removed and the inner surface of the cast treated with an aqueous solution of diphenylamine and sulphuric acid. This reagent forms a blue colour with nitrates and nitrites; however, the test lacks specificity (it forms blue compounds with other nitrogen containing substances) and sensitivity. It was therefore found to give both false negative and positive results in test firings [58]. Accordingly in 1935 and 1940 the Federal Bureau of Investigation advised against the use of this test, although for want of any better technique it was employed for a further twenty to thirty years.

References pp. 460–462

TABLE 9

CHEMICAL COMPOSITIONS OF HANDGUN PRIMING MIXTURES

Component	Composition (%):				
	1	2	3	4	5
Lead styphnate	36	41	39	43	37
Barium nitrate	29	39	40	36	38
Antimony sulphide	9	9	11	–	11
Calcium silicide	–	8	–	12	–
Lead dioxide	9	–	–	–	–
Tetrazene	3	3	4	3	3
Zirconium	9	–	–	–	–
Pentaerythritol tetranitrate	5	–	–	–	5
Nitrocellulose	–	–	6	–	–
Lead peroxide	–	–	–	6	6

A possible alternative to the measurement of residues from the propellant charge of the cartridge was to determine residues from the primer. The primer charge is a shock-sensitive mixture which is detonated by the action of the hammer and subsequently fires the main propellant charge. Table 9 lists typical compositions of hand gun primer charges [59]. Harrison and Gilroy [60] developed colorimetric spot tests for the detection of traces of barium, antimony and lead removed from the hand. However, these tests also suffered from a lack of sensitivity and selectivity and were difficult to apply in the investigation of actual crimes as opposed to test firings.

Ruch et al. [61, 62] applied neutron activation analysis to the determination of barium and antimony deposited on the hands following the discharge of a gun. Table 10 lists the typical amounts of these elements removed from the back of the hand after a single firing of the weapon.

TABLE 10

TYPICAL AMOUNTS OF BARIUM AND ANTIMONY DEPOSITED ON THE HAND AFTER A SINGLE FIRING OF VARIOUS FIREARMS

Weapon	Barium (ng)	Antimony (ng)
0.22 calibre revolver	390	80
0.22 calibre automatic pistol	700	140
0.38 calibre revolver	1300	420
0.44 calibre revolver	1400	420
0.45 calibre automatic pistol	3600	600
0.25 calibre automatic pistol	4700	630
9 mm automatic pistol	7500	730

Electrothermal atomization and atomic absorption has sufficient sensitivity for these elements, and techniques were developed for the determination of gunshot residues by Sherfinski [63], Goleb and Midkiff [64], Kinard and Lundy [57] and by the major suppliers of atomic absorption equipment. For all these methods the residue is removed from the hand by rubbing the skin with cotton swabs moistened with 1 M nitric acid. The swabs are then leached with 2 ml 1 M nitric acid and aliquots of this solution, with or without neutralisation by ammonium hydroxide, are used for the analysis.

The main problem associated with the application of gunshot residue analysis is in the interpretation of the results. Lead is a ubiquitous contaminant of the environment, and little reliance can be placed on the presence of lead removed from the hand of the suspect. Barium and antimony are less commonly found in normal control samples; however these elements can be present in hand swabs due to metal working, painting and other occupations. Also the suspect must be apprehended soon after the incident and the swabs taken before the hands are washed. Booker et al. [65] examined the range of barium and antimony concentrations and ratios found in priming mixtures and concluded that it was not possible to establish a practical threshold for either or both metals.

Nesbitt [59] successfully applied scanning electron microscopy with an energy dispersive X-ray fluorescence detector to provide highly reliable identification of gun shot residue particles. This technique employed an examination of the characteristic morphology of the residue particles with direct simultaneous analysis of the distribution and quantities of lead, barium and antimony in these particles. This method would appear to be the most suitable for the unequivocal determination of the presence of gun shot residues, although the high cost of the instrumentation has limited its routine application in forensic laboratories.

B. Cautionary note

On the 22nd November 1963 the most infamous assassination in recent times took place when President J.F. Kennedy was shot in Dallas. One would expect that the most sophisticated techniques would have been employed in the examination of the suspect, Lee Harvey Oswald; this was not the case. In spite of the F.B.I. reports on the unsatisfactory nature of the diphenylamine test, this procedure was used on paraffin wax casts made from hands and right cheek of Oswald. The test was positive for both hands and negative for the right cheek. The Warren Commission Report [66] stated, "the test is completely unreliable in determining whether a person has recently fired a weapon or whether he has not ... (the reagent) will react positively with nitrates from other sources and most oxidising agents, including dichromates, permanganates, hypochlorates, periodates and some oxides. Thus, contact with tobacco, Clorox, urine, cosmetics, kitchen matches, pharmaceuticals, fertilisers or soils may result in a positive reaction." However, worse was to follow; the paraffin casts were examined by neutron activation analysis for the presence of barium and antimony. These elements were found on both surfaces of all the casts and in fact more barium was found on the outside surface of the cheek cast than on the surface next to the skin. It was clear that contamination of the paraffin wax had occurred prior to the analysis. That this could have happened during the most intensive scientific examination of evidence from a crime is a warning to all involved in forensic science.

VII. CONCLUSION

Forensic science is an exciting and interesting field of analytical chemistry. It would be accurate to state that the investigator can never be certain from day to day what analysis will be required and the significance of his evidence to the solving of a crime. Atomic absorption spectrometry has many benefits for the accurate analysis of small samples submitted for examination.

REFERENCES

1   J. Glaister, The Power of Poison, C. Johnstone, London, 1954.
2   H.T. Delves, Ann. Clin. Biochem., 24 (1987) 529.
3   Y. Schmitt, J. Trace Elem. Electrol. Health Dis., 1 (1987) 107.
4   C.H. Collins and D.A. Kennedy, J. Appl. Bacteriol., 62 (1987) 385.
5   B.W. Haynes, At. Absorpt. Newsl., 17 (1978) 49.
6   I.C.R.P.–23 Report of the Task Group on Reference Man, Pergamon Press, London, 1975.
7   J.J.M. de Goeij, K.J. Volkers and P.S. Tjiol, Anal. Chim. Acta, 109 (1979) 139.
8   J. Locke, D.R. Boase and K.W. Smalldon, Anal. Chim. Acta, 104 (1979) 233.
9   C.E. Gleit and W.D. Holland, Anal. Chem., 34 (1962) 1454.
10 P.E. Paus, At. Absorpt. Newsl., 11 (1972) 129.
11 B. Welz and M. Melcher, Anal. Chem., 57 (1985) 427
12 P. Aysola, P. Anderson and C.H. Langford, Anal. Chem., 59 (1987) 1582.
13 G. Schnitzer, C. Pelterin and C. Clouet, Lab. Pract., 37 (1988) 63.
14 R. Blust, A. Van der Linden, E. Verheyen and W. Declair, J. Anal. At. Spectrosc., 3 (1988) 387.
15 J. Versieck, C.R.C. Critical Reviews in Clinical Laboratory Science, 22 (1985).
16 W.E. Kollmer and G.V. Iyengar, Normalen Konzentrationen verschiedener Elemente in Organen und Körperflüssigkeiten. Gesellschaft für Strahlen- und Umweltforschung mbH, München, 1972.
17 G.V. Iyengar, W.E. Kollmer and H.J.M. Bowen, The Elemental Composition of Human Tissues and Biological Fluids, Verlag Chemie, Weinheim, 1978.
18 P. Lievens, J. Versieck, R. Cornelis and J. Hoste, J. Radioanal. Chem., 37 (1977) 483.
19 S. Kind and M. Overman, Science Against Crime, Aldus Books, London, 1972.

20 J.D. Cross, I.M. Dale, A.C.D. Leslie and H. Smith, J. Radioanal. Chem., 48 (1979) 197.
21 A.C.D. Leslie and H. Smith, Med. Sci. Law, 18 (1978) 159.
22 J.E. Clay, I.M. Dale, J.D. Cross, J. Soc. Occup. Med., 27 (1977) 102.
23 E. Rentoul and H. Smith, Glaister's Medical Jurisprudence and Toxicology, 13th ed., Churchill Livingstone, Edinburgh, 1973.
24 J. Davidson and J.B. Henry, Clinical Diagnosis by Laboratory Methods, W.B. Saunders, Philadelphia, Penn., 1974.
25 B.A. Fowler, in R.A. Goyer and M.A. Mehlan (Eds.), Toxicology of Trace Elements, Wiley, New York, N.Y., 1974.
26 M. A. Lovell and J.G. Farmer, Human Toxicol., 4 (1985) 203.
27 J.A. Fiorino, J.W. Jones and S.G. Capar, Anal. Chem., 48 (1976) 120.
28 S. Peats, At. Absorpt. Newsl., 18 (1979) 118.
29 E.T. Solomons and H.C. Walls, J. Anal. Toxicol., 7 (1983) 220.
30 Standards Association of Australia, Australian Standard, AS 3502-1987.
31 K.S. Subramanian, J. Anal. At. Spectrom., 3 (1988) 111.
32 J.M. McKenzie and J.D. Neallie, Trace Substances in Environmental Health, Univ. Missouri, U.S.A., 8 (1974) 45.
33 N.A. Larsen, B. Neilson, H. Packenburg, P. Christoffersen, E. Damsgaard and K. Heydorn, Nuclear Activation Techniques in the Life Sciences, I.E.A.E., Vienna, 1972.
34 A.A. Gordus, J. Radioanal. Chem., 15 (1973) 229.
35 J.D. Cross, I.M. Dale and H. Smith, Forensic Sci. Int., 13 (1979) 25.
36 D. Barrowcliff, Med. Legal J., 39 (1971).
37 L. Gerhardsson, E. Dahlgren, A. Eriksson, B.E.A. Lagerkvist, J. Lundstrom and G. Nordberg, Scand. J. Work Environ. Health, 14 (1988) 130.
38 B. Castel, Johns Hopkins Med. J., 142 (1978) 27.
39 S. Moeschlin, Clin. Toxicol., 17 (1980) 133.
40 H. Aoyama, M. Yoshida and Y. Yamamura, Human Toxicol., 5 (1986) 369.
41 D.A. Hickman, Proc. Anal. Div. Chem. Soc., 16 (1979) 186.
42 C.D. Wall, Clin. Chim. Acta, 76 (1977) 256.
43 D.C. Paschal and G.G. Bailey, J. Anal. Toxicol., 10 (1986) 252.
44 A. Holden, The St. Albans' Poisoner, Hodder and Stoughton, London, 1974.
45 C.Y. Bhatkhande and V.D. Joglekar, Forensic Sci., 9 (1977) 33.
46 A.K. Chaturvedi, N.G.S. Rao and M.D. Moon, Human Toxicol., 5 (1986) 377.
47 J.F. Cade, Med. J. Aust., 2 (1949) 349.
48 I.M.B. Scott, J. Forensic Sci. Soc., 22 (1982) 41.
49 K.C. Thompson and R. Reynolds, in Atomic Absorption Fluorescence and Flame Emission Spectroscopy, Griffin, England, 1978.
50 B. Dawson, in D.L. Williams, R.F. Nunn and V. Marks (Eds.), The Scientific Foundations of Clinical Biochemistry, Heinemann, London, 1978.
51 D.A. Hickman, Anal. Chem., 56 (1985) 844A.
52 D.A. Hickman, G. Harbottle and E.V. Sayre, Forensic Sci. Int., 23 (1983) 189.
53 D.A. Hickman, Forensic Sci. Int., 23 (1983) 213.
54 T. Catterick and C.D. Wall, Talanta, 25 (1978) 573.
55 C.M. Hoffman, R.L. Brunelle, M.J. Pro and C.E. Martin, J. Assoc. Off. Anal. Chem., 51 (1968) 380.
56 F.W. Lima, C.M. Silva and R. Guimaraes, J. Radioanal. Chem., 15 (1973) 157.
57 W.D. Kinard and D.R. Lundy, Am. Chem. Soc. Symp. Ser., 13 (1975) 97.

58 V.P. Guinn, Ann. Rev. Nucl. Sci., 24 (1974) 561.
59 R.S. Nesbitt, J.E. Wessel and P.F. Jones, Aerospace Report No. ATR-75(7915)-2, El Segundo, Calif., 1974.
60 H.C. Harrison and R. Gilroy, J. Forensic Sci., 4 (1959) 184.
61 R.R. Ruch, V.P. Guinn and R.H. Pinker, Trans. Am. Nucl. Soc., 5 (1962) 282.
62 R.R. Ruch, V.P. Guinn and R.H. Pinker, Nucl. Sci. Eng., 20 (1964) 381.
63 J.H. Sherfinski, At. Absorpt. Newsl., 14 (1975) 26.
64 J.A. Goleb and C.R. Midkiff, Appl. Spectrosc., 29 (1975) 44.
65 J.L. Booker, D.D. Schroeder and J.H. Propp, J. Forensic Sci. Soc., 24 (1984) 81.
66 Warren Commission Report, Report of the President's Commission on the Assassination of President John F. Kennedy, U.S. Government Printing Office, Washington, D.C., 1964.

Chapter 4m

Fine, Industrial and Other Chemicals

L. EBDON and A.S. FISHER

Plymouth Analytical Chemistry Research Unit, Polytechnic South West, Drake Circus, Plymouth PL4 8AA (Great Britain)

I. INTRODUCTION

While many samples arrive in the analytical laboratory in a form ready for analysis by atomic absorption spectrometry, this is not usually so for the chemicals discussed in this chapter. Often the requirement is for quality control on trace metal impurities in a troublesome matrix (e.g. reagents, chemical starting materials, electronic components and medicines) to meet strict limits. The major component in the sample may cause chemical interference, non-specific absorption or salting up of the burner, as well as potential dissolution problems. Thus, although some samples can be analysed by a straightforward dissolution and determination, a recurrent theme in this chapter will be methods of circumventing matrix problems. An attempt will be made to highlight particularly useful methods of matrix destruction and analytical procedures which avoid matrix problems. On occasion it is necessary to isolate the analyte from the matrix, and a number of approaches to doing this will be described with an emphasis upon those which are simple and not time-consuming.

II. CHEMICALS

A. Inorganic

1. *Fine chemicals and analytical reagents*

In a number of cases, simple dissolution of a solid sample in an appropriate solvent is possible and some laboratory reagents may even be analysed without further treatment. Prior to flame analysis, the best solvent is dilute hydrochloric acid, provided of course that the major matrix elements are not silver, lead or another element which forms a sparingly soluble chloride. If additional oxidising ability is required, concentrated nitric acid may be added to the solvent. This acid is the preferred solvent when the analysis is to be completed by electrothermal atomization. If the material contains large amounts of silica, it may be necessary to add hydrofluoric acid after preliminary digestion with hydrochloric acid. Care should of course be taken

References pp. 507–514

with this reagent to avoid all skin contact, to use plastic laboratory ware and to avoid losses of silicon in the form of volatile silicon tetrafluoride. It should be noted that such losses can be eliminated, provided an excess of hydrofluoric acid is used and the solutions are not heated too strongly. In some cases, the silicon may be deliberately removed by evaporation to dryness of the fluoride medium, although this will cause problems with other elements such as calcium and the lanthanides, which form insoluble fluorides. It is often preferable to use aqua regia as the solvent, and to promote the precipitation of silica in those cases where the determination of silicon is not required. Provided the solution is kept highly acidic, and the precipitate aged by prolonged boiling, it can be demonstrated that there is very little co-precipitation of analyte species onto the silica. The subsequent filtration of the sample removes the troublesome silica and a number of potential chemical interferences in the flames. Ways of digesting samples which require more oxidising treatments than outlined here, can be treated as described in Section B below, with the appropriate precautions.

Acid dissolution is a particularly favourable approach for carbonates and sulphides, where the matrix anion will be removed during the evolution of carbon dioxide or hydrogen sulphide, and for salts of organic acids, where the anion seldom causes interference problems. Conversely, sulphates can cause problems during flame atomization and chlorides during furnace atomization; ways of dealing with such problems are discussed below.

Inorganic chemicals not amenable to acid dissolution are unlikely to be successfully attacked by ashing procedures, but in these cases fusion procedures may be useful. Unfortunately, fusion increases still further the dissolved-solid content and may introduce contamination. Various fusion mixtures have been used by different workers. Silicates may be attacked using lithium metaborate (or a mixture of lithium carbonate and boric acid) [1]. In this example, silicates were fused with $LiBO_2$ and LiB_4O_7. The residue was dissolved in 3% nitric acid, and 10% oxine was added. A typical fusion is to mix 0.2–0.5 g of finely ground sample with 2 g of lithium metaborate in a platinum crucible, and heat to 900°C in a muffle furnace for 15 min or until a clear melt is obtained. After cooling, the crucible is placed in a beaker containing concentrated nitric acid (8 ml) and water (150 ml). The dissolution is best completed at room temperature. Sodium carbonate fusions over standard Bunsen burners are favoured by many workers, but obtaining a sufficiently high temperature for a true fusion may prove problematical on an air/natural gas flame. Fusions using sodium peroxide and sodium carbonate mixtures in a zirconium crucible are much simpler, and more reliable (zirconium crucibles have considerably longer life-times than nickel crucibles). The method is to mix 0.5 g of sample with 1.5 g of sodium carbonate (which acts to moderate the fusion) and 4 g of sodium peroxide. After fusion over a Bunsen burner, the zirconium crucible is cooled, and immersed in a beaker containing concentrated hydrochloric acid (35 ml) and water (100 ml). Dissolution of the melt is completed by gentle warming if necessary and the final solution is made up to 500 ml. The standards are best prepared with a matching sodium chloride content (15 g l^{-1}) to minimise errors from different viscosities and ionisation suppression.

Blank determinations following any of the above dissolution procedures are important and may prove significant at the trace level. While it is always advisable to prepare the standards in the same concentration of acid and major matrix elements as the sample, to overcome variation in uptake rate and atomization, it is also preferable to take a blank through the whole dissolution procedure. Losses of trace metals will be minimised if dilute solutions are kept below pH 2 and analysis not delayed.

Little need be said about analytical procedures directly following dissolution. Using a flame as the atom cell, care needs to be taken that large contents of dissolved solids do not cause the burner or nebuliser to partly block. A small PTFE cup which accepts samples up to 100 μl attached directly to the metal capillary of the nebuliser, the so-called injection-cup technique [2], can be of use here. Samples in the range 20–100 μl are pipetted into the cup using a precision micropipette, and the resulting absorption peaks recorded using the peak heights or area mode on the instrument or a chart recorder. In this way, total dissolved salt contents of 15% may be tolerated compared with the 5% normally recommended in the conventional flame mode. Using this technique, also known as discrete sample nebulisation, surprisingly little deterioration in detection limits is observed. An alternative approach to handling samples with high dissolved solids or where only small samples are available is to use flow injection [3, 4]. The sample is introduced as a plug into a flowing carrier stream. The flow injection technique is readily automated and is rapidly growing in popularity as a tool for enhancing the versatility and speed of analysis by atomic absorption spectrometry.

A warning note has already been sounded about possible interferences which may be encountered in the flame. Suppression of ionisation of easily ionised elements by large amounts of alkali metals (and alkaline earth metals in a nitrous oxide/acetylene flame) in the matrix may be overcome by adding excess potassium or caesium to all samples and standards (1000 μg ml^{-1} usually suffices). Physical interferences are commonly encountered in inorganic chemical analysis because of the high viscosity of the samples. This can be overcome by buffering the standards with equivalent amounts of high-purity major matrix compounds and acids. Certain highly refractory matrices form stable compounds, usually oxides, in the air/acetylene flame, e.g. zirconium, and trace elements otherwise fully atomized in this flame may be occluded in the microscopic particles and thus a negative error results. Therefore, the flame should be chosen with the matrix as well as the analyte in mind, and in such cases of occlusion the use of a hotter and reducing nitrous oxide/acetylene flame may be preferable. The addition of ammonium salts may, by rapid sublimation, also aid the breakup of clotlets in the flame. Chemical interferences, the formation of actual chemical compounds in the flame which atomize more slowly than other forms of the analyte, can be encountered. The classical example is the calcium/phosphate interference which may greatly reduce the population of calcium atoms in the flame, compared with calcium in a chloride medium. This interference can be overcome by the use of releasing agents (such as lanthanum which preferentially complexes with the phosphate), sequestering agents (such as EDTA which preferentially complexes with calcium, but does not retard atomization), or a hotter flame (i.e. nitrous oxide/acetylene). Similar approaches can be used to overcome many other chemical interferences, such as those of sulphate, phosphate, silicate and aluminate ions on alkaline earth metals. The method of standard additions will provide useful information on such interferences when results are compared with those obtained by direct calibration. Standard additions will, however, provide no information about the extent of non-specific or molecular absorption interferences. Fortunately, these can be minimised by the use of automatic background correction. A full description of such techniques can be found in Chapter 2, Section II. In the analysis of solutions with high dissolved salt contents, automatic background correction is advisable whenever measurements are made at wavelengths below 300 nm.

The development of modern electrothermal atomizers has proved to be of immense value in the determination of low levels of trace elements in pure chemicals and reagents. Many analyses previously possible only after tedious, contamination-prone, preconcentration techniques are now viable. This is especially so for involatile elements in volatile matrices, where the majority of the matrix can be removed during the ashing step without loss of the analyte element. Some elements which may be lost during a vigorous or prolonged ashing step can be retained by the addition of a stabilising agent. Examples of such stabilisations are the addition of nickel, or more universally, palladium to retain arsenic, selenium, tellurium, lead, etc; phosphate to retain cadmium and sulphide to retain mercury. Others may be identified from the literature or from a study of the melting points of compounds of the analyte. While the use of coated furnace tubes, sophisticated temperature control and platform, or isothermal atomization, have reduced the extent of interferences observed in graphite furnace atomization, these still exceed those observed in flames. Optimisation of the temperature programme, particularly the ash or char cycle, and the use of background correction are essential. The use of peak integration facilities and standard addition may also prove useful. Many interferences occur when the analyte atoms first leave the hot walls of the atomizer and condense with the cooler vapour phase species; such problems may be ameliorated by isothermal atomization. A practical approach to such conditions when using a modern small graphite furnace is to break up an old tube, and use the fragments as small boats, which can be pushed into the centre of the furnace.

The sample can then be pipetted onto the boat from which atomization will not take place until the tube is hot (i.e., the atoms enter a hot environment). A tube clean cycle is always advisable on the completion of atomization to prevent the build up of matrix.

Chloride media present a particular problem in electrothermal atomization. Manning and Slavin [5] have described the determination of lead in magnesium and sodium chlorides at the 0.1 μg g^{-1} level, and their paper contains many points of relevance to these analyses. Pyrolytic carbon-coated tubes were further coated with molybdenum using ammonium molybdate dissolved in ammonia solution. Ammonium nitrate was added to the sample as matrix modifier. This is a particular useful agent, as it promotes the following reaction:

$$\text{NaCl} \quad + \quad \text{NH}_4\text{NO}_3 \quad \rightarrow \quad \text{NaNO}_3 \quad + \quad \text{NH}_4\text{Cl}$$

| boiling point 1413°C | decomposes at 210°C | decomposes at 380°C | sublimes at 335°C |

The temperature programme used was dry for 20 s (110°C), ramp 15 s for a 15-s char (550°C) and ramp 9 s for a 9-s atomization (2500°C). With determinations in chloride media, the ash or char stage is particularly critical in furnace atomization, as hydrogen chloride may be formed which aids the removal of chloride. This might otherwise interfere by vaporising the analyte as the chloride before the atomization temperature is reached.

Nearly all analyses undertaken on a graphite furnace nowadays utilise most or all of the conditions of the stabilised temperature platform furnace (STPF) concept, devised by Slavin. This states that platforms, matrix modifiers, Zeeman background correction, maximum heating rate to the atomization temperature, peak area integration and instrumentation with fast electronics should all be used. In addition to that, argon purge gas and a cool stage just before atomization is also recommended.

Despite these recent advances, for many determinations the level of interest is so low that the only feasible approach is to separate and perhaps preconcentrate the analyte prior to determination in a flame or furnace. Some of the possible methods of doing this for high-purity mineral acids, such as volatilisation, precipitation and complexation, have been discussed by an IUPAC Commission [6]. An alternative way of categorising these possibilities is presented in the discussion below.

(i) *Evaporation*

An approach particularly suited to liquid samples, such as mineral acids, ammonia solutions and volatiles such as silicon tetrachloride, is to evaporate the matrix carefully from a silica crucible. An inverted filter funnel can usefully be placed over the evaporating dish to reduce aerial contamination. For very stringent requirements, a clean air cupboard should be used. The residue should be taken up in 1–10 ml of hydrochloric acid for flame

TABLE 1

WAVELENGTHS AND ATOMIZATION TEMPERATURES FOR DETERMINATION OF TRACE METALS IN ACIDS AND AMMONIA [8]

Element	Wavelength (nm)	Atomization temperature (°C)
Cadmium	228.8	1800
Copper	324.7	2500
Iron	248.3	2540
Lead	283.3	1800
Manganese	279.5	1900
Zinc	213.9	1700

analysis, or 0.5–1 ml of nitric acid for graphite furnace work. By evaporating large amounts of acid, considerable concentration factors can be achieved. Kolodko and Babai [7] determined copper and lead in sulphuric acid to the ng g^{-1} level using an air/acetylene flame. A 50-g sample was evaporated to dryness and the residue dissolved in 3 ml HCl (1:1) and 2 ml water. The solution was then analysed by AAS using an air/acetylene flame at 324.8 nm for copper and 217.0 nm for lead. Copper (0.02 ppm) and lead (0.05 ppm) could be determined with a relative error of <10%.

If suitable, samples can be evaporated directly in a graphite furnace. An elegant method has been described by Langmyhr and Hakedal [8] for the determination of cadmium, copper, iron, lead, manganese and zinc in reagent and technical-grade acids and ammonia solution. Ammonia was evaporated directly in the furnace by adding 50–250 μl of the ammonia to a furnace preheated to 70 to 80°C. This procedure prevented the sample spreading. After evaporation to dryness, the furnace was heated for 60 s at 200°C before proceeding to the atomization stage. The acids were analysed by injecting 20–150 μl onto a glassy carbon boat made to fit directly into the furnace. The boats were evaporated to dryness on a thermostatically controlled hot plate and were covered by an inverted filter funnel attached to a water pump. This latter arrangement removed the acid fumes as well as reduced the possibility of contamination. Sulphuric acid was evaporated initially at 180°C for 5 min with a finish at 220°C for 5 min, whereas hydrofluoric acid was evaporated at 115°C. The boats were inserted into the furnace using PTFE tipped forceps. The analytical conditions shown in Table 1 were used. A 60-s atomization stage was used, followed by a clean cycle of 30 s at 1980°C. Calibration was by the method of standard additions.

(ii) *Ashing and pyrolysis*

Although more widely applicable to organic chemicals, the matrix of some inorganic chemicals may be reduced or modified by heating to an elevated temperature. Care must be exercised not to lose volatile trace metals such as arsenic, selenium, cadmium and zinc. Mercury will almost invariably be lost to some extent. Matrices which might be amenable to this treatment, include ammonium salts, sulphates, nitrates and the salts of organic acids such as the oxalates.

Gan and Gu [9] dry-ashed samples of sodium alkanesulphonates or sodium arenesulphonates. The residues were dissolved in water, and aspirated into an air/acetylene flame. The absorption at 330.2 nm was measured. The relative standard deviation of 8 determinations for samples containing 8–11% sodium, was ±0.57%.

Dry ashing may be performed as part of an electrothermal atomization cycle. Headridge et al. [10] determined antimony dopant, and some other ultra trace elements in semi-conductor silicon by this method. For the determination of antimony, 0.2 to 1.2 mg of silicon was placed in a micro-boat in a graphite furnace. The temperature was ramped to 70°C over 15 s, then to 120°C over 15 s and to 1600°C over 15 s (5 s hold), and stepped to 2250°C (10 second hold). The absorption was measured at 231.2 nm. In the determination of silver, cadmium, iron, lead, manganese and zinc, a larger sample weight was used (1–5 mg). Responses were found to be linear for all elements studied except cadmium.

(iii) Electrodeposition

Electrodeposition onto solid electrodes or mercury cathodes is a long established pretreatment capable of large concentration factors, and provided the cathode potential is carefully controlled, it is also of considerable selectivity. When atomic absorption is used as the finish, selective deposition is not usually required. Holen et al. [11] have developed a technique whereby selenium can be determined in sulphuric acid. Dilute sulphuric acid (1 + 4) (25 ml), was heated to 50°C, and electrolysed for 5 min at −0.8 V versus a silver/silver chloride reference electrode, using a platinum wire filament electrode. The filament was then rinsed with water and acetone, and connected to an electrical circuit. Selenium was then atomized by simultaneous heating of the filament (4.05 V, 10.18 A), and by an argon–hydrogen flame.

Trace elements may also be separated from the matrix by electrodeposition. Hiraide et al. [12] have separated aluminium from high-purity iron. The iron was dissolved by anodic dissolution, and redeposited on a flowing mercury cathode. The aluminium stayed in the sodium acetate–potassium chloride electrolyte, which was then injected into a graphite furnace.

The use of a solid graphite cathode has several advantages over a mercury electrode or a platinum wire cathode. One of these advantages has been exploited by Thomassen et al. [13]. Analytical-grade potassium and magnesium chloride were analysed for bismuth, cadmium, copper, indium, lead, mercury and thallium by acidifying the solution to pH 2, to prevent a rapid rise in pH during electrolysis, and carrying out electrolysis with a

References pp. 507–514

cathode of spectral-grade graphite (previously cleaned by heating for 30 min at 2000°C) at a potential of -1.0 V, versus a silver/silver chloride reference electrode for 15 h. The electrode was then ground down, yielding about 250 mg for subsequent solid sample analysis in a graphite furnace. The sample (2–3 mg) was introduced into the furnace using a tantalum scoop, dried for 30 s at 300°C and then atomized at 1900°C using a 50°C s^{-1} ramp. The sample can usefully be moistened with 100 μl water to aid reproducibility.

(iv) *Adsorption*

Adsorption of elements, ions or complexes, onto active carbon has been used for several years by numerous workers as a means of separation and preconcentration. Trace amounts of several elements were determined in sodium perchlorate by Kimura and Kawanami [14]. The pH of the sodium perchlorate solution was adjusted before filtration through activated charcoal. The metals (silver, bismuth, cobalt, copper, cadmium, iron, indium, lead, manganese, and nickel) were then de-sorbed using nitric acid. A more popular approach has been to adsorb trace metals as a complex. Cadmium, cobalt, copper, iron, lead, nickel and zinc have been determined in high-purity selenium [15]. After dissolution of the selenium in hot 14 M HNO$_3$, the solution was adjusted to pH 6.5 by the addition of 8 M NH$_3$. A 1% sodium diethyldithiocarbamate solution was added, and the mixture was filtered through activated carbon. The diethyldithiocarbamate complexes of the analytes were released with 0.1 M HNO$_3$ and determined by flame AAS, by the injection-cup sampling technique. Detection limits ranged from 0.04 μg g^{-1} for Cd and Ni, to 0.4 μg g^{-1} for Pb.

As well as activated carbon, several other adsorption media have been used. Satake et al. [16] have determined cobalt by adsorption of its acenaphthenequinone dioximate complex on microcrystalline naphthalene. To a cobalt solution was added 3 ml of 0.2% acenaphthenequinone dioxime solution in 0.5 M NaOH and 2 ml of acetate buffer (pH 6.5). After 10 min, 2 ml of 20% naphthalene solution in acetone was added. The precipitated naphthalene was filtered off and dissolved in DMF–1 M HNO$_3$ (24 : 1). The solution was diluted to 10 ml for atomic absorption analysis in an air/acetylene flame. Adsorption of the Co complex was found to be quantitative in the pH range of 5.9–7.9.

Adsorption of DDC complexes onto polymeric thioethers has been advocated by Nazarenko et al. [17]. Again desorption was achieved using HNO$_3$.

(v) *Ion exchange*

An obvious way to separate metal ions from acids, bases, digests and alkali metal salt solutions is to use an ion-exchange column or merely a batch of resin added to the analysis sample. The sample may be passed down a strong acid cation exchange resin, when divalent and trivalent metals will be strongly adsorbed, or a chelating resin (such as Chelex-100), when transition metals will be adsorbed. Desorption can be by elution with an acid, or, the resin may be directly injected into a graphite furnace.

Ion exchange and chelation have become extremely popular methods of separation and preconcentration. Foerster and Lieser [18] have determined traces of copper, iron, lead, nickel and zinc in analytical-grade sodium chloride, calcium chloride, aluminium nitrate, zinc chloride, manganese nitrate and chromium nitrate. One litre of an aqueous 10% solution of each salt (pH 5.75, 5.25, 2.8, 5.4, 3.15 or 2.4 for the respective matrix) was passed (at 5 ml min^{-1}) through a 2-cm diameter column packed with 1 g of a cellulose exchanger substituted with 4-(2-pyridylazo) resorcinol. The collected trace metals were then eluted with 100 ml of 1 M HCl, and determined by electrothermal AAS. Recovery was quantitative except for iron from aluminium nitrate and chromium nitrate, and for nickel from manganese nitrate.

Burba and Willmer [19] have determined traces of bismuth, calcium, cadmium, iron, magnesium, manganese, nickel, strontium and zinc in high-purity aluminium. The aluminium was dissolved in 25% NaOH, and the resulting solution diluted with water. The mixture was treated with cellulose Hyphan at 90°C, and the adsorbed trace metals were eluted with 2 M HCl. The eluate was analysed by flame AAS, using the injection cup technique. Detection limits were reported to be 0.02–0.1 μg g^{-1} (Bi 1 μg g^{-1}). Recently, Strelow [20] has determined trace amounts of lead in gram amounts of bismuth, cadmium, indium and tin, by using a column of AGMP-50 cation exchange resin, and 0.5 M HCl–50% acetone as the mobile phase. Elution of the lead was by 3 M HCl. After dilution of the eluate in 0.5 M HNO$_3$, lead was determined in an air/acetylene flame. In a similar report [21] lead was determined in large amounts of zinc, indium and gallium and various other elements, using an AGI-X4 anion exchange resin (Br form). The lead was retained from a 0.2 M HBr solution, while the other elements were eluted. Lead was then desorbed using 2 M HNO$_3$.

(vi) *Co-precipitation and precipitation*

While trace levels of impurity elements may not be precipitated quantitatively from solutions of reagents, the addition of a suitable carrier usually ensures complete recovery. Such coprecipitation is a very simple and apt method for atomic absorption spectrochemical analysis. Some selectivity is possible since the valency of the many impurities has considerable influence; for example, vanadium and chromium form oxyanions in strongly oxidising media, and are not precipitated. Alkali metals form soluble hydroxides, and are therefore not precipitated by the addition of ammonia solution even in the presence of a carrier such as lanthanum or iron hydroxides.

Traces of bismuth, cadmium, cobalt, iron, lead, indium, nickel, thallium and zinc have been separated from high-purity aluminium by precipitation using ammonium pyrrolidine dithio-carbamate, and Cu^{2+} as a co-precipitant [22]. The precipitate was filtered off on a cellulose nitrate membrane filter, which was then destroyed using acid. The residue was dissolved in a small volume of HNO_3 and analysed by flame AAS.

Iron hydroxide is an extremely popular co-precipitant. It has been used during the determination of sub-microgram amounts of bismuth [23], and antimony [24], in waters. It has also been used in conjunction with hydrated TiO_2 at pH 6 to determine V in sea water [25] and in river water [26].

Traces of silver, gold, copper and palladium in refined lead have been enriched by partial precipitation of the matrix with sodium borohydride [27]. Lead (10 g) was dissolved in HNO_3 (1:1), and the solution evaporated to dryness. The residue was dissolved in water, and a small portion of the lead, together with the trace metals, was precipitated by a solution of $NaBH_4$ in 0.01 M NaOH. The precipitate was then filtered off, and dissolved in HBr containing bromine. The solution was then concentrated so that excess bromine was evaporated. The remaining solution was diluted with 20% HBr, and the trace metals determined in an air/acetylene flame.

Trace metals have also been determined in ammonium fluoride and hydrofluoric acid [28]. The sample (4.7 g of ammonium fluoride, or 5 ml of 38–40% hydrofluoric acid, adjusted to pH 4) was dissolved, together with the appropriate standard additions, in 25 ml of water. Calcium (6 mg as 0.1 M $Ca(NO_3)_2$) was added. The mixture was shaken, allowed to stand for 30 min and then centrifuged in a polyethylene tube for 2 min. The supernatant was decanted. The sediment was dried by infra-red heating, dissolved in 5 ml HCl (1:3) and analysed in an air/acetylene flame. Co-precipitation was nearly complete for iron, but only partial (but reproducible) for cobalt, copper, manganese, nickel and zinc. Limits of detection were reported to be 10 ng g^{-1}.

(vii) *Solvent extraction*

The most popular separation and preconcentration technique for atomic absorption analysis is solvent extraction. In this case, it is easy to identify extraction systems which will remove a broad range of impurities from matrices such as acids, bases and alkali metal salt solutions. In addition to the advantages to be gained from separation, especially valuable in furnace work, and the concentration factors available, solvent extraction confers an additional advantage (typically a factor of 3–5) in flame analysis arising from the favourable nebulisation characteristics of several organic solvents.

The most widely applied reagents have been chelating agents which will complex with many metals, e.g. dithizone and the various thiocarbamate derivatives, such as diethyldithiocarbamate and pyrrolidine dithiocarbamate. The latter agent as the ammonium salt (APDC) has been shown to complex with thirty elements [29], most of which can be readily extracted into various solvents. Methyliso-butyl ketone (MIBK or 4-methylpentan-2-one) is usually the favoured solvent, because of its excellent compatibility with flames. The solubility of MIBK in water is not negligible and this limits the available concentration factor to ten. Higher molecular weight ketones (e.g. decan-2-one) offer better concentration factors, and chloroform up to 50 times. Chloroform is, however, only really suitable for electrothermal atomization, although using microlitre amounts and flow injection it can, with care, be used in conjunction with flames.

APDC complexes are formed over a wide pH range and generally extracted from solutions of pH 2–3. A wide range is, however, permissible and a summary of typical limitations is given in Table 2. It should be noted that extractions should always be made at reproducible pH values. A suitable solution of APDC can be made by dissolving 1 g in 100 ml of water. This solution is not very stable, and is best prepared freshly every few days, and filtered before use. For a typical 50 ml sample, 5 ml of this APDC solution is adequate. The pH can be adjusted as necessary with acetic acid or ammonia. Inspection of Table 2 shows that pH 5 is adequate for most metals, and an indicator such as bromocresol green can be used to indicate this. In the case of manganese, a two-minute excursion to pH 12 aids complexation, and certain metals, notably chromium and molybdenum, need warming to 80°C for 5 min to overcome kinetic problems in complex formation. The solution

TABLE 2

RECOMMENDED pH LIMITATIONS FOR APDC–METAL CHELATE EXTRACTIONS

Limit	Formation	Extraction
No limits, pH range 1–14	Ag, Au, Bi, Cd, Co, Cu, Fe, Hg, Ir, Ni, Os, Pb, Pd, Pt, Re, Rh, Ru, Ti, Zn	Ag, Au, Cu, Ir
pH not less than:		
2	Cr, Ga, In, Mn, Mo, Nb, Sb, Se, Te, U, V	In, Nb, Sb, Ti
3	Sn	Cr, Ga, Se, Te, U
4	Th	Mn, Sn, Th, V
pH not greater than:		
3	W	Mo, W
4	Nb	Nb, U
5	U	As, Sb, Te
6	As, Mo, Te, V	Mn, Se, Sn, Th, V
7	Sn	Cr
8	Th	Ga, Pb
9	Cr, Sb	In
10	Ga, In, Se	Bi, Co, Fe, Hg, Ni, Os, Pd, Pt, Re, Ru, Zn
11		Cd
12	Mn	Rh, Ti

References pp. 507–514

is transferred to a 100-ml separating funnel, and the complex (which may have precipitated) is extracted into 5 ml of MIBK. This is best achieved by vigorous inversions for 30 s. The extraction should be repeated with a second 5 ml aliquot to ensure quantitative recovery. The combined extracts can be aspirated into the flame, or injected into the furnace. For electrothermal atomization, chloroform may be used as the extractant, and a larger volume of sample, or smaller volume of solvent used. Standards must be taken through the same procedure.

Extremely good detection limits are obtainable, especially if the analytes are back-extracted into a minimum of acid. Micro-traces of bismuth, cobalt, iron, indium, manganese, nickel, lead, copper, antimony and zinc have been determined in phosphoric acid (as used for etching in microchip manufacture) and ammonium phosphate [30]. High-purity phosphoric acid was neutralised with aqueous ammonia, and APDC was added to form the complexes with the trace metals. The complexes were extracted into a solution of 4-benzoyl-3-methyl-1-phenylpyrazolin-5-one in MIBK, and analysed by flame atomic absorption spectrometry. Alternatively, the complexes were extracted into quinolin-8-ol in butyl acetate and then analysed by flame atomic absorption. Calibration was by standard additions.

Other complexing agents may be used for elements that do not form complexes with APDC. Phosphorus was indirectly measured in high-purity silica and chlorosilanes by the formation of an antimoniomolybdophosphate complex [31]. This was then extracted into MIBK and analysed by flame atomic absorption. Since the antimony:phosphorus ratio is 1:1 in the complex, then by measuring the antimony signal at 206.8 nm, the phosphorus was indirectly determined. A sensitivity of 0.02 ppm was obtained. Another complexing agent is 8-hydroxyquinoline (oxine). This has the advantage of forming chelates with a number of elements, such as Al, Ca, Mg and Sr which do not react with APDC. Oxinates may be extracted into MIBK or chloroform, typically at pH 11 for calcium and magnesium. Traces of calcium have been determined in phosphoric acid and its salts by extraction as the oxine complex [32]. To a sample of phosphoric acid (9.8 g) was added 6 M hydrochloric acid (1 ml) and 10% sodium tartrate solution (20 ml). This was neutralised with 6 M sodium hydroxide and diluted to 100 ml. To an aliquot (10 ml), saturated sodium chloride solution (10 ml), butyl cellosolve (2 ml), oxine (3 ml of a solution of 5 g in 100 ml ethanol) and potassium chloride/sodium hydroxide pH 13 buffer (5 ml) were added. The complex was extracted in 3-methyl-1-butanol, which was aspirated into an air/acetylene flame. Measurement was at 422.7 nm.

More exotic complexing agents that have been used, include 2,9-dimethyl-4, 7-diphenyl-1, 10-phenanthroline [33]. This was used to extract trace impurities of copper and iron from high-purity bismuth, into benzene. Analysis was performed using a tungsten ribbon furnace atomizer.

Arsenic (III) has been extracted from MOS-grade sulphuric acid by extraction into hexane by ammonium di-S-butylphosphorodithioate [34]. It was then back-extracted into water, and analysed by electrothermal atomization. The detection limit was 0.2 ppb for a 500-ml sample.

Some elements can be extracted as simple salts into organic solvents, e.g., arsenic trichloride into benzene. Cresser [35] has written a useful guide to solvent extraction for flame spectrometry which can be consulted when developing new methods.

(viii) *Vapour generation*

Mercury and those elements (principally antimony, arsenic, bismuth, germanium, lead, selenium, tellurium and tin) which form volatile covalent hydrides, may be separated from the matrix by vapour generation. The use of tin(II) chloride to generate elemental mercury and its subsequent aeration into a long path absorption cell with silica windows has been described elsewhere in this book, as has the use of sodium borohydride to produce hydrides which are swept into a flame, or heated tube for atomization. This approach is extremely successful for these elements, and sub ng g^{-1} detection limits are obtainable for several. The hydride generation technique is, however, subject to interference from transition metals such as nickel and some oxyanions.

Mercury has been determined in silver and silver nitrate by adding potassium bromide to the nitrate solution (for silver dissolve 0.1–1 g in 6 ml 8 M nitric acid) adjusted to pH 4–6 with ammonia (as indicated by the purple form of bromocresol purple) [36]. The silver bromide formed can be removed by filtration, the mercury remaining in the solution as the $[HgBr_4]^{2+}$ complex. Tin(II) chloride was used to generate the mercury vapour. As the recovery obtained was only 70–85%, because of co-precipitation a calibration curve was used. Arsenic has been determined in solutions with a 10^4 excess of selenium [37]. Hydrazine was used to reduce selenium to its elemental state. Any As(V) present would be reduced to As(III) but not As^0. To arsenic standards (0–200 μl of a 16 mg l^{-1} standard), 4.0 g l^{-1} Se(IV) (20 ml) and 12 M hydrochloric acid (20 ml) were added. Finally, 0.5 M hydrazine (20 ml) was added. One hour after mixing, the flasks were heated on a water bath at 80°C for one hour. After cooling, the contents of the flasks were diluted to 100 ml with HCl (4M). The As(III) was then determined using hydride generation with sodium borohydride (2% m/V) as the reductant.

Nickel has been determined in various certified reference materials by generation of the volatile nickel carbonyl [38]. For the analysis of steel, a 20-mg sample was dissolved using 1 : 1 HNO$_3$/HCl (5 ml). After dissolution and evaporation, the residue was taken up in concentrated HNO$_3$ (5 ml), and diluted to 100 ml with water. The aliquots used were in the range 0.5–1 ml. The calibration volume was 10 ml, and the reductant used was 3% NaBH$_4$ in 1% NaOH solution. Gas flow rates were 0.5 l min^{-1} N$_2$ and 0.5 l min^{-1} CO. A detection limit of 10 ng in the aliquots used was claimed.

2. *Industrial chemicals*

Similar general remarks about inorganic industrial chemicals can be made as for Section A above. Additionally, such chemicals and technological chemical solutions may be more complex and their major constituents less well defined. Thus interferences may constitute a greater problem. Fortunately, however, the levels of trace metals present may be higher than those of interest in laboratory chemicals, and flame analysis is often appropriate. Again there is need to exercise care over blanks and to match standards for major constituents and acid content.

Aluminium ($0.01-4$ mg l^{-1}) has been determined in hydrothermal reaction fluids by electrothermal atomization [39]. The reaction fluids contained up to 0.1 M KCl. Difficulties due to the enhancement of the Al signal by the KCl were overcome by matching the KCl concentration of the standards and samples. Nitric acid (1%) was preferred to ammonium nitrate as matrix modifier. A micro-extraction system based on the extraction of the aluminium-quinolin-8-olate complex into MIBK was also described.

Solid samples may frequently require dissolution in strong and sometimes hazardous acids. Sodium and potassium have been determined in molecular sieves by Niu and Li [40]. The sample (0.03 to 0.05 g) was dissolved in 2 ml of HCl (1 : 1), and the solution diluted with water to 100 ml for determining potassium and 1 l for determining sodium. The solutions were aspirated into an air/liquefied petroleum gas flame. This flame was found not to cause ionisation interferences, and hence no ionisation buffer was required. An alternative approach is to use fusions. These have been advocated by other authors [41].

The analysis of brines deserves special mention, because the high chloride content is extremely unfavourable for electrothermal atomization, and most troublesome for flame analysis. One approach is solvent extraction with either oxine or APDC to remove the trace metals into a small volume of MIBK. Care must be taken to avoid interference from chloro-complexes in the extraction, and if this is suspected, an ion-association extraction of these complexes might be preferable. Preconcentration of brines has also been performed using ion exchange [42]. A cooling brine containing 22% m/m potassium chloride solution and 4% m/m glycerol was diluted with a sodium acetate buffer (0.1M, pH 5.5) and spiked with 1 μg g^{-1} of each of arsenic, barium, cadmium, copper, lead and zinc. Then 5, 15 and 25 ml of sample was preconcentrated onto each of four different columns (Chelex-100, oxine cellulose, CPPI resin and Hyphan Cellulose). In each case, elution was with 5 ml of nitric acid (2M). The Hyphan Cellulose bound only copper, but the other three ion-exchangers gave recoveries of between 90 and 1–103% for the elements cadmium, copper, lead and zinc. Arsenic and barium were not retained on any column.

B. Organic

1. *Organometallic*

The determination of the metallic components in organometallic compounds is an important adjunct in characterisation for the synthetic chemist and for quality control in industry. Atomic absorption offers an excellent means of doing this through a combination of good sensitivity with good precision. Often only small samples are available and electrothermal atomizers are preferred, but when precision is an important requirement, the flame is likely to be optimal. Three approaches to organometallic analysis are frequently encountered: (a) dissolution in a suitable solvent, usually organic; (b) dissolution in a suitable solvent followed by ligand exchange; and (c) digestion and dissolution.

The first case is the simplest, but since different organo-metallic compounds will nebulise and atomize in different ways, and probably very differently to inorganic standards, standards made from similar organic compounds in the same solvent must be used.

Kudryashova et al. [43] determined sodium and silicon in organo-silicon and -fluorosilicon surfactants by atomic absorption. The sample (≈ 20 mg) was dissolved in water, and diluted to 100 ml. For sodium, the solution was atomized in an air/acetylene flame set at right angles, and measurement was at 589.0 nm. For silicon, a nitrous-oxide/acetylene flame was used, and measurement was at 251.6 nm. Standard solutions were prepared from sodium chloride and sodium silicate. The method was suitable for the analysis of solutions containing 5–20μg ml^{-1} sodium and 10–100 μg ml^{-1} silicon. Carbonyl complexes of ruthenium, e.g. [RuBr$_2$(CO)$_3$]$_2$ and RuBr$_2$(CO)$_2$[P(C$_6$H$_5$)$_3$]$_2$ have been analysed for ruthenium content by dissolving the organometallic compound (20–40 μg) in MIBK, so as to give solutions in the range 1–10 μg Ru ml^{-1} [44]. Ruthenium tris-acetylacetonate in MIBK was used for standards in the subsequent assay with an air/acetylene flame.

After dissolution in a suitable solvent, the analyte may be displaced from the organometallic compound by a stronger ligand. This has considerable advantages when either suitable standards are not available, or a wide range of organometallic compounds is being received for analysis. This ligand exchange method is not suitable for highly stable compounds. By dissolving nickel complexes in MIBK and then adding chloroform and diethyldithiocarbamate, Leonard and Swindall [45] developed a method independent of the bonding in the organo-nickel complexes. This method also had the advantage of giving more stable solutions, and allowing the use of an air/acetylene flame. Previously, the nitrous oxide/acetylene flame had to be used to overcome the effect of the bonding on absorbance.

Organometallic compounds may also be acid-digested. Shah and Kulkarni determined palladium in various organometallic compounds [46]. Samples (5–8 mg) were digested in concentrated nitric/sulphuric acids (0.3 ml of each), containing ≈20 mg of silver nitrate, at 280°C for 4 h. After cooling and dilution with water to 15 ml, the contents of the micro-Carius tube were heated at 85°C for 5 min. The silver chloride precipitate was then filtered off. The filtrate was diluted to the desired levels of 2–20 ppm, and the absorption was measured at 247.64 nm using an air/acetylene flame. Ashing of the matrix, either by wet digestion or by dry ashing, is the preferred method of sample preparation, if there is uncertainty about the formulation. Organorhodium complexes have been analysed by Hartley et al. [47]. The organic material was destroyed using a mixture of concentrated sulphuric acid and hydrogen peroxide. After dissolution, the sample was diluted to volume using an aqueous solution of 2% lanthanum nitrate. The resulting solution was aspirated into an air/acetylene flame. Standards were prepared by digesting [RhH(CO)(PPh$_3$)$_3$] in the same way as the sample. At 343.5 nm, the calibration graph was linear between 0 and 20 ppm. Involatile organoantimony compounds and organic antimony salts have been analysed by Marr et al. [48]. A sample (5–10 mg) was dissolved in butan-2-one (5 ml) in a 50 ml flask. Concentrated hydrochloric acid (2.5 ml) was added to aid dissolution, and ethanol (35 ml) was also added. The solution was diluted to the mark with water. The antimony signal was measured at 217.6 nm in either an air/acetylene, or preferably, an air/hydrogen flame. For more volatile compounds, digestion should be performed in a PTFE-lined digestion vessel. This method has been used to digest organo-mercurial compounds of the type (CH$_3$)$_3$SiCH$_2$HgX [49]. The complex was dissolved in water (2–5 mg in 50 ml), and an aliquot of this solution (0.1 ml) sealed in a bomb with 16 M nitric acid (1.3 ml) and 18 M sulphuric acid (0.7 ml). The bomb was heated at 125°C for 4 h. After cooling, the contents were transferred to a flask and 4% potassium permanganate (7 ml), followed by 5% potassium sulphate (2 ml), were added. The solution was allowed to stand overnight before 6% hydroxylamine hydrochloride (10 ml) was added. The mercury vapour was generated by 20% tin(II) chloride solution (5 ml) for determination by the cold vapour technique.

Dry ashing of the matrix has been performed by Cai et al. [50]. The organo-silicon compound was combusted and the combustion products were absorbed in a solution of sodium hydroxide (3–6M) and hydrogen peroxide. The resulting solution was warmed to 90°C in a water bath for 30 min. The silicon content could then be determined by aspirating the diluted solution into a nitrous oxide/acetylene flame. Calibration was by the method of standard additions.

2. *Impurities in organic compounds*

At first sight, the determination of trace metal impurities in organic compounds may appear to be refreshingly simple. The matrix can be more readily destroyed by wet digestion or ashing than some of the substances previously considered, and thus, several complications may be avoided. This is often the case, but words of warning must be given. Elements such as arsenic, cadmium, lead, selenium and zinc may be lost upon ashing at even moderate temperatures (e.g., 500°C is often recommended in the older literature) and there may be mechanical losses of analyte in any smoke evolved. If suitable precautions are taken, this can be a very useful approach. Schoenberger et al. [51] determined selenium, at the part per million range, in several organic matrices. The organic matrix was ashed at a low temperature (450°C for 2 h) in the presence of ethanolic 10% magnesium nitrate as an ashing aid; and nickel nitrate. The nickel nitrate was used to form the thermally stable Ni–Se complex, and hence prevent loss due to volatilisation of selenium. Determination of the selenium, at 196 nm, was by electrothermal atomization. Recoveries were very dependent on the nature of the matrix. For chrysene and fluoranthene, recoveries were between 90 and 103%, but for bis-(3,3,5-trimethylcyclohexyl) phthalate, 3,4-dimethoxybenzoic acid and 3,5-dihydroxybenzoic acid, recoveries were only 23–50%.

Wet-ashing needs to be considered equally carefully. No one mixture will suffice for all compounds, sometimes organic sulphur may be precipitated as a troublesome suspension, and above all, care must be taken with oxidising mixtures. The great value of peroxide and perchloric oxidants ensure their continued use, but when used to attack pure compounds, additional care must be taken. As the temperature rises, the oxidation potential of different compounds is reached. This causes fewer problems with mixtures, and more warning is given of the need to remove the source of heat temporarily; but with a pure, single compound this point may be reached rapidly and explosively. The reports of the Analytical Methods Committee of the Analytical Division of the Royal Society of Chemistry on the determination of small amounts in organic matter are authoritative guides to digestion procedures; these are published in *The Analyst* (e.g. for nickel [52] and selenium [53]). The use of hydrogen peroxide (50% m/m) and concentrated sulphuric acid or nitric acid is often advocated, and can be successful, but the reader is referred to warnings on the use of peroxide published by the above Committee [54, 55]. Vigler et al. [56] determined phosphorus in petrol containing tritolyl phosphate. With this method, a sample was added to magnesium sulphonate (1 g) and concentrated sulphuric acid (0.1 ml). The mixture was heated until white fumes were evolved, and then ignited at 650°C. The residue containing magnesium pyrophosphate was dissolved in nitric acid (2 ml of 1:1), and diluted with water. Analysis was by electrothermal atomization, and the signal measured at 213.6 nm. Lanthanum nitrate (1%) was used as matrix modifier.

Ashing with mixtures of perchloric acid and nitric and/or sulphuric acid are again widely used, but extreme caution must be exercised. Few substances withstand such an attack, but a large excess of acid to sample must be used, and the sample should be well wetted before heating, which should be applied gently and cautiously, e.g., with a micro-burner. Martinie and Schilt [57] have published the report of a very useful investigation of the effectiveness and hazards of two perchloric acid digestions of 87 organic substances. In one digestion, the sample (1 g) was attacked with a 2:1 perchloric acid/nitric acid mixture (15 ml). After observation for violent behaviour, the temperature was slowly raised to 120°C to distil off the nitric acid. If necessary, the heat was removed to moderate the reaction. Over a period of 15 min, the temperature was raised to 140°C, and then more rapidly to 203°C, where it was boiled for 30 min, provided this was safe. A protective screen and fume hood were of course used. In the second digestion, a sample (1 g) was boiled with sulphuric acid (15 ml) for 15 min. After cooling, nitric acid (15 ml) was added and the temperature raised to 320°C slowly to allow for distillation of the nitric acid. After cooling, perchloric acid (15 ml) was added, and the solution slowly heated so as to distil the perchloric acid from the sulphuric acid solution. Most of the 87 organic compounds investigated were completely oxidised by this treatment. For the compounds studied, some were found to have 50% residual carbon in the nitric/perchloric digest; these were: alanine, proline, methionine, 2, 2′-bypyridine, 8-hydroxyquinoline, 5-nitro-1,10-phenanthroline, 1,10-phenanthroline, pyridine, quinoline and 2,4,6-trimethyl pyridine. The following compounds gave violent reactions during the nitric/perchloric acid procedure: lanoline, cottonseed oil, lecithin, thiophene, furan, cholesterol, squalene, tygon, latex rubber and Amberlite XAD-2. Other compounds gave vigorous reactions; the addition of more nitric acid should be used. These were: pyrrole, sodium tetraphenylborate, tetraphenylarsonium chloride, 1-nitroso-2-naphthol, 2-hydroxyquinoline coumarin, Amberlite CT-120, quinazoline and anthranilic acid.

Organic solvents have been analysed by Manoliu et al. [58]. The elements calcium, copper, iron, magnesium, manganese, nickel, potassium, silver, sodium and zinc were determined in methanol and iso-propyl alcohol. A sample (200–250 ml) was concentrated by evaporation in a silica vessel at 8–10 ml h^{-1} under an IR lamp in a stream of nitrogen. The resulting concentrate was diluted with water to 25 ml and this solution was analysed by atomic absorption using an air/acetylene flame.

C. Phosphors

Phosphors used in the production of fluorescent lamps are of the calcium–halogenophosphate type, and are prepared by firing together calcium hydrogen phosphate, calcium carbonate and a lesser amount of calcium fluoride. The phosphor (0.25 g) can be dissolved in a small amount of concentrated hydrochloric acid, made up to 10 ml, and impurities such as sodium determined using a flame [59].

Scott [60] has determined europium in yttrium phosphors. Yttrium vanadate (0.1 g) was fused with potassium carbonate (2 g) in a platinum crucible. Yttrium oxide and yttrium oxysulphide were dissolved directly in the same solvent as the above melt, 50% hydrochloric acid. Standard flame conditions were used. Ametani et al. [61] analysed zinc sulphide–silver phosphors for several trace elements. Samples (20–100 mg) were dissolved in hydrochloric acid with heating. After dilution, the analytes (magnesium, silver, sodium and zinc) were determined in an air/acetylene flame at 285.2 nm, 328.1 nm, 589.0 nm and 213.8 nm, respectively.

III. PHOTOGRAPHIC FILM AND CHEMICALS

The measurement of various metals in photographic film is important in quality control, as the presence of some metals at given concentration levels may be vital or deleterious, especially in colour photography. Often the levels in solutions of photographical chemicals can be determined by simple dilution and aspiration into a flame. Care must be taken to avoid the build-up of silver in the spray chamber and drain tube, as explosive silver acetylide may be formed. Thorough washing out procedures are therefore advisable. Andronov et al. [62] by-passed these dangers by determining silver in chemical photographical solutions using an air/propane flame. Using the silver line at 328.068 nm, they obtained a sensitivity of 0.1 μg ml^{-1}. Potassium cyanide has been used to strip trace metals from photographic emulsions. Silver has been stripped from a photographic emulsion by 0.2 M [63], and cadmium by 0.04 M potassium cyanide [64]. Enzymes have also been used to digest photographic film [65] prior to the determination of gold, and to extract palladium from films [66]. Two dm^2 of film were extracted at 37°C by the enzyme serizyme. The film was then removed, and the extract evaporated to dryness. The residue was treated with nitric acid/hydrogen peroxide, re-evaporated, treated with a nitric/sulphuric/perchloric acid mixture and again evaporated before adding potassium bromide/hydrobromic acid. The palladium was extracted into toluene using 0.2 M dibutyl sulphide solution. The toluene was injected into a graphite furnace for ashing at 500°C and atomization at 2700°C. This enabled very low levels of palladium, e.g. 1 ng ml^{-1}, to be determined in the presence of many other metals. Such enzyme digestion procedures to remove the film are likely to be generally applicable and more meaningful as the metal content of the paper or polymer backing may be included in procedures involving ashing.

IV. CATALYSTS

The importance of catalysts in our energy- and pollution-conscious age is growing. Many catalysts depend for their activity on low levels of rather exotic metals, while even trace surface levels of elements such as lead, may impair their activity. Thus there is plenty of scope for atomic absorption spectrometry. Many homogeneous catalysts are deposited on an alumina base, thus obtaining dissolution is not always easy, and some interference in the air/acetylene flame may be encountered. A leaching procedure (e.g. with nitric acid) to dissolve adsorbed trace metals may be used to circumvent these problems.

Catalysts are widely used in the petroleum industry, and various dissolution procedures, usually involving hydrofluoric acid, are used. Fabec and Ruschak [67] determined titanium in a silica–alumina based fluid catalytic cracking catalyst by flame atomic absorption. Approximately 0.4 g of catalyst was weighed into a PTFE dish, 10 ml of water, 2 ml of sulphuric acid, and an excess of hydrofluoric acid was added to remove the silica. The contents of the dish were evaporated to dryness on a hot plate (230°C). Upon cooling, 30 ml water and 5 ml hydrochloric acid were added. These were heated until dissolution was complete (340°C). The solution was then cooled and diluted to 50 ml with water. Analysis was performed using a nitrous oxide/acetylene flame. Interference was eliminated by the addition of 1000 mg l^{-1} Al and 1000 mg l^{-1} La to both standards and samples.

Various approaches have been proposed for the analysis of catalysts. Dissolution in strong acids seems to be the most popular. Ruthenium has been determined by Fabec [68] in alumina–silicate catalysts and also titanium in a similar way. Other workers that have used hydrofluoric acid for dissolution, include Urbain and Chambosse [69] for the determination of zirconium, Westerman et al. [70] who used H_2SO_4/HF for the determination of vanadium and alkali metals in vanadium pentoxide catalysts, Chen and Yao [71] who determined chromium in a catalyst used to convert ethylbenzene to styrene, and Madar and Vlcakova [72] who determined palladium in alumino-silicate-based catalysts.

Other acids are also used for dissolution. Aqua regia was used by Petrosyan et al. [73] to decompose a silica-based catalyst. In this method, powder (0.1 g) was heated with 40 ml of aqua regia. The solution was then filtered and the insolubles washed. The washings were combined with the filtrate. Gold was then extracted from 50 ml of solution by 10 ml of iso-amyl alcohol. This solution was then analysed at 242.8 nm using flame atomic absorption.

Concern about air pollution has led to extensive investigations of catalysts for automobile exhausts. Elements such as lead, palladium and platinum can be determined after dissolution of the catalysts in acids such as aqua regia. A review of methods used for analysis of automobile catalysts has been made by Kallmann and Blumberg [74].

Fusions techniques also seem to be a popular method for the preparation of catalysts for analysis. Sinha and Gupta [75] determined minor quantities of iron, and macro quantities of nickel in NiO–Al_2O_3 type catalysts by fusing the sample with potassium bisulphate, and dissolving the flux cake in hydrochloric acid. The solution was then analysed using an air/acetylene flame and a double capillary aspiration system. Igoshina and Talalaev [76] have determined 4–40% ruthenium in iron- or aluminium-based catalysts used in the nitrate industry. A 70–150 ng sample was mixed 2:3:3 with sodium peroxide and sodium hydroxide. The mixture was fused at 400°C for 50 min and the melt dissolved in 3 M hydrochloric acid. This solution was treated with aqueous lanthanum nitrate and diluted with 3 M hydrochloric acid, to contain 10–75 μg ml^{-1} Ru(IV), and 2 mg l^{-1} La. Analysis was in an air/acetylene flame at 349.9 nm.

Chiricosta et al. [77] determined palladium in a polymer-supported catalyst by forming a homogeneous suspension in an organic solvent. Powdered samples of poly-(p-phenylene terephthalamide) supporting ≤2% Pd were prepared as homogeneous suspensions (≈0.004% w/v) in triethylene glycol. Aliquots (25 μl) were placed in a graphite furnace and dried at 150°C (30 s), ashed at 400°C (25 s) and atomized at 2600°C (6 s).

V. SEMI-CONDUCTORS

These represent a class of generally easily dissolved chemicals in which even very low levels of trace metals may be deleterious. There is also some interest in determining major element levels, often to distinguish different, semi-conductor types. The low levels are best determined using electrothermal atomization and most interest centres around possible matrix interferences. It is very important to avoid contamination during digestion, and for this reason enclosed PTFE-lined pressure digestion vessels may be used. Various methods have been proposed for layer by layer determinations of trace metals, and this has been achieved by etching, electrochemical oxidation followed by dissolution, ion-beam bombardment and the use of organic solvents. Yudelevich and Beizel [78] etched semi-conductor silicon with nitric acid–hydrofluoric acid (50:1 to 250:1) for layer by layer analysis. The matrix was removed by repeated evaporation with nitric acid, and the residue was dissolved in 0.3% or 1% nitric acid for determination of indium and thallium, respectively. A 20 μl aliquot was then injected into a graphite furnace. Down to 70 pg of indium and 50 pg of thallium could be determined. Very thin layers were analysed by anodic oxidation, and successive stripping with aqueous 1–2% hydrofluoric acid. Beizel et al. [79] have also analysed semi-conductors of the types Ga P$_x$ As$_{1-x}$, Al$_x$ Ga$_{1-x}$ As and In$_x$ Ga$_{1-x}$ As by etching, using both appropriate acids, and by anodic oxidation in a silica cell (for layers of 0.1–0.3 μm). The electrolyte for anodic oxidation was 3% citric acid solution in ethane-diol. The oxidation product was treated with 6 M hydrochloric acid before analysis. Etching using both chemical and mechanical means has been reported by Yudelevich et al. [80]. Chemical etching involved the use of nitric and hydro fluoric acids. Thin layers (>1

μm) of silicon and germanium were cut using a diamond knife in a solid state microtome. A similar technique was used by Dittrich et al. [81] for layer by layer analysis of AIIIBV semi-conductors. Some interferences due to the formation of diatomic molecules between the trace and matrix components were observed but these were minimized by the use of a platform, and nickel nitrate matrix modifier during the analysis by electrothermal atomization. A chemical etching method using hydrobromic acid–nitric acid was also reported.

Non-metals have also been determined in semi-conductors. Dittrich et al. [82] have determined Br, Cl, F, I and S by using diatomic molecular absorbance in a graphite furnace. The sample solution and an additional solution containing the metal component of the analyte species were added simultaneously or successively to a conventional graphite furnace. A temperature programme was chosen for drying, ashing and molecular evaporation. Gaseous molecular spectra for diatomic molecules were obtained for alkali-metal halides, alkaline earth halides and halides of aluminium, gallium, indium and thallium. Chloride has also been determined by Gladysheva et al. [83]. A single crystal (5–25 mg) of cadmium selenochromite or sulphochromite was dissolved in concentrated nitric acid (2.5 ml) containing 50 μg ml^{-1} silver and 6% barium nitrate solution (1 ml). The solution was gently heated to about 60°C and mixed with 5% sodium sulphate solution (1 ml) and sufficient 20% sodium hydroxide solution to give pH 2. After 30 min the mixture was cooled and filtered. The precipitate was treated with aqueous 4% ammonia solution (5 × 3 ml). The portions were combined and then diluted to 25 ml with water and filtered. The silver content was then determined at 328.1 nm on a propane–butane–air flame.

Headridge et al. [10] have determined antimony dopant and some other trace metals (cadmium, iron, lead, manganese, silver and zinc), in semi-conductor silicon by introducing solid samples into a graphite furnace. For the determination of antimony, 0.2–1.2 mg of sample was placed on a micro-boat in a graphite furnace. For the other metals, 1–5 mg was required. Responses for each were rectilinear, except for cadmium, which was curvilinear. Busheina et al. [84] have determined various elements in gallium arsenide with a graphite furnace, after pre-treatment of the sample in gas streams. Bismuth, indium, lead, manganese and silver were determined after pre-treatment to remove gallium and arsenic by heating at 250°C for 4 min in Ar–Cl (5 : 1). Results were similar to those obtained after dissolution and determination by spark source mass spectrometry.

VI. ELECTROPLATING SOLUTIONS

In electroplating baths, there is often a need to check the concentration of major metals, any metal containing additives (such as brighteners) and trace impurities, to ensure good deposition. Impurities may build up in the baths following introduction of the items to be plated, commercial salts and tap waters, and these need to be monitored carefully. A number of ancillary applications such as determining plating thickness by leaching off the plate, or controlling the composition of alloys (e.g. the lead/tin ratio in a tinning bath by dissolving the alloy in hydrofluoric and nitric acids) are useful in the same industry. Silver-plated ware used for foodstuffs is subject to legislation in a number of countries to control toxic metals being leached into consumables. The legislation usually requires that acetic acid (e.g. 4% v/v in water) be stood in the food container for 24 h, and then arsenic, antimony, cadmium and lead contents determined. Levels around 1 ppm are typically quoted, although higher levels of lead may be allowed. This analysis can be performed by flame atomic absorption.

When major components of plating solutions are determined, large dilutions may be required (e.g. a factor of 5000) to bring the sample into the normal working range for flame analysis. Such dilutions will, however, minimise any interference effects and viscosity effects from additives, and are thus to be preferred to the use of less sensitive lines or burner rotation. The above interference effects may be important in the determination of trace metal levels, and attempts should be made to match the standards for major component levels, or to use the method of standard additions. Solvent extraction to remove the analyte from the matrix may be necessary.

Yu and Ye [85] have determined impurities such as copper, iron, lead and zinc in various electroplating solutions. The concentration of the analytes ranged from several tens of ppm for copper, lead and zinc, to 50 or several hundred ppm for iron. The sample (5 ml) was treated with nitric acid (4 ml of 1:1), and diluted with water to 50 ml. The solution was nebulised into an air/acetylene flame. Free cobalt has been determined in acid gold-plating solutions by Goolen and Verstraelen [86]. In this method, a sample (25 ml) was mixed with 1 M hydrochloric acid (20 ml) and immediately passed through a column containing 50 g of Amberlite IRA-900 anion exchange resin, which retains anionic complexes of cobalt, but not Co^{2+}. The column was then rinsed with 400 ml of water. The washings and the eluate were then combined, and diluted to a volume of 500 ml. This solution was then analysed for cobalt. The average cobalt concentration over ten analyses was found to be 55 mg l^{-1}. Battistoni et al. [87] determined copper and cobalt from nickel-plating baths. Satisfactory results were achieved for Ni/Cu ratios of 5×10^5 and Ni/Co of 4×10^4. A sample was filtered through a 0.8 μm millipore disc. The sample was then mixed with potassium thiocyanate to give a solution 10^{-3} M with respect to the thiocyanate, and acidified to pH 1.7–1.8 with sulphuric acid. The resulting solution was then passed through a column containing 1.2–1.8 g of 3% Adogen 464 anion exchange resin adsorbed on silica gel. The column was then washed with water, and the analytes stripped off the column by nitric acid (0.5 ml) and hydrogen peroxide (0.5 ml of 30%). Determination of the analytes was by electrothermal atomization. Ashing was at 700°C (40 s) and atomization at 2500°C (5 s). Budniok [88] analysed samples of a bath solution containing ethylene diamine, used for the electrolytic preparation of a copper–cadmium alloy. A sample of the solution was decomposed by the addition of sulphuric acid. The solution was then diluted with water and analysed for copper and cadmium on an air/acetylene flame. Coatings (of thickness 2–3 μm) of this alloy, on single crystal silver, were dissolved in nitric acid, and the copper and cadmium contents determined similarly. Deng and Zhou [89] determined cobalt, copper, iron, lead and zinc in nickel electrolytic solutions by directly nebulising the sample into an air/acetylene flame. Davey [90] has also directly determined elements in electroplating baths. The technique of flow injection-flame atomic absorption spectrometry was used to determine zinc in galvanising solutions. Measurements were at 213.9 nm for the mg l^{-1} range, and 307.6 nm for the g l^{-1} range. Zinc concentrations of up to 110 g l^{-1} were determined with an RSD of less than 7%. High levels of potassium, sodium and ammonium chlorides were found not to interfere.

Some workers have determined trace amounts of anions in electroplating solutions. Ozawa [91] developed a technique in which chloride was determined indirectly in chromium plating solutions. Samples (10 ml) were mixed with silver nitrate to precipitate silver chloride. The precipitate formed was separated by centrifugation, and the silver remaining in the solution was determined by atomic absorption spectrometry. The calibration graph was linear for chloride up to 6 μg ml^{-1}. Trace amounts of thiocyanate in plating solutions have also been indirectly determined by atomic absorption [92]. A waste electroplating solution was mixed with 0.5 M potassium dihydrogen phosphate (1 ml), 25 mM thiourea (0.25 ml) and 10 mM neocuproine–CuII complex (0.25 ml), and then diluted with water to 8.5 ml. The solution was then extracted with chlorobenzine (1.5 ml). After agitation for 2 min, the organic layer was sprayed into an air/acetylene flame for determination of thiocyanate by measurement of copper at 324.8 nm.

VII. COSMETICS, DETERGENTS AND HOUSEHOLD PRODUCTS

Soaps consist essentially of organic compounds (see Section II.B.2), detergents contain high concentrations of phosphates and sulphates, while cosmetics may contain further inorganic species. Although problems may arise from the presence of large amounts of phosphate and sulphate in flame analysis, where the use of lanthanum as releasing agent may be essential, the presence of oxyanions is favourable for graphite furnace analysis. In a furnace, lengthy ashing times may be essential to remove non-specific absorption, arising from the pyrolysis of orthophosphates. Unless care is taken, this may result in loss of volatile metals.

Titanium dioxide has been determined in soaps by Lercari et al. [93] using electrothermal atomization. Soap samples were dispersed in water, and calibration standards were derived from aqueous $TiCl_4$ solution. No matrix effects were reported.

Zinc has been determined in 37 samples of dyes and cosmetics by Maslowska and Legedz [94]. The organic samples were digested with HNO_3–H_2SO_4–$HClO_4$ (70:7:23), and inorganic pigments were extracted with concentrated hydrochloric acid at 50°C for 30 min. The use of $LaCl_3$ (1%) prevented interference from iron(III). The linear working range was found to be 0.1–20 μg ml^{-1} zinc. Coefficients of variation ranged from 9.6 to 15.3%. In a similar report by the same authors, bivalent copper, lead, zinc and cadmium were determined in cosmetics and dyes used in cosmetics [95]. The samples were digested using the same HNO_3–H_2SO_4–$HClO_4$ (70:7:23) mix, which was then evaporated. The residue was ignited at 550°C for 5 h, and the ash dissolved in concentrated hydrochloric acid. This solution was evaporated, and the residue ignited as before. The ash was dissolved in HCl–HNO_3–H_2O (1:1:1). This solution was then extracted with concentrated hydrochloric acid at 50°C for 30 min, and the extract was subjected to photo-oxidation with aqueous 30% H_2O_2 under a mercury discharge lamp. The metals were then determined by AAS.

Bismuth oxychloride is used in a number of preparations and Okamoto et al. [96] have determined lead in bismuth-oxychloride-based hair, deodorant and shaving cream sprays and lipstick. Spray (5 g for hair and deodorant, 1 g for shaving cream) was dissolved in ethanol (50 ml). The bismuth oxychloride (1 g or 5 g for low levels) was dissolved in 6 M HCl (15 ml) and diluted to 100 ml with 0.5 M HCl. The lipsticks were ashed at 500°C before extraction with first 20 ml and then 10 ml of 2 M HCl. The lead content was determined in an air/acetylene flame using background correction. For the spray samples, the standards should be dissolved in ethanol also. Gladney [97] developed a simple graphite furnace procedure to determine Bi in eyeshadow. Nickel was used to prevent loss of the volatile bismuth. The sample was dissolved in hydrochloric/nitric/sulphuric acids, and diluted with a nickel solution. A 50 μl aliquot was dried at 100°C for 30 s, charred at 1000°C for 40 s and atomized at 2400°C for 7 s.

Titanium has been determined in a commercial sunscreen formulation by Mason [98]. Samples were digested with hot concentrated H_2SO_4 and $(NH_4)_2SO_4$, then 30% H_2O_2 solution was added dropwise until decomposition of the organic matter was complete. The solution was diluted with water, and Ti was determined in a nitrous oxide/acetylene flame using the 364.3 nm line. The calibration curve was linear up to 120 ppm Ti, and recovery was found to be 99.8% for six determinations. The use of a matrix blank was found to be essential due to the enhancement effects of iron.

Trace metals are required to be monitored in toothpastes for quality control and a particular check must be kept on lead levels. According to

British Standard 5136 [99] there should not be more than 5 mg lead kg^{-1} of toothpaste. The B.S. involves slurrying the toothpaste (2 g) with ethanol (10–15 ml) and then evaporating the solvent. The dried paste is then placed in a muffle furnace at 100°C and the temperature raised to 450°C in 50°C steps to avoid ignition. A few drops of concentrated nitric are added to the dark ashes and these reheated at 450°C for 0.5–1 h. Then water (1 ml) and 5 M nitric acid (10 ml) are added, the mixture boiled for 5 min, and filtered. The extraction is repeated a second time. For a flame atomic absorption finish, the pH is adjusted to 3–4 with ammonia solution, APDC solution added (2 ml of 1% w/v in acetone/water) and after 5 min MIBK added (10 ml). After shaking for 1 min, the organic layer may be aspirated into an air/acetylene flame with measurement at the 283.3 nm lead resonance line. The standards should be prepared the same way.

Trace metals (cadmium, chromium, copper, iron, lead and nickel) have been determined in detergents by simple dissolution in water [100]. The use of low detergent concentrations was necessary to overcome severe background absorption interferences.

VIII. PHARMACEUTICAL PRODUCTS AND DRUGS

Most modern therapeutic drugs are organic-based compounds. Such pharmaceuticals may contain metallic ions as impurities, arising from starting materials, reagents, catalysts or contamination, and these have to be monitored in the products and raw materials to ensure that control limits are not exceeded. American [101], British [102] and European [103] pharmacopoeia identify metals such as: aluminium, arsenic, barium, calcium, chromium, copper, iron, lead, lithium, magnesium, manganese, mercury, potassium, silver, sodium, palladium, selenium, strontium, tin, zinc and other heavy metals as potential impurities. Additionally, trace metal impurities may be indicative of the origin of preparations, this is of particular significance with regard to drugs of abuse. Some mineral salts and organometallic derivatives are of pharmacological interest, and several metallic compounds are in pharmaceutical use. Elements which may be of therapeutic use include: aluminium, antimony, arsenic, barium, bismuth, calcium, chromium, copper, gold, iron, magnesium, mercury, platinum, potassium, silicon, silver, sodium, titanium and zinc. A large number of reports have also been made of indirect determination of organic compounds used in drugs via complexation and atomic absorption spectrometry. Thus there is considerable scope for atomic absorption applications in this field, and clearly use in the pharmaceutical industry is increasing. Several informative reviews have appeared [104–112].

Very often the drugs are indirectly soluble in water, simple organic solvents or dilute acids. If an organic solvent is used, the standards must be prepared in the same solvent. Hydrochloric acid at concentrations similar to that found in the stomach is the preferred acid solvent. On occasion, direct determination using in-situ sample treatment in a graphite furnace (e.g. for oily injectable solutions) is used.

A. Determination of metallic concentrations of significance

Sodium and potassium salts occur in a great number of drugs and pharmaceutical preparations. These include: anti-convulsants (phenytoin sodium), antacids (sodium bicarbonate), anti-microbial preservatives (sodium benzonate), anticoagulants (sodium acid citrate), vasodilators (sodium nitroprusside), antibiotics (phenoxymethyl penicillin potassium), antiseptics (potassium permanganate), and various electrolyte solutions (chlorides etc.). Sodium and potassium may usually be determined with greater precision by flame atomic absorption than by atomic emission. Infusion fluids and dialysis solutions have been analysed by Hazebroucq [113] and by Smith [114]. Samples should be adjusted to 0.1 to 20 μg ml^{-1} for sodium, and between 0.1 and 6 μg ml^{-1} for potassium. An air/acetylene, air/propane or entrained air/hydrogen flame may be used. To prevent suppression of the signals by ionisation interference, an excess (1000 μg ml^{-1}) of an ionisation buffer (often caesium) should be added to both standards and sample solutions.

Lithium carbonate is commonly used in the treatment of manic depression. Lithium carbonate tablets may be dissolved in 10% hydrochloric acid. After addition of an appropriate ionisation buffer (e.g. potassium), the solution may be aspirated into an air/acetylene flame. Drugs containing rubidium and caesium have also been used to treat mental illness. Ionisation problems are again severe and hence an ionisation buffer is required. In addition to this, problems may be encountered due to the poor instrumental sensitivity of the resonance lines (780.0 nm and 852.1 nm, respectively). A red sensitive photomultiplier may be used to overcome these problems.

Magnesium salts also have a wide range of applications in drugs. They are used for the treatment of hypermotivity and spasmophilia, as antacids (oxide, carbonate, hydroxide and trisilicate), as purgatives (hydroxide and carbonate in large doses, and sulphate), or excipients (stearate). Magnesium may also be used, in the chloride form, in peritoneal dialysis or haemodialysis solutions. Samples may be dissolved in hydrochloric acid, concentrated if necessary. In the case of preparations with high organic content, gentle ashing prior to dissolution may be useful. An air/acetylene flame, and standard conditions can be used. Rousselet and Thuillier [104] recommended the use of a spectral buffer of lanthanum chloride (50 g l^{-1}), sodium chloride (10 g l^{-1}) and potassium chloride (2 g l^{-1}), which will eliminate chemical and ionisation interferences. Calcium salts are also widely used in tablets, syrups, suspensions and injections. They too are used in haemodialysis solutions. Generally, these preparations are soluble in hydrochloric acid, and the use of the above buffer, or 5% lanthanum nitrate [115], with standard air/acetylene conditions is recommended. Calcium pantothenate, a B vitamin, may be determined individually by its calcium content at a level of 12 mg per capsule, by the method of standard additions [105], provided calcium is not otherwise present in the formulation.

Aluminium salts are used as antacids (phosphate and glycinate), astringents (aluminium potassium sulphate), and antiseptics. Organic salts, alumina as the hydroxide and phosphates may be attacked with concentrated hydrochloric acid and then diluted to bring the aluminium concentration into the range of 10–50 μg ml^{-1}. Alternative procedures for antacids using hydrochloric/nitric acid [116] and extraction with 4 M hydrochloric acid [117] have been proposed. For silicates, the sample is best taken up in perchloric/hydrofluoric acid, evaporated to dryness to remove silica, and then the residue dissolved in warm hydrochloric acid [107]. In each case, the nitrous-oxide/acetylene flame is the preferred atom cell, and the method of standard additions may be used to minimise any errors arising from lateral diffusion interference in the flame.

Often the chemical speciation of iron, and hence its availability, is significant in iron supplements. Thus, atomic absorption may be preceded by separation procedures [118]. To measure total iron in multi-vitamin preparations, the sample can be dissolved in hydrochloric acid, and determined in an air/acetylene flame. Care must be taken with high mineral content samples to avoid inter-element interferences and the method of standard additions may be employed. Similar comments may be made about the determination of manganese and copper [119, 120], except that interferences are less likely, but there may be a requirement for lower levels to be determined.

Vitamin B_{12} (cyanocobalamin) can be determined indirectly by its cobalt content (4.35%) and given the low concentrations (0.2 μg cyanocobalamin per tablet has been reported [105]), sometimes encountered, atomic absorption is a useful method for this determination. Care must obviously be taken if cobalt was used to prepare the vitamin, in which case standard additions and a conventional air/acetylene flame can be used. If the concentration permits, the tablet may be dissolved in hot water, ethanol or preferably hydrochloric acid [121, 122]. Lower concentrations can be determined by electrothermal atomization [123, 124]. Peck [123] determined cobalt in the range 15–20 ng ml^{-1} on a 20 μl sample using the programme: dry 150°C (30 s), ash 675°C (30 s) and atomize 2700°C (40 s). Whitlock et al. [125] determined vitamin B_{12} in dry feeds by dissolving the sample in 200 ml water, adding EDTA (5 g), and adjusting the pH to 7, using ammonia solution. The cobalt was then concentrated on charcoal (5 g) which was filtered and ashed. The ash was taken up in 5 M nitric acid (3 ml) before aspiration into an air/acetylene flame. Organic components present may enhance the absorbance of cobalt in the flame, so either the method of standard additions, or standards prepared from cyanocobalamin should be used.

Zinc oxide (a mild astringent), zinc stearate (a soothing cream) and zinc undecanoate (a fungicide) are encountered in a variety of creams, ointments and pastes. Moody and Taylor [126] dissolved the residue from such samples after ether extraction (1 g in 5 ml ether) in concentrated hydrochloric acid. After dilution, the determination can be completed at the 213.9 nm line in an air/acetylene flame, where interferences are not normally encountered. Zinc is also a constituent of different forms of insulin. A review of analytical methods for determining zinc in insulin concluded that atomic absorption was the preferred method because of its accuracy, speed and precision [127]. Spielholtz and Toralballa [128] determined zinc in insulin by dissolving a sample (5 mg) in 6 M hydrochloric acid (1 drop) and water (10 ml). After dilution to 50 ml, the solution was aspirated into an air/acetylene flame.

The anti-rheumatic activity of gold salts (e.g. sodium aurothiomalate) ensures their continued therapeutic use. Preparations such as injectable solutions can be analysed directly using an air/acetylene flame and the 242.8 nm line. Standards can be made from tetrachloroauric acid. Conflicting successes have been reported for gold determinations in sodium aurothiopropanol. Complex platinum salts, e.g. cis-platin, are frequently used in chemotherapy. The drugs are generally water-soluble, and therefore extended dissolution procedures are unnecessary. It is advisable to use a similar platinum compound for the standards to overcome any differences in atomization efficiency. Alternatively, the hotter nitrous oxide/acetylene flame may be used.

The use of mercury-based drugs has come under considerable scrutiny recently. They are used as antiseptics, diuretics, disinfectants and bactericides. Acid digestion may be used to release the mercury by the cold vapour reduction/aeration method. Unless great care is exercised, mercury may be lost during the hot stages of the digestion. Miller [105] has proposed digesting ointments and creams (100 mg) in nitric acid (3 ml) using a PTFE-lined bomb for 1 h at 120°C. Organo-mercurial preservatives used in eye drops and multi-dose injections to prevent microbial growth (e.g. phenylmercuric acetate and nitrate, mersalyl and thiomersal), can be acid-digested apparently without loss. When four digestion procedures were compared [129], a mixture of sulphuric/perchloric acids (5:2) was preferred as the most rapid and interference free. Girgis-Takla and Valijanian [130] determined mercury in organo-mercurial preservatives by extracting the solution (100 ml) with 0.005% dithizone in chloroform (2 ml). The dithizone/chloroform layer was then analysed by electrothermal atomization using a graphite rod.

Various other metals are used in drug preparations. These include: silicon (as antacids, silicone oil, etc.), which may be determined after either digestion with hydrofluoric acid (provided that the solution is not heated strongly), or by fusion; chromium (found in disinfectants); arsenic (in arsenamide preparations); selenium (as the sulphide for treating dandruff); bismuth (subgallate) and barium (as a radio-opaque material).

B. Determination of impurities

The quality control of pharmaceuticals is particularly important. Care must be taken to limit the levels of toxic materials in the final product. The acid dissolution procedures described above, (e.g. 6 M hydrochloric acid), are often equally applicable for the determination of impurities. Complete destruction of the matrix by wet oxidation or dry ashing may be necessary to obtain a completely independent method. Raw materials, catalysts, preparative equipment and containers are all possible sources of contamination. Lead, arsenic, mercury, copper, iron, zinc and several other metals may be subject to prescribed limits. Greater sensitivity is often required for lead and arsenic determinations, and this can be achieved by electrothermal atomization, or hydride generation. Akguen and Pindur [131] determined trace amounts of lead, cadmium and thallium in tablets, and their active ingredients. Samples were heated with small amounts of 65% nitric acid until they were decomposed. The residue was dissolved in 0.2–0.5% nitric acid. Aliquots of this solution were analysed using electrothermal atomization. Signals were measured for lead, cadmium and thallium at 283.3 nm, 228.8 nm and 276.8 nm, respectively. In a similar report by the same authors [132], samples were wet-ashed using 65% nitric acid in a Kjeldahl flask. Lead and cadmium were determined in the resulting solution by either flame or electrothermal atomization atomic absorption. Mohay et al. [133] determined traces of lead, zinc, iron and copper in various pharmaceutical materials (barbitone, caffeine and amidopyrine), by direct flame AAS, after dissolution of the sample in water, or a mineral acid, and filtration of the solution. In a report by Iwasa et al. [134], lead was determined in the drugs bismuth subnitrate, bismuth subgallate, silver nitrate and zinc oxide. Concomitants such as sulphate, bismuth, silver and zinc were found not to interfere.

Lead and cadmium have been determined in traditional Chinese medicines by Zhu [135]. A sample was dried for 3 h (105°C), and powdered. The powder (1 g) was treated with concentrated nitric acid (10 ml), perchloric acid (1 ml) and 1.7% phosphoric acid (1 ml). The mixture was allowed to stand overnight. After heating to digest the material, the residue was dissolved in 0.3% nitric acid/potassium iodide. A 5 ml aliquot was then treated with 0.8 M potassium iodide (2 ml), 10% ascorbic acid (1 ml) and MIBK (2 ml). After extraction, the organic layer was analysed for cadmium and lead using electrothermal atomization. The furnace programme used was: dry 120°C (30 s), ash 550°C (Pb) or 300°C (Cd) for 30 s, and atomization at 2300°C (8 s).

Andersen and Helboe [136] developed a method for the determination of aluminium in antihaemophilia preparations by electrothermal atomization. Blood-coagulation factors VIII and IX were obtained from human plasma by precipitation with $Al(OH)_3$. A check was therefore required to ensure that patients were not exposed to unacceptably high levels, and the consequent toxic effects of aluminium.

Mercury has been determined in pharmaceutical products by several workers. Girgis-Takla and Valijanian [130] determined traces of mercury in numerous drugs including potassium chloride, citric acid, sulphacetamide and nicotinamide. After dissolution of the sample in 0.1 M hydrochloric acid (100 ml), the mercury was extracted by shaking with 0.005% dithizone in chloroform (2 ml). The mercury in the extract was then determined by electrothermal atomization, using a carbon rod. Kovar et al. [137] determined mercury in pharmaceutical preparations by pyrolysing the sample in argon, and then completely combusting it in oxygen. The specially designed apparatus was described in their paper. The mercury released during the combustion was passed into silver wool, forming an amalgam. The wool could then be heated in a stream of argon, and the liberated mercury was swept to the light beam in an atomic absorption spectrometer. Recovery was found to be 99.9%, and the sensitivity for 1% absorption was 0.3 ng of mercury. Lau et al. [138] determined mercury in Chinese medicinal pills by cold vapour atomic absorption spectrometry. Samples were digested using concentrated sulphuric acid and 35% hydrogen peroxide. After reduction with 6% tin(II) chloride, the mercury signal was measured at the 253.7 nm line.

C. Indirect methods

Many modern drugs form extractable complexes with metals. This effect has been exploited in a large number of published indirect methods for the determination of such drugs. While the final measurement of the metals so complexed provides a rapid and simple finish, many of the methods are likely to be non-specific. Often related drugs will react with the same metal, and some metals, such as copper, have indeed been advocated for many different species. Careful control of pH must be exercised during complexation to optimise recoveries and selectivity. Thus, such methods are often of academic rather than practical interest. If however, the identity of the drug is known, either via production information or specific tests, and interferents are absent, such tests may have a role to play when speed and simplicity are of the essence. Several of the published procedures are presented in Table 3. A review of indirect methods of determining organic substances, including drugs, has been made by Clark and Yacoub [139].

IX. FUNGICIDES

Organo-metallic-containing fungicides are used in a number of treatments. Given the associated toxicity problems, close analytical control is advisable. Wood preservatives may contain copper, chromium, tin and ar-

TABLE 3

INDIRECT METHODS FOR THE DETERMINATION OF DRUGS

Drug or compound	Analyte metal	Notes	Ref.
Alkaloids	Fe	Extract $Fe(SCN)_6^{3-}$ ion complex into 1,2-dichloroethane.	[140]
Alkaloids (e.g. brucine, strychnine)	Mo	Extract phosphomolybdate complex into MIBK.	[141–144]
Cinchona alkaloids	Hg	Mix sample with 0.02 M K_2HgI_4. Wash precipitate. Dissolve in ethanol. Aspirate ethanol.	[145]
Alkaloids	Co	Extraction of ion pair of drug $Co(SCN)_4^{2-}$ into 1,2-dichloroethane.	[146]
Amino-acids	Cu	Extract ion association complex in to MIBK.	[147–149]
p-Amino benzoic acid	Cu	Extract ion association complex in to MIBK.	[150]
Amino-quinoline antimalarials	Co	Extract ion association complex $(Co(SCN)_4^{2-})$ into nitrobenzene.	[151]
Azepines	Cu	Extract ion association complex in to MIBK.	[152]
Barbiturates	Cu	Precipitate complex. Redissolve.	[153]
Barbiturates	Cu	Mix barbiturate with $CuSO_4$ and diethylamine. Filter, measure filtrate.	[154]
Bromohexine	Co	Extract ion association complex in to 1,2-dichloroethane.	[155]
Benzylpenicillin	Cd	Extract ion association complex in to nitrobenzene.	[156]
Caffeine	Mo	Phosphomolybdate complex dissolved in NH_3–NH_4Cl buffer.	[157]
Chinoform (clioquinol)	Zn	Complex formed in MIBK.	[158]
Chlorpheniramine maleate	Cu	Extract complex into chloroform.	[159]
Chorprothixene	Cr	Precipitate reineckate.	[160]
Ethambutol	Cu	Extract complex with ketone.	[161]
Flufenamic acid	Cu	Extract complex with isopropyl acetate.	[162]
Folic acid (vitamin M)	Ni	Oxidise, extract ion association complex into MIBK.	[163]
Halogen containing compounds	Ag	Acidify, precipitate Ag halide.	[164]
Isonicotinyl hydrazine (isoniazid)	Cu	Extract complex into MIBK.	[165]
Methamphetamine	Bi	Precipitate complex.	[166]
Nitrogenated drugs	Bi	Extract ion association complex (BiI_4^-) into dichloromethane.	[167]
Noscapine	Cr	Extract reineckate into chloroform.	[168]

TABLE 3 (continued)

Drug or compound	Analyte metal	Notes	Ref.
Pilocarpine	Hg	Precipitate complex with Mayer solution. Aspirate supernatent.	[169]
Quinoline	Zn	Extract chelate into MIBK.	[158]
Sodium EDTA	Ni	Add Ni. Precipitate excess.	[170]
Sulphonamides	Ag or Cu	Precipitate complex. Measure unconsumed cation.	[171]
Vitamin B$_1$	Pb	Desulphurise with plumbite.	[172]

senic compounds as part of their formulation. Atomic absorption procedures are recommended in British Standard 5666 [173] for the analysis of wood preservatives and treated timber. Arsenic, chromium or copper can be leached from the wood or the preservative after digestion with sulphuric acid/hydrogen peroxide at 75°C. After filtration, sodium sulphate solution is added and chromium determined in a nitrous oxide/acetylene flame, arsenic in an argon/hydrogen entrained air flame, and copper in an air/acetylene flame. Di- and tri-butyl-tin containing fungicides have been extracted from samples of synthetic resin emulsion paints by shaking with methanolic 1.7% hydrochloric acid [174]. The extract was diluted four-fold with phosphate-citrate buffer solution at pH 2. The compounds were then extracted into hexane. This extract was evaporated, and the residue was taken up in dichloromethane, and passed through a column of alumina to retain any di-butyl- or phenyl-tin compounds. The eluate was evaporated, and the residue was dissolved in hot nitric acid. Tin in the resulting solution was determined by electrothermal atomization using the programme: dry 110°C (20 s), ash 500°C (50 s) and atomize 2500°C (10 s). Measurement was at the 286.3 nm tin line.

Fungicides containing zinc, iron and manganese were analysed by Gudzinowicz and Luciano [175]. The fungicides were acid-hydrolysed with hydrochloric acid. Standards were prepared from the corresponding cyclohexane butyrates.

X. PAINTS AND PIGMENTS

Metals may be introduced into paints as the principal constituents in pigments, in drying oils, in pesticides or antifouling compounds and as markers in security paints. The need to monitor metal levels arises from the effects trace metals may have upon the colour or quality of the paint, its toxicity or in forensic applications. Many countries have legislation specifying permissible levels, and analytical methodology for determining toxic elements in paints. The British Standards Institution has published

References pp. 507–514

a series of reports for both total [176] and soluble amounts of a range of elements [177–180]. Total lead may be determined in several ways [176]. The paint is first tested for the presence of antimony and nitrocellulose. The solvent from a 5 g sample is then evaporated using a hot-plate. If nitrocellulose was found to be present, the sample must first be mixed with liquid paraffin (2 g), and then evaporated. The resulting residues are then mixed with magnesium carbonate (2 g) and ashed at 475°C in a muffle furnace. If antimony is absent, this ash is extracted with hot 18% (w/v) hydrochloric acid (100 ml). The mixture is then filtered, and the combined filtrate and washings diluted to 250 ml with water. If antimony is present, the ash is ground with sodium carbonate–sulphur (10 g of 1 : 1), and heated until SO_2 is no longer evolved. The residue is treated with a small amount of water, and then collected on a filter paper from 1% sodium sulphide solution. The filter paper and residue are boiled with 31.5% (w/v) nitric acid (15 ml). The resulting mixture is mixed with hot hydrochloric acid, and filtered as before. An alternative method is to treat the sample (0.5 g) with either sulphuric acid–hydrogen peroxide or sulphuric acid–nitric acid. The residue is treated with a 3.7% solution of the disodium salt of EDTA (50 ml), aqueous 8.5% (w/v) ammonia (10 ml) and water (50 ml). The mixture is then boiled for 15 min and filtered. The filtrate is diluted to 250 ml with water. The filtrate solutions may then be aspirated into an air/acetylene flame, and the lead signal measured at the 283.3 nm line. The methods are applicable to lead concentrations of 0.01–2% (w/w). An extremely rapid method of estimating lead in paints has also been reported more recently by the British Standards Institution [181]. In this case the paint sample was spotted onto a hardened filter paper, which was then placed in an air/acetylene flame. The range of the method covers lead concentrations of 0.01–0.1%.

Hautbout et al. [182] have determined aluminium, calcium, magnesium, silicon, titanium and zinc oxides and barium sulphate in some pigments for road paints. The paint sample was fused with $Li_2B_4O_7$, and the resulting melt was dissolved in dilute sulphuric acid or dilute nitric acid. Silicon, aluminium, titanium and barium were determined using a nitrous oxide/acetylene flame, and calcium, magnesium and zinc were determined using an air/acetylene flame. Fusions have also been performed by Meranger and Somers [183]. They fused titanium dioxide pigments with potassium bisulphate and glucose, and then dissolved the melt in sulphuric acid. Chromium in pigments from art paintings has been determined by flame atomic absorption following fusion with sodium peroxide/potassium carbonate/sodium carbonate (2 : 1 : 1), and extraction with hydrochloric acid [184].

Dry ashing of paint samples can lead to losses of volatile elements. Hausknecht et al. [185] reported losses of lead of up to 97% after ashing samples at 500°C. They also reported loss of 30% lead from simple acid digestion. To overcome the problem of analyte loss, they developed a technique whereby the sample was ashed at 300°C, and the resulting residue was taken up in dilute nitric acid (1:1). The lead signal was measured at 283.3 nm in an air/acetylene flame.

Acid digestion of paints is also a popular procedure. Takahashi et al. [186] have published a report in which several wet-digestion methods were compared for the determination of copper, zinc and tin in antifouling paints. In an earlier report by Takahashi et al. [187], copper and tin were determined in antifouling paints, used on ship-bottoms, by decomposing samples (0.2 g) in 3:1 nitric acid–sulphuric acid with gentle heat, filtering, and diluting the filtrate with 1 M hydrochloric acid. Mercury has been determined in paints by James and Judson [188]. The antifouling paint was dried and ground and then samples were digested with potassium permanganate at 70°C, or with concentrated nitric acid in a PTFE bomb at 140°C. The latter was the preferred method because it was both faster, and less liable to analyte loss. Mercury was determined by atomic absorption spectrometry using the cold vapour technique. An alternative method has been advocated by Ebdon et al. [189]. Non-oxidative pyrolysis of the sample and collection of the vapours in a strongly oxidising solution of permanganate/sulphuric acid, followed by cold vapour generation produced good, reliable results.

Oils and thinners may be diluted with white spirit or MIBK, usually by at least a factor of ten to promote efficient nebulisation. Metal naphthenates can be used as standards [116]. Paints as well as the oils above may be diluted with MIBK for direct injection into a graphite furnace.

Cobalt and lead have been determined in paint driers and varnishes [190]. Toluene solutions of the paint drier or varnish were mineralised with concentrated nitric acid, and then emulsified by the addition of Triton X-100 (10%). The resulting emulsions have been analysed by flame absorption using an air/acetylene flame and the 240.7 nm line for cobalt and the 283.3 nm line or 261.4 nm line for lead.

If only analysis of the pigment is required, the sample may be dissolved in an organic solvent. Centrifugation allows the collection of the insoluble pigments for subsequent acid dissolution or alkali fusion. If necessary, solvent extraction may be used to remove the trace metals from salts introduced during fusion.

Lead has been determined in wax crayons by direct insertion of the sample (0.15 mg) into a graphite furnace [191]. The programme used was dry 100°C (60 s), ash 500°C (120 s to eliminate all smoke), atomize 2300°C (10 s). The results using the 283.3 nm line compared well with a more conventional procedure of dry-ashing at 500°C, followed by dissolution in nitric acid.

XI. PAPER AND PULP

Paper and pulp, being based on plant materials represent particularly favourable matrices for atomic absorption analysis by flame or electrothermal atomization. Differences in opinion exist as to whether wet or dry ashing is to be preferred for flame analysis, but increasingly, paper samples are being added to graphite furnaces without any sample pretreatment.

Trace metals have been determined in paper additives, such as starch, casein, carboxymethylcellulose, alginate and mannogalactin by Knezevic [192]. A sample (2 g) was decomposed by heating with 65% nitric acid (8 ml) at 160°C, overnight, in a PTFE bomb. The resulting solution was analysed for cadmium, chromium and lead using a graphite furnace, and for copper and zinc using an air/acetylene flame. Calibration was by the method of standard additions. Arsenic has been determined in the same additives, by the same author [193]. The sample was decomposed with 30% nitric acid at 160°C, for 3 h, in a PTFE-lined autoclave. The mixture was then evaporated at 80°C, and a solution of the residue in 1.5% hydrochloric acid was analysed for arsenic by the addition of sodium borohydride in sodium hydroxide solution and measurement of the arsenic signal at 193.7 nm. Knezevic has also determined heavy metals in paper and paperboard [194]. Samples were either decomposed by acid digestion under pressure, or by combustion with oxygen in a "Trace-O-Mat" apparatus. Copper and iron were determined directly in the digest by flame atomic absorption; cadmium was determined by standard additions and lead by the use of a L'vov platform in an electrothermal atomizer. Mercury and arsenic were determined using vapour generation. Detection limits ranged from 0.001 to 3 $\mu g\ g^{-1}$.

Cadmium has been determined in a number of food packaging papers, e.g. sweet wrappers, tea bags, flour sacks and kitchen towels [195]. In this method, samples (100–200 mg) were digested by 1 ml of nitric acid and 1 ml of sulphuric acid at 140°C for 30 min. Solutions were then cooled, diluted with water, and analysed by electrothermal atomization. In a similar report, lead was determined in confectionery wrappers [196]. The wrapper was wiped clean with a damp tissue, cut into 0.5 × 0.5 mm pieces and dried (110°C for paper, 80°C for plastic), for 1 h. The sample (0.5 g) was heated with concentrated nitric acid (1 ml) at 90–100°C for 10–20 min. Ammonium nitrate (1 ml of 0.1 M) and aluminium nitrate (1 ml of 0.1 M) were added, the mixture heated at 110°C for 2 h until dry. The residue was ashed for 12 h at 450°C, and the ash dissolved in hot nitric acid (1 ml). Lead levels up to 30,000 $\mu g\ g^{-1}$ were found in wrappers with yellow or orange pigments. The determination of trace metals in packaging materials has recently been reviewed [197].

George et al. [198] extracted cadmium from cellulosic materials with hot acetic acid. The cadmium was then extracted into an organic solvent as its diethyldithiocarbamate complex.

The direct determination of trace metals in paper and paper products following solid sample introduction into the graphite furnace has been reported by several workers. Knezevic and Kurfuerst [199] compared the methods of acid digestion and solid sampling for the determination of the trace elements cadmium, chromium, copper, lead and mercury in wrapping paper, cellulose and greaseproof paper. For the solid sampling technique, paper strips of 1 mm × 0.5–10 mm (0.2–10 mg) were introduced in a graphite boat into the furnace of an atomic absorption spectrometer with Zeeman background correction. The results were in good agreement with those obtained by digestion. A review by Kumar et al. [200] includes a discussion of the analysis of solid samples of paper and pulp by electrothermal atomization.

XII. TEXTILES, FIBRE AND LEATHER

There is interest in trace metal levels in both natural and synthetic fibres and fabrics, but perhaps most interest in synthetic fibres as these may contain residues of catalysts, treatments or stabilising agents. Reviews have been published of trace metal analysis of rayon, polyamide, polyester and polypropylene fibres [201] and of cotton fabrics, especially for flame retardants [202]. A more recent review of metal determinations in textiles has been made by Tonini [203]. Some elements may be determined by simple acid extraction, or by ashing the fibres and dissolving the ash in hydrochloric or nitric acid [204]. The latter procedure is especially useful for synthetic fibres. Certain synthetic fibres can be dissolved in an organic solvent which may be sprayed directly into a flame, provided it does not harm the spray chamber and the standards are made up in the same solvent, e.g. cellulose acetate in MIBK, polyacrylonitrile in dimethylformamide, nylon in formic acid, wool in 5% sodium hydroxide, cotton and cellulose in 72% sulphuric acid.

Voellkopf et al. [205] determined copper, manganese and rubidium in polyester, nylon and Perlon by atomic absorption using the cup-in-tube technique. The results were in good agreement with those obtained by standard AAS and XRF procedures. In a more recent report by the same authors [206], cadmium, copper, manganese and rubidium were determined in the same substances using both flame atomic absorption and the cup-in-tube method. For the flame method, samples of Perlon and nylon (1 g) were decomposed in concentrated sulphuric acid (1 ml) and nitric acid (3 ml) with heating to 300°C. The solution was warmed (2 s) and cooled (20 s) alternately for 15 min. The solutions were then diluted to 100 ml with water and analysed. Polyester was similarly digested but in 5 ml of concentrated nitric acid, with the use of a percolator. For the analysis of the solid samples, 5% sulphuric acid with 3% nitric acid was used as matrix modifier for copper, manganese and rubidium, and 5% sulphuric acid with 3% nitric acid and 2% di-ammonium hydrogen phosphate was used for cadmium. Calibration was by aqueous standards.

Polypropylene has been analysed by Brzozowska et al. [207]. Samples (5 g) were ignited at 480° ± 20°C. The ash was dissolved in concentrated hydrochloric acid (5 ml), and the mixture evaporated to dryness. The residue was dissolved in 1 M hydrochloric acid (5 ml). The solution was then diluted to 50 ml with water, and cadmium, lead and zinc were determined by atomic absorption with the use of an air/acetylene flame. Recoveries ranged between 95 and 110%, and the limits of detection were 2, 6 and 1 ppm, respectively.

Korenga [208] determined arsenic in acrylic fibres containing antimony oxide. A sample (10 mg–1 g) was decomposed by treatment in a Kjeldahl flask with nitric acid, perchloric acid and sulphuric acid. The product was dissolved in hydrochloric acid. The solution was treated with TiCl$_3$, and after 30 min at 60°C and cooling to room temperature, the solution was extracted with benzene. The arsenic(III) was back-extracted into water and reduced to arsine with the use of hydrochloric acid, potassium iodide tin(II) chloride and zinc. Determination was by atomic absorption using a nitrogen–hydrogen diffusion flame. Trace metals have been determined in cotton by Tonini [209] and by Erb [210].

The use of chromium salts in the tanning of leather has been reflected in the interest in the determination of chromium in leather. Huang and Chen [211] determined chromium oxide in chromium tanned leather, by dissolving the chromium out of the leather by heating with sulphuric acid–nitric acid. The method was simple, rapid and appeared accurate. Knecht [212] has determined chromium in leather dust. The sample was digested with a mixture of nitric acid–perchloric acid (2:1) until white fumes appeared. The residue was dissolved in water. A nitrous-oxide/acetylene flame was recommended, and calibration was by standard additions.

Metal salts may be used in the treatment of wool. Flame methods for the determination of aluminium [213], barium, chromium, copper, mercury, strontium, tin, zinc [214] and zirconium [215] in wool have been published. Standard addition to wool, cleaned by soaking and washing it with disodium EDTA (800 ml of 0.5 M per 30 g of wool soaking for 3 days and double washing), was used as the calibration technique. This compensated for interferences from hydrochloric acid and amino acids. The samples were equilibrated to a constant humidity for 24 h and then 0.3 g was sealed with 5 ml of constant boiling point hydrochloric acid in a glass tube. The tubes were placed in an oven at 110°C for 20 h. A nitrous-oxide/acetylene flame was used in the determination of aluminium and zirconium.

XIII. POLYMERS AND RUBBER

Many of the remarks made in the previous section concerning fibres can be applied to the analysis of plastics. Some polymers are soluble in organic solvents and samples may be prepared for direct aspiration into a flame in this way, e.g. MIBK is a suitable solvent for polyesters, polystyrene, polysiloxanes, cellulose acetate, and butyrate, dimethylformamide for polyacrylonitrile, dimethyl acetamide for polycarbonates and polyvinyl chloride, cyclohexanone for polyvinyl chloride and polyvinyl acetate, formic acid for polyamides, and methanol for polyethers. These organic solutions may alternatively be injected into a graphite furnace. Otherwise polymers may be wet- or dry-ashed and the resultant ash dissolved in acid. Direct insertion of solid samples into a graphite furnace has also attracted interest. A very common method of digestion is to use a sulphuric acid–hydrogen peroxide mix [216–222]. Mendiola-Ambrosio et al. [216] reviewed several methods for determining lead in polyvinyl chloride, but concluded that the best method was to decompose the polymer with 95% H_2SO_4/30% H_2O_2. The precipitated lead sulphate was then dissolved in dilute nitric acid. Alternatively, soxhlet extraction of the plasticisers from the polymer with diethylether, followed by decomposition of the residue with perchloric acid and nitric acid, was recommended. In both methods, the lead was determined using an air/acetylene flame. Belarra et al. have determined aluminium, antimony, calcium [220], together with lead and magnesium [221] in samples of polyvinyl chloride. Samples (0.1 g) were heated for 15 min with concentrated sulphuric acid (2 ml), and, after dropwise addition of 30% hydrogen peroxide (5 ml), the mixture was boiled vigorously to eliminate excess hydrogen peroxide. The solutions were then cooled and treated with concentrated ammonia solution (10 ml) and 4% EDTA (10 ml). The analytes were determined using an air/acetylene flame. The procedure was modified very slightly by Belarra et al. [222] when cadmium, antimony and tin were determined in PVC. An extra stage was added in which fuming nitric acid was used to complete the digestion. Other acid digestions have also been reported. Rombach et al. [223] determined traces of arsenic, barium and mercury in samples of polycarbonate, polyethylene, polypropylene and PVC. A sample (0.4–1.1 g) was decomposed by heating with 65% nitric acid (10 ml) at 240°C for 8 h in a PTFE-lined bomb. The resulting solution was diluted to 100 ml with water and analysed. Mendiola-Ambrosio and Gonzalez-Lopez [224] made a comparative study of different acid digestion procedures for the determination of cadmium in PVC composites. They concluded that for cadmium, perchloric acid, nitric acid or a sulphuric–nitric acid mix was suitable, but hydrochloric acid was not. For zinc, nitric acid or sulphuric–nitric acid mix gave better results than did perchloric or hydrochloric acids.

Vladescu et al. [225] have determined iron in PVC by dissolving the sample in tetrahydrofuran or dimethylformamide. After treatment with sulphuric–perchloric acids (1 : 1), iron was determined by atomic absorption spectroscopy. Jaeger [226] has dissolved PVC in dimethylsulphoxide and concentrated nitric acid. The cadmium analyte was determined using an electrothermal atomizer. Aznarez et al. [227] have determined antimony in plastics by hydride generation AAS. They dissolved the plastic (0.3 g) in 20 ml of 2.5% (v/v) hydrochloric acid in dimethylformamide. Aliquots (2 ml) were mixed with 10% (v/v) sulphuric acid in DMF (2 ml) in the hydride generator, and the stibine was produced by the injection of 3% sodium borohydride in DMF.

Solid sampling has many advantages, including reduced sample pre-treatment times, minimal errors from contamination and avoidance of the dilution associated with dissolution. These advantages are particularly useful in polymer analysis where low levels may be significant and the large molecules involved may be difficult, or slow, to dissolve. Hence the considerable interest in the direct insertion of solid samples of plastics into graphite furnace atomizers. Voellkopf et al. [205] have determined trace metals in plastics by the cup-in-tube technique. Copper, manganese and rubidium were determined using the furnace programme: dry 100°C (20 s) and 200°C (10 s), ash 320°C (25 s) and 550°C (20 s) and atomize 2300°C (3 s) for copper, 1900°C (5 s) for manganese and 1900°C (5 s) for rubidium. Each procedure included a tube clean cycle at 2650°C (5 s). Janssen et al. [228] determined trace metals as additives (e.g. pigments and stabilisers) in polyethylene using solid sampling. Solid standards of similar composition to the sample were used for calibration because the matrix of the sample lowered the signal as compared with pure salt solutions. Due to the abundance of the analytes, less sensitive wavelengths were used, e.g. lead 364.8 nm, chromium 429.0 nm, copper 249.2 nm and nickel 339.1 nm. Background levels were high and Zeeman background correction was used.

Rubbers and adhesives are amenable to most of the treatments discussed above. By heating at 105°C, and then to constant weight at 800°C, samples have been analysed for cobalt, copper, iron, manganese, nickel and zinc [229]. The residue was fused with sodium carbonate/borax (2 : 1), dissolved in hydrochloric acid (1 : 1), and an air/acetylene flame used. An indirect method for determining free sulphur in rubber has been developed by Puacz [230]. Rubber was extracted at 20°C for 25 h, with mechanical shaking, by dimethylformamide. The extract was then treated with a 2% solution of sodium borohydride in a mixture of DMF and water plus 5% ammonium chloride (1 ml). The mixture was warmed at 40°C for 15 min. Unconsumed sodium borohydride was decomposed by careful addition of 5 M hydrochloric acid (1 ml). A known amount of mercuryII chloride solution (10 or 0.1 ppm of Hg) was added with stirring, and the solution was diluted to 50 ml with 10% hydrochloric acid. The solution was then vigorously stirred for 90 s with 1 ml of 20% tin(II) chloride solution in hydrochloric acid. The liberated Hg° was swept by air into an absorption cell for atomic absorption measurement at 253.7 nm. The calibration graph was linear for 0.35–4 μg of sulphur.

For a more detailed description of analytical applications of relevance to polymer and petroleum products the reader is directed to Chapter 4h.

XIV. THE NUCLEAR INDUSTRY

Atomic absorption is being increasingly applied to the energy industries including those associated with nuclear power. While radiochemical monitoring is frequently used, several metals at the trace level may be monitored by atomic absorption. Particular problems arise when the matrix is radioactive and stringent safety precautions may be required. Electrothermal atomization in an enclosed system, such as a glove box, is preferable to flame atomization.

Mathews [231] has determined traces of calcium, chromium, cobalt, copper, lead, lithium, magnesium, manganese and potassium in the sodium coolant of fast breeder reactors. The sodium was distilled in a tantalum or Al_2O_3 crucible. The residue was then dissolved in dilute acid. Handling techniques for the sodium were discussed.

The determination of daughter nuclides in fuels or fuel processing liquors is one potential application of AAS. In view of the high radioactivity of the solutions, and low levels of the metals of interest, selective solvent extraction to separate the analyte metals from the uranium matrix is advisable. Several workers [232–236] have extracted uranium with tri-butyl phosphate into MIBK, or some other similar solvent. The trace analytes remain in the aqueous phase, which can then be analysed by flame or electrothermal atomization. Cadmium, cobalt, copper and nickel have been determined in high-purity uranium by Bangia et al. [237]. In this method, the sample of uranium metal or oxide (approximately 100 mg) was dissolved in concentrated nitric acid, and then diluted to 5 ml. Uranium was removed by shaking the solution with three portions of a 0.2 M solution of dioctyl sulphoxide in xylene. The aqueous phase was evaporated to dryness and the residue dissolved in 1 ml of 0.1 M nitric acid. Aliquots (5 ml) of this solution were injected into a graphite tube for drying at 180°C (5 s), ashing at 225°C (5 s) and atomization at 1725°C (3 s) for Cd, 2700°C for Co, and 2400°C for Cu and Ni.

Ion exchange offers an alternative method of separating the analytes from their matrix. Lead has been determined in uranium products [238] by adsorption onto Dowex 1-X8 from 2 M hydrobromic acid. The lead was eluted from the Dowex column with 6 M hydrochloric acid prior to flame atomic absorption.

Gerardi and Pelliccia [239] have determined metallic impurities and rare earths in nuclear solutions containing uranium, thorium and fission products. Sample preparation was just a simple dilution with water. Analysis was performed using a graphite furnace. Background absorption was found not to be a problem provided that no more than 258 μg of the U and Th matrix was present.

References pp. 507–514

XV. ARCHAEOLOGY

One of the lesser known areas where atomic absorption is proving to be particularly useful, is archaeology. It has been used in survey work (e.g. to identify the sources of objects and trade routes, to identify the raw materials used, to study ancient technology), and to detect forgeries. Reviews on the role of atomic absorption spectroscopy in these areas have been published [240–242]. While non-destructive techniques are clearly preferred in this area of work, the high sensitivity and low sample requirements of atomic absorption commend themselves, particularly when it is recalled that techniques such as X-ray fluorescence requires clean surfaces. By comparison therefore, the drilling of a 1 mm diameter hole may well cause less damage than removing corrosion layers, or a patina, of several mm^2. Atomic absorption spectrometry is suitable for the analysis of several types of ancient inorganic materials, e.g. metals and alloys, silicates and minerals. Only a few milligrams of sample are required, typically 10 mg may be dissolved in 25 ml for analysis. Electrothermal methods may require even less sample, and are thus very attractive in this field. Often papers describing results obtained by atomic absorption give little or no analytical details. Table 4 lists some of these publications to illustrate the potential scope of atomic absorption spectrometry in archaeology.

Sampling presents the major difficulty in archaeological analysis for most techniques including atomic absorption. Szonntagh [251] has developed

TABLE 4

ARCHAEOLOGICAL APPLICATIONS OF ATOMIC ABSORPTION SPECTROMETRY

Material examined	Reference
Ancient bronzes	[241, 243, 244]
Ancient glasses	[245, 246]
Ceramics	[240, 244, 247, 248]
Coins	[249–251]
Copper ores and artefacts	[252–254]
Flint-mine products	[255, 256]
Glass-blowing tubes	[257, 258]
Gold torcs	[259]
Iron-age axes	[260]
Iron-age bronze collars	[261]
Iron artefacts	[262]
Iron swords	[263]
Roman horse trappings	[264]
Roman bricks and wall slabs	[265]
Romano–British pottery	[266]
Stained glass	[267–279]

a micro-drilling method and apparatus for representative sampling of ancient coins. By drilling through the cylindrical surface of coins, a more representative sample than that obtained by surface sampling techniques may be obtained. The process left barely visible holes on the surfaces of the coins, and the faces remained unaffected. In the review by Hughes et al. [240], it is reported that the British Museum also use hand-held drills for sampling. To obtain 10 mg of sample, about 5 mm penetration was required. Samples can usually be dissolved readily in 1 ml aqua regia, with gentle heating to 60°C. The solution may then be diluted with water, and analysed. Similarly, Aycik and Edguer [262] drilled holes in iron artefacts, e.g. arrowheads, and removed 0.17–0.50 g of non-surface powder. The elements bismuth, cobalt, copper, manganese, silver and zinc were then determined using an electrothermal atomizer.

Sample preparation often involves digestion by acids. Stratis et al. [248] made comparisons of six techniques involving acid decomposition in the analysis of Neolithic ceramic samples. Decomposition with HF–HCl and aspiration into an air/acetylene flame gave reliable results. Varma [254] also found that dissolution of archaeological samples in aqua regia, or HCl–HF prior to aspiration into an air/acetylene flame, gave good results for copper. Hughes et al. [240] recommended that various samples should be taken as chips (10–15 mg), which could be powdered in a mortar, or by abrasion using a manual or mechanical wheel. If the sample contains calcium carbonate, or organic matter, a dried sample should be evaporated to dryness with concentrated hydrochloric acid (1 ml) and concentrated nitric acid (1 ml). To such a sample, 40% hydrofluoric acid (2 ml) was added, and the crucible heated to 100°C until fumes appeared. After allowing all the fluoride to evaporate, 60% perchloric acid (2 ml) was added. With gentle heating, the sample dissolved and was then made up to 50 ml. Blanks and matched standards were of course essential. The analysis of alloys often requires two different solvents to determine the normal range of elements. Silver alloys (5–10 mg) may be dissolved in 50% nitric acid (1 ml). The silver content can then be determined. However, tin and gold appear as a black residue, which must then be taken up in concentrated hydrochloric acid (1.5 ml). This precipitates the silver as silver chloride. The precipitate may be removed by centrifugation.

Fusions have also been advocated by some workers. Tubb and Nickless [266] determined a number of metals in Romano-British pottery after fusing the sample with lithium metaborate. Lithium metaborate (800 mg) was mixed thoroughly with a dry, finely powdered sample (400 mg) in a platinum crucible. The mixture was fused for 20–30 min at 920°C in the oxidising flame of a Meker burner. The resulting flux was cooled rapidly by a cold air jet. The mass loss was determined, and the flux powdered in a stainless-steel mill. The powder (500 mg) was dissolved in 1.5 M nitric acid (50 ml). This solution was then aspirated into a nitrous oxide/acetylene flame. The

References pp. 507–514

results obtained were sufficiently precise to resolve pottery samples into groups correlating with their sources of origin. Stratis et al. [248] found that fusion of ceramic samples with sodium carbonate and lithium metaborate gave similar analytical results to those obtained by an HF–HCl digestion procedure.

Flint may be sampled with a hollow-cored trepanning drill [239]. A solid cylinder of about 1 g in weight can be obtained. The flint may be cleaned by brief immersion in aqua regia, washed, dried and then dissolved in 40% hydrofluoric acid (10 ml) at 70°C in a PTFE beaker. The excess fluoride was evaporated, and the dry residue taken up in concentrated nitric acid (0.5 ml) and evaporated again to remove any carbon. The residue was then dissolved in 60% perchloric acid (2 ml), and diluted to 50 ml. Such a procedure should ensure dissolution for most rocks.

XVI. ISOTOPES

Atomic absorption offers a more practical opportunity for determining isotopic composition than atomic emission. Useful reviews of the possibilities of the techniques have appeared in two books [270, 271]. With the advent of inductively coupled plasma–mass spectrometry, isotopic composition can now be determined in solutions rapidly and across the periodic table. Therefore, there has been a decline in applications of atomic absorption spectrometry for determining isotopes in solutions, but the technique may still offer advantages for some specialist applications.

Isotopic analysis is in theory possible, provided that highly enriched isotope sources are available, the absorption line width available is less than the isotopic displacement, and for a given isotope the nuclear spin hyperfine components must be partially resolved from the other isotopic components of the absorption line. In the simplest possible case, for an element with two isotopes, the lamp is prepared for the first isotope and only this isotope in the atom cell will absorb the radiation. The procedure can then be repeated with a lamp prepared with the second isotope. Effectively, this is an extension of the impressive selectivity of atomic absorption because of the classic lock and key effect, treating the different isotopes as different analytes. This very simple case is rare because of the small wavelength shifts involved; such an effect has been observed for mercury 202 and mercury 200 [272]. Lithium 6 and lithium 7 ratios have been determined in geological materials by Meier [273]. A double channel, double beam, instrument was used with an ^{6}Li lamp in one channel, and an ^{7}Li lamp in the other. Each lamp was operated at 4 mA, and each monochromator was set to 670.8 nm with a bandwidth of 0.5 nm. An oxidising air/acetylene flame was used. The absorption has non-linear dependence upon total lithium concentration and isotopic composition. A method using non-linear equations to describe the relation of ^{6}Li and ^{7}Li lamp radiation was proposed as a means of calculating isotopic composition that is independent of total lithium concentration. The isotopic abundances of lithium have also been investigated by Chapman et al. [274]. They used the more commonly reported technique of plotting the ratio of the absorbance measured using a ^{6}Li lamp to the absorbance measured using a ^{7}Li lamp against percentage ^{6}Li. This technique was also used by Goleb and Yokoyama [275] and Wheat [276].

Heavier elements such as lead [277] and uranium [278, 279] have also been investigated. Goleb [278, 279] determined uranium 235 and uranium 238 using a water-cooled sputtering cell as the atom reservoir, and water-cooled hollow-cathode lamps as sources. Samples could be placed as chips, compounds or oxides in the emission source, and standards were placed in the absorption source. The 415.3 and 502.7 nm uranium lines were used as uranium 235 has negligible fine structure at these lines, but appreciable isotopic shifts (0.07 and 0.010 nm, respectively) were observed. A linear calibration curve of the isotopic abundance of uranium 238 versus the percentage transmissions through the absorption cell was obtained.

The technique is limited to light elements, e.g. lithium and boron, and to heavy elements with appreciable nuclear charge, e.g. lead, mercury and uranium. This is because the technique requires elements where resonance lines show an isotopic shift comparable to, or greater than, the absorption profile under experimental conditions. Further complications arise when elements have more than two major isotopes, and in the accurate determination of minor constituents.

REFERENCES

1 I.A. Voinovitch and M. Druon, Analusis, 14 (1986) 87.
2 E. Sebastiani, K. Ohls and G. Riemer, Fresenius Z. Anal. Chem., 264 (1973) 105.
3 J. Ruzicka and E.H. Hansen, Flow Injection Analysis, Wiley, Chichester, 1981.
4 J.F. Tyson, Analyst, 110 (1985) 419.
5 D.C. Manning and W. Slavin, Anal. Chem., 50 (1978) 1234.
6 IUPAC Analytical Chemistry Division, Commission on Micro-chemical Techniques and Trace Analysis, Pure Appl. Chem., 49 (1977) 893.
7 T.M. Kolodko and A.N. Babai, Khim. Prom-St., Ser. Metody Anal. Kontrolya Kach. Prod. Khim. Promsti, 8 (1980) 46.
8 F.J. Langmyhr and J.T. Hakedal, Anal. Chim. Acta, 83 (1976) 127.
9 Z. Gan and T. Gu, Shiyou Huangong, 15 (1986) 309.
10 J.B. Headridge, D. Johnson, K.W. Jackson and J.A. Roberts, Anal. Chim. Acta, 20 (1987) 311.
11 B. Holen, R. Bye and W. Lund, Anal. Chim. Acta, 131 (1981) 37.
12 M. Hiraide, P. Tschoepel and G. Toelg, Anal. Chim. Acta, 186 (1986) 261.
13 Y. Thomassen, B.V. Larsen and F.J. Langmyhr, Anal. Chim. Acta, 83 (1976) 103.
14 M. Kimura and K. Kawanami, Nippon Kagaku Kaishi, 1 (1981) 1.
15 E. Beinrohr and H. Berndt, Mikrochim. Acta, 1 (1985) 199.

508

16   M. Satake, M. Katyal and B.K. Puri, Fresenius Z. Anal. Chem., 322 (1985) 514.
17   I.I. Nazarenko, I.V. Kislova, L.I. Kashina, G.I. Malofeeva, O.M. Petrukhin, Y.I.
     Murinov and Y.A. Zolotov, Zh. Anal. Khim., 40 (1985) 2129.
18   M. Foerster and K.H. Lieser, Fresenius Z. Anal. Chem., 309 (1981) 355.
19   P. Burba and P.G. Willmer, Fresenius Z. Anal. Chem., 322 (1985) 266.
20   F.W.E. Strelow, Anal. Chem., 57 (1985) 2268.
21   F.W.E. Strelow, Anal. Chim. Acta, 183 (1986) 307.
22   H. Berndt and J. Messerschmidt, Fresenius Z. Anal. Chem., 299 (1979) 28.
23   S. Nakashima, Fresenius Z. Anal. Chem., 303 (1980) 10.
24   S. Nakashima, Bull. Chem. Soc. Jpn., 54 (1981) 291.
25   T. Shimizu and K. Sakai, Nippon Kagaku Kaishi, 1 (1981) 26.
26   T. Shimizu, Y. Uchida, Y. Shijo and K. Sakai, Bunseki Kagaku, 30 (1981) 113.
27   E. Jackwerth, R. Hoehn and K. Musaick, Fresenius Z. Anal. Chem., 299 (1979)
     362.
28   P. Heininger and G. Henrion, Z. Chem., 25 (1985) 73.
29   C.A. Watson, Ammonium Pyrrolidine Dithiocarbamate, Monograph 74, Hopkin
     and Williams, Chadwell Heath, England, October 1971.
30   I. Khavezov, E. Ivanova, P. Koehler, N. Iordanov, R. Matchat, K. Reiher, G. Emrich
     and K. Licht, Fresenius Z. Anal.Chem., 326 (1987) 536.
31   D.C. Parashar and A.K. Sarkar, Anal. Lett., 17 (1984) 1269.
32   M. Yanagisawa, M. Suzuki and T. Takeuchi, Talanta, 14 (1967) 933.
33   A. Hioki, N. Fudagawa, M. Kubota and A. Kawase, Bunseki Kagaku, 36 (1987) 91.
34   Y. Chen, Z. Hu and X. Jia, Huaxue Shiji, 8 (1986) 310.
35   M.W. Cresser, Solvent Extraction in Flame Spectroscopic Analysis, Butterworth,
     London, 1978.
36   W.W. White and P.J. Murphy, Anal. Chem., 49 (1977) 255.
37   R. Bye and W. Lund, J. Anal. At. Spectrom., 4 (1989) 233.
38   J. Alary, J. Vandaele and C. Escrieut, Talanta, 33 (1986) 748.
39   P.C. Lindahl, K.C. Voight, A.M. Bishop, G.M. Lafon and W.L. Huang, At. Spectrosc.,
     5 (1984) 137.
40   R. Niu and Q. Li, Fenxi Huaxue, 14 (1986) 158.
41   C. Manoliu, B. Tomi and F. Petruc, Rev. Chim., 24 (1973) 991.
42   G. Knapp, K. Mueller, M. Strunz and W. Wegscheider, J. Anal. At. Spectrom., 2
     (1987) 611.
43   L.M. Kudryashova, O.P. Trokhachenkova and E.A. Bondarevskaya, Zh. Anal.
     Khim., 42 (1987) 1036.
44   G. Braca, G. Sbrana, G. Scandiffio and R. Cioni, At. Absorpt. Newsl., 14 (1975) 39.
45   M.A. Leonard and W.J. Swindall, Analyst, 98 (1973) 133.
46   V.G. Shah and S.Y. Kulkarni, Anal. Chem., 59 (1987) 1375.
47   F.R. Hartley, S.G. Murray and P.N. Nicholson, J. Organomet. Chem., 231 (1982)
     369.
48   I.L. Marr, J. Anwar and B.B. Sithole, Analyst, 107 (1982) 1212.
49   D.T. Burns, F. Glockling, V.B. Mahale and W.J. Swindall, Analyst, 103 (1978) 985.
50   Q. Cai, C. Xue and C. Liu, Youji Huaxue, 4 (1983) 270.
51   E. Schoenberger, J. Kassovicz and A. Shenhar, Int. J. Environ. Anal. Chem., 18
     (1984) 227.
52   Analytical Methods Committee, Analyst, 104 (1979) 1070.
53   Analytical Methods Committee, Analyst, 104 (1979) 778.
54   Analytical Methods Committee, Analyst, 92 (1967) 403.

55 Analytical Methods Committee, Analyst, 101 (1976) 62.
56 M.S. Vigler, A. Strecker and A. Varnes, Appl. Spectrosc., 32 (1978) 60.
57 G.D. Martinie and A.A. Schilt, Anal. Chem., 48 (1976) 70.
58 C. Manoliu, O. Popescu, T. Balasa, V. Tamas and I. Georgescu, Rev. Chim., 31 (1980) 291.
59 J. Perkins, Analyst, 88 (1963) 324.
60 R.L. Scott, At. Absorpt. Newsl., 9 (1970) 46.
61 K. Ametani, E. Shima and T. Oka, Microchem. J., 28 (1983) 37.
62 Y.G. Andronov, V.K. Efimov, S.I. Polyakov and V.A. Cherkasov, Zh. Nauchn., Prikel, Fotogr. Kinematogr., 27 (1982) 129.
63 N. Kunimine and H. Kawada, Bunseki Kagaku, 16 (1967) 185.
64 N. Kunimine and H. Kawada, Bunseki Kagaku, 16 (1967) 189.
65 K. Dittrich and W. Mothes, Talanta, 22 (1975) 318.
66 P. Ambrosetti and F. Librici, At. Absorpt. Newsl., 18 (1979) 38.
67 J.L. Fabec and M.L. Ruschak, At. Spectrosc., 5 (1984) 142.
68 J.L. Fabec, At. Spectrosc., 4 (1983) 46.
69 H. Urbain and A. Chambosse, At. Spectrosc., 3 (1982) 143.
70 D.W.B. Westerman, I.E. Ruffio, M.S. Wainwright and N.R. Foster, Anal. Chim. Acta, 117 (1980) 285.
71 X. Chen and L. Yao, Huaxue Shiji, 8 (1986) 120.
72 J. Madar and A. Vlcakova, Ropa Uhlie, 23 (1981) 342.
73 R.A. Petrosyan, O.S. Eginyan, V.K. Boyadzhyan and R.M. Kachatryan, Arm. Khim. Zh., 34 (1981) 209.
74 S. Kallmann and P. Blumberg, Talanta, 27 (1980) 827.
75 R.C.P. Sinha and P.C. Gupta, Fert. Technol., 17 (1980) 181.
76 E.V. Igoshina and B.M. Talalaev, Zh. Anal. Khim., 38 (1983) 1648.
77 S. Chiricosta, G. Cum, R. Gallo, A. Spadaro and P. Vitareli, At. Spectrosc., 3 (1982) 185.
78 I.G. Yudelevich and N.F. Beizel, 1ZV Sib. Otd. Akad. Nauk SSSR, Ser. Khim. Nauk, 6 (1978) 94.
79 N.F. Beizel, C.A. Kozhanova, I.G. Yudelevich and L.M. Buyanoa, Izv. Sib. Otd. Akad. Nauk SSSR, Ser. Khim. Nauk, 5 (1978) 113.
80 I.G. Yudelevich, N.F. Beizel, T.S. Papina and K. Dittrich, Spectrochim. Acta, Part B, 39 (1984) 467.
81 K. Dittrich, W. Mothes, I.G. Yudelevich and T.S. Papina, Talanta, 32 (1985) 195.
82 K. Dittrich, B. Vorberg, J. Funk and V. Beyer, Spectrochim. Acta, Part B, 39 (1984) 349.
83 V.P. Gladysheva, A.K. Nurtaeva, T.A. Mal'Tseva and G.I. Vinogradova, Zavod. Lab., 48 (1982) 1.
84 I.S. Busheina, J.B. Headridge, D. Johnson, K.W. Jackson, C.W. McLeod, and J.A. Roberts, Anal. Chim. Acta, 197 (1987) 87.
85 S. Yu and G. Ye, Fenxi Huaxue, 10 (1982) 416.
86 W. Van Goolen and L. Verstraelen, Plat. Surf. Finish., 65 (1978) 50.
87 P. Battistoni, S. Bompadre, G. Fara and G. Gobbi, Talanta, 30 (1983) 15.
88 A. Budniok, Microchem. J., 25 (1980) 531.
89 B. Deng and X-H. Zhou, Ch'ing Hua Ta Hsueh Pao., 20 (1980) 79.
90 D.E. Davey, Anal. Lett., 19 (1986) 1573.
91 T. Ozawa, Tokyo-toritsu Kogyo Gijutso Senta Kenkyu Hokuku, 9 (1980) 115.
92 Y. Hou, X. Chi and S. Liu, Fenxi Huaxue, 16 (1988) 242.

93   C. Lercari, B. Sartorel, L. Sedea and G. Toninelli, J. Am. Oil Chem. Soc., 60 (1983) 856.

94   J. Maslowska and E. Legedz, Rocz. Panstw. Zakl. Hig., 33 (1982) 149.

95   J. Maslowska and E. Legedz, Rocz. Panstw. Zakl. Hig., 35 (1984) 431.

96   M. Okamoto, M. Konda, I. Matsumoto and Y. Miya, J. Soc. Cosmet. Chem., 22 (1971) 589.

97   E.S. Gladney, At. Absorpt. Newsl., 16 (1977) 114.

98   J.T. Mason, J. Pharm. Sci., 69 (1980) 101.

99   British Standard 5136, Specification for Toothpastes, British Standards Institution, 2 Park Street, London W1A 2BS, 1981.

100 S.I. Pardhan and J.M. Ottaway, Proc. Anal. Div. Chem. Soc., 12 (1975) 291.

101 United States Pharmacopoeia, XIX edn., Mack Printing Co., Easton, PA, 1975.

102 British Pharmacopoeia, HMSO, London, 1980.

103 European Pharmacopoeia, Vol. III., Maisonneuve, France, 1975.

104 F. Rousselet and F. Thuillier, Prog. Anal. At. Spectrosc., 1 (1979) 353.

105 J.H.M. Miller, Int.Lab. July/August 1978, 37.

106 P.B. Bondo, Guidelines Anal. Toxicol. Prog., 2 (1977) 171.

107 V.V. Tkachuk, Farm. Zh. (Kiev), 6 (1978) 35.

108 J. Pawlaczyk and M. Makowska, Pol. Pharm., 36 (1979) 59.

109 J.B. Smith and S. Ahuja, Chem. Anal. (N.Y.), 85 (1986) 217.

110 S.G. Schulman and W.R. Vincent, Drugs Pharm. Sci., II (1984) 359.

111 S. Ali, Pharm. Ind., 46 (1984) 391.

112 S. Ali, J. Pharm. Biomed. Anal., 1 (1983) 517.

113 G.F. Hazebroucq, in 3rd International Conference on Atomic Absorption and Atomic Fluorescene Spectrometry, Hilger, London, 1971, 577.

114 R.V. Smith and M.A. Nessen, J. Pharm. Sci., 60 (1971) 907.

115 B.A. Dalrymple and C.T. Kenner, J. Pharm. Sci., 58 (1969) 604.

116 W.J. Price, Spectrochemical Analysis by Atomic Absorption, Heyden, London, 1979.

117 P.P. Kharkhanis and J.R. Anfinsen, J. Assoc. Off. Anal. Chem., 56 (1973) 358.

118 S.L. Ali and D. Steinbach, Pharm. Z., 124 (1979) 1422.

119 E. Van Den Ereckout and P. de Moerloose, Pharm. Weekbl., 106 (1971) 749.

120 Y.S. Chae, J.P. Vacik and W.H. Shelver, J. Pharm. Sci., 62 (1973) 1838.

121 Y. Kidani, K. Takeda and H. Koike, Bunseki Kagaku, 22 (1973) 719.

122 F.J. Diaz, Anal. Chim. Acta, 58 (1972) 455.

123 E. Peck, Anal. Lett., B11 (1978) 103.

124 P.O. Kosonen, A.M. Salonen and A.L. Nieminen, Finn. Chem. Lett., 1978, 136.

125 L.L. Whitlock, J.R. Melton and T.J. Billings, J. Assoc. Off. Anal. Chem., 59 (1976) 580.

126 R.R. Moody and R.B. Taylor, J. Pharm. Pharmacol., 24 (1972) 848.

127 K. Szivos, L. Polos, L. Bezev and E. Pungor, Acta Pharm. Hung., 43 (1973) 90.

128 G.I. Spielholtz and G.C. Toralballa, Analyst, 94 (1969) 1072.

129 I.T. Calder and J.H.M. Miller, J. Pharm. Pharmacol., 28 (1976) Suppl. 25P.

130 P. Girgis-Takla and V. Valijanian, Analyst, 107 (1982) 378.

131 E. Akguen and U. Pindur, Pharm. Ind. 44 (1982) 930.

132 E. Akguen and U. Pindur, Pharm. Acta. Helv., 58 (1983) 130.

133 J. Mohay, M. Veress and G. Szasz, Magy. Kem. Foly., 85 (1979) 465.

134 S. Iwasa, M. Morishita, H. Yamada, H. Iizuka, H. Ichiba and M. Katayanagi, Iyakahin Kenkyu, 14 (1983) 819.

135 Y. Zhu, Yaowu Fenxi Zazhi, 3 (1983) 175.

136 J.R. Andersen and P. Helboe, J. Pharm. Biomed. Anal., 4 (1986) III.
137 K.A. Kovar, G. Jarre, W. Lautenschlaeger and J. Maassen, Arch. Pharm., 47 (1982) 173.
138 O-W. Lau, P.K. Hon, C.Y. Cheung and M.H. Chau, Analyst, 110 (1985) 483.
139 E.R. Clark and E.-S.A.K. Yacoub, Talanta, 31 (1984) 15.
140 C. Nerin de la Puerta, A. Garnica, J. Cacho Palomar, Mikrochim. Acta, III (1986) 117.
141 S.J. Simon and D.F. Boltz, Microchem. J., 20 (1975) 468.
142 I. Mitsui and Y. Fujimura, J. Hyg. Chem., 21 (1975) 183.
143 T. Minamikawa, K. Matsumara, A. Kamei and M. Yamakawa, Bunseki Kagaku, 20 (1971) 1011.
144 T. Minamikawa and N. Yamagishi, Bunseki Kagaku, 22 (1973) 1058.
145 M.M. Ayad, S.E. Khayyal and N.M. Farag, Spectrochim. Acta, Part B, 40 (1985) 1205.
146 C. Nerin, A. Garnica and J. Cacho, Anal. Chem., 57 (1985) 34.
147 Y. Kidani, S. Uno and K. Inagaki, Bunseki Kagaku, 25 (1976) 514.
148 Y. Kidani, S. Uno and K. Inagaki, Bunseki Kagaku, 26 (1977) 158.
149 Y. Minami, T. Mitsui and Y. Fujimura, Bunseki Kagaku, 32 (1983) 206.
150 Y. Kidani, T. Saotome, K. Inagaki and H. Koike, Bunseki Kagaku, 24 (1975) 463.
151 S.M. Hassan, M.E.S. Metwally and A.A. Abou Ouf, Analyst, 107 (1982) 1235.
152 J. Alary, A. Villet and A. Coeur, Ann. Pharm. Fr., 34 (1976) 419.
153 T. Mitsui and Y. Fujimura, Bunseki Kagaku, 24 (1975) 575.
154 Y. Minami, T. Mitsui and Y. Fujimura, Bunseki Kagaku, 31 (1982) 604.
155 C. Nerin de La Puerta, J. Cacho Palomar and A. Garnica, Anal. Lett., 18 (1985) 1887.
156 Y. Kidani, K. Nakamura, K. Inagaki and H. Koike, Bunseki Kagaku, 24 (1975) 742.
157 A. Bazzi, J. Montgomery and G. Alent, Analyst, 113 (1988) 121.
158 Y. Kidani, K. Inagaki, N. Osugi and H. Koike, Bunseki Kagaku, 22 (1973) 892.
159 Y. Kidani, T. Saotome, M. Kato and H. Koike, Bunseki Kagaku, 23 (1974) 265.
160 S. Tommilehto, Acta Pharm. Fenn., 88 (1979) 25.
161 A.V. Kovatsis and M.A. Tsougas, Arzneim-Forsch., 28 (1978) 248.
162 T. Minamikawa, K. Sakai, N. Hashitani, E. Fukushima and N. Yamagishi, Chem. Pharm. Bull. (Tokyo), 21 (1973) 1632.
163 Y. Kidani, K. Nakamura and K. Inagaki, Bunseki Kagaku, 25 (1976) 509.
164 Y. Kidani, H. Takemura and H. Koike, Bunseki Kagaku, 22 (1973) 187.
165 Y. Kidani, K. Inagaki, T. Saotome and H. Koike, Bunseki Kagaku, 22 (1973) 896.
166 T. Mitsui, Y. Fujimura and T. Suzuki, Bunseki Kagaku, 24 (1975) 244.
167 C. Nerin de la Puerta, A. Garnica and J. Cacho Palomar, Anal. Chem. 58 (1986) 2617.
168 T. Minamikawa and K. Matsumura, Yakugaku Zasshi, 96 (1976) 440.
169 M.M. Ayad, S.E. Khayyal and N.M. Farag, Microchem. J., 33 (1986) 371.
170 R. Hartubise, J. Pharm. Sci., 63 (1974) 1131.
171 S.S.M. Hassan and M.H. Eldesouki, J. Assoc. Off. Anal. Chem., 64 (1981) 329.
172 S.S.M. Hassan, M.T. Zaki and Eldisouki, J. Assoc. Off. Anal. Chem., 62 (1979) 315.
173 British Standard 5666, Methods of Analysis of Wood Preservatives and Treated Timber, Part 3, Quantitative Analysis of Preservatives and Treated Timber Containing Copper, Chromium, Arsenic Formulations, British Standards Institution, 2 Park Street, London W1A 2BS, 1979.

512

174   S. Kojima, A. Nakamura and M. Kaniwa, Eisei Kagaku, 25 (1979) 141.
175   B.J. Gudzinowicz and V.J. Luciano, J. Assoc. Off. Anal. Chem., 49 (1966) 1.
176   British Standard, BS 3900, Part B4, 1986, 8.
177   British Standard, BS 3900, Part B6, 1986, 9.
178   British Standard, BS 3900, Part B7, 1986, 9.
179   British Standard, BS 3900, Part B9, 1986, 9.
180   British Standard, BS 3900, Part B12, 1986, 11.
181   British Standard, BS 3900, Part B15, 1987, 6.
182   R. Hautbout, G. Legrand and I.A. Voinovitch, Analusis, 14 (1986) 139.
183   J.C. Meranger and E. Somers, Analyst, 93 (1968) 799.
184   O. Beniot and G. Geiger, Bull. Liaison Lab. Ponts Chaussées, 79 (1975) 83.
185   K.A. Hausknecht, E.A. Ryan and L.P. Leonard, At. Spetrosc., 3 (1982) 53.
186   K. Takahashi, M. Shigematsu and Y. Ohyagi, Shikizai Kyokaishi, 59 (1986) 410.
187   K. Takahashi, S. Minami and Y. Ohyagi, Shikizai Kyoaishi, 54 (1981) 606.
188   A.D. James and L.P. Judson, Chem. N.Z., 46 (1982) 60.
189   L. Ebdon, J.R. Wilkinson and K.W. Jackson, Analyst, 107 (1982) 269.
190   J. Garcia-Anton and J.L. Guinon, Analusis, 14 (1986) 158.
191   V. Bartocci, P. Cescon, F. Castellani, R. Riccioni and M. Gusteri, At. Absorpt.
       Newsl., 13 (1974) 121.
192   G. Knezevic, Papier (Darmstadt), 34 (1980) 226.
193   G. Knezevic, Papier (Darmstadt), 36 (1982) 534.
194   G. Knezevic, GIT Fachz. Lab., 28 (1984) 178.
195   W. Griebenow, B. Werthman and B. Schwarz, Papier (Darmstadt), 39 (1985) 105.
196   D. Watkins, T. Corbyons, J. Bradshaw and J.D. Winefordner, Anal. Chim. Acta, 85
       (1976) 403.
197   G. Knezevic, Verpack-Rundsch, 37 (1986) 39.
198   J. George, D. Hubner, E. Muller and G. Trojna, Zellst. Pap., 35 (1986) 216.
199   G. Knezevic and U. Kurfuerst, Fresenius Z. Anal. Chem., 322 (1985) 717.
200   A. Kumar, M.Z. Hasan and B.T. Deshmukh, Indian J. Pure Appl. Phys., 25 (1987)
       49.
201   C. Tonini, Tinctoria, 75 (1978) 160.
202   B. Piccolo and V.W. Tripp, U.S. Dept. Agric., Agric. Res. Serv. (Rep), ARS 72-98
       (1972) 45.
203   C. Tonini, Tinctoria, 77 (1980) 358.
204   C. Tonini, Tinctoria, 76 (1979) 117.
205   U. Voellkopf, R. Lehmann and D. Weber, Labor Praxis, 9 (1985) 990.
206   U. Voellkopf, R. Lehmann and D. Weber, J. Anal. At. Spectrom., 2 (1987) 455.
207   B. Brzozowska, H. Mazur, J.K. Ludwicki and I. Lewandowska-Mahinowska, Rocz.
       Panstw. Zakl. Hig., 36 (1985) 197.
208   T. Korenga, Analyst, 106 (1981) 40.
209   C. Tonini, Tinctoria, 77 (1980) 119.
210   R. Erb, Fresenius Z. Anal. Chem., 322 (1985) 719.
211   S. Huang and L. Chen, Pige Keji, 3 (1984) 36.
212   J. Knecht, Fresenius Z. Anal. Chem., 316 (1983) 409.
213   F.R. Hartley and A.S. Inglis, Analyst, 92 (1967) 622.
214   F.R. Hartley and A.S. Inglis, Analyst, 93 (1968) 394.
215   A.S. Inglis and P.W. Nicholls, J. Text. Inst., 64 (1973) 445.
216   J.M. Mendiola-Ambrosio, A. Gonzalez-Lopez and S. Arribas Jimeno, Afinidad, 37
       (1980) 39.

217 J.M. Mendiola-Ambrosio and A. Gonzalez-Lopez, Rev. Plast. Mod., 41 (1981) 550.
218 A. Gonzalez-Lopez, Quim. Ind. 29 (1983) 801.
219 J.M. Mendiola-Ambrosio, A. Gonzalez-Lopez and S. Arribas-Jimeno, Afinidad, 37 (1980) 251.
220 M.A. Belarra, F. Gallarta, J.M. Anzano and J.R. Castillo, J. Anal. At. Spectrom., 1 (1986) 141.
221 M.A. Belarra, F. Gallarta, J.M. Anzano and J.R. Castillo, J. Anal. At. Spectrom., 2 (1987) 77.
222 M.A. Belarra, M.C. Azofra, J.M. Anzano and J.R. Castillo, J. Anal. At. Spectrom., 3 (1988) 591.
223 N. Rombach, R. Apel and F. Tschochner, GIT, Fachz. Lab., 24 (1980) 1165.
224 J.M. Mendiola-Ambrosio and A. Gonzalez-Lopez, Rev. Plast. Mod., 41 (1981) 413.
225 L. Vladescu, R. Lerch and E. Diacu, Bul. Inst. Politech. Gheorghe Gheorgiu-Dej Bucuresti, Ser. Chim.-Metal., 43 (1981) 43.
226 H. Jaeger, Labor Praxis, 8 (1984) 345.
227 J. Aznarez, J.C. Vidal and J.M. Gascon, At. Spectrosc., 7 (1986) 59.
228 A. Janssen, B. Brueckner, K.H. Grobecker and U. Kurfuerst, Fresenius Z. Anal. Chem., 322 (1985) 713.
229 I.G. Putov, S.A. Popova and A.G. Brashnarova, Khim. Ind. (Sofia), 49 (1977) 16.
230 W. Puacz, Acta Chim. Hung., 124 (1987) 293.
231 C.K. Mathews, Pure Appl. Chem., 54 (1982) 807.
232 J.G. Chang and R.L. Graff, ASTM Spec. Tech. Publ., 1981, 747.
233 K.H. Henn, R. Berg and L. Hoerner, Atomspektrom. Spureanal., Vortr. Kolloq. Verlag Chemie, 1981, 553.
234 T.R. Bangia, K.N.K. Kartha, M. Varghese, B.A. Dhawale and B.D. Joshi, Proc. Nucl. Chem. Radiochem. Symp., 1981, 302.
235 O. Andomie, L.A. Smith, L.A. and S. Cornejo, Nucleotecnica, 4 (1985) 28.
236 C.R. Walker and O.A. Vita, Anal. Chim. Acta, 43 (1968) 27.
237 T.R. Bangia, K.N.K. Kartha, M. Varghese, B.A. Dhawale and B.D. Joshi, Fresenius Z. Anal. Chem., 310 (1982) 410.
238 J. Korkisch and H. Gross, Mikrochim. Acta, II (1975) 413.
239 M. Gerardi and G.A. Pelliccia, At. Spectrosc., 4 (1983) 193.
240 M.J. Hughes, M.R. Cowell and P.T. Craddock, Archaeometry, 18 (1976) 19.
241 W.A. Oddy, Eur. Spectrosc. News, 34 (1981) 31.
242 K. Zimmer, Fresenius Z. Anal. Chem., 324 (1986) 875.
243 P. Clayton, J. Egypt. Archaeol., 58 (1972) 167.
244 M.J. Hughes, Nucl. Instrum. Methods Phys. Res., Sect. B, 14 (1986) 16.
245 R.H. Brill, A chemical analytical round-robin on four synthetic ancient glasses, Ninth International Congress of Glass: Artistic and Historic Communications, Paris, 1972, 93.
246 M.R. Cowell and A.E.A. Werner, Analysis of some Egyptian glass, Annales de 6e Congres de l'Association Internationale pour l'Histoire du Verre, Liège, 1974, 295.
247 N.M. Magalousis and V. Gritton, Occas. Pap., Br. Mus., 19 (1981) 103.
248 J.A. Stratis, G.A. Zachariadis, E.A. Dimitrakoudi and V. Simeonov, Fresenius Z. Anal. Chem., 331 (1988) 725.
249 W.A. Oddy and F. Schweizer, in E.T. Hall and D.M. Metcalf (Eds.), Methods of Chemical and Metallurgical Investigation of Ancient Coinage, Royal Numismatic Society, London, 1972, 171.
250 H. McKerrell and R.B.K. Stevenson, in E.T. Hall and D.M. Metcalf (Eds.), Methods

of Chemical and Metallurgical Investigation of Ancient Coinage, Royal Numismatic Society, London, 1972, 195.

251 E.L. Szonntagh, Mikrochim. Acta, 1 (1981) 191.
252 P.R. Fields, J. Milstead, E. Henrickson and R. Ramette, in Science and Archaeology, Cambridge, MA, 1971, 131.
253 E. Pernicka, Nucl. Instrum. Methods Phys. Res., Sect. B., 14 (1986) 24.
254 A. Varma, Talanta, 28 (1981) 785.
255 G. de G. Sieveking, P.T. Craddock, M.J. Hughes, P. Bush and J. Ferguson, Nature, 228 (1970) 251.
256 G. de G. Sieveking, P. Bush, J. Ferguson, P.T. Craddock, M.J. Hughes and M.R. Cowell, Archaeometry, 14 (1972) 151.
257 J. Lang and J. Price, Archaeol. Sci., 2 (1975) No. 4.
258 J. Price, Some Roman glass from Spain, Annales de 6e Congres de l'Association Internationale pour l'Histoire du Verre, Liège, 1974, 65.
259 J. Brailsford and J.E. Stapley, Proc. Prehistoric Soc., 38 (1972) 219.
260 J.V.S. Megaw, Proc. Prehistoric Soc., 35 (1969) 358.
261 J.V.S. Megaw, P.T. Craddock and A.E.A. Werner, Br. Mus. Quart., 37 (1973) 70.
262 G.A. Aycik and E. Edguer, J. Radioanal. Nucl. Chem., 82 (1984) 319.
263 J. Lang and A.R. Williams, J. Archaeol. Sci., 2 (1975) 199.
264 P.T. Craddock, J. Lang and K.S. Painter, Br. Mus. Quart., 37 (1973) 9.
265 E. Blasius, H. Wagner, H. Braun, R. Krumbholz and B. Thimmel, Fresenius Z. Anal. Chem., 310 (1982) 98.
266 A. Tubb and G. Nickless, Anal. Proc., 19 (1982) 334.
267 R.H. Brill, J. Glass Stud., 12 (1970) 185.
268 C.D. Vassass, Chemical thermal analysis and physical study of glasses of mediaeval stained glass windows, Ninth International Congress of Glass: Artistic and Historical Communications, Paris, 1972, 241.
269 G. Rauret, E. Casassas and M. Baucells, Archaeometry, 27 (1985) 195.
270 J. Ramirez-Munoz, Atomic Absorption Spectroscopy, Elsevier, Amsterdam, 1968, 387.
271 K.C. Thompson and R.J. Reynolds, Atomic Absorption, Fluorescence and Flame Emission Spectroscopy, 2nd ed., Charles Griffin, London, 1978, 306.
272 K.R. Osborn and H.E. Gunning, J. Opt. Soc. Am., 45 (1955) 552.
273 A.L. Meier, Anal. Chem., 54 (1982) 2158.
274 J.F. Chapman, L.S. Dale and H.J. Fraser, Anal. Chim. Acta, 116 (1980) 427.
275 J.A. Goleb and Y. Yokoyama, Anal. Chim. Acta, 30 (1964) 213.
276 J.A. Wheat, Appl. Spectrosc., 25 (1971) 328.
277 H. Kirchhoff, Spectrochim. Acta, Part B, 24 (1969) 235.
278 J.A. Goleb, Anal. Chem., 35 (1963) 1978.
279 J.A. Goleb, Anal. Chim. Acta, 34 (1966) 135.

Chapter 4n

Analysis of Polluted Soils

M. CRESSER

Aberdeen University, Department of Plant and Soil Science, Old Aberdeen AB9 2UE (Great Britain)

I. INTRODUCTION

To the man in the street, soil is unlikely to be a topic which stimulates much excitement. He tends to see it as something which, with a bit of tinkering, will allow the growth of crops for food and trees, and ornamental flowers and shrubs. He might even see it as playing an important role in supporting buildings and roads. He is most unlikely to perceive soil as a vital resource upon which all life as we know it ultimately depends, but that is exactly what it is. Soil must provide water, nutrient elements and physical support for plants, but also the soil microbial population must be capable of dealing with continuous inputs of natural plant and animal detritus. In breaking down these organic residues, it releases nutrients for re-use in the biosphere. It also releases carbon dioxide back into the atmosphere, allowing photosynthesis and the growth of plants to continue at a reasonable rate. Thus soil biological activity plays a major role in regulation of the composition of the air we breathe, since photosynthesis generates oxygen.

Surface freshwaters originate primarily from rain and snow, but their chemistry is usually regulated to some extent by contact with soil. Thus the quality of drainage water also depends upon the chemical, physical and biological properties of the soil from which it originates.

Over the centuries, mankind has rather tended to take soil for granted as a natural resource. He has been able to get away from this abuse because the resource was vast compared with the population it supported. Scientific progress, however, has imposed new burdens upon soil, mostly linked to population explosion and increasing demands for comfort and material goods. Soil has increasingly become a global dustbin. At a local level, effects of sludge spreading or use of land for industrial and domestic garbage dumping may be all too obvious to the local residents or their visitors. On a global scale, however, pollution effects may be slower and change almost imperceptible without painstaking scientific investigation. Table 1, for example, shows the changes in lead concentration in samples of surface organic soils from a forest in northeast Scotland, collected in 1949/50, and again in 1987, and analysed in 1989 by flame AAS [1]. It also shows the estimated lead accumulation rates in these soils, taking into account changes

References pp. 525–526

TABLE 1

CHANGES IN LEAD CONCENTRATION OVER 38 YEARS AND LEAD ACCUMULA-
TION RATES FOR SOME FOREST SITES IN NORTHEAST SCOTLAND

Site	Pb concentration (mg kg^{-1})		Accumulation rate (g ha^{-1} yr^{-1})
	1949/50	1987	
1	72	81	230
2	51	66	30
3	34	94	345
4	62	98	668
5	67	96	306
6	61	52	120

in the thickness of the surface soil layers.

Changes such as that described above occur over decades rather than
years, as a consequence of minute pollutant inputs from the atmosphere.
They may produce no obvious visible effects either in the soils or in the
biota which they support for several decades, or even centuries. Their
reliable quantification from field studies requires very precise relocation
of old experimental sites, because of complications imposed by natural
variability. For the unambiguous elucidation and quantification of long-term
pollution effects upon soil chemical composition, storage of old samples for
decades under contamination-free conditions is also highly desirable. These
requirements are rarely met in practice, however [2–4]. The use of old
analytical data, rather than the re-analysis of stored samples, requires a
facility for exact reproduction of the procedures used in early studies. This,
too, may limit reliability of the comparison of data sets, because often the
details preserved of original methodology are insufficient.

Many environmental scientists try to avoid the problems associated with
quantification of the effects of long-term, low-dose inputs of atmospheric
pollutants by examination of the fate of pollutants added in simulated
precipitation to miniature ecosystems. These may range from small, re-
packed cores to large intact-core soil lysimeters [5]. By understanding the
mechanisms of retention and fixation (chemical or biological) by soil at
different depths, they are able to predict longer-term effects from shorter-
term studies.

The concentrations of pollutant elements used in this way are invariably
less than 1 μg ml^{-1}, making flame-, electrothermal-, mercury cold vapour-
and hydride generation-AAS excellent investigatory tools. Usually, the
concomitant matrix species are primarily calcium, magnesium, sodium,
potassium, iron, manganese, aluminium, ammonium, chloride, phosphate,
nitrate, sulphate and bicarbonate, each generally only being present at
the mg ml^{-1} concentration or less. Thus matrix effects are rarely an
insurmountable problem.

Where pollutants are added to soil with waste disposal over much shorter periods, the problems posed for the analyst are more straightforward in many respects. However, a major question which needs to be answered before deciding upon the best analytical approach in any investigation of possible pollution effects is: "For what exact purpose are the results to be obtained?" The environmentalist must decide whether total pollutant concentrations are of real interest. If his concern is with food chains, then the plant-available components only may need to be studied, and the sample preparation techniques will be different. He may be interested in drainage-water contamination or in possible adverse effects upon plant growth or upon soil microbial activity. In each instance, however, atomic absorption analysis could have a role to play. The environmental scientist must also decide whether he is interested in short-term effects (days to weeks) or long-term effects (months to years). The latter is always a problem for systems like soil which are biologically active and therefore subject to long-term biological, chemical and physical change, and to mixing effects. The latter may be ameliorative (i.e. beneficial through dilution over a greater soil volume), but they also may change the physico-chemical environment to which the added materials are being subjected. This, in turn, could influence the mobility of potentially toxic species. Clearly, then, the raison d'être for the analysis must be given very careful consideration when deciding upon sampling strategy and upon sample preparation technique.

II. TOTAL ELEMENT ANALYSIS OF POLLUTED SOILS

Total element analysis of soils is reasonably straightforward, but tends to be rather time-consuming. In many pollution studies, simple oxidising acid digestions are used, rather than the more rigorous alkaline flux fusions or digestions with acid mixtures which include hydrofluoric acid. The justification is based (often tacitly) upon the concept that any heavy metals which are so firmly locked up in mineral matrices that they cannot be dissolved by boiling aqua regia or some similar mixture, are of no practical consequence to plant growth, food chains or drainage-water chemistry.

If true total analysis is deemed essential, fusion with sodium carbonate or lithium metaborate is one possible approach. However, not all minerals are attacked by such fluxing agents. Chromite, for example, may be better dissolved by fusion with sodium hydrogen sulphate [6]. It is not attacked by a boiling sulphuric and hydrofluoric acid mixture, or by sodium carbonate fusion, except on prolonged heating at 1200°C. Zircon and rutile also require this high temperature treatment. Lithium metaborate leaves a range of minerals only partly attacked [7, 8].

Prior to carrying out a sodium carbonate fusion, it is desirable to pre-ignite soil samples containing substantial amounts of organic matter to minimise risk of damage to platinum ware. Soils with high concentrations of free iron oxides are best pre-digested with acid for the same reason. Fusion of the residue is then carried out with at least 4-fold excess of flux at 900–1000°C for 20–60 min [7]. The cooled flux is usually dispersed with water and slightly acidified with hydrochloric acid before dilution to mark.

Various alkali element hydroxides and peroxides have also been recommended over the years as fusion fluxes, but none has found as widespread popularity as sodium carbonate, except in specialised applications [7]. All fusion fluxes suffer from similar drawbacks in the context of atomic absorption spectrometry and trace analysis in general. They provide a high concentration of matrix solute, which necessitates careful and regular use of blanks, quality control on flux reagents, and matrix matching of samples and standards. They partially dissolve the crucible metal, be it platinum, gold, silver, nickel or zirconium. Their use leads to some loss of volatile constituents. The high salt concentration may lead to erratic nebuliser operation, for example, as a result of partial blockage, making regular checks for drift very important, and to scatter signals in both flame and furnace AAS. The latter problems may be overcome by background correction, or by using solvent extraction to separate determinants from the matrix components, but only at a high cost in terms of time of analysis [9].

Acid digestions offer a number of potential advantages over fusion techniques, and many argue that these more than outweigh their drawbacks with respect to safety. All hot oxidising acids must be treated with care, however, and perchloric acid especially offers potential fire and explosion risks. No analyst should use hydrofluoric acid without being aware of the nature of the burns it can cause and, more importantly, the first aid treatment to be used in the event of an accident (washing with copious amounts of water and treatment with monosodium glutamate gel [7]). If perchloric acid is to be used regularly, a wash-down fume hood must be regarded as essential.

Most acid digestion procedures for total element analysis of soils involve the use of hydrofluoric acid, because of its ability to attack silicate via the formation of $[SiF_6]^{2-}$. If silicon is not to be determined, the hydrofluoric acid is usually used with perchloric acid to attack dry-ashed soil in a platinum crucible. The digestion is often carried out with slow, controlled heating on a sand bath so that silicon (a common interference in AAS procedures) is slowly volatilised off as SiF_4, together with excess hydrofluoric acid. Sulphuric acid, in spite of its high boiling point, is best avoided for digestion purposes because of the limited solubility of some common element sulphates and its depressive effects on the absorption signals for many elements determined by AAS.

Although the open digestion procedure described above has the advantage of eliminating silicon from the sample solutions, it should not be forgotten that several other elements of interest in the present context also have volatile fluorides which may be partially or completely lost. These include arsenic, boron, germanium and antimony. If such elements are to be determined, it is preferable to use a closed digestion system (a bomb). In this case, typically ca. 200 mg of finely ground sample is treated with nitric acid (0.2 ml) and 40% hydrofluoric acid (5 ml) in a sealed polytetrafluoro-ethylene (PTFE) container enclosed in a steel bomb, which is then heated for one hour in an oven at 110–150°C [7, 10]. At the elevated pressures generated (up to 100 atm) most minerals are attacked, but volatile elements are retained, provided the PTFE cap seal holds. The bomb and contents are cooled thoroughly prior to opening with care, and boric acid is added to dissolve any precipitate formed.

Over recent years, several authors have used domestic microwave ovens to facilitate digestion of soils in inert plastic vessels. This has now led to the development of safer and more suitable purpose-built systems. Although these are much more expensive, they give more precise control and have a pressure relief system incorporated.

Whatever digestion procedure is employed, the sample solutions ultimately obtained will contain high and often variable concentrations of matrix components. Even if silicon has been volatilised off during the digestion process, the solutions will invariably contain large amounts (relative to trace elements) of aluminium, iron, magnesium, calcium, sodium, potassium, sulphate and phosphate. It may contain other potential problem elements such as titanium and zirconium. While careful matrix matching may go a long way towards eradication of matrix or contamination effects, it does not help greatly with interference effects from the soil matrix components. The analyst must therefore be satisfied that his or her proposed conditions of determination cope adequately with all conceivable concomitant elements.

III. USING THE LITERATURE FOR METHOD DEVELOPMENT

One of the major problems confronting the scientist when scanning published literature on AAS is knowing how far one can safely extrapolate from the paper to real samples when assessing potential interference effects [11]. Right from the early days of flame AAS, papers have tended to include rudimentary interference studies. Usually these involved selection of a single concentration of determinant (one which gave a signal which could be measured with reasonable precision (e.g. ±2%), and examination of the effect of selected concentrations of potential interfering elements on the measurement of this concentration of determinant. Generally, only up to ca. 100- or 1000-fold mass or molar excesses of potential interferant have been studied and very rarely the effect of possible additive interference has been considered. Even more rarely, interferences were compared on different instruments, or under diverse analytical conditions. Thus the results obtained, and the conclusions which could be reliably drawn, were specific to a particular nebuliser and aspiration capillary in a particular condition, at particular fuel and oxidant flows, at a particular height above a particular burner design, with the impactor in a particular position. Nowadays we know that a change in any one of these "particulars" may result in a change in the extent of interference.

As a general rule, it is accepted that the probability and extent of chemical interference is reduced under conditions which favour the production of finer aerosol (and hence finer solid particulates which are more readily atomized). Thus the greatest dilution [12] and the lowest aspiration rate [13] compatible with adequate precision are desirable. It is also preferable, from the point of view of interferences, to work high up, in as hot a flame as possible [12]. This sometimes involves compromise in terms of working in a fuel-lean flame at sub-optimal sensitivity.

Confronted by this lengthy list of interactive variables associated with flame AAS, the analyst has two options. The first is to conduct a systematic study of interferences which brackets all conceivable variability. This is very tedious and time-consuming, but has the advantage that it also provides information on how rigidly analytical conditions must subsequently be controlled. The second is to make use of a selection of certified reference materials, to make sure that any interpolation from the published literature is indeed sound. A third possible approach, but a slightly less satisfactory one, is to vary conditions slightly, for example, by changing the fuel flow or height of observation, and see if the same results are obtained for a selection of samples. Significant differences between two sets of results invariably indicate interference problems in one or both sets, and warrant further investigation.

For many years, the problems associated with soil matrix effects greatly restricted the use of electrothermal atomization (ETA) for the determination of total trace elements in soils by AAS. However, improvements in furnace design, and the development of simultaneous background correction, the platform-in-furnace technique, and the use of matrix modifiers to prevent volatilisation of determinant at too low a temperature have recently made such determinations possible, if not common place [14–16]. Indeed, for volatile metallic elements such as cadmium, copper, lead and thallium, direct atomization from finely powdered solid samples, often in suspension form, may be possible [14]. This eliminates high blanks from reagents, although of course grinding may still introduce problems.

For the determination of hydride-forming elements by hydride-generation AAS, the soil matrix components present at relatively high concentrations tend not to be the ones which cause interference problems. These elements are therefore often determined in this way [14–16].

IV. DETERMINATION OF WATER-SOLUBLE AND PLANT-AVAILABLE ELEMENTS

The fact that a soil is polluted, makes little difference to the extraction technique used at the sample preparation stage for assessing these labile components. Where it has been decided that water-soluble or plant-available elements are of interest, flame AAS very rarely provides adequate sensitivity for direct determination, although it may do so for iron, manganese and zinc, and even for copper and nickel in some very heavily polluted soils. Thus if only direct flame AAS is available, large samples must be extracted, and trace elements such as arsenic, silver, nickel, lead, chromium, cobalt, etc., concentrated by solvent extraction prior to nebulisation [9, 16]. Care must be taken that the extraction is sufficiently quantitative and reproducible, especially if a complexing agent such as EDTA or DTPA has been employed to extract plant-available metals.

Often several different extractants are recommended for the determination of the "available" amount of a single element. For example, Table 2 lists five procedures commonly used for the determination of zinc in soils [17–21]. Of these extractants, the dilute hydrochloric acid generally extracts the most zinc, and may remove one hundred times more zinc than water does. DTPA, however, is usually regarded as giving the best indication of plant-available zinc. The sensitivity of analytical techniques ideally should never be allowed to dictate the choice of extractant. In practice, however, where resources are limited it often does.

Water-soluble concentrations of trace elements are generally lower than those in extractant solutions such as ammonium acetate, ammonium hydro-

TABLE 2

EXAMPLES OF EXTRACTION PROCEDURES RECOMMENDED FOR THE DETERMINATION OF AVAILABLE ZINC IN SOIL

Extractant	Ratio soil : extractant	Shaking time (min)	Ref.
0.1 M hydrochloric acid	1 : 10	60	[17]
1 M ammonium acetate at pH 4.8	1 : 5	30	[18]
EDTA/ammonium carbonate at pH 8.6	1 : 2	30	[19]
DTPA/triethanolamine/calcium chloride at pH 7.3	1 : 2	120	[20]
Water	Field capacity water content, equilibrated for 48 h	No shaking, centrifuging for 120 min	21

References pp. 525–526

gen oxalate, EDTA, DTPA, etc., although concentrations in the rhizosphere soil (immediately adjacent to plant roots) tend to be greater than those in bulk soil. Over recent years, solutions in both categories have been analysed most commonly by ETA-AAS, ideally with a L'vov platform, a matrix modifier and automatic simultaneous background correction [14–16], or by hydride generation AAS. Many different matrix modifiers have been suggested; some of the more common and successful are mentioned in Section VI.

V. SPECIATION

Over recent years there has been growing interest in the distribution of trace elements between the various soil components. This has been partly prompted by concern over the fate of pollutants, and partly by the increasingly widespread occurrence of trace element deficiencies as a result of element removal season after season in high yielding crops. Environmental chemists are usually interested in the native forms of elements in soil minerals, and in the quantitative assessment of their fixation on inorganic and organic cation exchange sites, by soil biota, or in secondary minerals such as the amorphous or microcrystalline hydrous oxides of iron or aluminium or, in more arid climates, in carbonates [22–23].

Usually, the exchangeable trace elements are extracted first, for example, by successive shake-and-centrifuge steps with 1 M magnesium nitrate at pH 7 [23, 24]. If carbonates are present, the carbonate fraction is then extracted with sodium acetate at pH 5. It may be necessary, for highly calcareous soils, to add, in small increments, sufficient acetic acid to dissolve the carbonate and maintain the pH at 5, otherwise this fraction may be seriously underestimated, and the next overestimated [22]. The organically bound fraction may then be estimated by oxidation of the organic matter with 0.7 M sodium hypochlorite at pH 8.5 over 30 min in a boiling water bath.

Trace metals bound in the manganese oxides fraction may be dissolved by treatment with 0.1 M hydroxylamine hydrochloride at pH 2, and the amorphous iron oxides fraction is then removed by treatment of the residue with 0.4 M ammonium hydrogen oxalate at pH 3 [24]. Heating on a water bath for 30 min with the same extractant, but in the presence of added ascorbic acid, separates the crystalline iron oxides fraction. Finally, the residue may be separated into residual clay-, silt- and sand-size fractions by standard sedimentation techniques.

The cadmium, copper, manganese, iron and zinc in each of the above fractions may be determined by flame AAS, although sometimes the concentration found is near the detection limit and ETA-AAS may give better precision. It is important to use matrix-matched standards whatever technique is used, and an appropriate matrix modifier if a furnace is employed.

For lead, nickel and chromium, ETA-AAS is essential to attain adequate sensitivity, unless the soil is severely polluted.

The above separation/speciation procedure is very time-consuming, and simplified procedures with less fractions are sometimes preferred. The precise choice of technique ultimately depends upon the purpose of the analysis.

VI. NOTES ON THE DETERMINATION OF INDIVIDUAL ELEMENTS

It is appropriate at this point to look briefly at the determination of individual elements, and the problems which may be encountered when polluted soils are analysed. The comments are based upon the author's experience of soil analysis accumulated over more than 20 years. The reader may also wish to refer to Chapter 4g for specific extraction procedures suitable for geological-based materials.

A. Aluminium

There has been considerable interest in the solubilisation of aluminium in acidified soils over recent years because of its proven link to Altzheimer's disease. The element is only soluble at concentrations above 0.1 mg l^{-1} in highly acidified soils when a mobile anion such as sulphate is present (for example, from acid rain), or when it is organically complexed in water draining from highly organic soils. Thus flame AAS using nitrous oxide–acetylene and an ionisation buffer is rarely sufficiently sensitive for precise determination. The element may be determined using ETA if a matrix modifier such as magnesium nitrate is used.

B. Arsenic

Although platform ETA-AAS has been successfully employed for the determination of arsenic in soils, using nickel as a matrix modifier, hydride generation is generally regarded as a more satisfactory approach. The latter technique has been extensively used in the author's Department, for example, in studies of the adsorption isotherms of arsenate in soils. Although there has been some interest in the speciation of organo-arsenic compounds in aquatic sediments over recent years, this interest does not appear to have been extended to soils at the time of writing. Arsenic is sufficiently volatile to have been considered as a suitable determinant for slurry atomization techniques to be applied.

C. Barium

Although not a particularly common pollutant, barium is worth including because of its toxicity. The determination in a fuel-rich nitrous oxide–

References pp. 525–526

acetylene flame in the presence of an ionisation buffer is very sensitive, but care is necessary to make sure that the buffer adequately copes with other easily ionisable elements present. Usually samples contain a large excess of calcium, and background correction may be necessary to eliminate the resultant molecular spectral interference.

D. Cadmium

The determination of cadmium in an air–acetylene flame is very sensitive if source, burner, nebuliser and flame conditions are carefully optimised, and therefore flame AAS may sometimes be used for the analysis of polluted soils, in spite of the low concentrations usually present. The element sometimes accumulates in soils because of regular trace additions with poor-quality phosphate fertilizers. If better sensitivity is required, ETA may be used with a suitable matrix modifier; phosphate is sometimes used. Slurry atomization is possible in both flames and furnaces. In the latter instance, palladium and ascorbic acid have been recommended as a matrix modifier [15]. Atom trapping techniques may also be appropriate for the determination of cadmium extracted from soils, as discussed elsewhere in this volume (Chapter 2 Section II).

E. Chromium

Even if conditions are very carefully optimised, the determination of chromium in a fuel-rich air–acetylene or nitrous oxide–acetylene flame is not usually sufficiently sensitive for the direct determination of soil chromium. In part this is because the chromate ion is strongly adsorbed by most soils. ETA-AAS may be suitable, but has rarely been applied in practice. For speciation purposes, chromium(III) and (VI) may be separated by anion or cation exchange prior to AAS determination.

F. Copper

Although for unpolluted soils the concentrations of copper in soil extracts are sometimes close to the detection limit of its determination by flame AAS, leading to poor precision, this is less of a problem when soils have been contaminated with copper. Although the element is tightly bound to soil organic matter and not significantly soluble in water for most soils, determinable amounts are usually found when DTPA is used as an extractant. Determination by ETA-AAS is straightforward and sensitive for soil extracts. Direct atomization from slurries is possible.

G. Lead

There has been great interest in lead accumulation in soils for many years now, because the element is known to accumulate near the surface, especially in acidic, organic soils. The effect is so pronounced that it is often possible to determine the lead in acid digests of such soils by direct flame AAS, in spite of the rather poor sensitivity of the method. Water-soluble and extractable lead concentrations are invariably too low for flame determination. For such samples, ETA-AAS is most commonly used, with phosphate as a matrix modifier. Direct slurry atomization is possible, and atom-trapping is also occasionally used to enhance the sensitivity of flame determinations.

H. Mercury

Considerable interest has been shown in the determination of mercury in soils. Cold-vapour AAS has been employed in the majority of studies.

I. Nickel

The determination of nickel by flame AAS is not especially sensitive unless the element is pre-concentrated by solvent extraction [9]. Most determinations of extractable nickel are therefore performed by ETA-AAS, using a matrix modifier such as phosphate.

J. Thallium

Thallium is not a common pollutant, but due to its toxic nature, its determination should be mentioned briefly. Because of the low concentrations invariably involved, ETA-AAS must be used, with an appropriate matrix modifier such as a mixture of palladium and magnesium.

K. Other elements

In a general text of this nature, it would not be useful to discuss the determination of every element which could conceivably occur as a soil pollutant, and the above list is therefore of necessity highly selective. Hopefully, sufficient information has been presented to give some insight into the sort of problems likely to be encountered in, and the methodology of, the analysis of polluted soils by AAS. More information may be readily found in the annual review of applications of atomic spectrometry in environmental chemical analysis published in the *Journal of Analytical Atomic Spectrometry* (see, for example, refs. [14–16]).

REFERENCES

1   M.F. Billett, Long-Term Changes in the Chemistry of Soils from the Alltcailleach Forest, North-East Scotland, M.Sc. Thesis, University of Aberdeen, 1989.

526

2   M.F. Billett, E.A. Fitzpatrick and M.S. Cresser, Soil Use Manage., 4 (1988) 102.
3   M.F. Billett, F. Parker-Jervis, E.A. Fitzpatrick and M.S. Cresser, J. Soil Sci., 41 (1990) 133.
4   L. Hallbacken and C.O. Tamm, Scand. J. For. Res., 1 (1986) 219.
5   U. Skiba, A.C. Edwards, T. Peirson-Smith and M.S. Cresser, Chemical Analysis in Environmental Research, NERC, ITE Merlewood, (1987) 16.
6   M.S. Cresser and R. Hargitt, Anal. Chim. Acta, 82 (1976) 203.
7   P.J. Potts, A Handbook of Silicate Rock Analysis, Blackie, Glasgow, 1987.
8   M. Cremer and J. Schlocker, Am. Mineral., 61 (1976) 318.
9   M.S. Cresser, Solvent Extraction in Flame Spectroscopic Analysis, Butterworths, London, 1978.
10 B. Bernas, Anal. Chem., 42 (1968) 1682.
11 M.S. Cresser, Lab. Pract., 26 (1977) 171.
12 M.S. Cresser and D.A. MacLeod, Analyst, 101 (1976) 86.
13 I.L. Garcia, C. O'Grady and M.S. Cresser, J. Anal. Atom. Spectrom., 2 (1987) 221.
14 L. Ebdon, M.S. Cresser and C.W. McLeod, J. Anal. Atom. Spectrom., 2 (1987) 1R.
15 M.S. Cresser, L. Ebdon and J.R. Dean, J. Anal. Atom. Spectrom., 3 (1988) 1R.
16 M.S. Cresser, L. Ebdon and J.R. Dean, J. Anal. Atom. Spectrom., 4 (1989) 1R.
17 D.C. Martens, Soil Sci., 106 (1968) 23.
18 R.C. Tiwari and B.M. Kumar, Plant Soil, 41 (1974) 689.
19 J.R. Trierweiler and J.R. Lindsay, Soil Sci. Soc. Am. Proc., 33 (1969) 49.
20 W.L. Lindsay and W.A. Norvell, Soil Sci. Soc. Am. J., 42 (1978) 421.
21 F. Adams, C. Burmesten, N.V. Hue and F.L. Long, Soil Sci. Soc. Am., J., 44 (1980) 733.
22 L.M. Shuman, Soil Sci., 140 (1985) 11.
23 E. El-Sayad, M.S. Cresser, M.Abd. El-Gawad and E.A. Khater, Microchem. J., 38 (1988) 307.
24 A. Tessier, P.G.C. Campbell and M. Bisson, Anal. Chem., 51 (1979) 844.

SUBJECT INDEX

Printed and bound by CPI Group (UK) Ltd, Croydon, CR0 4YY

03/10/2024

01040329-0011